W9-BBM-101

Thesis

PRIMARY
ANATOMY

PRIMARY

SEVENTH EDITION

Baltimore

ANATOMY

John V. Basmajian, M.D., F.A.C.A.

Professor of Anatomy and Physical Medicine and
Director of Rehabilitation Research & Training Center
Emory University, Atlanta, Georgia

The Williams & Wilkins Company

First Edition, 1948

Second Edition, 1951
  Reprinted January 1953

Third Edition, 1955
  Reprinted January 1956
  Reprinted September 1957

Fourth Edition, 1960
  Reprinted May 1962

Fifth Edition, 1964
  Reprinted August 1965
  Reprinted April 1967

Sixth Edition, 1970
  Reprinted January 1972
  Reprinted January 1973
  Reprinted May 1974
  French Edition, 1972
  Spanish Edition, 1973
  International Student Edition, 1973
  Braille transcription, 1974
  Italian Edition, 1975
  Portuguese Edition, in preparation
  Japanese Edition, in preparation

Seventh Edition, 1976
  Reprinted May 1977
  Reprinted June 1978
  Reprinted February 1979

Library of Congress Cataloging in Publication Data

Basmajian, John V   1921-
    Primary anatomy.

    Seventh ed. of H. A. Cates' Primary anatomy, first pub-
lished in 1948.
    Includes index.
    1. Anatomy, Human. I. Cates, Harry Arthur,
1890–1953. Primary anatomy. II. Title. [DNLM: 1. Anatomy. QS4 B315p]
QM23.2.B37 1976   611   75-11861
ISBN 0-683-00549-9

Copyright, © 1976
The Williams & Wilkins Company
428 E. Preston St., Baltimore, Md., U.S.A. 21202
Made in United States of America

Composed and Printed at the
Waverly Press, Inc.
Mt. Royal & Guilford Aves.
Baltimore, Md., U.S.A. 21202

TO THE MEMORY OF

Professor H. A. Cates

Preface to Seventh Edition

Primary Anatomy has entered an explosive phase in its growth that has surprised and gratified both its author and its advocates. We must assume that the reason can be traced to our attempt to make this textbook both as free of jargon and as scientifically sound as precise words and drawings can make it. A feature that has been unique among small textbooks always has been its profuse special illustrations. These have been augmented further with more than 50 new or revised text-figures and a 26-page color atlas.

A new Chapter 16 has been written in response to widely expressed needs. This chapter is added for more advanced students than those who were the intended readers of early editions. More and more medical and dental students are now regular users—they require at least a general section on regional anatomy.

This edition has gone metric rather rapidly and thoroughly, not "inch by inch." While students in the U.S.A. number more than half of its regular users, even they are supposed to be at ease with the metric measures; those that are not should be! Non-U.S. readers now read this book in French, Spanish, Italian, and other foreign editions, and many of them are not familiar with obsolete "English" measures.

I am particularly pleased that we are able to include, as Plates 5 to 30 in Chapter 16, some of the regional anatomy illustrations from *Stedman's Medical Dictionary*, 23rd edition.

Once more, I thank those who gave me ideas, and welcome further suggestions for the improvement and correction of this book. While improvements may take several years to show up in a future edition, they will help to keep this book alive and lively. As noted in previous editions, anatomy is not the science of the cadaver, it is concerned with the fabric of living man. A living subject deserves a lively treatment.

J. V. B.

Emory University
Atlanta, 1975–1976

From the Prefaces of the First to Sixth Editions

Even the most experienced teachers will never be in complete agreement as to the exact content of an elementary course. Some would omit what appears vital to others; some would include what appears unimportant to others. In this connection it should be remembered that even the beginner possesses scientific curiosity, and that it is proper to foster such legitimate curiosity provided the general balance is not lost.

This book embodies the experience of many years and it expresses certain convictions as to the questions raised; first, that a systematic approach involves the beginner in fewer difficulties than does a regional one; secondly, that discrimination must be exercised both as to the amount of detailed instruction given for each system, and as to the amount of 'embryological' instruction given; thirdly, that access to anatomical material is highly desirable if not absolutely vital.

Complex illustrations, such as—very properly—are to be found in larger works, are of little help to the beginner, but simple, schematic, illustrations often profitably replace the written word.

• • •

Those who are familiar with previous edi-

tions will notice the very marked increase in
the use of headings and subheadings, which is
in accordance with my belief that this makes
the organization and study of anatomy more
pleasant and the plan of the book more apparent. Many other devices have been employed
to place correct emphasis on the information
supplied (e.g., the use of small type for unimportant details which need not be memorized
or reread). A large number of new tables have
been scattered through the text in the hope
that this may prove to be useful for summary
and review.

Wherever a new term is introduced, its deri-
vation is now explained briefly—usually in one
word—not necessarily for the student to mem-
orize but rather to make his reading and learning easier and more pleasant.

•   •   •

Although considerable numbers of medical
and dental students (and even graduates) are
known to be using *Primary Anatomy* to gain
preliminary concise but authoritative information, the temptation to change its character by
introducing minute details has been actively
resisted. At the other extreme, certain phrases
and similes originally used here and there to
enliven or simplify the descriptions for the dull
witted, having proved unnecessary, have been
largely pruned out.

•   •   •

Although *Primary Anatomy* has changed
greatly both in format and in content, is has
never changed its basic aim. This aim is to provide non-medical university students studying
human anatomy with a professional textbook of
gross and functional anatomy that neither en-
tangles them in a web of details nor deceives

them with a style so bare and basic as to be boring and misleading. There are already enough books, both good and bad, that are loaded with details, and there are more than enough puerile anatomy books intended for the beginner which are simply awful when viewed at the university level. The growing success of *Primary Anatomy* in recent years suggests that it has filled a real need.

•   •   •

While this book is for students and not specialists, I hope that teachers of anatomy will find here new material of solid worth. As before, I count on them and on readers everywhere to provide me with their comments and criticisms. Such 'feed-back' has helped keep this book as lively a fabric as is its subject, living man! Anatomy is a study of a living organism, not a cadaver—this must never be forgotten.

Contents

Contents

1

Introduction

PROTOPLASM AND THE CELL

An **element** is a gas, liquid, or solid that cannot by any chemical means be broken down into a simpler one. Although there are little more than 90 of them—some abundant and well known, others so rare as to be the objects of scientific curiosity—yet it is these elements, and endless combinations of them, that make up the whole animate and inanimate world. They are the building blocks for the fashioning of Nature's products.

One of these products is unlike any other and is known as **protoplasm.** When the earth was still very young, certain of the elements combined—perhaps by accident —to form this new substance which had the unusual characteristic of possessing *life*. From protoplasm were fashioned the myriads of living things, some extremely simple, others very complex.

Twelve elements at least enter into the composition of a protoplasm. They are: hydrogen, oxygen, nitrogen, chlorine, calcium, carbon, iron, sodium, potassium, and phosphorous, with traces of magnesium and sulfur. These are combined in innumerable patterns to produce an equally innumerable variety of protoplasms, each of which has defied all attempts to arrive at its exact formula. This is in part owing to the fact that protoplasm is very unstable, i.e., it easily breaks down into simpler compounds, a little too much heat or a little drying so altering it that it becomes inert and loses its magic property of life.

What is life? To say that it is the essential condition of existence, or that it is the fundamental property of protoplasm does not tell us anything. There are those who say that it is merely protoplasmic activity resulting from changes in the environment. There is something to be said for such a definition. Even the most complex animal is essentially an aggregation of protoplasmic units and their products.

Life is manifested by inherent movement. The opossum 'plays dead' by remaining absolutely motionless. Inherent motion is the commonest criterion of life. But motion implies energy. Whence comes the original energy residing in a piece of primordial protoplasm? Is it stored there by the processes of chemical building up, or **synthesis** as it is called, that elaborated the protoplasm? If so it is but a minute amount; it is soon expended and, unless renewed, the protoplasm dies or is broken down into more stable and simpler compounds. Life has gone.

Tissue Fluid

Since protoplasm will not survive drying, water is the medium necessary for its unstable existence. Life, of course, is either

1

aquatic or terrestrial, but the fact that terrestrial life demands water for its continued existence must not be forgotten. Every living cell of which even the human body is composed survives only because it is bathed in water or rather, in what is known as **tissue fluid,** and it has been estimated that about 85 per cent of the human body actually consists of tissue fluid. In that tissue fluid, therefore, the living protoplasm must find the materials that it can utilize to renew the energy it is constantly expending. These necessary materials are **oxygen** and **food.** By the utilization of oxygen the chemical activities of protoplasm proceed and by the ingestion of food renewal of the protoplasm is assured. Deprived of oxygen most cells soon die; deprived of food, a cell is doomed to die a slow but inevitable death.

The Cell

Before progress toward a higher and more complex form of life can be made, certain changes in the protoplasm must occur. A portion of the protoplasm becomes condensed and more highly specialized than the rest and, taking up a central position in the mass, acts as an executive presiding over the activities of the whole. This condensed and specialized portion is known as the **nucleus** (nux, L. = nut), and the mass of protoplasm containing it and whose activities it supervises is called the **cytoplasm,** i.e., cell-plasma. Nucleus and cytoplasm, contained within an extremely thin limiting membrane, constitute the basic unit of life—the **cell** (fig. 1).

The body of the lowest form of animal life, the *amoeba*, is just such a single cell. (Although in fig. 2 it is shown greatly enlarged, actually a magnifying lens is required to see an amoeba.) This tiny animal exhibits all the characteristics of 'purposeful' life—a mystic drive to remain alive and to keep its race alive.

Is there than a conscious utilization of oxygen and a seeking of food? Are the responses made by the amoeba to environmental changes conscious responses? Not as we ordinarily think, but the presence in the environment of beneficial or deleterious conditions so influences the nucleus

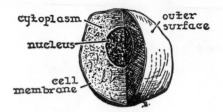

Fig. **1.** Three-dimensional diagram of a cell with a segment removed (greatly enlarged).

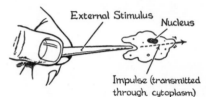

Fig. **2.** Amoeba responding negatively to stimulus.

that it initiates responses that may be called positive or negative (fig. 2). The amoeba will respond positively to the presence of food particles in its vicinity by changing the shape of its protoplasm and engulfing the particles. It will then 'digest' that food and convert it into energy, rejecting as a result certain waste products that it cannot use. Some of the food it will use to replace wornout parts of its own body.

By promptly changing its shape it will respond negatively and withdraw itself from harmful stimuli (stimulus, L. = a goad). Thus oxidation, digestion, elimination, reception of stimuli, and the locomotor responses called forth are essential activities of the protoplasm controlled by its nucleus. They foreshadow the great 'systems' of the complex body and it is these systems that will so fully engage our attention.

There comes a time in the life history of the amoeba, when there occurs a remarkable series of changes, first in the nucleus and then in the remainder of the protoplasm. Thus two objectives are accomplished: first, a complete renewal of the vitality and youthful vigor of the protoplasm; and second, a cleavage into two whereby from an elderly and decrepit amoeba two young and lusty progeny result. Each separates from the other to take up independently the struggle for existence. What the effective stimulus is whereby these changes are set in motion and brought to fruition is unknown, but this phenomenon of **cell cleavage** or division is vital to the survival of the amoeba as it is to most cells.

Cell Colony

The beginning of life for the individual of higher animal species is the result of the merging or fusion into one, of two cells derived from two parents. When, by the process of cleavage, this original cell divides, the resulting two cells do not separate from one another, as is the case in the amoeba, but continue to exist together. Subsequent divisions, occurring in rapid succession, lead to the formation of a cell mass or **cell colony.**

Life now becomes community life and the work to be performed in order to maintain it may be shared among the individual members. The cells begin to specialize and a division of labor takes place. This phenomenon is called **differentiation.** Eventually, about 25,000,000,000,000,000 cells form the human body.

We cannot hope in the space of a few pages to follow in any detail the changes whereby the single cell becomes the adult animal, but must content ourselves with a bare outline (page 8).

It has been noticed that the protoplasm of a cell controlled by its nucleus reacts to its environment. When, as the result of the formation of a cell colony, some cells find themselves at the surface of the mass and others in the interior, it is apparent that the environment for some cells becomes altogether different from that for others, so that the duties each group undertakes are determined by the advantages resulting from the peculiarities of its situation (fig. 3). Further, the duties assumed by a cell markedly influence its form and appearance (fig. 4). Thus it comes about that a whole science known as **histology** (Gr. = tissue study) is devoted to the investigation and recognition of the function of a cell by a study of the peculiarities of its form and appearance.

A group of cells collected together to

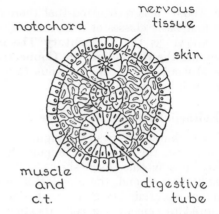

Fig. **3**. A cell colony produces specialists

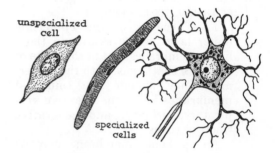

Fig. **4**. Specialized cells—from muscle and from nerve.

perform a specific function or functions is called an **organ.** The cells composing the essential structure of that organ are known as the **parenchyma.** It must not be supposed, however, that there is any organ consisting only of parenchymal cells. These essential cells are supported by a framework of less specialized cells. Insinuated among them are many tiny blood vessels carrying blood cells and many nerve fibers to regulate the activity of the organ.

CLASSIFICATION OF THE TISSUES OF THE BODY

The word 'tissue' is derived from the latin word 'texere'—a texture or fabric. [The word 'text' means a weaving together of words into a composition.] Even before the introduction of the microscope, anatomists observed that the parts of the body were woven together into a variety of patterns or textures and the use of the expression 'tissues of the body' came into general use. Today, it is recognized that there are but four basic classes of tissues and these are known as the **Elementary Tissues.** They are the **Epithelial,** the **Connective,** the **Muscular,** and the **Nervous** tissues. Each will be discussed briefly.

I. Epithelial Tissues

Two characteristics distinguish this group: first, the cells by arranging themselves in a mosaic, form sheets or lining membranes; second, the amount of intercellular material, being no more than is required to cement the cells together, is reduced to a minimum. The group is subdivided into three subsidiary groups:

1. EPITHELIAL TISSUES PROPER

These consist of the **epidermis** (outer layer of the skin) and those lining membranes that at some site or other are continuous with the epidermis. They include the **(mucous) membranes** lining the digestive, respiratory, urinary, and generative tracts or tubes.

An epithelial membrane may be only one cell thick (*simple*) or it may consist of several layers (*stratified*) (fig. 5). Furthermore, the shape of the cells is variable, being anything from flat and pavement-like (*squamous*) to tall (*columnar*). Thus, for example, an epithelial membrane consisting of flat cells in several layers is spoken of as a *stratified squamous epithelium* (fig. 6).

An epithelial cell may manufacture a substance that it secretes onto the surface of the membrane; it is then a *secretory cell.* Or, a bud-like or tubular extension from an epithelial surface may occur, forming what is known as a **gland.** A gland may remain in this simple form or, by secondary extensions of the bud or tube, may become compound. Moreover, the secretory cells may come to lie at a distance from the epithelial surface as a more-or-less independent organ. Such a secretory organ is then usually connected to its original epithelial surface by a nonsecretory

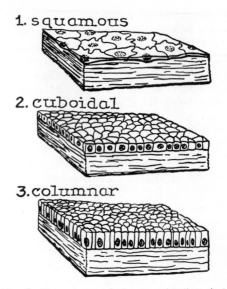

Fig. **5.** Three types of simple epithelium (schematic three-dimensional diagrams, modified after Ham).

tube—the **duct** of the organ—which serves to conduct the secretions to the epithelial surface. These conditions are illustrated in figures 7 and 347.

2. ENDOTHELIAL TISSUES

These are almost entirely confined to the inner lining of the walls of blood (and lymph) vessels (including the heart); they have no continuity anywhere with the epidermis. They consist of a lining of flattened cells that is only one layer thick. It presents an extremely smooth surface to the blood and so reduces friction and prevents blood clots.

3. MESOTHELIAL TISSUES

Mesothelium is a special lining membrane for the four great cavities of the body. These are the *peritoneal cavity* in the abdomen, the two *pleural cavities* in which the lungs are enclosed, and the *pericardial cavity* in which the heart lies. Mesothelium is also referred to as **serous membrane.** It consists of a sheet of areolar tissue whose free surface is covered by a single layer of flat cells—not unlike those of endothelium. The free surface is extremely smooth and slippery. The three serous membranes are

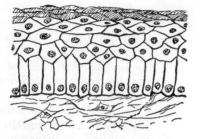

Fig. **6**. Stratified squamous epithelium (greatly magnified) showing the layers (strata) of cells as seen on raw surface of vertical cut into ordinary skin.

called **peritoneum, pleura,** and **pericardium.**

II. Connective Tissues

Widespread throughout the body and extremely diversified, connective tissues range from low-grade primitive tissue to such highly specialized tissue as bone. They are characterized by possessing a great deal of inert intercellular substance which may assume many forms (inter, L. = between).

An intercellular substance is the product of the cells in whose neighborhood it is found. It is of considerable importance, and, when abundant, gives a specific character to the tissue possessing it. In an epithelial tissue, such as a mucous or a serous membrane, only a minimal quantity of intercellular substance exists—enough merely to cement together the closely packed cells—and the tissue as a whole is characterized by the abundance of these living cells. In a connective tissue, intercellular substance is so abundant as to cause the cells producing it to lie scattered widely in it. It is by the character of the non-living intercellular substance rather than by that of the living cell that the following subdivisions of connective tissue are recognized.

1. AREOLAR TISSUE (fig. 8)
   In this the cells lie in a loose, irregular network of scattered fibers.*

---

*The word 'fiber' here denotes a strand of nonliving intercellular substance, i.e., a connective tissue fiber. The word is also used in association with muscle or nerve when it denotes a living cell (muscle fiber) or a part of one (nerve fiber).

This is a primitive form and due to its 'open' nature it is often called *loose areolar tissue*. Nature uses it as a bed for skin and mucous membranes and as one would use 'excelsior,' viz., for surrounding more important structures and filling in what might otherwise be empty spaces. It is the packing of the body.

2. ADIPOSE TISSUE
   This is areolar tissue that is impregnated with cells loaded with fat. It is the presence of many *fat cells* that gives the tissue its name (adeps, L. = fat). The combination of the two allied tissues is often called *fatty areolar tissue*. The areolar or adipose

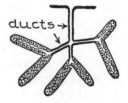
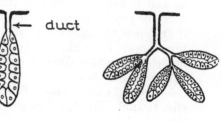

tubular

duct

alveolar

SIMPLE   COMPOUND

Fig. **7**. Types of glands (greatly magnified and diagrammatic). On left: three examples of simple (i.e., unbranched) glands. On right: three examples of compound glands.

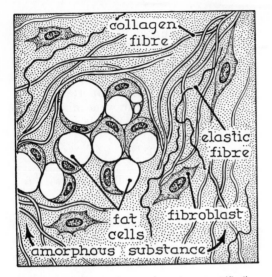

Fig. **8.** Areolar tissue (greatly magnified)—scattered cells lying in a loose meshwork of non-living fibers.

tissue deep to the skin is called **superficial fascia.**

3. DENSE FIBROUS TISSUE (fig. 9)

The intercellular substance predominates and consists of closely packed fibers or elongated strands. According to the nature of these fibers the tissue is variously designated. The fibers may be white in color and unyielding **(white fibrous tissue or collagen),** or they may be yellow in color and elastic **(yellow fibro-elastic tissue).** To this sub-group belong:

(a) **Tendon**—a tough cord or band composed of many closely packed white fibers and always part of a muscle, usually forming an attachment of muscle to bone.

(b) **Aponeurosis**—in reality, merely a sheet-like tendon.

(c) **Ligament**—resembling tendon except that it possesses varying amounts of elastic fibers and joins bone to bone. Ligaments occur in the neighborhood of joints whose movements they restrict.

(d) **Fascia**—dense fibrous tissue arranged in a sheet (not to be confused with an epithelial membrane) and most commonly found in association with muscles where, as **deep fascia,** it intervenes between muscles and the overlying superficial fascia. It forms envelopes for individual muscles, binds muscles together in groups, and, as **intermuscular septa** passing as partitions from the deep fascia to bone, separates groups of muscles from one another.

(e) **Reticulated connective tissue**—the very delicate trellis-work of loose areolar tissue provided for the support of the parenchyma of organs.

4. CARTILAGE OR GRISTLE (fig. 10)

In this the cells are separated from one another by wide intervals filled with what looks like a homogeneous matrix. If the matrix retains this structureless appearance—actually it contains many invisible fibers—the tissue is called **hyaline cartilage** and this is the most rigid variety (e.g., the 'Adam's apple'). If greater numbers of connective tissue fibers are present in the matrix, the tissue is called **fibro-cartilage** or **elastic cartilage** according to the chief type of fibers present (e.g., the elastic cartilage of the external ear).

5. BONE [AND DENTINE OF THE TEETH] (fig. 11)

This is a highly specialized connective tissue whose mode of origin will be considered later (see page 21).

6. HEMOPOIETIC TISSUE

This tissue, whose name means 'blood-making,' forms the *red marrow*

Fig. **9.** Dense fibrous tissue (greatly magnified)—scattered cells lying between dense bundles of non-living fibers.

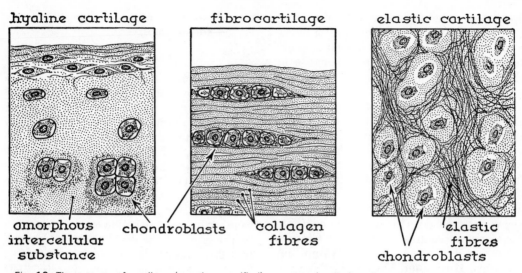

hyaline cartilage   fibrocartilage   elastic cartilage

amorphous   chondroblasts   collagen   elastic
intercellular   fibres   fibres
substance   chondroblasts

Fig. **10**. Three types of cartilage (greatly magnified)—scattered cells in a homogeneous matrix with vary-ing amounts and types of fibers.

Fig. **11**. Bone (greatly magnified)—many cells arranged in concentric circles in the inert material (cf. fig. 32).

of the bones (page 22) and is found also in several internal organs (e.g., liver, spleen) and in many lymph nodes scattered throughout the body. Unlike other connective tissues, he-mopoietic tissue is highly cellular since it produces millions of red and white blood cells every minute.

III. Muscular Tissues

The tissue known as muscle consists of highly specialized elongated cell masses possessing the inherent power of contrac-tion (see page 113).

IV. Nervous Tissues

Nervous tissues consist of highly special-ized cells of innumerable forms capable of

receiving stimuli and of initiating or trans-mitting impulses (see page 296).

The following summary is given for reference:

Classification of Tissues

I. EPITHELIAL
  1. Epithelial proper (and glands)
  2. Endothelial
  3. Mesothelial
II. CONNECTIVE
  1. Areolar }
  2. Adipose } superficial fascia
  3. Dense fibrous
    (a) *Tendon*
    (b) *Aponeurosis*
    (c) *Ligament*
    (d) *Fascia (deep and intermuscular septa)*
    (e) *Reticular*
  4. Cartilage
    (a) Hyaline
    (b) Fibro-
    (c) Elastic
  5. Bone and Teeth
  6. Hemopoietic
III. MUSCULAR
IV. NERVOUS

ANATOMY DEFINED. MAN'S KINSHIP

Anatomy has been defined authorita-tively as the **"Science of the Structure of Organized Bodies."**

Such a definition of anatomy may be applied to the investigation of any form from the lowest to the most highly developed. Restricted to animals that possess that very characteristic structure, a backbone, the science becomes known as **Vertebrate Anatomy**—vertebrate because in the language of the anatomist a backbone is a **vertebral column.**

Man is a vertebrate animal and it is only the emphasis placed on this or that structure that distinguishes him from other vertebrates for, after all, a common environment demands a fundamental similarity of structure best suited to live in that environment. When the environment shows major differences the structure of the vertebrates does likewise. Hence, there have arisen five great divisions or **Classes** of the vertebrates, viz., **Fishes, Amphibia, Reptiles, Birds,** and **Mammals.** They reflect major differences in structure demanded by major differences in environment. "A bird is known by its feathers, a beast (mammal) by its hair, a fish by its fins;" while amphibia and reptiles, representing as they do a transition from an aquatic to a terrestrial existence, are "neither fish, flesh nor fowl" (Gadow).

Each of these five great **Classes** is further subdivided into **Orders,** which are, in turn, divided into **Genera,** and each genus contains various **Species.** Finally, there are different **Varieties** in most species. Each subdivision is based on structural differences less and less important and pronounced as we proceed, so that ultimately it may tax the ingenuity of a trained observer to distinguish two species of the same genus; but no one can mistake a mischievous monkey for his favorite dog.

Modern man, with the great Apes and Monkeys, belongs to the order **primate** of the class **mammal.** He is the only living species, **sapiens** (L. = wise), of the genus **homo (man).** Anthropologists have found remains of several extinct—apparently not very wise—species of man, e.g., Neanderthal man, Rhodesian man, Heidelberg man. They also have found bones of animals, now extinct, which have characteristics midway between human and simian, e.g., the South African 'man-apes', the 'erect ape-man' of Java. Based upon many and various findings and upon a flood of related data, the theory that man's antiquity dates back millions of years has received added proof.

EMBRYOLOGY

The life of an individual begins with the union of two cells—one from the mother, the other from the father. The maternal cell is the **ovum** (L. = egg), the paternal cell a *spermatazoon* or **sperm** (Gr. = seed). The union, which is known as **fertilization,** takes place deep within the mother's body. The fertilized ovum, remaining within the mother's body, becomes adherent to the inner lining of the *uterus* or womb, the special hollow organ that is to be the home of the developing child for its first 280 days of life (ten lunar months).

When a sperm and an ovum unite, their substances become intermingled to form a single cell which possesses potentialities inherited from both parents. The single-cell stage is probably a momentary one followed by **cleavage,** a rapid series of divisions into more and more cells. At first, these cells obtain their nutrition directly from the tissue fluids of the inner lining of the uterus.

Cleavage produces a rounded mass of closely packed cells, the *morula* (fig. 12). In this mass of cells, **differentiation** soon becomes evident, one part of the mass becoming differentiated into cells that are destined to become the developing child or **embryo.** Another part differentiates to become the outer membrane (the *chorion*) that surrounds the embryo (fig. 13). A part of this membrane differentiates further to form a special and wonderful structure, the **placenta** (see below).

The **embryonic mass** of cells further differentiates into two adjoining masses of cells. One is destined to cover the body and is known as the **ectoderm** ('outer skin').

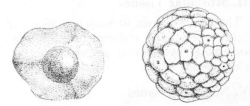

Fig. **12.** A single fertilized cell splits and resplits to form a clump of cells, the morula.

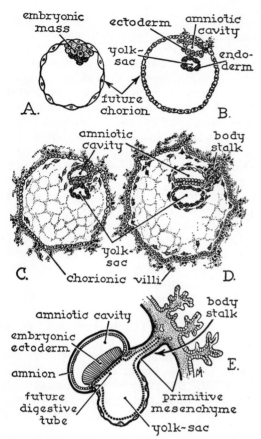

Fig. 13. Development of early embryo and its membranes (schematic; greatly enlarged). A. The morula becomes hollow, one part forming embryonic mass. B. Mass differentiates into amniotic sac (ectodermic vesicle) and yolk-sac. C. Outer surface of chorion proliferates. D. Body stalk now forming. E. At about 2 weeks after conception (magnified × 10).

individual—the skin—and, by a special process, the nervous system (page 12 and fig. 17). The rest of the ectoderm forms the wall of a 'balloon' filled with fluid which expands and soon envelopes the whole embryo (including the yolk-sac), thus forming a second (inner) surrounding membrane. This membrane is the *amnion*, and its contained fluid the *amniotic fluid*. At this stage the embryo is still only a few weeks old and less than two mm (1/16″) long (fig. 13).

Some primitive cells, forming a tissue known as **mesoderm,** appear quite early in the region between endoderm and ectoderm. Mesoderm increases rapidly in quantity and importance. It is destined to form the connective tissues (including bone), the muscles, the blood, and the vessels of the body. The *notochord*, or primitive backbone, develops as a condensation of cells in the mesoderm (page 11, and figs. 3 and 16). At this stage, the body of the embryo becomes organized into a longitudinal series of about 35 blocks of tissue known as *somites* (fig. 13.1). Each of these body segments develops independently for a time but gradually the segmental appearance is lost. However, many souvenirs of the somites are apparent even in the adult, e.g., vertebrae and ribs.

In the second month of life in the uterus, the embryo grows rapidly and comes more and more to resemble the final form. The head end becomes enlarged; the other end forms a definite tail which persists for some months. On the sides of the embryo two pairs of lumpy projections appear. These

The other eventually lines, in a sense, the inner surface of the body and is known as the **endoderm** ('inner skin').

Within the endodermic mass a space appears and enlarges to form the *yolk-sac*. The yolk-sac eventually becomes a tube that runs the length of the embryo and is known as the digestive tube or gut.

The ectoderm also becomes a hollow mass of cells known as the ectodermic *vesicle* (L. = little bladder). The portion of the ectodermic vesicle adjoining the endoderm increases in size and importance with a great proliferation of cells. It eventually forms the outer surface of the

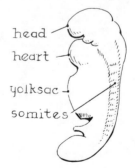

Fig. 13.1. A 25-day embryo shows about 14 somites along its trunk. (Modified after Langman.)

limb buds soon grow longer and form the upper and lower limbs (fig. 14).

At the end of eight weeks, the embryo looks quite human, having a large head with a well formed face, hands and feet with fingers and toes, a functioning heart, and a circulatory system. But the entire length is only 30 mm (about 1″), and the embryo floats in *amniotic fluid* surrounded by two cellophane-like membranes, the inner *amnion*, and the outer *chorion*. The belly of the embryo is connected by a cord —the **umbilical cord**—to a specialized part of the chorion known as the **placenta.** The cord contains blood vessels which carry blood from the embryo to the placenta and back to the embryo.

The placenta is to the unborn child the equivalent of a combination of lungs, kidneys, and food passage. Here, the blood of the child meets but does not mingle with the blood of the mother. Thin partitions separate the blood of mother and child, and there is a rapid interchange of oxygen, carbon dioxide, food, and wastes which pass readily through the barrier.

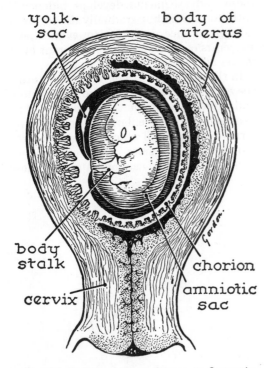

yolk-sac

body of uterus

body stalk

cervix

chorion

amniotic sac

Fig. **14**. Embryo and membranes at 2 months. About actual size.

In the third month, development of all the organs is completed and thereafter the chief process is one of enormous growth in size. The first two (or three) months are called the **Embryonic Period,** the remaining (seven or) eight months the **Fetal Period.** It is during the embryonic period that almost all congenital abnormalities (e.g., of the heart) become apparent. Indeed, it is now known that many congenital abnormalities are caused by external factors affecting the child at this stage, such as toxic drugs or German measles in the mother.

The **fetus** grows very rapidly. At three months the total length is less than 8 cm and the weight, less than ½ kg (1 lb). Of course the uterus grows to accommodate the increase in size of the fetus.

At the fifth lunar month, the fetus begins to exercise skeletal muscles, moving the trunk and limbs. The mother 'feels life' in her womb.

In the last months of the pregnancy the child becomes bigger and better able to cope with the external environment (fig. 15). If, however, birth occurred prematurely, it would take very special care to keep the child alive.

At the end of about ten lunar months, a train of processes known as **labor** begins. The protective uterus changes its rôle and, by contractions of its muscular wall, expels the child. Birth is a momentous occasion. Now the lungs assume their adult function. Important changes occur in the circulatory system as now more blood must pass through the lungs and none through the umbilical cord. The kidneys, too, begin to function in earnest. The baby is hungry and must be fed.

SYSTEMS OF THE BODY
Their Origins and Locations

The following very brief and general survey indicates how the various systems came to occupy their relative positions. It is more particularly designed to give the beginner a basic anatomical vocabulary so that he can follow subsequent discussion without feeling the need for constant explanatory digression.

The number of systems assigned to the

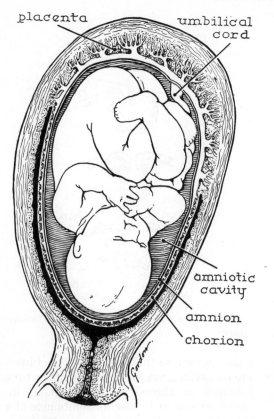

placenta

umbilical cord

amniotic cavity

amnion

chorion

Fig. **15.** Fetus at full term lies in amniotic fluid, surrounded by membranes and muscular wall of uterus.

7. **Generative or Reproductive,** closely allied in development to the foregoing and often included with it under the term **Urogenital (Genito-urinary) system.** It exists for the perpetuation of the species.

8. **Circulatory (and lymphatic),** conveying nutrition to all tissues and renewing tissue fluids.

9. **Nervous,** for the regulation of the activities of all the tissues of the body.

10. **Integumentary,** or the skin and its derivatives—a protective covering with a variety of functions.

11. **Endocrine glands,** producing chemical 'messengers' (hormones) that greatly influence the chemistry of the body.

Notochord and Its Fate. If one were to examine the embryo of any mammal while its structure was still simple and relatively uncomplicated, one would be struck by the fact that it takes form around a rod-like condensation of tissue (cells) running along the length of the back of the embryo (see fig. 3). This rod, the **notochord** (noton, Gr. = back), is the forerunner or precursor of the adult backbone or vertebral column. It is comparable to the keel of a ship and just as the 'ribs' of a ship are fastened to the keel so the ribs of the animal are fastened to the vertebral column. These rib elements are at first very short but they grow around the developing body until they ultimately meet in front and so enclose a chest cavity. With the ribs grow muscles in similar fashion. Where the ribs fail to appear muscles alone form the body wall, as they do in the abdomen or lower half of the trunk (see fig. 19).

When bone surrounds, incorporates, and replaces the notochord it is laid down as a series of blocks or segments united to one another by discs of gristle or cartilage; it thus retains some of the flexibility of the softer notochord.

As a result of this segmentation of the vertebral column, each block of a bone (a **vertebra**) supports typically a pair of ribs and on each side of it lie one block of muscle—a myomere (Gr. = muscle share), one artery and one nerve; to the composite

body depends on how certain organs, such as the skin and the ductless glands, are regarded. There is no need for rigidity in this matter and different authors give different numbers. For our purposes eleven systems will be listed:

1. **Skeletal,** comprising the bones and certain cartilaginous parts.

2. **Articular,** or the system of joints and ligaments.

3. **Muscular,** which, operating with 1 and 2, produces locomotion.

4. **Digestive,** for the assimilation of food.

5. **Respiratory,** for the provision of oxygen and the elimination of carbon dioxide.

6. **Urinary,** for maintaining constant the chemical composition of the blood by eliminating waste products.

unit the name *somite* or **body segment** is applied, the wall of the trunk being but a succession of such segments applied to one another. This arrangement may later become considerably modified and obscured but it is desirable to be familiar with the basic pattern (fig. 16).

Neural Tube—Brain and Spinal Cord. The vertebral column not only gives rigidity to the body, but also gives protection to an extremely important structure that develops and lies behind it. This, unlike the notochord, is a hollow tube and, being the beginning of the nervous system, is known as the **neural tube** (neuron, Gr. = nerve). It results from an infolding of a strip of surface cells lying immediately behind the notochord. Figure 3 illustrates its position, figure 17 its mode of origin. Notochord and neural tube extend parallel and adjacent to one another from end to end of the developing embryo. The front end of the tube becomes the **brain,** the remainder, the **spinal cord.** While the cord is destined to remain in actuality a tube (although its

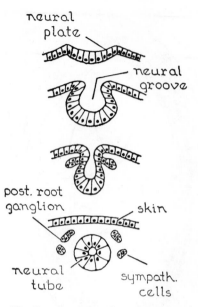

Fig. **17**. Development of neural tube (cross-section).

walls become so thick as almost to obliterate its cavity), yet the front end of the tube becomes so altered by its enormous increase in size as to lose all semblance of a tube. However, the cavities or **ventricles** in its interior proclaim its hollow origin.

The factors responsible for the marked increase and expansion of the fore-end of the neural tube are associated with the fact that it is at the fore-end of the animal that the special senses (smell, sight, hearing, etc.) develop. It is by these senses that the brain is informed of very many important changes in the external environment. The sense of smell is the first of these to appear since, while the animal body remains close to the ground (as in the reptiles), smell is the sense most informative with respect to environmental changes. When the body is lifted off the ground by the increasing length and strength of the limbs, the animal is able to command a wider view of his surroundings and sight becomes of increasing value. This is particularly true of the primates, who, forsaking the ground for an arboreal existence, can still further widen their horizon. It is still more true of the flying birds. Man's upright posture demands keen vision even though he, alone among the primates, has forsaken the trees

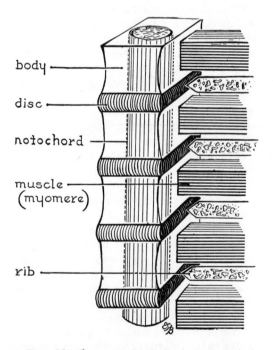

Fig. **16**. The notochord is incorporated in bony blocks and there disappears. Between the blocks it persists as the pulpy nucleus of the cartilaginous disc (see text).

for the ground. In all mammals hearing is well developed and remains acute.

From all of these senses impulses constantly reach the fore-end of the neural tube which increases in size to receive, interpret, and co-ordinate these impulses, to initiate suitable motor responses, and to store up a memory of them resulting in experience.

But this is not all. The body is constantly receiving impressions of touch and they too demand suitable representation. Then, too, the animal gradually becomes aware of his own body. None of us finds it necessary, for example, to look at his limb to see whether the knee is bent or straight. Impulses are forever reaching the brain from the joints and muscles to tell us of their position and state of tension. Lastly, impressions reach the brain from the organs of the body keeping it informed of their state of well-being and activity though, fortunately for us, these impulses rarely seem to reach a conscious level. Thus, the brain primarily develops to receive and sort out sensory impulses from the external and internal environments and to initiate appropriate motor responses. These are 'animal' functions. How do the higher functions arise? How do we 'will' to do things? How do we exercise judgment and lastly what is consciousness? These are vital and important questions but they are beyond the scope of our present enquiry.

The spinal cord, unlike the brain, remains primitive and, in a very general way, exhibits that segmentation we have noted whereby each cord segment exercises some control over its own body segment. It is, of course, dominated by 'higher centers' in the brain.

From the brain 12 pairs of **cranial nerves,** and from the spinal cord 31 pairs of **spinal nerves** issue. These nerves constitute the **peripheral nervous system** in contradistinction to the brain and spinal cord themselves which are the **central nervous system.**

The close proximity of the spinal cord to the developing vertebral column allows each vertebra to afford protection to the cord by throwing an arch of bone around it (fig. 18). Each arch is known as a **vertebral**

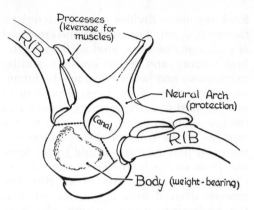

Fig. **18.** Parts of a vertebra (with associated ribs).

(or **neural) arch** and the longitudinal canal formed by the succession of arches is known as the spinal or **vertebral canal.** At the sides, between adjacent neural arches, a series of paired **intervertebral foramina** (pl. of foramen, L. = hole) permits the escape of the series of paired segmentally disposed spinal nerves (fig. 47).

The bony protection of the brain offers a greater problem. It is so voluminous that vertebrae, strive how they may, cannot keep pace with its increasing size. The floor of the brain-case does consist in part of highly modified vertebrae fused together and expanded sideways, but that is the best the vertebrae can offer. How then is the brain protected at the sides and above?

The vault of the cranium develops from the subcutaneous membranes covering the brain—hence the bones are called *membrane bones.* They meet the modified vertebrae of the base (*cartilage bone*) and so complete the brain-case or **cranial cavity.**

In front of the brain-case in most animals but below it in man lie the bones of the face and the nasal and oral (mouth) cavities. Between the brain-case and the face lies a pair of spaces, the orbital (eye) cavities or **orbits.**

The Myomeres. In the mesoderm close beside each vertebral body, paired blocks of muscle tissue make their appearance; they are known as **myomeres** or **myotomes** (Gr. = muscle 'slices') and are the source of the musculature of the trunk of the body.

Each myomere divides in a characteristic fashion (fig. 19). The back or posterior part of each remains associated with the vertebral column and by union with similar parts above and below becomes the intrinsic musculature of the back. The front or anterior part of each myomere undergoes a further division. Part gives rise to the muscles of the front of the vertebral column and part migrates round to the sides and front of the body. Thus, the walls of the *thoracic cavity* consist of ribs and muscles derived from myomeres, whereas the *abdominal cavity*, where ribs have failed to grow, is enclosed by myomeric parts only.

Limb Buds. During the second embryonic month, when the embryo is 30 mm long (about 1″), the upper and lower limbs appear as little buds. These rapidly grow in length and become more and more like the adult's arms and hands, legs and feet.

Joints. Development described on page 75.

The Digestive Tube. In front of the notochord another hollow tube appears (fig. 20). It is the **digestive tube** or **alimentary canal**, and its upper end is fixed to the outside of the base of the skull behind the nose and mouth. It acquires an opening forward into the mouth and, until the palate develops, the mouth or oral cavity is continuous with the nose or nasal cavity (fig. 21). In the neck and upper part of the thoracic cavity, the digestive tube, here known as the *esophagus* or gullet, lies in front of the vertebral bodies; but as it

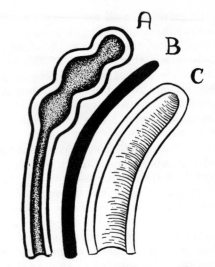

Fig. **20**. The notochord (B) is a stiffening rod for the support of neural tube (A) and digestive tube (C) (cf. fig. 3).

proceeds downward it gradually takes up a more forward or anterior position until, in the lower part of the thorax and in the abdomen, it leaves room behind itself for other systems. In the *pelvic cavity* (enclosed between the two hip bones) the digestive tube returns to a midline position in front of the lower end of the vertebral column. It finally breaks through the surface and opens to the exterior.

A striking feature of the abdominal portion of the digestive tube is the remarkable increase in its length and the considerable variation in its caliber; the numerous coils into which it is thrown possess a degree of mobility not granted to other parts, and the variation in caliber permits its division into the **stomach, small intestine,** and **large intestine.**

Furthermore, if the **digestive tract** (as it is more usually called) is to serve the purpose of digestion, it must be provided with digestive enzymes. Some enzymes are provided by the cells lining the interior of the tube; others are provided by organs which have grown bud-like from the wall of the tube. The latter are the **liver** and the **pancreas,** and the sites of their budding are denoted by the openings of their ducts by which they empty their secretions into the digestive tube. The liver ultimately becomes the largest organ in the abdomen

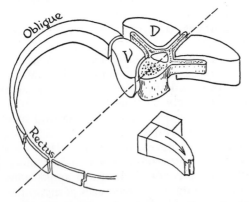

Fig. **19**. Splitting of a myomere. D. = dorsal; V. = ventral.

and is situated immediately below the **diaphragm**—the transversely disposed muscular partition separating the abdominal cavity below, from the thoracic cavity above. Because the diaphragm is domed upward, the liver enjoys the shelter and protection of the lower ribs. The pancreas lies below the liver and behind the stomach; it is placed transversely across the back of the upper part of the abdominal cavity.

The close relationships of the **spleen** to the stomach, pancreas, and diaphragm justify mentioning it now even though it has no functional relationship to the digestive tract. The spleen is about the size of a fist and is a ductless organ that has important functional relationships to the blood. It lies well around to the back below the extreme left part of the diaphragm. It is attached to the stomach by a membrane and is sheltered by the lower ribs.

The Respiratory System. This develops as a tube growing out of the digestive tube in the region later to be identified as the voicebox or **larynx.**

Below the larynx, it is known as the **trachea** or windpipe, and it lies directly in front of the gullet or esophagus. In the thorax, the trachea divides into two branches (right and left **bronchi**) which divide and subdivide like the growing

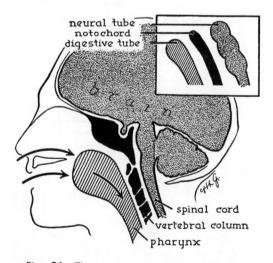

Fig. **21.** The upper end of the notochord is incorporated in the middle of the base of the skull.

branches, twigs, and leaves of a tree to form the two **lungs.** The lungs are separated by a midline partition, the **mediastinum,** occupied by the developing heart and great vessels.

The Circulatory System. Very early in the life of the embryo, small collections of blood cells appear known as *blood islets.* These coalesce to form tiny vessels which join in turn to form larger and larger vessels. The tiny vessels remain among the tissues as **capillaries** (L. = hairs). The blood enters the capillaries from **arteries** and gives off oxygen and food products which pass through the walls of the capillaries into the tissue fluid to nurture the cells. The reverse process of taking up wastes from the tissue fluids is accomplished at the same time and the blood is returned to **veins.** The largest veins join the largest arteries at the developing heart and so the blood makes a 'circle,' giving us the term *circulatory system.*

Developmentally the **heart** is a greatly enlarged and modified blood vessel. It lies immediately behind the sternum or breastbone and between the two lungs. But the main stems of the blood vessels lie in front of the vertebral column in the space left for them by the forward shift of the digestive tube. In order to gain this position from the heart in front, the great artery **(aorta)** arches backward and finally resembles a cane with a curved handle. Many branches arise from the aorta and reach all parts of the body. A great vein and the principal lymph duct (see page 292) accompany the aorta where it lies in front of the vertebral bodies.

The Urinary and Reproductive Systems. Closely associated in their development, they consist primarily of paired structures. The chief (paired) organs of the urinary system (fig. 22), the **kidneys,** develop and lie in the upper part of the abdomen one on each side of the great vessels and close to the vertebral column.

Each kidney connects by a tube, the **ureter,** to the **urinary bladder.** This, at first, is not an individual structure but merely the front part of a chamber or reservoir at the lower end of the digestive tube and known as the *cloaca.* When the

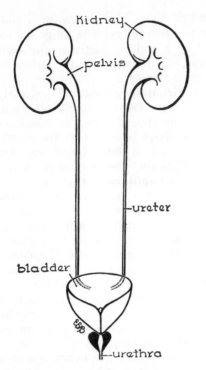

Fig. **22.** Scheme of the urinary system. (The pelvis of the ureter must not be confused with the part of the abdomen called the pelvis.)

cloaca splits into two independent reservoirs early in embryonic life, the front or anterior one (the bladder) receives the ureters (fig. 23).

The bladder occupies the more anterior position in the pelvic cavity; the ureters, consequently, must pass from a posterior position in the abdomen to an anterior one in the pelvis. Finally, the bladder discharges to the exterior by way of a tube or duct known as the **urethra.**

The essential organs of generation, the paired **testes** (male) or **ovaries** (female), originate beside the kidneys and close to the vertebral column, but, by migrating downward, they leave the kidneys. Each testis migrates so far that it actually makes its way through the muscles of the anterior abdominal wall and takes up its position in a sac formed by an outpouching of the skin of the abdomen. This sac is the **scrotum** (L. = a skin) (fig. 24).

The first part of the tube or duct conveying the secretion of the testis is known as the **epididymis.** Then, as the **ductus (vas)**

deferens (L. = vessel carrying down), it passes down the back of the bladder to empty into the urethra just beyond. When the testis migrates to the scrotum it carries its duct with it, and this duct must, in order to reach the back of the bladder, retrace its course from the testis back through the abdominal wall. On the lateral side of each ductus deferens, where it lies against the back of the bladder, is a reservoir or sac joined to the ductus and known as the *seminal vesicle* (semen, L. = seed). Surrounding the first 3 cm of the urethra is a small solid secretory organ known as the **prostate gland.** This median unpaired organ is about the size of a chestnut and adds its secretion to those of the testes and seminal vesicles.

The **ovaries,** not migrating so far downward, enter the pelvic cavity where they lie near its side wall. Each is provided with a tube, the **uterine tube,** which runs horizontally medialward to enter the side of the **uterus** or womb. The uterus is situated in the middle of the pelvic cavity, behind the urinary bladder and in front of the rectum (fig. 25). The uterus is a thick-walled, hollow, muscular organ whose tapering lower end, the neck or **cervix,** projects into, and is embraced by, the upper end of a canal leading to the exterior and known as the **vagina.**

HOW ANATOMY IS STUDIED

An anatomist thinks it absurd and profitless to consider form and structure without reference to function.

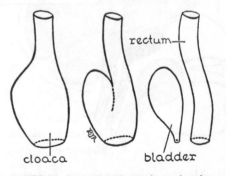

Fig. **23.** In birds the cloaca is a chamber common to digestive and urinary systems. In man it splits into bladder and rectum.

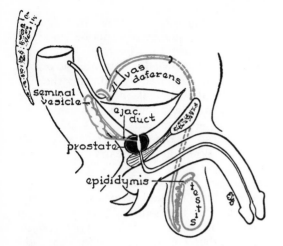

Fig. 24. Scheme of male pelvis (from right side). Reproductive organs in red.

The reverse is also true: function depends on structure.

There are several avenues of approach open to the student of human anatomy; each, by the particular view it affords, can reveal an aspect of the subject that other avenues leave obscure. It will be helpful to note briefly some of these avenues and to point out the special value of each.

1. Gross dissection, or the process of cutting up the cadaver in order that some insight may be gained into how the living body is constructed. This is the approach that gave its name to the science, for the literal meaning of the word 'anatomy' is to cut up or to dissect. It is the most direct approach and, on the whole, the most authoritative; our textbooks are chiefly descriptive accounts of organs and parts as they are revealed by this method. It is, therefore, important to observe that the human cadaver lying before the anatomist on the dissecting table is the highest authority to whom he can appeal. Vesalius, the father of modern anatomy, wrote in 1561, "I shall examine that true book of ours, the human body of man himself."

2. Microscopic anatomy (histology), or the study of the tissues that make up the gross organ or part. It is of inestimable value and today is one of the most important links uniting anatomy and physiology. It has become a major division of the subject and a science in its own right

demanding its own separate treatment. In this book, an attempt has been made to outline briefly the microscopic structure of the important tissues.

3. Developmental anatomy, or the study of the development of the adult form from its earliest beginnings as a fertilized ovum or egg. It is divisible into two periods, **pre-natal** and **post-natal** (natus, L. = birth). The study of pre-natal development is called **Embryology;** it is a period of intra-uterine development which, in man, lasts for ten lunar months, and is subdivided for convenience into an **embryonic** and a **fetal period.** The more important post-natal changes are described in Chapter 15.

4. Surface anatomy, i.e., the study of the living body. It is remarkable how much information may be gained by investigating intelligently the living human body, and it is still more remarkable how little inclined is the beginner to take advantage of it (see Plates 1, 2, 3, and 4, preceding page 19).

Bones can be felt (palpated) and their shapes and positions appreciated. Muscles can be made to act and their forms, attachments, and uses demonstrated. Joints can be moved and their ranges and limitations observed. Arteries, by their pulsations, reveal their situations. Veins can be seen through the skin and their courses traced. Finally, many nerves can be rolled under the finger and their positions located.

5. Comparative anatomy, or the study of form and structure in other animals. By

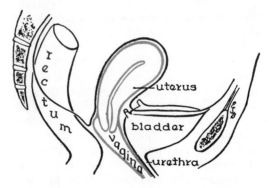

Fig. 25. Scheme of female pelvis (in sagittal— i.e., vertical, front to back—section). Note relative simplicity compared to male.

its means information may be obtained that is unavailable when one animal alone is studied. The study of comparative anatomy is of the utmost value to the advanced student and will often serve to gratify legitimate scientific curiosity.

These are some of the avenues of approach open for study and none should be neglected; there are many others and some will be mentioned as occasion requires.

An elementary understanding of some of the so-called *pure sciences* permits a better appreciation of how the body performs its daily tasks. For example, the properties of bone are understood because the chemist has revealed that bone is composed of inorganic salts and organic matter; the uses of bones, joints, and muscles are interpreted according to the principles of the actions of levers; some enlightenment on the conduction of the nerve impulse, if not on its nature, results from a familiarity with the phenomena of electricity; an understanding of how hearing is possible follows from a knowledge of the conduction of sound; and the anatomy of the eye is illuminated by an acquaintance with the rudiments of optics.

Terminology. He who desires to read a book in a foreign language must first learn that language. Anatomy is an ancient science and it, too, has its own language. Most of the scientific words and terms used come from Latin or Greek so that some familiarity with these two languages is a considerable asset. If however, 'in Latin and Greek we are sadly to seek' we must do the best we can. In this book the derivations of most of the technical words are given when they are first encountered.

A good rule to follow is one that is employed in every day life. No one ever uses a word whose meaning he does not know. That would be absurd, for language is a vehicle for the transmission of thoughts and ideas. The same rule applies in anatomical studies; thus, never use a word or term whose meaning is not understood, or conversely, **always look up in a good dictionary the meaning of every new word or term.** It is astonishing how much lighter the task then becomes.

If it is remembered that anatomy is a descriptive science and that the names given to structures are very often descriptive of some quality or function they possess, it is easy to appreciate how much anatomical knowledge can, almost unconsciously, be acquired by following the simple rule above mentioned. For example, when a muscle is found bearing the name *extensor carpi radialis longus*, instead of being awed by the designation, we do well to ask, "What does it mean or why did the muscle get such a name?" Reference to a dictionary reveals that *extensor* is a noun derived from the verb to extend or straighten; *carpi* is the genitive ('*of——*') of the latin noun *carpus* meaning a wrist; *radialis* is an adjective agreeing with *carpi* and refers to the radius—the outer of the two bones of the forearm; *longus* qualifies *extensor* and of course means long, and at the same time it implies that there is also a *brevis* or short extensor. The Latin term then becomes—*the muscle on the radial side of the forearm that is the longer of two and that extends or straightens the wrist.* (One might even become a little critical and say that the more desirable form of the term is *extensor longus carpi radialis*.) The situation and function of the muscle are recalled simply by knowing the meaning of the term.

Anatomical Position. In order to avoid confusion, anatomists the world over have agreed to adopt the convention of referring all descriptive accounts to what is known as the **anatomical position** (fig. 26). It is the position of the living body standing erect with the arms by the sides and the palms facing forward. *Anatomy is a science of the living, not of the dead,* and although our knowledge is, perforce, acquired from the cadaver, it is in terms of the living that we must think. Therefore, all descriptive speaking and writing are in terms of the living and refer to the above-mentioned position.

It requires a little practice and self-discipline to conform to this convention but science *is* a discipline and its good habits must be acquired. The greatest difficulty lies in overcoming the tendency to misuse prepositions—and that seems to be true in the learning of any new language. It is necessary to be wary and to remember that prepositions such as *over, under, behind, in*

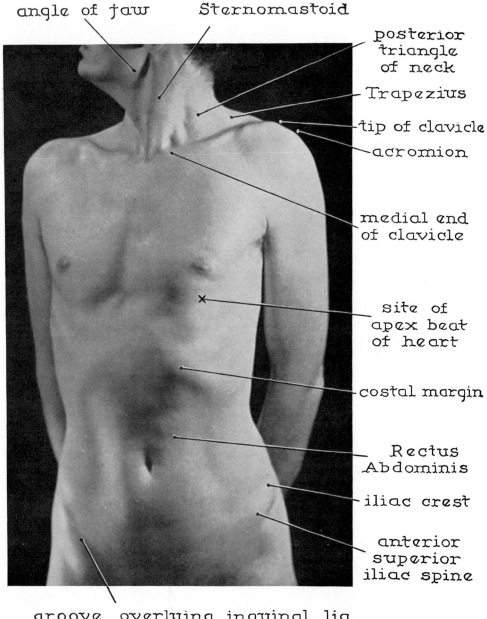

angle of jaw   Sternomastoid

posterior
triangle
of neck

Trapezius

tip of clavicle

acromion

medial end
of clavicle

site of
apex beat
of heart

costal margin

Rectus
Abdominis

iliac crest

anterior
superior
iliac spine

groove overlying inguinal lig.

Plate 1
SURFACE ANATOMY

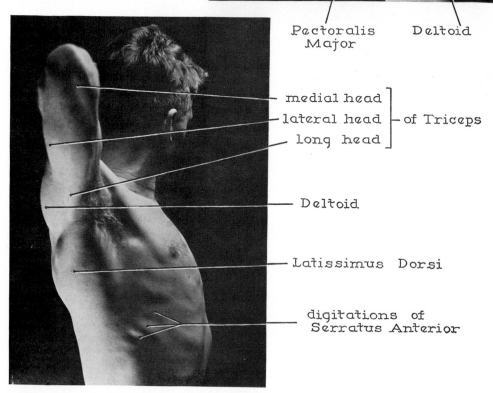

pisiform and
Flexor Carpi Ulnaris

Abd. Pollicis Longus

Flexor Carpi Radialis

Brachioradialis

basilic vein

olecranon process

medial head of Triceps

Biceps

long head of Triceps

Pectoralis
Major

Deltoid

medial head
lateral head
long head

⎤
⎬ of Triceps
⎦

Deltoid

Latissimus Dorsi

digitations of
Serratus Anterior

Plate 2
SURFACE ANATOMY

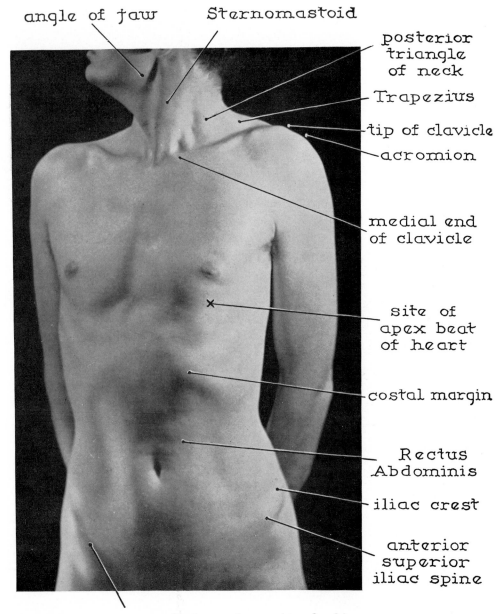

angle of jaw   Sternomastoid

posterior
triangle
of neck

Trapezius

tip of clavicle

acromion

medial end
of clavicle

site of
apex beat
of heart

costal margin

Rectus
Abdominis

iliac crest

anterior
superior
iliac spine

groove overlying inguinal lig.

Plate 1
SURFACE ANATOMY

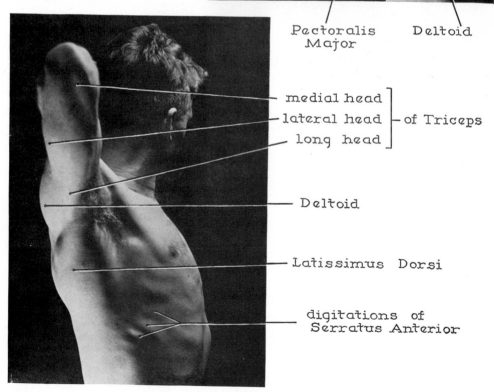

pisiform and
Flexor Carpi Ulnaris

Abd. Pollicis Longus

Flexor Carpi Radialis

Brachioradialis

basilic vein

olecranon process

medial head of Triceps

Biceps

long head of Triceps

Pectoralis
Major

Deltoid

medial head
lateral head
long head

of Triceps

Deltoid

Latissimus Dorsi

digitations of
Serratus Anterior

Plate 2
SURFACE ANATOMY

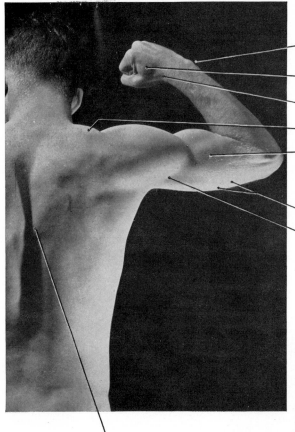

head of Ulna

1st Dorsal Interosseus

snuff box

Trapezius

Brachialis

long head of Triceps

post. border of Deltoid

spine of 7th cervical vert.

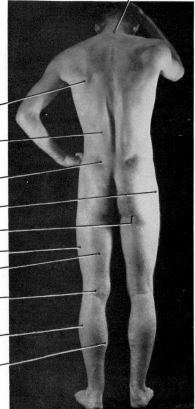

inferior angle of scapula

Sacrospinalis

post. superior iliac spine

greater trochanter

fold of buttock

Vastus Lateralis

hamstrings

popliteal fossa

Gastrocnemius ⎡ lateral head
⎣ medial head

Plate 3
SURFACE ANATOMY

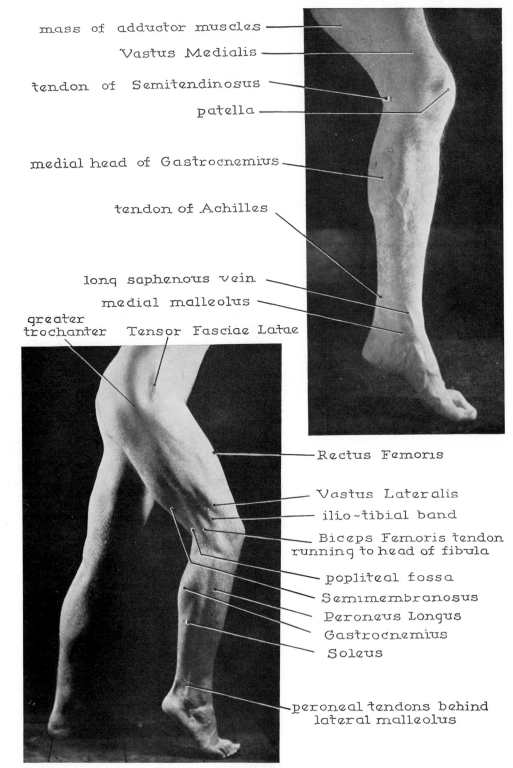

mass of adductor muscles

Vastus Medialis

tendon of Semitendinosus

patella

medial head of Gastrocnemius

tendon of Achilles

long saphenous vein

medial malleolus

greater trochanter   Tensor Fasciae Latae

Rectus Femoris

Vastus Lateralis

ilio-tibial band

Biceps Femoris tendon running to head of fibula

popliteal fossa

Semimembranosus

Peroneus Longus

Gastrocnemius

Soleus

peroneal tendons behind lateral malleolus

Plate 4
SURFACE ANATOMY

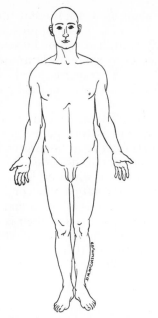

Fig. **26**. The anatomical position

older textbooks and some physicians incorrectly using 'inner' and 'outer' to mean medial and lateral because many years ago this was acceptable.

Planes of Body. One can divide the human body along many different planes. The plane which divides the body into right and left parts is a *sagittal plane* and many million sagittal planes are possible. The most useful is the *median sagittal plane* which divides the body into perfect halves in the midline (fig. 27). Any vertical plane passing from side to side and dividing the body into anterior and posterior parts is a *coronal (or frontal) plane* (fig. 28). Any plane cutting across the long axis of a structure at right angles is a *transverse plane* and, of necessity, transverse planes of the human body as a whole are all *horizontal* planes (fig. 29). However, transverse planes—or *cross-sections*—of structures or organs other than vertical ones are not horizontal.

Use of Specimens. Many students in courses related to medicine are today ac-

front, *above*, *below*, etc., have different interpretations according to whether the body is erect or recumbent. In brief, the danger of misunderstanding a written account is real if the accepted convention that is always implied is not remembered.

Comparative Terms. Terms of relative position come in pairs. Thus we have *anterior* (toward or at the front) and *posterior* (toward or at the back). To avoid confusion, many anatomists use *ventral* (for anterior) and *dorsal* (for posterior). These terms are often used in comparative anatomy and zoology. *Lateral*, meaning away from the midline of the body, is the opposite of *medial* or toward the midline. *Median*, of course, means midline and is used in many other sciences, including mathematics. *Superior* means higher or above and is the opposite of *inferior*. The beginner often confuses these two terms with *deep* and *superficial* which refer to depth from the surface—any surface—be it on the front, back, sides, or bottom of the body. Referring to the limbs, *proximal* and *distal* mean closer to and farther from the trunk, i.e., superior and inferior.

Internal and *external* refer only to body cavities, and *inner* and *outer*, to be correct, must conform to this definition as well. However, the student will sometimes find

Fig. **27**   Fig. **28**

Fig. **27**. Median sagittal plane
Fig. **28**. Coronal (or frontal) plane

Fig. **29**. Transverse (or horizontal) planes

corded the privilege of entering a dissecting room. The contact they make with human material should never engender a feeling of repugnance but rather one of gratitude for the opportunity offered. Before a 'subject' is dissected it is properly prepared for scientific study and there is, therefore, not the slightest danger in handling freely material offered for investigation; indeed it is the cadaver that is likely to suffer under the hands of the inept and clumsy.

The chief method of study adopted in these pages is that made possible by gross dissection and it will be presumed that the actual work of dissection has already been performed and that dissected specimens are available.

Two plans of attack present themselves: The first is a functional approach whereby the body is studied system by system. This is spoken of as **systematic anatomy.** The second is a topographical approach whereby the body is studied by regions in each of which more than one system is represented. This is spoken of as **regional anatomy** and is the method particularly suited to the needs of the surgeon and those who work under his direction. Neither plan can be dispensed with but, by studying the body first systematically, a comprehensive view of the whole is afforded and, with this as a background, the beginner is less likely to become lost in the complexities of the regional approach (Chapter 16).

2

Skeletal System

DEVELOPMENT OF BONE

In the embryo, the forerunner of bone is either cartilage or a membrane of fibrous tissue. Limb bones, for example, are formed out of cartilage; the bones of the vault of the skull are formed out of membrane. Each variety will be considered in turn.

The limbs develop as buds from the sides of the body; the buds are filled with a tissue of high potentiality, usually known as *mesenchyme*. In the axis of the bud where bone ultimately is found, this mesenchyme becomes condensed and finally cartilaginous. [The fate of the mesenchyme in the area between two contiguous cartilaginous units will be considered when the origin of joints is discussed.] When the mesenchyme becomes converted into cartilage, a miniature and discrete cartilaginous model, having the shape and appearance of the adult bone and united to its neighbors by remaining mesenchyme, is soon observable. This cartilage is invaded by salts (calcium phosphate) and the cartilage so becomes *calcified*.

At a point somewhere near the middle of the shaft of the calcified cartilage, bone-forming cells known as **osteoblasts** make their appearance and proceed to replace the calcified cartilage with bone (fig. 30). The site at which the osteoblasts begin

their work is known as the **primary center of ossification,** and in each cartilaginous model this center appears at some time between the fifth and the 12th week of intra-uterine life, i.e., mainly in the embryonic period.

By the time of birth, the process of ossification has progressed and has reached almost to the ends of the cartilage, at which time provision is made for growth. This is accomplished as follows: a new center of ossification appears in each expanded cartilaginous end, so that the bone of the shaft becomes capped at its ends with bone developing from these new centers. But, in order that growth in length may go on until the adult length of the bone is attained, the bone formed from the original or **primary center** of ossification does not fuse with that formed from the **secondary centers** or **epiphyses.** Instead, a plate of cartilage—known as the **growth or epiphyseal plate**—intervenes between them.

The growth in length of a long bone is accomplished by the epiphyseal plate drifting farther and farther from the middle of the shaft. This steady migration of the plate occurs as the result of new cartilage being formed on the epiphyseal surface while the surface on the shaft side is turned into bone. Growth ceases when the entire plate ossifies. This is known as *fusion* of the

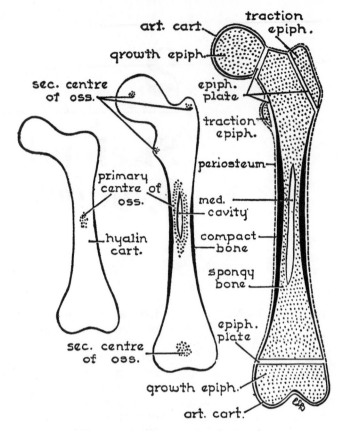

Fig. **30**. Development and parts of a long bone

epiphysis with the shaft. The fusion of the many epiphyses in the body begins at the time of puberty and is completed as the youth becomes an adult.

At all times, a phenomenon of modeling goes on such as occurs when the sculptor in clay adds a little here and removes a little there, so that, by this process, as the bone grows in length and size it retains the form and appearance of its miniature cartilaginous precursor.

Thus three important periods occur. They may conveniently be styled:

1. **The formative period,** from the fifth to the 12th week of intra-uterine life, during which nearly all primary centers of ossification appear;
2. **The growth period,** lasting to puberty, during which secondary growth centers appear—almost all after birth;
3. **The consolidation period** from pu-

berty to the attainment of the adult stature (14th–25th yr.).

The growth phenomenon is of more than academic interest. Fractures, etc., may involve one or more epiphyses; as a result growth may cease with a resulting serious and permanent shortening deformity.

A membrane bone (e.g., the cranial vault) results from a very similar but shortened process. Osteoblasts simply invade directly the membranous tissue without the intervention of cartilage and, from a primary center of ossification, situated near the center of the membranous tissue, bone radiates toward the periphery until a plate of bone of adult size is formed.

STRUCTURE AND FUNCTION OF BONE

Long Bones

The shafts or bodies of long bones are hollow (fig. 30). An outer shell known as compact bone is almost ivory-like in its hardness and density, but within, the bone has a much more trellis-like appearance and is called **spongy** (or **cancellous**) **bone** (fig. 31). In the middle of the shaft, the space which is known as the **medullary cavity** is occupied by yellow marrow (chiefly fat). A long bone is particularly spongy at its expanded ends and here the compact bone is exceedingly thin and

shell-like, while the interstices of the spongy bone are filled with **red marrow**—red because red marrow manufactures red blood cells.

Covering the outside of a bone, is a tough fibrous membrane known as **periosteum.** It is intimately adherent to the bone although it can be stripped off as a sheet. Its inner surface is almost entirely constituted of innumerable **osteoblasts** which are the important cells in growth and repair of bone. The periosteum also is richly furnished with capillaries (fig. 32) and these are responsible for the nourishment of the bone; they account for the pink hue of fresh bone. For these reasons a bone denuded of its periosteum dies.

At the expanded ends of a long bone, the periosteum is replaced by smooth and slippery cartilage taking part in the formation of the joint (fig. 126, page 78). The cartilage is known as *articular cartilage* and is of the hyaline variety since it is a persistent portion of the original cartilaginous precursor of the bone; such articular cartilage possesses no blood vessels because it requires little nutrition.

Near the middle of the shaft there is a small hole called the **nutrient foramen** by means of which a nutrient artery can enter and reach the spongy bone and the marrow.

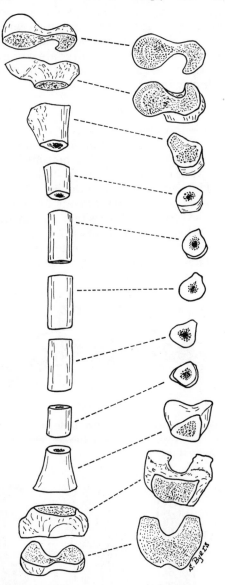

Fig. **31**. Changing structure (both internal and external) of a long bone revealed by serial sections of a right femur.

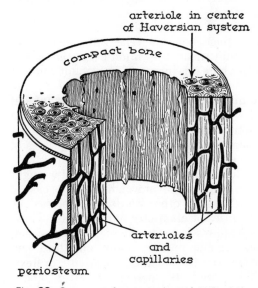

Fig. **32**. Structure of compact bone (cf., fig. 11)

Although the details of the microscopic structure of bone have been avoided, from what has been stated it is apparent that bone is a highly specialized tissue that is alive and capable of growth and change in form.

Flat Bones

In the flat or rather curved bones of the cranium the compact bone forms an **outer** and an **inner 'table'**; between them the spongy bone is rich in veins and is known as **diploë** (fig. 33).

A blow on the head may fracture the outer table leaving the inner table and the under-lying brain intact. Fracture of the inner table with involvement of the underlying brain is obviously far more serious. Paradoxically, the brain may be seriously damaged without any fracture of the skull.

Many of the bones adjacent to the nasal cavities are hollow; their interiors, instead of being occupied by spongy bone, have been excavated to form air sinuses (fig. 34). These excavations extend from the nasal cavity as **paranasal air sinuses.** They are described with the respiratory system on page 224.

Architecture of Bone

Bone possesses a definite architecture which reflects the duties it is called upon to perform. A perfection of function is achieved with a minimum of material. Compact bone is relatively greatest in amount near the middle of the shaft; elsewhere it is increased in amount where it is liable to buckle, particularly in the concavity of bends. Spongy bone is not laid down in a haphazard fashion but lies in thin delicate plates or lamellae which arrange themselves like struts along lines of force transmission (fig. 35).

Bone will 'give' with the ease of wrought iron and it will bear weight with the ease of cast iron. In other words, it possesses the qualities of **elasticity** and **rigidity.** These it acquires from its chemical composition; its organic matter, about one-third, accounts for its toughness, elasticity, or flexibility, and its inorganic matter, about two-thirds, for its hardness or rigidity (fig. 36).

The relative amount of organic to inorganic matter varies with age. Organic matter is relatively greatest in childhood, and during that period bones will bend—hence the deformities from which children suffer (e.g., in rickets) when deprived of correct amounts of calcium salts. Furthermore, fractures in childhood are infrequent compared to the risks that the bones of childhood are forced to assume, and when fractures do occur, the bones often break like a green stick. With increasing years the amount of salts increases relatively; fractures of middle life and after are common, and the bones, losing their elasticity, break with a snap.

Functions of Bone

The important functions of bone are:
1. They give **protection** by forming the rigid walls of cavities housing important organs, e.g., skull.
2. They give **rigidity** to the body.
3. They provide attachments for muscles, so serving as **levers in pulley systems** whereby movements can be produced by muscles and permitted by joints.
4. They are **factories** for the manufacture of blood cells.

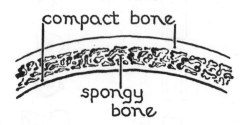

Fig. **33.** The compact flat bone that encloses brain has two 'tables.' The spongy bone (diploë) between is rich in veins.

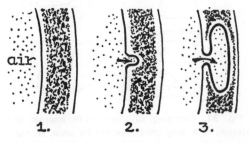

Fig. **34.** Scheme of development of an air sinus (pneumatic bone).

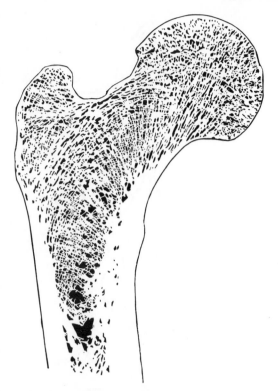

Fig. **35.** Architecture of bony trabeculae or struts in upper end of femur. (Photograph of sawed surface slightly retouched to improve reproduction.)

5. They are **storehouses** for minerals and
   chlorides.

'Markings' on Bone

The student will soon learn that the surface of a bone tells a story. The lumps (tuberosities, or, if small, tubercles) and the projections (processes) indicate localized strong attachments of fibrous cords—tendons or ligaments. The ridges and lines indicate attachments of broad sheets of fibrous tissue—aponeuroses or intermuscular septa. The smooth surfaces declare that they gave attachment to fleshy fibers during life or that nothing was attached to them (i.e., they were 'bare'). Grooves (furrows or sulci), holes (foramina), notches, and depressions (fossae) usually suggest the location of important structures.

CLASSIFICATION OF BONES

For purposes of classification, bones are divided into four categories, **long, short,**

flat, and **irregular.** They are also classified as **axial** and **appendicular,** i.e., those that contribute to the formation of the walls of the body cavities and are therefore largely protective; and those that are the skeleton of the limbs and are levers for the actions of the muscles of the limb (fig. 37).

Fig. **36.** A fibula decalcified in a bath of acid for several days is easily tied in a knot (Anatomy Museum, Queen's University).

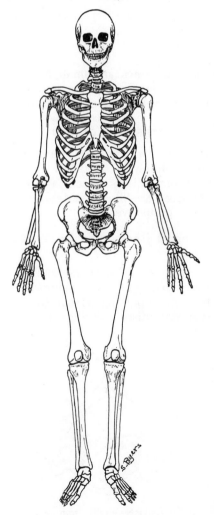

Fig. **37.** Human skeleton

The **axial skeleton** is made up of:
1. The vertebrae (including the sacrum and coccyx).
2. The bones of the skull (and certain others associated with it).
3. The mandible or lower jaw.
4. The ribs and the sternum.

The **appendicular skeleton** of each limb consists of four parts:
1. The bones of the girdle, or the part of the limb skeleton that is buried amid muscles in the wall of the body proper. The girdle bones unite the limb to the axial skeleton. They are known as the pectoral or upper limb girdle, and the pelvic or lower limb girdle.
2. The bone of the (upper) arm and that of the thigh.
3. The bones of the forearm and those of the leg.
4. The bones of the hand and those of the foot.

Sesamoid bones are small bones that develop in certain tendons and have a fancied resemblance to sesame seeds.

VERTEBRAL COLUMN

There are two essential parts to the vertebral column: (1) vertebral bodies and intervertebral discs, which are weight-bearing (figs. 38–43); (2) vertebral arches, which are protective.

1. Vertebral Bodies and Intervertebral Discs

In describing the vertebral column, it is profitable to discuss also the joints between the bodies of adjoining vertebrae known as intervertebral discs.

Fig. **38**. Cervical vertebra from above

Fig. **39**. Cervical vertebra from side

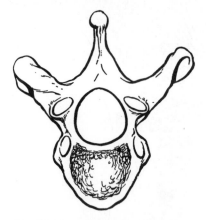

Fig. **40**. Thoracic vertebra from above

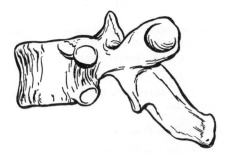

Fig. **41**. Thoracic vertebra from side

Each **body** is a short, cylindrical block of bone flattened at the back and possessing a slight 'waist.' Many vertical lamellae of spongy bone in its interior enable it to resist compression; its outer covering of compact bone is very thin.

Adjacent bodies are firmly united to one another by an **intervertebral disc** roughly one-fifth to one-third as thick as the neighboring bodies. This disc is composed of concentric rings of fibro-cartilage and a central mass of pulpy tissue, the **nucleus pulposus,** which represents the remains of the notochord. The disc, being under pressure, bulges, i.e., it is convex at its periphery. (See figures 137–139 on page 83.)

Discs are 'shock-absorbers' giving resilience to the column and they are relieved of pressure only when the body is recumbent. Being the non-rigid portion of the column, they also give it its flexibility. When the body is erect and in the 'normal' position the various parts of each disc are under uniform pressure; but when the vertebral column is flexed, extended, or bent sideways, one part of a disc is under increased compression whereas another part of the same disc is under tension (fig. 44).

An **anterior** and a **posterior longitudinal ligament** extend the length of the column, one down the fronts, the other down the backs of the vertebral bodies; they are firmly attached to the discs and they guard against excessive movement of the flexible column. An abnormal stress may tear the surrounding fibers of a disc with a resulting outward bulge of the

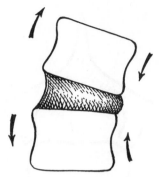

Fig. **44**. The same disc may be under compression and tension at the same time.

enclosed nucleus pulposus. This mass may then press on neighboring spinal nerves causing various distressing symptoms.

2. Vertebral Arches

A vertebral arch springs from the upper part of the back of its vertebral body (fig. 45) and each half is made up of two parts: (1) a very short rounded bar projecting backward from the body and known as the **pedicle** (L. = little foot); and (2) an oblong plate with sloping surfaces known as the **lamina** (L. = plate).

The lamina is continuous with the pedicle and meets its fellow of the opposite side in the midline. The hole thus framed in behind each vertebral body is the **vertebral foramen;** the succession of foramina makes up the **vertebral canal** in which the spinal cord and its coverings reside.

A typical vertebra begins to ossify before birth from three centers of ossification; these ultimately unite to form the adult bone (fig. 46).

A projection of bone is called a **process.** At the angular junction of pedicle and lamina on each side there are three processes, one upward (superior), one downward (inferior), and one lateralward (transverse). The superior and inferior processes meet and form joints with similar processes from adjacent vertebral arches; they are, consequently, called the **superior** and **inferior articular processes** and their chief function is to prevent undesirable movements between adjoining vertebrae (fig. 47). The lateral-projecting process is

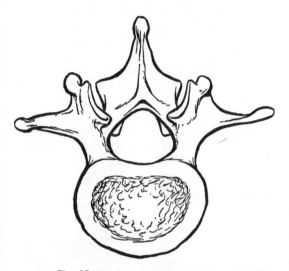

Fig. **42**. Lumbar vertebra from above

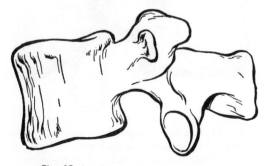

Fig. **43**. Lumbar vertebra from side

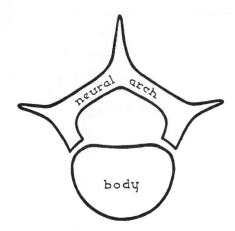

Fig. **45**. The scheme of a vertebra is simple—body for weight; vertebral (neural) arch for protection; processes for movements.

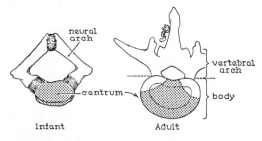

Fig. **46**. Three centers of ossification joined by cartilage form the vertebra in the infant. These unite to form an adult vertebra, but the 'body' is not the same as the centrum.

called the **transverse process;** it is chiefly for attachments of muscles although in the thoracic region a rib abuts against the transverse process and is steadied by it.

Where the two laminae meet in the midline posteriorly there projects backward a **spinous process** which is for attachment of muscles.

The succession of sloping laminae united by ligaments looks somewhat like a shingled roof and completely closes in the vertebral canal posteriorly.

Because a vertebral arch springs from the *upper* part of its vertebral body, a deep notch is visible below the pedicle when the vertebra is viewed from the side. When two adjacent vertebrae are in position, the notch becomes a hole—the **intervertebral foramen**—which gives exit and entrance to spinal nerves and vessels (fig. 47). The

intervertebral foramen is bounded above and below by pedicles; it is bounded in front and behind by joints (intervertebral disc and bodies in front, articular processes behind).

There are 24 'free' vertebrae distributed in three regions: seven in the neck—**cervical vertebrae;** 12 in the thorax—**thoracic vertebrae;** five in the loin—**lumbar vertebrae.** Below the last lumbar vertebra lies the sacrum.

Sacrum

Five vertebral bodies, united by four ossified intervertebral discs, are easily distinguishable on the concave anterior surface of the curved triangular bone that comprises the sacrum (sacrum, L. = sacred; the origin of the term is obscure). The body of the first of these vertebrae has a prominent, oval, upper surface with a distinctly forward slope. To it is attached a thick disc that unites it to the body of the fifth lumbar vertebra (see figs. 48–51).

Two parallel rows of four **pelvic** (*anterior*) **sacral foramina,** in line with intervertebral foramina above, serve to separate the sacral vertebral bodies from the parts of the bone lateral to these openings and known as the **lateral masses.** The thick upper part of each lateral

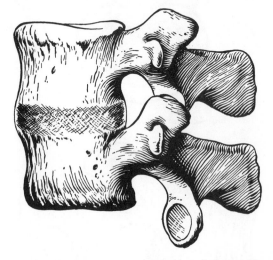

Fig. **47**. Two lumbar vertebrae in articulation. The lower pair of processes grasps the upper and prevents rotation. Note the boundaries of the intervertebral foramen, the exit for a spinal nerve.

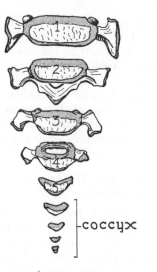

Fig. **48**. Young child's sacrum is made up of five vertebrae which later unite to form one bone.

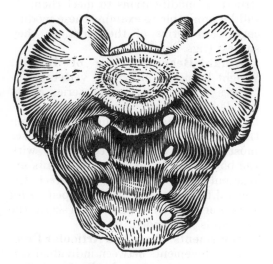

Fig. **49**. Sacrum from front

tuberosity; it is for a mass of strong, short ligaments that further unite the pelvic girdle to the sacrum and that constitute the **fibrous sacro-iliac joint.** Below the level of the auricular surface the whole sacrum tapers rapidly since the lower part of the bone bears no weight.

Behind, the bone is convex and much rougher; here the two rows of **dorsal sacral foramina** (in the same vertical plane as the anterior ones) serve to separate the laminae, which are all fused together, from the lateral mass on each side (fig. 51).

At the upper end of the bone, there projects a pair of large superior articular processes that face one another and embrace the inferior ones of the fifth lumbar vertebra. The triangular upper opening of the *sacral canal* lies between them and is bounded behind by the paired laminae which are easily recognizable.

An irregular vertical ridge down the back of the lateral mass of each side represents transverse processes, while a similar ridge medial to the dorsal foramina represents articular processes. Four bony tubercles in the midline represent spinous processes.

At the lower end of the bone, there is a pair of small processes called the *sacral cornua* (sing., cornu, L. = horn). Between

mass, smooth above and in front where it is continuous with the side of the first sacral vertebral body, is called the **ala** (wing) of the sacrum. On the side of each ala are two areas. In front, is a large ear-shaped area by means of which the sacrum articulates with the hip bone on each side. This area is called the **auricular surface** and the articulation is the *synovial sacro-iliac joint* (auricle, L. = external ear) (see page 98). Behind (above, when the bone is correctly oriented) the auricular surface is an equally large rough area known as the

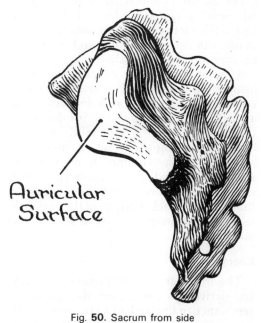

Auricular Surface

Fig. **50**. Sacrum from side

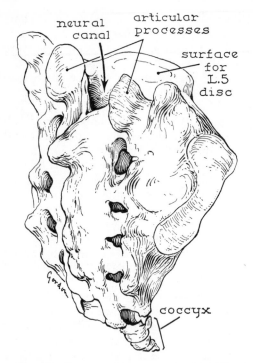

neural canal   articular processes

surface for L.5 disc

coccyx

Fig. **51**. Sacrum and coccyx viewed obliquely from behind and right side.

them is the lower end of the vertebral canal which is closed in life by a tough fibrous membrane.

The extent to which the vertebral (sacral) canal is closed by bone in the lower half of the sacrum is very variable; in other words, the lowest vertebral arches are frequently incomplete or absent.

A vertebra has three primary centers of ossification. One forms most of the body; each of the other two forms one-half of the arch (fig. 46). Failure of the two halves of the arch to meet in the midline, while commonest in the sacral region, may occur anywhere in the column. It is known as **spina bifida** and it may allow the coverings of the spinal cord with their contained fluid to bulge backward out of the vertebral canal and so produce a tumor known as a meningocele.

Coccyx

The irregular coccyx (pronounced cock-six) derives its name from its fancied resemblance to the beak of a cuckoo (fig. 51). It consists usually of four fused vestigial vertebrae of which the first is by far the largest. The body of this first coccygeal vertebra is united to the lower end of the sacrum by a fibrocartilaginous intervertebral disc; the vertebral arch is represented by two upwardly projecting horns or *cornua* which meet those of the sacrum. Below this vertebra the bone is nodular and represents the vestigial human tail. It can be felt quite easily in the living person.

Functions and Differences in the Vertebral Column

An isolated vertebra can be assigned to its correct region with assurance in spite of the fact that all vertebrae are built on the same general plan. This is because various situations impose various tasks upon the vertebrae which, in consequence, exhibit structural modifications to meet them. It will be profitable to examine these modifications in terms of the functions they suggest (figs. 38–43).

Weight-Bearing; Size of Bodies. Weight borne by individual vertebrae increases progressively as the series is descended; vertebral bodies, in consequence, become more massive as they proceed from the cervical to the lumbar region; discs must increase in size in conformity with the bones (fig. 52). [These statements are true only in animals that walk erect; the general tendency among quadrupeds is for vertebrae actually to decrease in size as the column is 'descended.']

Movements; Discs and Articular Processes. Movements between individual vertebrae take place (1) at the discs, and (2) at the joints between the (paired) articular processes of the vertebral arches. Movements at the discs are greatest where the discs are thickest; movements at articular processes are greatest where the joint surfaces are largest.

In the lumbar region, both of the above conditions exist and here movements are freest. In these lumbar movements, lumbar discs accept considerable strain. If to the normal lumbar movements of flexion, extension, and side-bending, rotary movements were added, the resulting torsion of the discs might be more than they could stand. For this and other reasons, the dispositions of the joints between lumbar articular processes restrict rotation (fig. 47). These joints are 'interlocking,' i.e., the superior articular processes of the vertebra below grip the outsides of the inferior articular processes of the vertebra above. In this region, bending forward and backward can be freely performed; sideways bending is much less free; and rotation is severely limited. Even with this safeguard lumbar discs rupture more readily than any others.

In the thoracic region, the discs are relatively thin, the opposing surfaces of the articular processes on the vertebral arches are small and flat and face backward and forward; no type of movement is entirely prohibited, yet movements in all planes are slight. Nevertheless, because there are 12 vertebrae in this region, the total mobility between the first and the last is considerably greater than might be thought. The transition from the more mobile lumbar vertebrae to the less mobile thoracic vertebrae is unfortunately rather abrupt, and it is the 11th or the 12th thoracic vertebra that is most commonly fractured in a broken back.

In the cervical region, the discs are relatively thicker than those in the thoracic region, and the articular surfaces of the processes—facing at first upward and downward but gradually changing to a forward and backward direction—are small but relatively larger also. Thus the seven cervical vertebrae permit movement in all planes as do the thoracic ones, but the range is considerably greater. In addition to their rather free movements of flexion, extension, and side-bending, the cervical vertebrae indulge also in movements of rotation (twisting or torsion). But it has been observed that such a combination of movements throws excessive strain on the discs. Perhaps it is for this reason that the periphery of the upper surface of a typical cervical vertebral body is built up markedly at the sides and slightly at the back; it slopes away in front (fig. 53).

The first and second cervical vertebrae are very specially modified, for they carry the head and aid in its movements. They will be discussed shortly (page 33).

Vertebral Foramina. The caliber of the spinal cord varies very little from end to end, but it has two local enlargements associated with the sites of origin of the great nerves destined for the upper and for the lower limbs; these are the cervical and the lumbar enlargements. With these facts in mind it is easy to understand why it is that: (1) the vertebral foramina are relatively and actually large in the cervical region, where they are also triangular; (2) they are small and circular in the thoracic region; (3) they are actually, but not relatively, large in the lumbar region where, again, they are triangular.

Transverse Processes; Joints for Ribs. A very special feature distinguishes the **thoracic vertebrae:** they carry the ribs.

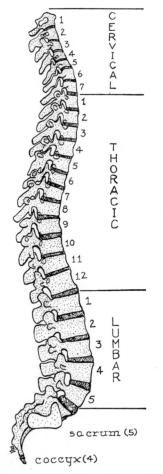

Fig. **52.** Vertebral column from right side. Note intervertebral discs between bodies.

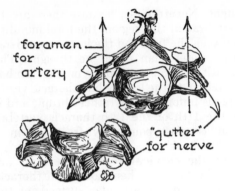

foramen
for
artery

"gutter"
for nerve

Fig. **53**. A cervical vertebra from above and from in front (see text).

The head of a rib meets the vertebral column at the side of a disc and at the adjacent edges of two vertebral bodies. In its excursion around the trunk, the rib at first sweeps backward to abut against the front of the tip of a transverse process. The presence, therefore, of little joint surfaces, the one on the body and the other on the transverse process, proclaims a vertebra to be rib-bearing and thoracic.

A special feature distinguishes **cervical vertebrae** also. Each transverse process has a hole or foramen in it, the bar of bone in front of the foramen being, in effect, an undeveloped rib (fig. 53).

Observation of the articulated column will reveal this 'rib' to be in line with the true ribs in the thoracic region, so that the foramen obviously is the space between the rib and the transverse process which, too, is readily seen in the thorax. Through the series of foramina, vessels (vertebral artery and vein) thread their way upward to aid ultimately in the blood supply of the brain. Lateral to the foramen, the issuing spinal nerve rests in a bony gutter formed by the two elements of the compound transverse process.

Sometimes one of these normally undeveloped ribs (usually the seventh) becomes greatly enlarged and like a true rib. The condition is known as a 'cervical rib.' It may cause symptoms if it presses on the large vessels and nerves that pass from the neck into the upper limb.

Spinous Processes. The 'spines' are distinctive for each region: cervical ones have double tips, i.e., they are bifid; thoracic ones are long, slender, pointed, and tend to project more downward than backward; lumbar ones are massive, square-cut, and project straight backward.

Curves of the Vertebral Column

During intra-uterine life the body of the fetus is noticeably flexed, and this applies to the vertebral column as well as to other structures. Soon two primary curves, both with their concavities forward, are recognizable, one involving the pre-sacral vertebrae, the other the sacrum itself (fig. 54).

In a structure such as the vertebral column which is built up of superimposed bodies and discs, these curves must be expressions of differences in thickness between the fronts and the backs of the individual bodies and discs. Furthermore, when it is the bones (bodies) that are chiefly concerned, the curve is likely to be permanent; when the discs also or alone are concerned, the resulting curve can be temporarily abolished.

The two permanent or **primary curves** persist as thoracic and sacral (pelvic) (figs. 54, 55). **Secondary curves** develop in the cervical and lumbar regions, and there the relatively thick discs that these regions possess take a considerable part. These secondary curves are convex forward and are subject to temporary elimination. The cervical curve would seem to develop in response to the need for holding the head up. The lumbar curve develops so that the center of gravity of the body will not lie in a plane in front of the hip joints during the sitting or standing positions. Everyone is familiar with the stance of the child learn-

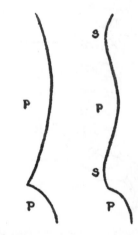

Fig. **54**. Curves of the vertebral column. P. = primary; S. = secondary.

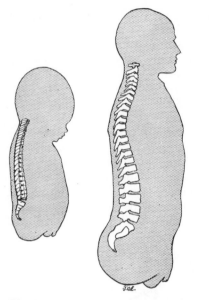

Fig. **55**. Curves of spinal column in infant and adult compared.

ing to walk—shoulders held far back and lumbar curve greatly exaggerated.

Where the last lumbar vertebra (fifth) meets the sacrum there is an abrupt change; this is provided for by the last lumbar vertebral body and the last lumbar disc being much thicker (or deeper) in front than behind, i.e., they are wedge-shaped.

Occasionally an accident occurs here and the lumbar body (fourth and fifth) slips forward; the accident is said to be due to a developmental failure of the neural arch to be fixed to its body but, since the separation is rarely, if ever, at the site of union of arch and body, it seems more probable that it is due to a bilateral fracture of the arch. The condition is known as spondylolisthesis (Gr. = vertebral slipping).

The normal curvatures of the column are rather more pronounced in the female than in the male, and in both sexes there is usually a slight lateral curvature to one side said to be associated with right- or left-handedness.

An excessive curvature to one side is known as scoliosis. It may be caused by muscle imbalance (e.g., poliomyelitis) in which the muscles of one side of the column are paralyzed and cannot counteract the pull of those of the healthy side. An excessive backward curvature is known as kyphosis (Gr. = hump-back). It is commonly, but not always, due to tuberculosis of one or more vertebral bodies (Pott's disease of the spine). The diseased bodies are eaten away and are crushed by the superimposed

weight. An excessive forward curvature in the lumbar region is known as lordosis. It occurs—particularly in women who wear excessively high-heeled shoes—as a result of faulty posture and is a common cause of backache.

A disease such as *arthritis* which attacks joints and, in the vertebral column, limits the range of movements of the joints of the vertebral arches, produces a rigidity of the column with limitation of movement in every direction.

Atlas

The first cervical vertebra, the atlas, supports the skull. It has lost its body and consists simply of a ring of bone made up of two **lateral masses** joined in front and joined behind by an arch—the **anterior** and the **posterior arch of the atlas** (fig. 56). The upper surface of each lateral mass is elliptical, concave, and articular. The long axis of the ellipse runs mainly antero-posteriorly, but the two ellipses are so placed that they are nearer one another in front than behind, i.e., the anterior arch is shorter than the posterior. On the concave surfaces of the lateral masses the head rocks in the nodding movement aptly described as the 'Yes' movement. The lower surface of each lateral mass is circular, flat, and articular and rests on a similar surface of the axis.

Lateral to each lateral mass is the very large and widely set transverse process, containing the foramen previously noted as transmitting the vertebral vessels. This process projects so far laterally that it may

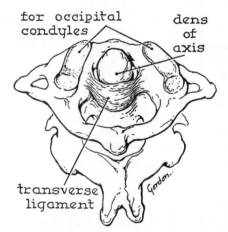

for occipital condyles   dens of axis

transverse ligament

Fig. **56**. Atlas and axis in articulation viewed from behind and above.

be felt on deep pressure behind the jaw bone just below the ear.

Medial to each lateral mass and encroaching somewhat on the anterior part of the very large vertebral foramen, there is a pronounced tubercle to which is attached the very strong **transverse ligament of the atlas** stretching to the tubercle on the other side. The vertebral foramen is divided by this ligament into a smaller anterior compartment and a larger posterior one.

Axis

The second cervical vertebra, appropriately named the axis, is distinguished by a blunt tooth-like process, the **dens** (L. = tooth) or *odontoid process*, projecting upwards from the body (figs. 57, 58). It is in reality the body of the atlas which has become divorced from the atlas and has joined the body of the axis. This dens projects upward into the anterior compartment of the vertebral foramen of the atlas and provides a mechanism, consisting of a pivot and a collar, whereby the head and atlas can rotate around the dens in the so-called 'No' movement. On each side of the dens is a large, flat, and circular joint surface for the support and rotation of the atlas.

THE SKULL

General Appreciation. There are two essential parts of the skull of any animal, the brain-case or cranial cavity, and the skeleton of the face. In most animals the cranial cavity is quite small and is situated behind the much larger facial region which projects forward as a prominent muzzle (fig. 59). In man, two factors greatly modify this state of affairs. The first is the very great increase in size of the brain accommodated in a greatly expanded cranial cavity; it is so pronounced that, when the human skull is viewed from above, nothing but the brain-case, except perhaps the tip of the nasal region, can be seen, everything else being completely overshadowed by it. This is an entirely human feature and exists in no other animal. The effect of this expansion in a forward direction is particularly noteworthy, for it has resulted in the

conversion of the significant and shallow depression that houses the eye of other animals, into a deep and conical socket in which the roof of the orbit is also the floor of the brain-case in this region. Similarly, the floor of the orbit is also the roof of the face which is no longer in front of, but below the cranial cavity.

The second factor is the foreshortening of the face, possibly associated with the lessening importance of the nose as an organ of smell and of the jaws as organs of prehension. Man depends much more on his sense of sight than on his sense of smell and, because he has assumed the erect

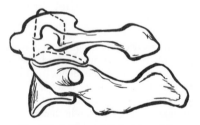

Fig. **57**. Axis from side

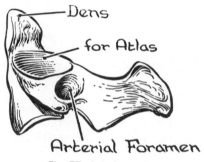

Fig. **58**. Atlas and axis from side

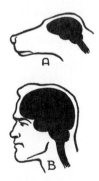

Fig. **59**. Brain-cases of dog and man compared

posture, he uses his upper limbs for duties performed in other animals by the jaws and teeth.

In order to sustain this position, the face is buttressed against the brain-case by abutments which the box provides for it. These abutments or buttresses will be examined later.

BONES OF SKULL

Some two dozen individual bones have lost their identity in the common task of enclosing the brain and the openings of the face. Except for the mandible or jaw-bone, which often is not included under the title 'skull,' these bones are immovably united together by sutures (page 76). Therefore, except as an aid to clear description, the individual bones will not be described at length but will be treated in the way that a geographer treats a group of counties forming a single state. Thus, 'key' bones will receive the most attention.

FACE

Most of the bones of the face can be seen from the front. It is obvious that their chief function is to surround and protect the openings of the face—mouth, nose, and eyes (fig. 60). The **mandible,** with the lower arch of teeth along its upper edge, forms the lower border of the mouth; it receives special attention on page 44. The horseshoe-shaped upper **alveolar** (or *dental*) **arch** is carried by the immovable, paired **maxillae** or 'upper jaw.' A horizontal plate of bone with a free posterior edge fills in the arch, forms the roof of the mouth or floor of the nose, and is known as the **hard palate** (fig. 64).

Each maxilla bounds the opening for the nose and the lower **margin** of the **orbit** or **orbital cavity.** A stout projection upward and medially (frontal process of maxilla) helps to form a narrow isthmus between the orbital margin and the **naris (anterior choana*)** or **nasal aperture;** here it meets the **nasal bone** of its own side and a downward-running process of the **frontal bone.**

---

* Pronounced Ko-anna; pl., choanae.

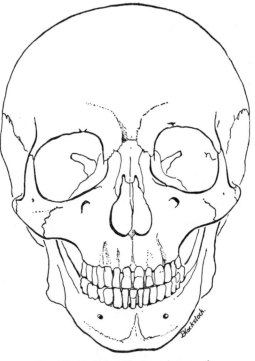

Fig. **60**. Skull from in front (see text)

The paired nasal bones are small, and wedged side by side between right and left maxillae; each forms the upper boundary of the anterior nasal aperture of its own side. They bulge outward to form the bridge of the nose where a blow may buckle them inward (fig. 61).

The unpaired frontal bone, the broad rounded bone of the forehead, is really more concerned with the brain-case but contributes to the face by forming the upper boundary of the orbital cavities.

Not only does the maxilla present a surface to the front but it also forms the floor of the orbital cavity and the greatest part of the lateral wall of the nasal passage. It is not, however, a big heavy block of bone, because a balloon of air (**maxillary sinus**) is 'blown' into it from the nasal cavity and so the maxilla in the adult is a light hollow shell.

Buttresses. During biting and chewing, enormous pressure is exerted on the upper alveolar (dental) arch by the lower arch. This would crush the face upward. To prevent this, three buttresses of bone run upward from the upper teeth.

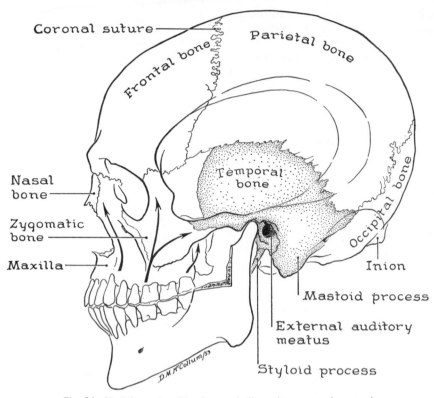

Fig. **61**. Skull from the side. Arrows indicate buttresses (see text)

One of these, already described, runs up the medial side of the orbit. The second runs up to the lateral aspect of the orbit and splits into an upper and lower limb. One limb passes horizontally backward as a free arch of bone to just above the external opening of the ear and is known as the **zygoma** (L. = yoke). The upper limb forms the lateral boundary and wall of the orbital cavity and reaches the frontal bone. The greatest part of both of these limbs is formed by the **zygomatic bone** which is diamond-shaped. The third buttress (the pterygoid process of the sphenoid bone) runs upward from behind the upper last molar tooth to the base of the brain-case (see fig. 61).

Orbit. One can see all of the roomy interior of this 5-cm long (2″) hollow pyramid in which the eyeball and its nerves are protected. The floor and lateral walls have been mentioned above. Since one can look through foramina at the apex into the interior of the brain-case, it is obvious that

its roof is also the floor of the brain-case (see Anterior Cranial Fossa). The medial wall is formed by the lateral surface of the oblong, hollow, box-like **ethmoid** bone whose medial surface helps to form the lateral wall of the nose.

The **orbital margin** is made up of parts of three bones—frontal, maxillary, and zygomatic—each forming about one-third of the margin.

SKULL FROM THE SIDE

Many of the features to be seen when the skull is viewed from the side have already been discussed. The posterior end of the zygomatic arch has been seen to end just above the **external acoustic meatus,** the deep bony canal which leads inward from (and gives attachment to) the elastic cartilage of the external ear (see fig. 56).

Immediately behind the meatus a large rounded lump of bone just downward; it is the **mastoid process.** This process, produced by the attachment of the Sternomastoid muscle, is hollowed out by a honey comb of air spaces (*mastoid air cells*) which communicate with the middle ear cavity (see page 354). Immediately in front of the external acoustic meatus is the oval depression into which the head of the mandible fits.

The Outline of the Brain-Case. Seen from the side this may be traced most conveniently from the **foramen magnum,** the enormous hole through which the brain communicates with the spinal cord. The foramen magnum is at the lowest point of the brain-case and lies midway between the mastoid processes. From it the outline runs obliquely upward and backward to a midline lump of bone, the *external occipital protuberance* or **inion.** From this lump the outline of the brain-case is quite obvious even in a living person. It ends anteriorly in a depression (above the nasal bones) which is on the same horizontal plane as the inion. (The *vertex* is the highest level of this outline and is not specially marked.)

The outline of the brain-case is the outer of three curved concentric lines (fig. 61). 'Parallel' to it and several centimeters (1–2″) away is a raised line on the surface of the bone arching from the posterior end of the zygomatic arch to its anterior end. It encloses the **temporal fossa** from the surface of which the fan-shaped Temporalis muscle arises. (This muscle passes deep to the zygomatic arch and attaches to the mandible.)

The innermost concentric line is a suture line which outlines the **squama** of the **temporal bone.** The vault of the skull above this suture line consists chiefly of the **parietal bone** which meets its companion of the opposite side at a midline suture, the **sagittal suture.** The two parietal bones are separated from the (single) frontal bone by a suture that runs like a diadem—thus, **coronal suture**—across the top of the anterior part of the vault.

Posteriorly the sagittal suture ends by dividing into a right and left limb; each of these runs downward, forward, and laterally on its own side to reach the suture outlining the temporal squama near the mastoid process. All of the area of bone

below and between the limbs is the **occipital bone** and includes inion, foramen magnum, and the large area of muscle attachments at the nape of the neck (see figs. 62 and 63).

SKULL FROM BELOW

Most of the features of the face have already been discussed. The *palate, upper alveolar* (*dental*) *arch*, and the *zygomatic arch* are all very obvious from below when the mandible is removed.

Very prominent both in size and position is the **foramen magnum** (fig. 64). Situated in the midline half-way between the posterior edge of the palate and the inion, it is large enough to accommodate the 'root' of an inserted thumb. Its outline is made rather pear-shaped by the encroachment of the paired **occipital condyles** on the sides of its anterior half. These oval lumps of bone have a smooth inferior surface which is convex from front to back, and give the impression of a pair of rockers from a rocking chair. They rest in concavities on the upper surface of the first cervical vertebra and by sliding back and forth produce the nodding motion of the head signifying 'Yes.'

Although the foramen magnum is almost at the same horizontal level as the hard palate, the base of the skull slopes rapidly upward and forward from the foramen magnum to a level 2 to 3 cm (1″) above the hard palate leaving room between the palate and the base of the skull for the *posterior choanae* of the nasal cavities. The sloping central area of bone between the foramen magnum and the roof of the nasal cavity forms the roof of the **pharynx.** Generally referred to vaguely as the 'throat,' the pharynx is the space behind the nasal and mouth cavities.

The paired nasal cavities are separated by a midline partition (**septum**) of bone and cartilage, the posterior part consisting of the (unpaired) *vomer* bone. The upper boundary of the posterior nasal openings is formed by the body of the **sphenoid** bone (which is seen best in the interior of the brain-case). Running downward and forming the lateral boundaries of the openings are the buttresses previously mentioned,

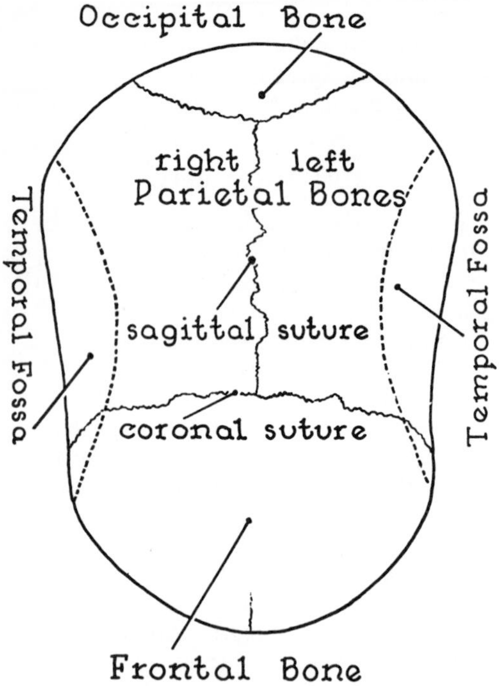

Fig. **62**. Skull from above

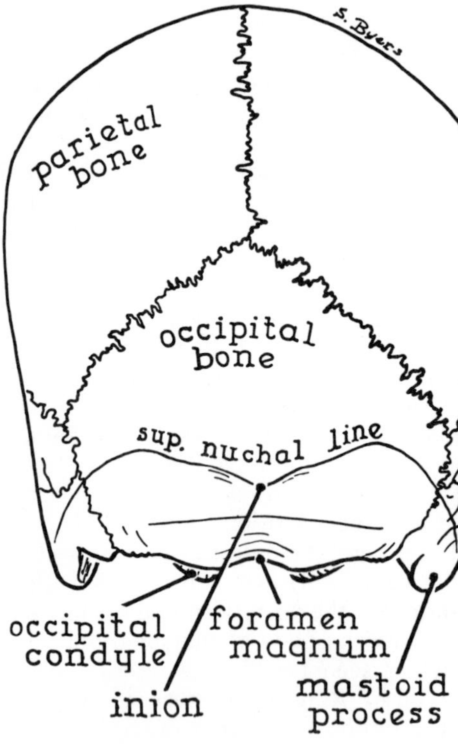

Fig. **63**. Cranium from behind

the paired *pterygoid processes of the sphenoid bone.*

The *pterygoid processes* each consist of two posteriorly projecting free plates. To the edges of the *medial plate* the walls of the pharynx are attached. The *lateral plate* gives origin to two muscles of mastication (or chewing) that run to the mandible and help fill the extensive area (**infratemporal region**) between the pterygoid process and the zygomatic arch.

We have already noted the oval depression at the posterior end of the zygomatic arch for the head of the mandible. Between it and the central area that forms the roof of the pharynx is a bewildering collection of holes for the passage of many important nerves and vessels. They will be seen again from the interior of the skull and the important ones will be mentioned later. They are limited postero-laterally by a long, stylus-like process, the **styloid process,** that lies just one centimeter medial to the outer opening of the ear canal. Rough handling of laboratory specimens soon shears off this slender process.

The large area of bone behind the foramen magnum has been described earlier as giving attachment to muscles in the nape of the neck (fig. 63).

INTERIOR OF BASE OF SKULL

Cranial Fossae

The brain has a frontal lobe which grows forward above the orbit, and a temporal lobe which grows forward at a lower level and behind the orbit. From the brain stem just above the foramen magnum the cerebellum grows and expands. So we have three 'steps,' the first above the orbit, the second behind the orbit, and the third surrounding the foramen magnum. These are known as the **anterior, middle,** and **posterior cranial fossae** (fig. 65). In a skull whose 'cap' has been removed these three steps downward from front to back in the floor of the brain-case constitute the most obvious feature.

Anterior Cranial Fossa. Irregular in surface, the thin bone in the floor of this fossa separates the brain from the orbital and nasal cavities. The chief contribution

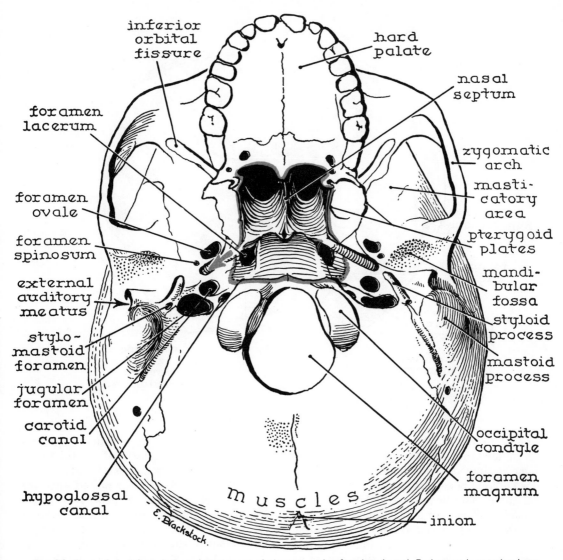

Fig. **64.** Base of skull from below. Attachment of pharynx and soft palate in red. Red arrow is entering bony part of right auditory (pharyngo-tympanic) tube which leads to middle ear.

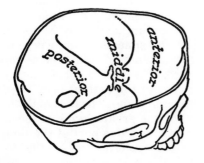

Fig. **65.** The three cranial fossae (cf., fig. 66)

is made by the frontal bone which sends back an extensive plate of bone from the region of the eyebrows (fig. 66). A narrow central strip, the roof of the nose, is slightly depressed and perforated with many small holes; this is the **cribriform (sieve-like) plate** of the ethmoid bone. Springing upward in the midline of the cribriform plate is a crest of bone fancifully called the *crista galli* (cock's comb). To it a membrane (falx cerebri) is attached.

The **ethmoid bone** will be described

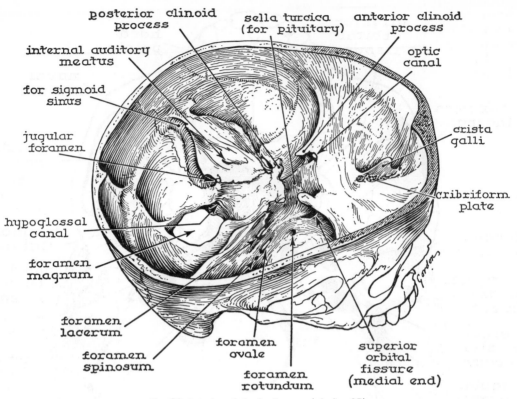

posterior clinoid process
sella turcica (for pituitary)
anterior clinoid process
internal auditory meatus
optic canal
for sigmoid sinus
crista galli
jugular foramen
cribriform plate
hypoglossal canal
foramen magnum
foramen lacerum
foramen ovale
foramen spinosum
foramen rotundum
superior orbital fissure (medial end)

Fig. **66**. Interior of the brain-case (cf., fig. 65)

briefly, here, since it is sometimes difficult to understand and 'fit in.' This (single) bone is shaped like a businessman's desk. Two oblong boxes are joined by a horizontal table. The boxes are hollowed out by about a dozen paranasal air sinuses and separate the orbital and nasal cavities (fig. 68). The plate of bone joining the upper surfaces of the boxes is the cribriform plate or roof of the nose; the crista galli is represented below the plate by a vertical continuation that helps to form the nasal septum.

Middle Cranial Fossa. It is the rare anatomist who does not describe the middle cranial fossa as shaped like a butterfly, because its central small area expands into large wing-like areas on both sides.

The central area is above the **body of the sphenoid** which is more or less a small square cube whose interior is hollowed out by the paired *sphenoidal air sinuses* that open into the nasal cavity. The body of the sphenoid helps to form the posterior part of the roof of the nose.

Since the vitally important *hypophysis cerebri* (pituitary gland) rests on the top of the central area, a bed is provided for it here. Four bed-posts—the paired **anterior** and **posterior clinoid processes**—have also stirred the imagination of anatomists (kline, Gr. = bed). [But the metaphors are mixed since the bed is called the *sella turcica* or turkish saddle.]

Just medial to each anterior clinoid process, the **optic canal** (for the optic nerve) passes into the back of the orbit.

The lateral areas are extremely concave and sharply demarcated fore and aft. The sharp boundary anteriorly is the **lesser wing of the sphenoid** from whose medial end the anterior clinoid process projects backward. The posterior boundary is the upper edge of the petrous bone (a part of the temporal bone). The floor of the fossa is formed by the **greater wing of the sphenoid** and the **petrous part of the temporal bone** (often referred to as the *petrous bone*).

At the 'root' of the wing of the butterfly

are found several large important holes for the passage of important vessels and nerves to and from the brain-case. These are described later for students who require additional information (see also figs. 64–69).

Posterior Cranial Fossa. The petrous bone on each side forms an 'anterior wall' for this fossa. Between right and left petrous bones is a sloping surface (for the brain stem) leading down to the foramen magnum, the dominant feature of the posterior fossa. The occipital bone surrounds the foramen magnum and makes up most of the concave floor occupied by the rounded cerebellum.

From the side wall, a deep groove that lodges a great blood vessel (sigmoid venous sinus) runs down to a large foramen (jugular foramen). Immediately above this is the internal acoustic meatus leading to the inner and middle ears encased in the center of the petrous bone.

Note on Development

The upper end of the notochord has been seen to be the core around which the bony foundation of the skull is laid (fig. 21). In a skull that has been split into right and left halves in the midline, this foundation is recognizable as a wedge of bone which continues upward and forward the line of the vertebral bodies (fig. 69). The wedge slopes upward from the front edge of the foramen magnum at an angle of about 60°, is about 4 cm (1½″) long, and is continuous in front

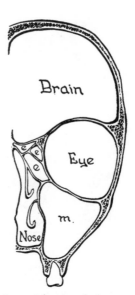

Fig. **68.** Coronal (i.e., vertical, transverse) section of skull to show maxillary (m) and ethmoidal (e) air sinuses.

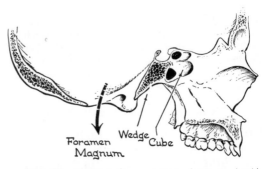

Fig. **69.** Sagittal (i.e., vertical, front to back) section of skull.

with the back of a hollow, bony, air-filled cube. The hollow cube is the *body of the sphenoid bone.* The front wall of the cube cannot be seen from inside the skull since it faces the back of the nasal cavity; its interior has been excavated from the front wall as one of the paranasal air sinuses (fig. 69).

Since the anterior or upper end of both the notochord and the neural tube terminated originally at about the same level, it is not difficult to appreciate that the enormous expansion of the neural tube that resulted in the brain started in the neighborhood of what is, in the adult, the upper surface of the sphenoidal body (fig. 21). Bone developed in the membranous covering that kept pace with this expansion and, as a result, to the 'foundation' wedge and cube (which are notochordal in origin) membrane bone was added in front, at the sides, and all over the surface of the expanding brain (fig. 70). And so the brain became completely contained in a bony cranial

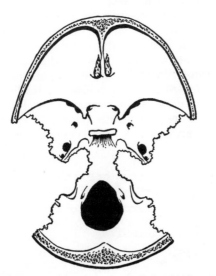

Fig. **67.** Interior of base of skull with temporal bones removed.

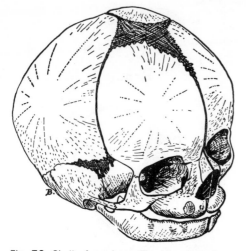

Fig. **70**. Skull of newborn. The kite-shaped anterior fontanelle and the suture lines are still filled by membranes which eventually will be replaced by bone. At birth the skull is pliable; this allows it to be forced through the relatively narrow birth canal.

cavity. Furthermore, it one may regard the bone at the sides and back of the foramen magnum as a highly modified vertebral arch, it is easy to understand how the backward extension of the growing brain was similarly protected by the expansion of this vertebral arch and by membrane bone added to it.

FORAMINA IN BASE OF SKULL

Group A, leading forward from the front of the middle cranial fossa:

1. *Optic canal*, a short round canal at the apex of the orbital cavity transmitting the nerve of vision—the optic nerve, and the artery of supply of the orbital contents—the ophthalmic artery.

2. *Superior orbital fissure*, a slit-like aperture under shelter of the lesser wing of the sphenoid, and also leading to the orbital cavity. It transmits:
   a. the sensory nerve of the cornea, upper eyelid, and neighboring skin of the forehead—the ophthalmic nerve.
   b. the three motor nerves for the eye muscles—oculomotor, trochlear, and abducent nerves.
   c. the veins that drain blood backward from the orbital contents—the ophthalmic veins.

3. *Foramen rotundum*, a round hole

(hence its name) below the medial end of the superior orbital fissure. It cannot be seen from the outside since it opens in a deeply buried space, the pterygopalatine fossa. It transmits the sensory nerve of the face—the maxillary nerve, which, swinging across the space like a rope across a chasm, enters the orbit in front. Running at first along the floor of the orbit, the nerve gradually sinks into the floor and appears on the face through a foramen, the infra-orbital foramen, below the orbital margin.

Group B, leading downward from the back of the middle cranial fossa:

1. *Foramen ovale*, an oval hole (hence its name) in the back part of the floor of the middle cranial fossa and leading downward to the masticatory region. It transmits the nerve of the muscles of mastication—the mandibular nerve.

2. *Foramen spinosum*, a little round hole just behind and lateral to the foramen ovale. It gives entrance to the chief artery supplying the meninges or coverings of the brain—the middle meningeal artery (see fig. 71).

3. *Foramen lacerum*, a large and irregular (lacerated) foramen closed below by cartilage; but since the apex of the petrous bone is channeled by a canal, the *carotid canal*, the internal carotid artery gains the cranial cavity by coming up through the bone and lying on the cartilage. It is the chief arterial supply of the brain.

Group C, in the posterior cranial fossa:

1. *Internal acoustic meatus*, leading to interior of the petrous bone. From it issues the eighth (vestibulocochlear) nerve which has come from the

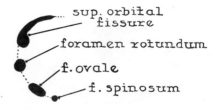

Fig. **71**. Four important holes in the middle cranial fossa are arranged in a semicircle.

organs of hearing and equilibrium. Into the foramen runs the seventh (facial) nerve which traverses the petrous bone and, after making a right-angled turn, issues at the base of the skull through the small *stylomastoid foramen* between styloid and mastoid processes. The nerve, as its name implies, supplies the facial muscles of expression.

2. *Hypoglossal canal*, channeling the bone at the side of the foramen magnum above the occipital condyles. It transmits the nerve that supplies all the tongue muscles—the hypoglossal nerve.

3. *Jugular foramen*, lying lateral to the foramen magnum and the hypoglossal canal. Approaching the foramen from the lateral side is a broad bony gutter in which lies the great vein that, draining blood from the interior of the skull, passes through the foramen to become the internal jugular vein. Issuing from the medial part of the foramen are three cranial nerves:
   a. ninth or glossopharyngeal nerve, the sensory nerve of the pharynx and back of the tongue.
   b. tenth or vagus nerve, running down with the internal jugular vein; it is a great nerve for the supply of internal organs.
   c. 11th or accessory nerve, a peculiar nerve made up of two parts; its cranial part is accessory to the vagus which it quickly joins.

Nasal Passages

When a skull is split into right and left halves (i.e., in the median sagittal plane) as in figure 69, special features of the nasal passages are revealed. These are considered with the nose, in Chapter 7.

BONES ASSOCIATED WITH SKULL

Mandible
Ossicles of Ear
Hyoid Bone

Development; Pharyngeal Arches

Respiration in fishes is performed by means of gills. The skeleton of a gill is a bony arch or bow hinged to the cranium dorsally and joined to its companion of the opposite side ventrally. It supports the delicate, pink, feathery, vascular mechanism whereby oxygen is extracted from the water as it flows between the gills (fig. 72).

Five or six similar arches, the **pharyngeal arches,** appear in the human embryo but, since man develops lungs for breathing air, the arches are modified for other uses. The beginner would be led too far afield if he attempted to trace the fate of each one in particular, but he will be amply repaid if he devotes a few moments to the understanding of the essential relations of each, since such a study will make clear the relations of their human derivatives.

The first pharyngeal arch is known as the mandibular arch and it gives rise to the mandible or lower jaw.

The first gill (or branchial) cleft still persists as the passageway, known as the auditory (Eustachian) tube, leading from the interior of the pharynx to the ear. At the site where the first arch is slung from the skull, two of the three tiny ossicles of the ear develop out of its extreme upper end. They are described on page 355.

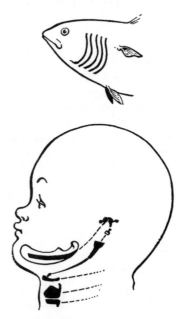

Fig. **72.** The gill or branchial arches in a fish compared with developmentally related structures in man.

At the skull, the second pharyngeal arch is represented by the **styloid process** of each side. Lower down, it is represented by the long, slender, stylohyoid ligaments and by the upper half of the (unpaired) hyoid bone.

The **hyoid bone** (fig. 73) is a slender U-shaped bone whose outer surface you can feel through the skin at the meeting place of the front of the neck and the floor of the mouth. Many muscles are moored to it (page 129 and fig. 214).

Although, normally, the second and the lower gill clefts close and disappear, occasionally one does persist and forms a discharging passage from the pharynx or throat to the skin.

The lower half of the hyoid bone and the chief cartilages of the larynx or voice-box suspended from it are the middle portions of succeeding pharyngeal arches which have lost all connection with the skull. In summary, the lower jaw, ear ossicles, hyoid bone, and chief cartilages of the larynx are, in man, derivatives of bony bows or arches functioning in aquatic animals as the skeletal framework of a respiratory system. Later, it will be desirable briefly to notice the fate of the muscles and nerves that are present also in these arches.

Mandible

The mandible (figs. 74, 75) results from a broadening of the anterior part of the first pharyngeal arch. The adult mandible consists of two vertical quadrangular plates, the **rami,** continuous at their lower ends with a U-shaped horizontal portion, the **body.** Each ramus is surmounted by two processes: (1) an articular **condylar process** behind, bearing a **head** (which resembles a little cylinder laid on its side) and forming the joint with the skull just in

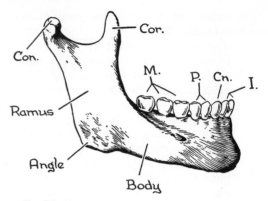

Fig. **74.** Mandible from outside. Con. = condyloid process; Cor. = coronoid process. M. = molar teeth; P. = premolars; Cn. = canine; I. = incisors.

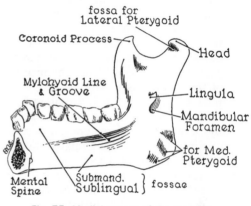

Fig. **75.** Medial aspect of the mandible

front of the ear; (2) a pointed and triangular **coronoid process** in front, receiving the insertion of the powerful biting muscle so evident on the side of the head. A second biting muscle clothes the outer surface of the ramus; a third parallels the second on the inner surface. Where ramus meets body a pronounced **angle** is present. The body not only carries the lower teeth but also forms the bony boundary of the mouth.

During the development of the human face a wing on each side grows medially; this wing meets a central downgrowth from the cranium and forms with it the upper jaw or **maxilla** (fig. 76). The maxilla carries the upper teeth and, by meeting its fellow of the opposite side behind the downgrowth from the cranium, forms also a bony **palate** which cuts off the mouth cavity from the nasal cavity. A failure of the two halves to meet results in an incomplete or **cleft palate.** A

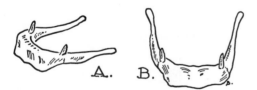

Fig. **73.** Hyoid bone. A. From left side. B. From in front. Greater and lesser horns are obvious.

similar failure of bilateral soft parts of meet results in a **hare-lip**—so called becaused this is the normal state of affairs in animals of the hare family. The section of maxilla that carries the incisor teeth and that represents the downgrowth from the cranium is a special midline element of bone known as the **pre-maxilla.** Therefore, when at the back, cleft palate is in the midline; at the front it and a hare-lip are never in the midline but to one side (or both sides) and along the line where maxilla and premaxilla have failed to meet.

RIBS

Each rib (fig. 77) has a **head** at its posterior end and a **costal cartilage** at its anterior end (costa, L. = rib). The heads of the ribs articulate with the vertebral column (bodies and discs). The costal carti-lages of the typical ribs articulate with the side of the sternum. Between its two ends each rib makes a sweeping arch helping to enclose the thoracic cage.

The greatest part of the rib is the **shaft** or **body** and, because it is flattened, it has upper and lower borders, and outer and inner surfaces. At the posterior end, the

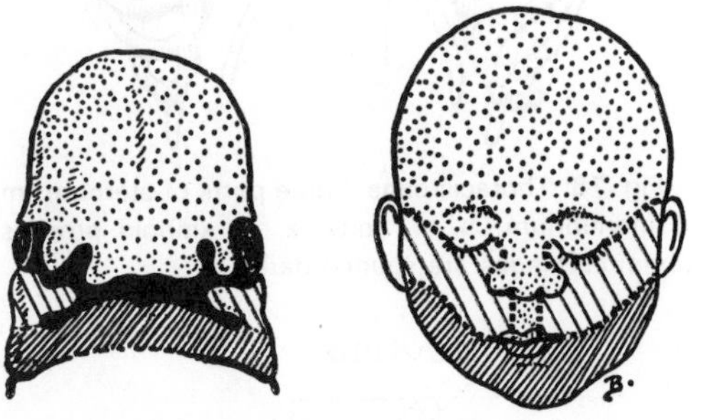

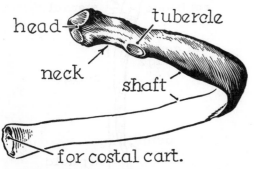

Fig. **76.** Development of face (see text)

Fig. **77.** A typical right rib from behind (costal cartilage absent).

head is separated by a short length of the rib, **the neck,** from a lump on the posterior surface known as the **tubercle.** The latter articulates (except in ribs 11 and 12) with the tip of the transverse process of the numerically corresponding vertebra.

A typical rib is joined to the vertebral column by means of a ligament which unites a horizontal crest on the **head** of the rib to an intervertebral disc; above and below this ligament the rib head articu-lates with adjacent vertebral bodies. For example, the crest of the head of rib number six is joined by an intra-articular ligament to the disc between vertebral bodies five and six, while the joint surfaces of the head articulate with the adjacent parts of these same vertebral bodies. Ribs 1, 11, and 12 are exceptions in that they are not united to a disc at all and each ar-ticulates with the side of the numerically corresponding vertebral body only.

In its excursion around the chest wall, a typical rib at first takes a direction down-ward and backward as well as lateralward until it abuts against the front of the tip of a transverse process. Head, neck, and tu-bercle are firmly anchored to the adjacent vertebrae by strong ligaments.

A short distance from the tubercle, the shaft ceases to pass backward and swings rather sharply forward; in consequence, the rib here presents an **angle.** Finally, as the shaft approaches the front of the chest it reaches its lowest point. At or near this point there is a sharp demarcation between the bone and the cartilage of the rib. Some anatomists consider the costal cartilage as a separate structure firmly united to the rib, but, in reality, it is only the unossified anterior part of the rib.

The lower border of the shaft possesses a **costal groove** protected laterally by a bony flange so that, while the upper border is smooth and rounded, the lower border is thin and sharp. Groove and flange are for the accommodation and protection of the intercostal vessels and nerve that accom-pany the rib. Ribs get progressively longer from the first to the seventh after which they once more get progressively shorter.

Costal cartilages four to ten soon turn abruptly upward to run to the side of the sternum where each articulates by structures very similar to those present at the head of the rib. The costal cartilage of the seventh rib, the last one to reach the sternum directly, is the longest. Costal cartilages eight, nine and ten do not reach the sternum but, in each case, join the cartilage next above (fig. 78).

Ribs 11 and 12 are distinctive. Each is short and articulates with the numerically corresponding vertebral body and not with a disc; each fails to articulate with a transverse process and, therefore, possesses no tubercle, each is tipped in front with a costal cartilage which, however, is free and does not reach its neighbor above, giving the name 'floating ribs' to 11 and 12.

STERNUM OR BREASTBONE

In the course of development, the costal cartilages (which are simply the unossified parts of the original rib elements that grow around the body from the vertebral column) approach the midline in front where their ends reach toward their fellows of the opposite side. Ossification occurs in the condensed mesenchyme between these ends and converts it into a sternum.

The sternum bears a vague resemblance to a dagger and consists of a **manubrium** (L. = handle), a **body,** and a **xiphoid** (Gr. = sword-like) **process** (fig. 78). The manubrium is wider than the rest of the sternum and, in consequence, gives better protection to the great vessels diverging from the heart.

At each upper lateral angle the manubrium articulates with the clavicle or collar bone, and this sternoclavicular joint bears the considerable responsibility of alone uniting the upper limb skeleton to the axial skeleton. Just below this important joint, the manubrium receives the first costal cartilage which is short and horizontal and, unlike the succeeding ones, is firmly fused to the side of the manubrium. Thus, manubrium and first ribs make a single 'unit.'

The body of the sternum is united to the manubrium by a movable joint which acts as a hinge allowing the sternal body to swing forward with each inspiratory effort. Viewed from the side, the sternum is seen to be slightly 'angled' at this joint, and here

a ridge can be felt in the living across the front of the bone. It is an important landmark known as the **sternal angle** (of Louis), and at this level the second costal cartilage joins the sternum (fig. 79).

The body of the sternum is 10 cm (4″) long and provides the greatest protection for the heart. Costal cartilages three to six

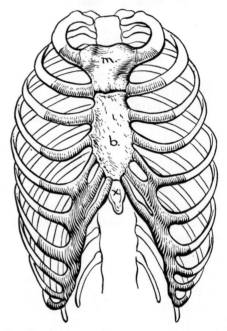

Fig. **78**. Thoracic cage. Three parts of sternum: m. = manubrium; b. = body; x. = xiphoid process. Costal cartilages are shaded dark.

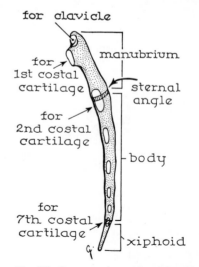

Fig. **79**. Sternum viewed for right side

join the side of the sternal body; cartilage number seven joins the sternum where the body meets the xiphoid process.

The irregular and pointed xiphoid process is cartilaginous in early life but soon starts to ossify. It can be felt between the seventh costal cartilages and is slightly movable up to middle life after which it becomes fused to the sternal body.

Complete failure of the ossified elements of the right and left sides of the sternum to meet in the midline may result in the heart lying outside the thorax, a condition that is usually fatal. Minor, insignificant holes due to partial failure of union may be diagnosed by the unwary as bullet holes.

Thoracic Cage and Respiration

When the ribs, costal cartilages, vertebrae, and sternum are in place the resulting composite structure is often spoken of as the **thoracic cage.** Besides containing the heart, lungs, and other important structures, it is a mechanism necessary to the respiratory act and it will therefore be profitable to examine it with this function in mind (fig. 78).

The thoracic cage has the general appearance of a truncated cone. It has an inlet above, an outlet below, a long posterior (vertebral) wall behind, and a short anterior (sternal) wall in front; the 12 ribs (with the intervening intercostal muscles) constitute its lateral wall and they contribute to its anterior and posterior walls; the costal cartilages are a conspicuous feature of its anterior wall.

Figure 80 illustrates the difference in shape of the cross-section of the human and non-human thorax. In man, the antero-posterior flattening has altered the heart-shaped appearance to a kidney-shaped one.

The vertebral bodies encroach on the thoracic cavity behind, but the consequent reduction in volume is more than compensated for by the backward sweep of the ribs. As a result of this, the thoracic cavity is particularly capacious on each side of the vertebral column.

The **thoracic upper aperture** (or **inlet**) is roughly equal in size to the area enclosed when the thumb and index finger of one hand are opposed to those of the other forming a kidney-shaped outline. This inlet is bounded by the body of the first thoracic vertebra, by the markedly curved, flat, broad, and short, first rib (fig. 81), and by the manubrium of the sternum. Its plane slopes toward the front. The structures bounding the inlet form a lever from which the remainder of the thorax moves in respiration.

The **lower aperture** is much larger; it is bounded by the body of the 12th thoracic vertebra, by the 12th rib, by costal cartilages 11, 10, 9, 8, and 7, and by the xiphoid process. To all of these, the circumference of a thin but strong fibromuscular partition, the diaphragm, is attached.

Diaphragm. The diaphragm is highly domed upward, its highest level reaching almost to the plane of the nipples. Although the diaphragm is the 'floor' of the thoracic cavity and the 'roof' of the abdominal cavity, its dome rises so high within the thoracic cage that the lower

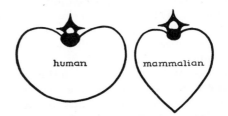

Fig. **80**. The human thorax is flattened; the general mammalian is keeled.

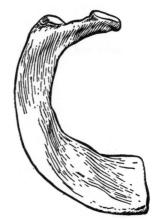

Fig. **81**. The first rib is the flattest, shortest, highest, most curved, and most fixed.

ribs give considerable protection to many abdominal organs (e.g., liver, stomach).

Movements of the Ribs in Breathing

The (voluntary) *act of inspiration* increases the capacity of the thoracic cage in the three principal directions. The transverse diameter is increased by the ribs swinging outward; the antero-posterior diameter is increased by the sternal body, hinged at the sternal angle, swinging forward; the vertical diameter is increased by the descent and flattening of the dome of the diaphragm.

A typical rib has a distinctive shape; it is curved and it is so twisted that it cannot be made to lie flat on a table. The reason for its peculiar shape must be sought in the movements it makes. A rib and its cartilage must be thought of as a unit fixed at one end to the vertebral column and at the other, to the sternum. At rest, this bow-shaped unit hangs at the side of the chest much as a bucket-handle hangs at the side of a bucket (fig. 82). Inspiration swings the unit up just as a bucket-handle can be swung up. This movement increases the width of the thoracic cavity and takes place about the two ends of the rib-cartilage unit.

At the same time, inspiration thrusts the sternum forward by causing the lowest point of the rib (its front end) to rise (fig. 83). This movement takes place around an axis that passes through the two posterior joints, viz., the joint at the head, and the joint at the tubercle of the rib. This axis is roughly at right angles to the one just described.

The lateral swing and the forward movement of the unit subject it to considerable strain. This, the 'springy' costal cartilage bears.

The inspiratory act is one in which the thoracic cavity in increased in volume by muscular effort, and, since nature will not permit a vacuum to exist, air flows in through the open respiratory passages and, by expanding the lungs, fills the increased space.

Although inspiration is a muscular effort, *expiration is an elastic recoil*. In this recoil the costal cartilages play a part

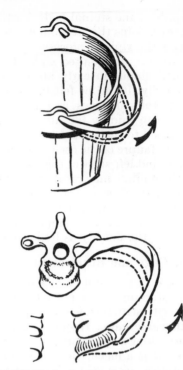

Fig. 82. The 'bucket-handle' inspiratory movement increases transverse diameter.

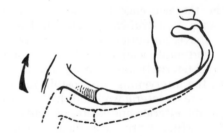

Fig. 83. The forward and upward inspiratory movement innreases front to back diameter.

second only to that played by the elastic tissue of the lung itself. Should the lungs lose their elasticity, as happens in certain diseases, the thoracic cage assumes a position of permanent inspiration and excessive muscular effort is required to squeeze the cage and so force air out of the lungs.

APPENDICULAR SKELETON

Development of Limbs

Each limb develops as a short bulbous outgrowth from the side wall of the embryonic body; it is known as a limb bud. Its interior is filled with a mass of unspecialized cells known as mesenchyme, out of which the bones, joints, ligaments, and muscles ultimately develop; nerves from the spinal cord invade the limb bud and supply it. As the limb bud elongates, a central core of condensed mesenchyme provides a certain degree of rigidity and, by division into three segments and conversion into cartilage and then into bone, this core of mesenchyme becomes the skeletal framework of three segments of the limb, viz., arm, forearm, and hand; thigh, leg, and foot.

The increasing size and development of the limb demand support at its 'root', i.e., in the side wall of the embryo. Here similar condensations of mesenchyme also occur which, starting in the vicinity of what is to be the shoulder or the hip joint, gradually extend toward the midline of the embryo both in front and behind, until, with those of the opposite side, they encircle the body and are the beginnings of an **upper limb (pectoral) girdle** or a **lower limb (pelvic) girdle.** When this condensed mesenchyme becomes converted into cartilage and then into bone, the degree to which it becomes anchored to the midline axial skeleton in front and behind determines the degree of mobility or fixity of the shoulder or hip region and, as a consequence, the freedom of movement of the limb concerned.

Stability vs. Mobility

In the upper limb, bony fixity to the vertebral column behind is never achieved, and the shoulder blade or **scapula** moves freely on the chest wall. Bony fixity to the sternum in front is, however, achieved and by means of the collar bone or **clavicle.** Whatever the real functional value of the clavicle may be—and that is still a subject of discussion among anatomists—it is the only bone uniting the upper limb to the axial skeleton.

The state of affairs in the lower limb is quite different. It is the limb of stability; the upper is the limb of mobility. The lower limb girdle is firmly anchored to the vertebral column by a large joint strengthened by strong ligaments; this is the *sacro-iliac* joint and it is not only a firm bond of union between axial and appendicular skeletons, but also a mechanism for the transmission of weight from the vertebral column to the lower limb girdle. Since there is no axial skeleton in front for the lower limb girdle to join, bony anchorage is secured by means of a firm midline union between the girdle bones of the two sides; this is known as the **symphysis pubis** and it too contributes to the fixity of the girdle.

Finally, there is a certain independence of movement between the two elements of the upper limb girdle, the scapula and clavicle, for they are united by a joint permitting movement; even this is denied to the lower limb girdle for the elements that comprise it are, on each side, entirely fused with one another into one solid composite hip bone.

UPPER LIMB GIRDLE

The girdle consists of the clavicle (L. = little key) in front, and the scapula behind. (Scapula is a modern word derived from scapulae, L. = shoulder blades.)

Clavicle

A man on the march swings his upper limbs and in doing so experiences no difficulty in clearing the side of his body. When it is remembered that the chest is cone-shaped and at its narrowest above, it is obvious that only by protecting the shoulder blade (which carries the arm at the shoulder joint) sufficiently far lateralward away from the chest wall can the swing be free. It is the collar bone or clavicle that accomplishes this for it is a strut from whose lateral end the scapula is suspended. A part of the scapula known as the acromion projects as a sort of hood above the shoulder joint; with this acromion the lateral end of the clavicle articulates and thus holds the scapula well laterally. A broken clavicle allows the whole shoulder region to collapse medialward.

The clavicle is situated at the root of the neck and can be palpated (felt) beneath the skin throughout its entire length. Its medial two-thirds is bowed with the convexity forward and is strong because of its having a prismatic cross-section. The lateral one-third is flat, relatively weak,

UPPER SURFACE

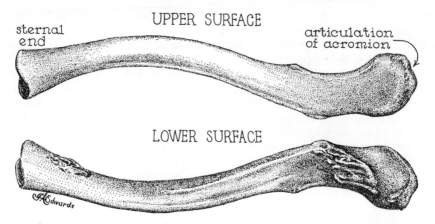

Fig. **84**. Right clavicle. The upper surface is smooth, the lower rough (for ligaments)

and bowed with a concavity forward (fig. 84).

The *medial (sternal) end* of the clavicle is enlarged and bulges above the concave facet on the upper lateral angle of the manubrium. Between the two prominent clavicles is the deep **jugular notch** noticeable at the front of the root of the neck.

As the clavicle passes almost horizontally lateralward, it crosses the flat first rib. The forward convexity of the medial part of the clavicle leaves room for the large vessels and nerves that cross the upper surface of the first rib to pass behind and below the clavicle on their way to the limb.

The *lateral (acromial) end* of the clavicle does not reach the 'point of the shoulder' which is entirely formed by the acromion; the small joint uniting the two bones can be felt 2 or 3 cm medial to the point of the shoulder. By swinging the arm slowly backward and forward, the gliding movement of this joint can be readily appreciated and, in the male, the site of the joint is often marked by a distinct prominence (see fig. 85 and Plate 1, *Surface Anatomy*, preceding p. 21).

The *upper surface* of the clavicle, being just deep to the skin, is smooth; the *lower surface* is much rougher, since strong ligaments bind it down to the first rib near its medial end and suspend the scapula from it near its lateral end.

Two muscles descend to the back of the clavicle; one to its medial end, the other to its lateral end. Two muscles ascend to the front of the clavicle; one to its medial two-thirds, the other to its lateral one-third. A

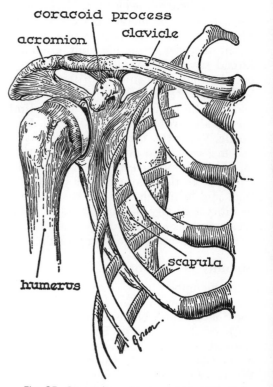

Fig. **85**. Orientation of bones of shoulder region. Note strong coraco-clavicular ligament that suspends scapula from clavicle.

small muscle occupies the lower surface of the clavicle between the two ligaments.

Scapula

The scapula (figs. 86, 87) is a thin, flat, triangular plate of bone clothed on both surfaces with muscles and applied to the upper part of the back of the thorax (fig.

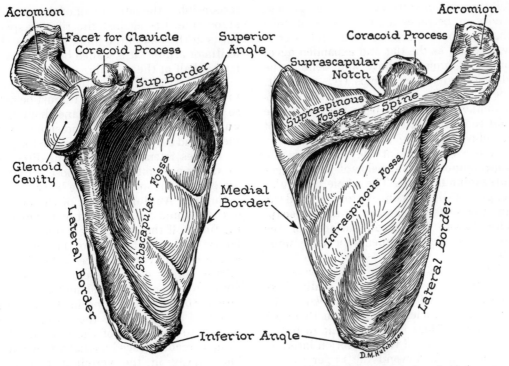

Fig. 86. Right scapula from in front

Fig. 87. Right scapula from behind

85); it covers the backs of ribs two to seven. When in the anatomical position, the longest side of the triangle runs parallel to, and about 5 cm (2″) away from, the vertebral spines; it is hence called the **medial** (or **vertebral**) **border,** and it is thin. At its lower end is the **inferior angle** of the scapula from which the thick and strong **lateral** (or **axillary**) **border** runs upward and in a lateral direction until it terminates in the 'lateral angle' of the scapula where a joint surface, taking part in the shoulder joint and known as the glenoid cavity is situated (glenoid, Gr. = socketlike). The heavy bar of bone in the axillary border (axilla = armpit) prevents buckling of this stress-bearing region.

The **glenoid cavity,** devoted to the reception of the head of the humerus or arm bone, is a shallow, concave, oval area about 4 cm (1½″) long and 2 to 3 cm (1″) wide; it faces forward, lateralward, and slightly upward. This region of the scapula is the thickest part of the whole bone.

Running almost horizontally medialward from the upper end of the glenoid cavity is the **superior border** which meets the vertebral border at the **superior angle** of the scapula. The superior border is extremely thin and sharp except near the glenoid cavity. Here, there springs from the superior border immediately medial to the upper end of the glenoid cavity, a stout, hooked projection of bone known as the **coracoid† process.** This resembles in size, shape, and direction, a bent little finger pointing down the arm. At the site of its bend it is firmly united by a strong ligament to the under surface of the clavicle as that bone crosses above it; to the tip of the coracoid process are attached muscles that run in the direction that it points (fig. 145, page 88).

Running across the posterior surface of the scapula is a ledge of bone that projects backward (fig. 87). It is known as the **spine** of the scapula and it begins about a third of the way down the vertebral border where it fades out into that border. As it passes lateralward toward the glenoid cavity, this spine gradually increases in prominence. A

---

† Although coracoid means 'like a crow,' no resemblance to a crow's body or beak is apparent.

full thumb's-breadth medial to the glenoid cavity, it becomes a free process of bone that sweeps forward, upward, and lateralward until, as the flat and expanded **acromion,** it projects above and behind the glenoid cavity (akros, omos, Gr. = summit, shoulder; cf., Acropolis) (fig. 88). This acromion has already been noticed as the part that articulates with the lateral end of the clavicle.

Above the spine of the scapula is the **supraspinous fossa;** below it is the larger **infraspinous fossa.** The whole triangular scapula is slightly concave forward and the anterior surface is known as the **subscapular fossa.** The broad surfaces of the fossae provide attachments for fleshy muscles.

The scapula is capable of very considerable movement on the chest wall; these movements will be analyzed when the numerous muscles that perform them are considered. Here it is sufficient to notice that, as a result of the projection laterally of the scapula beyond the upper ribs—a projection for which the clavicle is

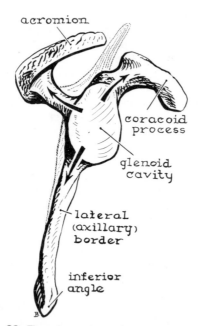

Fig. **88.** Three heavy bars of bone, indicated by arrows, radiate from the glenoid cavity to three areas of the scapula where important muscles and ligaments are attached. (Right scapula, antero-lateral view.)

responsible—the lateral half of the scapula at no time fits against the ribs. A space exists at the root of the limb, known as the **axilla** or armpit; it is walled behind by the lateral half of the scapula.

Finally, it should be noted that in order to orient a scapula correctly it must be held so that its surfaces lie in a plane about midway between front to back and side to side; the glenoid cavity will face as much forward as it does lateralward—a useful direction for the upper limb most of whose activities are carried out in front of the body.

Surface Anatomy. Most of the outline of the scapula can be traced by the exploring finger. If the clavicle be followed to its lateral end it will lead to the acromion whose outline can be traced; the tip and lateral border of the acromion limit the bony prominence of the shoulder, and the lateral border ends behind in the palpable *acromial angle* whence the finger is conducted to the spine of the scapula. The spine is readily traced medially to where it fades away in the vertebral border as already noted. The finger can follow the vertebral border throughout almost its whole length and, by passing round the inferior angle, it can trace the axillary border almost to the glenoid cavity. The superior angle and superior border are quite inaccessible to the finger except for the prominent coracoid process, which can be felt at the front in the visible depression situated below the clavicle near its lateral end.

If the fingers of one hand palpate the coracoid process and those of the other lift up the inferior angle, the scapula can be controlled by its two 'ends' and the whole shoulder girdle can be passively lifted up and manipulated.

HUMERUS

The humerus, like most long bones, consists of a shaft or body with two expanded ends (fig. 89, 90).

Upper End. At the upper end of the bone is the ball of the ball-and-socket shoulder joint. Actually, this ball or **head** consists of rather more than a third of a sphere carried on a cylindrical shaft. The head is not on the summit of the shaft but, in the anatomical position, faces medial-ward and upward; its surface is covered with articular cartilage and has an area more than three times as great as that of the glenoid cavity with which it articulates. (For epiphysis, see fig. 91.)

In close proximity to the head are two **tubercles**—a **greater** and a **lesser.** Both are for the insertions of certain of the numerous muscles that surround and move the shoulder joint. The **greater tubercle** faces lateralward and, in the living, is largely responsible for the roundness of the shoulder; the **lesser tubercle** faces forward and is separated from the greater by the **intertubercular** (or *bicipital*) **groove** prolonged down the shaft and in which lies a tendon (the long head of the biceps). The

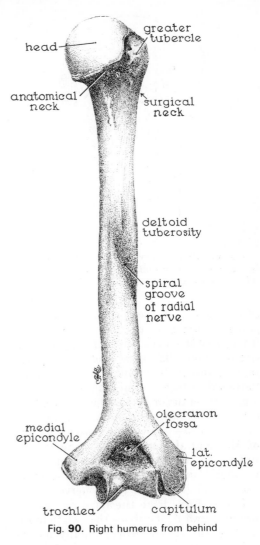

Fig. **90.** Right humerus from behind

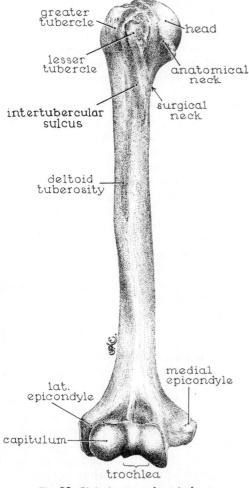

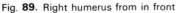
Fig. **89.** Right humerus from in front

lips of the groove give insertion to certain shoulder muscles.

At the circumferential margin of the cartilage-covered articular head lies the **anatomical neck;** it separates the head from the tubercles. Below the head and tubercles, where the shaft narrows, is the **surgical neck;** it is the site of most frequent fracture of the upper end of the shaft.

Shaft. The upper one-half of the shaft or body is cylindrical as one might expect, for it carries a head for a ball-and-socket joint. Half-way down its length, where a roughness on its lateral side is known as the **deltoid tuberosity,** the shaft begins to

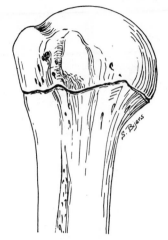

Fig. **91**. Upper epiphysis of the humerus shortly before uniting with the body of the humerus (age 19). (In life, cartilage fills the epiphyseal line.)

change until, at the lower end, it is expanded from side to side and flattened from front to back, although a midline strengthening bar gives the front of the shaft a distinctly rounded contour. Each flattened surface is completely occupied by a muscle—a flexor of the elbow in front, an extensor behind. A shallow, oblique—almost vertical—groove, the (**spiral**) **groove of the radial nerve,** limits the deltoid tuberosity behind and spirals from medial to lateral sides on the back of the shaft, lodging the important radial nerve.

Lower End. Here the bone curves distinctly forward; it carries the two bones of the forearm, viz., the ulna medially and the radius laterally. The two joint surfaces for these two bones are different from each other, since the functions of the two bones are different.

The humerus and ulna together form a secure hinge joint. To accomplish this, a pulley-like surface on the lower end of the humerus, known as the **trochlea** (L. = pulley), fits into a deep, rounded notch on the upper end of the ulna which swings round the pulley as the forearm is flexed and extended.

Immediately above the trochlea, two hollows or fossae exist, one in front and one behind; these are known as the **coronoid fossa** and the **olecranon fossa,** respectively, and are for the accomodation of the

corresponding parts of the ulna when the joint is fully flexed or fully extended.

Immediately adjoining the lateral part of the trochlea is a little rounded ball of bone, the **capitulum** (L. = little head). There is smooth articular cartilage covering its anterior and inferior aspects (but not its posterior aspect) because only anterior and inferior aspects articulate with the head of the radius (see Pronation and Supination, page 56).

A prominent process, the **medial epicondyle,** projects medially from the trochlea and provides origin for flexor muscles of the forearm. Similarly, the **lateral epicondyle** projects laterally from the capitulum and provides origin for extensor muscles.

From each epicondyle a ridge runs upward, contributing to the width of the lower half of the shaft of the humerus and disappearing half-way up in the cylindrical body. These ridges are the **medial** and the **lateral supracondylar ridges** and of them the lateral is the more pronounced.

Surface Anatomy and Orientation

The head of the humerus can be felt in the axilla (or armpit) when the arm is carried from the side (abducted); the greater tubercle can be grasped if the Deltoid muscle, which covers it, is relaxed. The shaft can be made out through the muscles that clothe it; the epicondyles (particularly the medial) are the easiest parts of the bone to palpate; the lateral (but not the medial) supracondylar ridge can be traced almost half-way up the arm.

When the limb is held in the anatomical position (palms facing forward) the head of the humerus looks medialward and upward but neither forward nor backward; the lesser tubercle looks directly forward and, at the lower end of the bone, the medial epicondyle lies slightly in advance of the lateral.

On the other hand, when the limb hangs comfortably by the side (palms facing medialward) there is some medial rotation of the humerus. Then the two epicondyles are about equally advanced and the head of the bone faces somewhat backward to make its best fit with the glenoid cavity.

ULNA

The ulna is the medial of the two bones of the forearm (figs. 92, 93).

Upper End. It looks rather like a spanner or wrench whose upper 'jaw,' known as the **olecranon process,** is simply the upper end of the shaft and whose lower 'jaw,' known as the **coronoid** (Gr. = crown-like) **process,** is a forwardly projecting ledge of bone. Between the two processes the large and deep semilunar 'mouth of the wrench' faces forward and is known as the **trochlear notch.**

The lateral side of the ledge or coronoid process carries a small, shallow, semilunar surface, the **radial notch,** in which the side of the head of the radius fits. The anterior surface of the coronoid process is rough and ends below in the **tuberosity of the ulna** which receives the insertion of the brachialis, the great flexor muscle of the elbow joint.

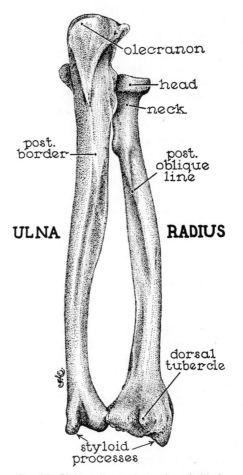

Fig. **93.** Right radius and ulna from behind

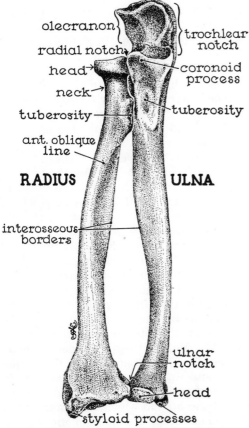

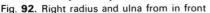

Fig. **92.** Right radius and ulna from in front

Shaft and Lower End. The **shaft** is triangular on section and tapers as it descends until, at its lower end, the bone consists merely of a small disc-like **head‡** from whose posterior part there projects a short blunt peg of bone known as the **styloid process.**

The sharp lateral border of the shaft faces the corresponding (medial) border of the radius; each is known as an **interosseous border** and they are united by a fibrous sheet—**the interosseous membrane.** The front or anterior surface of the shaft is for the origin of a muscle that extends uninterrupted to the medial surface of the bone. Hence the anterior border is smooth and rounded. The back of the

---

‡ Note: The head of the ulna is at the wrist.

shaft adjacent to the interosseous border gives origin to a succession of muscles devoted to the thumb. The sharp **posterior border** runs from the back of the olecranon down to the styloid process.

Surface Anatomy. The olecranon can be felt, and from it the whole length of the shaft of the bone can be followed just beneath the skin to the head and styloid process.

RADIUS

The radius is the lateral of the two bones of the forearm and whereas the important functions of the ulna lie at the elbow, those of the radius lie at the wrist; there the bone is charged with the important duties of carrying the hand and taking that structure with it in the movements of pronation and supination. For this reason, while the **head** of the radius is small, the **shaft** becomes progressively more massive as it is followed distally (downward) until, at the wrist, it is twice as wide as it is at the head.

Upper End. Just above the narrowest part of the radius, appropriately named the **neck,** is the **head,** a disc whose upper surface is slightly concave and articulates with the capitulum of the humerus, and whose circumference is grasped by a circular ring of bone and ligament. One-quarter of this ring is the radial notch of the ulna; the rest is a special ligament. The head of the radius rotates in its 'high collar' during pronation and supination. The **neck** of the radius is limited below (on the medial side) by the prominent **tuberosity of the radius.**

Shaft. There is an obvious lateral convexity or bowing of the shaft or body (fig. 92). Running obliquely across the front of the shaft from the region of the radial tuberosity to the point of greatest bowing is a distinct line known as the **anterior oblique line of the radius.** Above and below it the bone is clothed by muscles; along it a muscle arises; at each end a muscle is inserted. On the back of the radius is the corresponding, but much less important, *posterior oblique line.*

The sharpest border of the radius is the **interosseous border** which faces a similarly named border of the ulna and is joined to it by a membranous ligament.

Lower End. At the lower end, the interosseous border gives place to the concave **ulnar notch** into which the head of the ulna fits. On its lateral side, the lower end of the radius tapers abruptly to a **styloid process** which is larger than that of the ulna and projects farther distally. The inferior surface of the radius for articulation with the carpal bones is appropriately concave.

The anterior surface of the lower end is smooth and is terminated by a sharp and prominent anterior border separating it from the inferior surface; in the resultant hollow above this border lies the Pronator Quadratus muscle. The posterior surface of the shaft immediately above the lower end is convex and shows a series of more or less vertical grooves in which tendons play as they pass to the back of the wrist and fingers.

Pronation and Supination of the Forearm

As we have noted, the head of the radius can rotate freely within a ring or a collar formed by bone and ligament; also, the interosseous borders of the radius and ulna are joined by a membrane which is quite pliable. These two facts, coupled with the manner in which radius and ulna are joined at their lower ends, allow the lower end of the radius to swing right round the front of the lower end (head) of the ulna until the shafts are crossed.

This brings the lower end of the radius to the medial side, and, because the hand is attached to the radius, the palm faces backward (fig. 94). The movement which has been described is *pronation;* the reverse movement is *supination.*

The root of the ulnar styloid process is attached to the lower edge of the ulnar notch of the radius by a triangular fibrocartilaginous ligament which crosses below the head of the ulna (fig. 160 on page 94). This allows the radius to swing around the head of the ulna but prevents their separation. The fibro-cartilage excludes the head of the ulna from contact with the carpal bones and thus the ulna takes no part in 'carrying' the hand. Rather, it acts as the fixed bone around which the lower end of the radius swings in pronation and supination.

Fig. **94**. Forearm in pronation

the wrist; the posterior surface of the lower end of the shaft can be felt, from the styloid process laterally to the inferior radio-ulnar joint medially, and at this joint the movement of the radius can be appreciated as it swings round the ulna.

CARPAL BONES

The small bones that give flexibility to the wrist are known as the carpal bones (fig. 96), and sometimes, collectively, as the **carpus** (Gr. = wrist). They are eight in number arranged in two rows of four—a proximal or upper row, and a distal or lower row. In both rows each bone glides on its neighbors and is tied to them by interosseous ligaments. The two rows also glide on one another, and rather more movement is possible between the two rows

The head and upper part of the shaft of the radius are on a plane entirely in front of the shaft of the ulna, as can readily be observed by viewing the two articulated bones from the lateral side (fig. 95); this is because the head of the bone articulates with the side of the coronoid process which has already been observed to be a ledge of bone projecting forward from the shaft of the ulna. Furthermore, the shaft of the radius is not straight but is bowed laterally. These two features permit the radius to roll *en masse* across the ulna in spite of the fact that the head of the radius does not move in space as it revolves in its socket.

When the ulna moves on the humerus in flexion and extension of the elbow, the radius accompanies it and the rather concave upper surface of the head of the radius slides on the rounded surface of the capitulum.

Surface Anatomy. When the forearm is pronated and supinated, the head of the radius can be felt to rotate just below the lateral epicondyle of the humerus. Unlike the ulna, the shaft is not easily palpated but the sharp anterior border of the lower end can be felt on deep pressure in front of

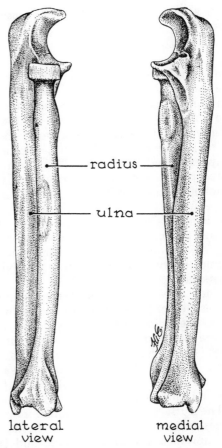

lateral view   medial view

Fig. **95**. Lateral and medial views of right radius and ulna.

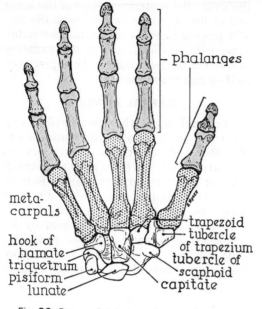

Fig. **96.** Bones of right hand from in front (palmar aspect).

head of the ulna is single, continuous, and oval in outline. It is markedly convex and fits into the concavity of the lower surface of the radius.

At this joint, the *radiocarpal* or *wrist joint*, the bones can glide from side to side on the radius and thus permit the hand to be abducted or adducted; and they can glide forward and backward so as to flex or extend the hand.

When the movement of abduction or adduction is performed at the wrist joint, the hand moves as a unit and the bones of the distal row must indulge in no independent movement, but rather must move with those of the proximal row. This is accomplished by the shape of the joint between the two rows; the joint is known as the *midcarpal joint* and it is sinuous in outline thereby locking the two rows together when abduction or adduction takes place.

The hand can be flexed and extended on the forearm to an equal degree, but two joints are involved—wrist and midcarpal. Most of the extension occurs at the wrist joint and flexion at the midcarpal joint.

In each row the bones are disposed so as to form a transverse arch whose concavity faces forward, and the marginal bones— thus already projected farther forward than their fellows—are rendered still more prominent by each possessing an eminence. To these four eminences are attached the four corners of a strong and broad postage-stamp-sized ligament that maintains the forward concavity and, at the same time, forms a bridge deep to which pass the numerous tendons that flex the wrist and fingers. It is the *flexor retinaculum* (fig. 268 on page 159).

Of the marginal bones, the scaphoid has a tubercle, the trapezium a ridge, and the hamate (L. = hooked) a hook.

Surface Anatomy. The tubercle of the scaphoid and the pisiform are easily palpable at the distal skin crease on the front of the wrist. The ridge of the trapezium and the hook of the hamate lie just below but are not felt as easily. The dorsum of the carpus presents a continuous rounded surface with no special features (fig. 97).

than is possible between adjacent bones in the same row.

From medial to lateral side the bones of the proximal row are named **pisiform, triquetrum, lunate,** and **scaphoid;** those of the distal row are **hamate, capitate, trapezoid,** and **trapezium.** The four marginal bones therefore are pisiform and scaphoid of the proximal row, hamate and trapezium of the distal row. The smallest is like a pea in size and shape and is therefore named the pisiform; it succeeds in projecting forward only by resting on the front of the neighboring triquetrum instead of lying by its side.

The names of the carpal bones are more or less descriptive of their appearance: thus, the triquetrum is three-sided, the lunate is similunar, the scaphoid is somewhat canoe-shaped, the hamate has a hook, the capitate has a rounded 'head,' the trapezium and trapezoid are rather irregular and four-sided.

The surface that the proximal row of bones (exclusive of pisiform) presents to the lower surface of the radius and the fibro-cartilaginous ligament below the

METACARPALS

The five bones in the palm of the hand are the metacarpals (meta, Gr. = beyond). They are miniature long bones and each possesses a more or less square-cut **base,** a rather stout **shaft,** and an expanded and rounded **head.** The bases are packed close together; the heads are separated from one another by distinct intervals. Therefore, the bones do not run parallel to one another but diverge like the sticks of a fan. Again, the rounded heads are not ball-shaped; their fronts are almost flat. The importance of these statements will be appreciated when finger movements are considered (pages 96, 169).

The metacarpal of the thumb is highly mobile because its base possesses an independent saddle-shaped joint surface; those of the little and ring fingers move slightly when the hand is firmly clenched; those of the middle and index fingers are practically immobile.

On the rounded heads of the metacarpals the free fingers move, and the joints at these heads or 'knuckles' are known as the **metacarpophalangeal joints.** The joint surfaces are prolonged on the fronts of the heads but do not reach the backs; the reason is apparent.

The capsule of each of these joints is united in front to its neighbor or neighbors by a short, strong, square-shaped ligament which prevents the knuckles from being farther separated from one another. The thumb being an exception, its 'knuckle joint' moves in space with the movements of the first or thumb metacarpal.

PHALANGES

Each of the four fingers possesses three phalanges, a **proximal,** a **middle,** and a **terminal or distal phalanx.**§ The proximal ones are the largest and the distal, the smallest. Like the metacarpals, the phalanges are miniature long bones and, although possessing somewhat expanded **bases,** the proximal and middle phalanges are little, if any, wider at their **heads** than at their **shafts.** Thus, while the knuckles

§ Note the spelling. The singular of phalanges is phalanx.

are relatively large and prominent, the fingers present no great enlargement at the *interphalangeal joints.*

The bases of the proximal phalanges are concave to fit the heads of the metacarpals; their heads, however, possess a pair of little condyles that fit into shallow depressions on the proximal aspect of middle phalanges. The heads of the middle phalanges and bases of the terminal ones are similarly made. These pulley-like surfaces permit, of course, flexion and extension but preclude entirely any other movement.

Each terminal phalanx is flattened, expanded, and triangular at its distal end where, of course, it is non-articular but is rather a bed for the nail.

The shaft of each phalanx is rounded or convex from side to side at the back, but it is flat or even slightly concave in front; it thus provides in front a suitable surface for the accommodation of the rounded tendons that flex the fingers. Extensor tendons on the backs of the fingers are flat bands, hug the sides of the shafts, and are not palpable.

The thumb possesses only two phalanges and except that they are rather stouter

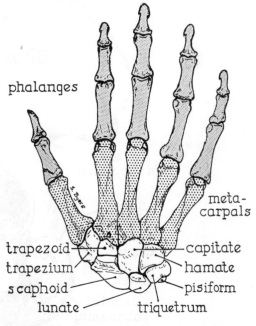

phalanges

meta-carpals

trapezoid
trapezium
scaphoid
lunate
capitate
hamate
pisiform
triquetrum

Fig. **97.** Bones of right hand—dorsal view

than those of the fingers they exhibit no other special features.

Surface Anatomy. The shafts of the metacarpals are felt at the back of the hand where they are close to the skin; the degree of mobility of each one is readily demonstrable. The heads of the metacarpals are the familiar knuckles and when the hand is clenched the joints of the knuckle row lie immediately distal to these heads and may be felt with the fingernail; they lie midway between the transverse palmar crease and the creases at the bases of the fingers. All phalanges can be readily felt and while the first row of interphalangeal joints corresponds to the level of the middle crease of the free finger, the second row of joints is just beyond the last crease.

LOWER LIMB GIRDLE

The *right* and *left hip bones* together constitute the pelvic or lower limb girdle. With the sacrum and coccyx the hip bones form the *bony pelvis* (page 62). It can thus be seen that the hip bones are equally important in relation to the abdominal cavity and to the lower limb. *Pelvis* is latin for *a basin* and refers to the general shape of the pelvic cavity.

HIP BONE

The hip bone (*os coxae*) is such a complex bone and so difficult to visualize that it is profitable to select its most salient feature and to reconstruct the bone around it. Somewhat below the center of the outer aspect of the bone, is a hollow hemisphere (fig. 98) about 5 cm (2") in diameter and known as the **acetabulum;** it is the socket for the reception of the spherical head of the femur or thigh bone and it faces lateralward, downward, and slightly forward.

At this acetabulum (L. = vinegar cup) three bony elements meet to form it and the composite bone. Until puberty the three elements are readily recognizable and, even in the adult, the lines along which they fused may be made out (fig. 99). The significance within the acetabulum, of a horseshoe-shaped articular area

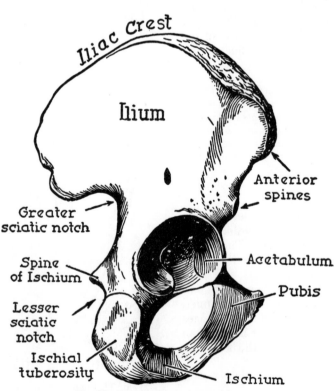

Fig. **98.** Outer aspect of right hip bone

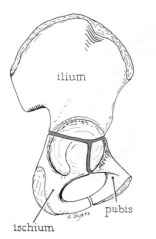

Fig. **99**. The three parts of the child's hip bone are united by cartilage along a Y-shaped line at the acetabulum.

and of the central non-articular area will be appreciated when the hip joint is studied (page 100).

Ilium. The upper two-fifths of the ace-tabulum and an expanded and curved plate of bone above it constitute the **ilium.** The ilium is roughly fan-shaped. Its upper mar-gin, the **iliac crest,** is thick and sinuous in outline when viewed from above. It is palpable from **anterior superior spine** to **posterior superior spine.** Below each of these spines is a corresponding *inferior spine.*

The iliac bone helps to protect the lower abdominal contents behind and at the sides, and on its inner concave and outer convex surfaces large and powerful muscles take origin. The anterior spines of the two sides are separated from each other by almost the total width of the trunk but the posterior spines are but 10 cm (4″) apart since, as the ilia pass backward, they approach one another.

Area of Sacro-Iliac Joint. The posterior part of the inner surface of each ilium meets the side of the sacrum and is firmly united to it over a large area. The union is a secure one and is known as the sacro-iliac joint. It consists of a large ear-shaped articulating surface known as the **auric-ular surface** (fig. 100), behind and above which is a very rough area known as the **tuberosity of the ilium;** the latter gives

attachment to strong interosseous and strong posterior sacro-iliac ligaments that bind the bones together and permit only a minimum of movement. Since the joint receives and transmits the weight of the body it is an important one.

Immediately below the sacro-iliac joint, the posterior border of the ilium suddenly sweeps forward and then downward to form an enormous notch, the **greater sciatic notch,** through which the sciatic nerve and other structures pass. Just below the angle of the notch the ilium meets the ischium.

Ischium. The lower lateral two-fifths of the acetabulum, and a stout but short prismatic column of bone that supports it, are known as the **ischium** (pronounced *iskium;* Gr. = hip). It descends almost vertically for 5 cm (2″) and terminates in a rough and large **ischial tuberosity** which receives the weight of the body when one sits up straight.

The greater sciatic notch of the hip bone is terminated below by the prominent, pointed **ischial spine.** Between it and the ischial tuberosity is a much smaller **lesser sciatic notch.** From the tuberosity a slen-der bar of bone, the *ramus* of the ischium, projects forward (to meet the inferior ramus of the pubis).

Pubis. The antero-medial one-fifth of the acetabulum belongs to the pubis, and this pubis is continued almost horizontally medialward as a bony bar—**superior** (or 'iliopubic') **ramus**—until it expands to form the **body.** The body of the pubis meets its fellow of the opposite side, where the union of the two bones is known as the **symphysis pubis.** By means of this symphysis the fixity and security of the hip bones are further enhanced. The joint sur-faces are oval and about 3 to 4 cm (1½″) long in the male.

From the lower part of this symphysis, a slender bar of bone, the **inferior ramus,** runs downward, backward, and lateral-ward to meet a similar one projected from the ischial tuberosity. Thus, the pubis is united to the ilium at the acetabulum by an 'ilio-pubic' ramus and to the ischium by an 'ischio-pubic' ramus.

Obturator Foramen. This large hole is ringed by the rami of the pubis and is-

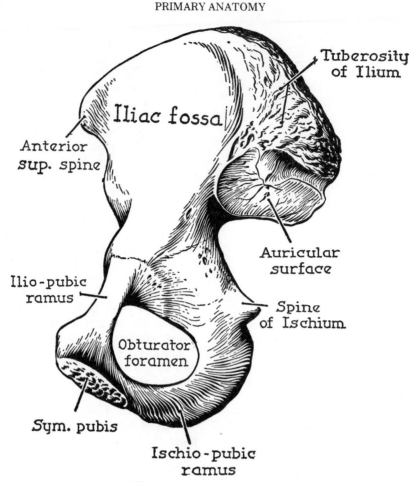

Fig. **100.** Inner aspect of the right hip bone

chium, and is a conspicuous feature of the bone in the region between the acetabulum and the pubic symphysis. The foramen is closed in life by a fibrous membrane except at its highest point where the spiral arrangement of its margin permits the passage of vessels and a nerve.

Lines of Force. The weight of the body received at the sacro-iliac joints is transmitted to the ischial tuberosities if one sits up straight, or to the hip joints if one stands erect. Hence the hip bones are thick and strong along both these lines of weight transmission (fig. 101). The symphysis pubis ensures that, under normal circumstances, the two bones shall be neither spread apart nor buckled inward.

Surface Anatomy. The iliac crest can be traced in the living from ant. sup. spine to post. sup. spine. Running from the ant.

sup. iliac spine to the pubis, a ligament lies deep to the crease in the groin; it marks the division between abdomen and lower limb and is the **inguinal ligament** (of Poupart) but it cannot be felt. The top of the symphysis pubis is readily felt in the midline in front.

Between the palpable posterior iliac spines the rough posterior surface of the sacrum lies close to the skin and is readily made out. The large muscle (Gluteus Maximus) of the buttock conceals the ischial tuberosity which, nevertheless, is easily felt, and from the tuberosity the ischio-pubic ramus can be traced to the lower part of the symphysis pubis.

Bony Pelvis

The composite, basin-like structure consisting of the sacrum and coccyx behind, and of the hip bones at the sides and in front, is known as the bony pelvis (fig. 102). It is divisible into: (1) a so-called **false** or **major pelvis** made up of the blades of the two iliac bones, and contributing to the abdominal wall; (2) a **true** or **minor pelvis** formed: (a) by the sacrum and coccyx; (b) by the part of the ilium in the neighborhood of the acetabulum; (c) by the ischium and pubis.

From what has been said it is perhaps realized that the hip bone is by no means all on one plane. Actually, the general plane of the blade of the ilium is such that its surfaces look chiefly medially and laterally, while the surfaces of the combined ischium and pubis look principally forward and backward; the ilium is thus almost at

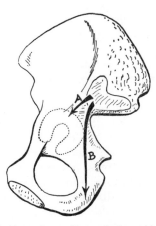

Fig. **101.** Heavy bars of bone, indicated by arrows, run from the sacro-iliac joint for weight-bearing: A. When standing, to the top of the acetabulum (dotted outline). B. When sitting, to the ischial tuberosity.

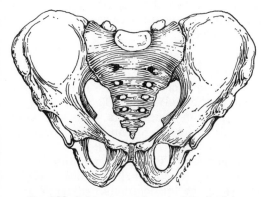

Fig. **102.** Pelvis viewed from in front (male)

right angles to the rest of the bone. This is an obvious feature when the bone is observed as it rests on the table. This difference in plane is marked on the inner aspects of the bone by a ridge or line that extends from the region of the top of the symphysis pubis to the most anterior point of the sacro-iliac joint. The line serves to divide the region above it, which takes part in the wall of the abdominal cavity, from the region below it, which takes part in the wall of the pelvic cavity.

The upper aperture of the pelvis minor (or **pelvic inlet**) is shaped like the conventional heart; it is encroached upon behind, by the front of the body of the first sacral vertebra. The prominence so produced is known as the **sacral promontory,** lateral to which the rounded and smooth anterior margins of the alae of the sacrum bound the inlet behind while the lines extending from the sacro-iliac joints behind to the symphysis pubis in front complete the boundary at the sides and in front (fig. 102).

The **lower aperture** or **outlet of the pelvis minor** is bounded on each side by the ischial tuberosities, anteriorly by the pubis, and posteriorly by the coccyx. The 'ischio-pubic' rami and a pair of ligaments (fig. 104) that join ischial tuberosities to coccyx and sacrum complete the sides of the 'diamond-shaped' area. The term **perineum** is used for the area when soft structures and skin are included (fig. 375, page 214).

When correctly oriented in the anatomical position, the plane of the pelvic inlet varies from 45° to 60° to the horizontal; the axis of the cavity is a curve with its concavity forward and is roughly parallel to the curve of the sacrum; the plane of the outlet is at about 10° to 15° to the horizontal and about 55° to 75° to that of the inlet.

The walls of the pelvic cavity consist of the floor of the acetabulum (in which all three parts of the hip bone partake), the back of the pubic bone, the medial surface of the ischium, the fibrous membrane that closes the obturator foramen, and the anterior surface of the sacrum.

The **female true** or **minor pelvis** is more capacious than that of the male because it surrounds and will limit the size of the birth canal, and as a result of the following characteristics (fig. 103):

1. The hip bones are set farther apart. This necessitates a relatively wider sacrum, longer pubic rami, and a greater distance between ischial tuberosities.
2. The ischial tuberosities are turned outward, thus increasing the subpubic angle of the *pubic arch*.
3. The sacrum is less curved and is set more horizontally. This increases the distance between coccyx and symphysis.

Many other, but subsidiary, sex differences may be made out but almost all of them are dependent on one or other of these three fundamental dispositions.

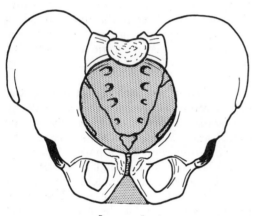

male pelvis

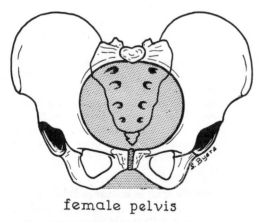

female pelvis

Fig. **103.** Male and female pelves compared (see text).

Mechanics of Pelvis

Essentially, the sacrum is slung more than it is wedged between the two iliac bones (see fig. 168, page 99) and it lies more horizontally than it does vertically; since it receives weight at its upper end (i.e., at the first sacral vertebra), it has a manifest tendency to rotate at the sacroiliac joints, a rotation that would result in the lower end of the sacrum and the coccyx tilting upward. On each side this tendency is resisted by two ligaments both of which are fixed to the side of the lower part of the sacrum and to the coccyx; one passes to the ischial spine and is known as the **sacrospinous ligament,** the other and much longer passes to the ischial tuberosity and is known as **sacrotuberous ligament** (fig. 104). Besides preventing sacral tilt, the ligaments help to complete the wall of the pelvis posteriorly, and they convert the greater and lesser sciatic notches into correspondingly named foramina. Through these foramina issue muscles, nerves, and vessels.

The superimposed weight on the sacrum tends to force it *en masse* forward and downward. This it would do easily except for the following mechanism. Joining the large rough areas behind the auricular surfaces of ilium and sacrum are many strong short fibers of the **interosseous ligament.** In addition the auricular areas have interlocking irregular surfaces. The interosseous ligaments are tightened by the downward pressure of the sacrum and draw the two ilia toward each other. The ilia squeeze the sacrum between them and the greater the weight the harder they squeeze the interlocking surfaces together (see fig. 168, page 99).

FEMUR

The femur or **thigh bone** is the longest bone in the body (figs. 105, 106) being 45 cm (18″) in length. Its ends are very specialized for the hip and knee joints but the shaft is relatively simple.

Upper End. The characteristics of the upper end are related chiefly to the peculiarities of the hip joint, its ligaments, and its muscles. It has been noted already that

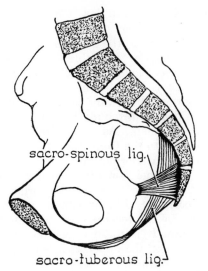

Fig. **104**. Two ligaments prevent sacral tilting

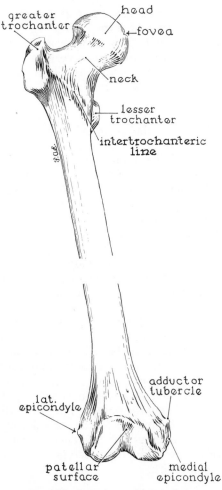

Fig. **105**. Right femur from in front

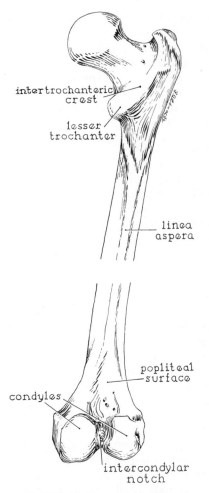

Fig. **106**. Right femur from behind

the sockets for the hip joints are separated by the width of the true pelvis; and it is obvious that the knees are normally in contact with each other. Therefore, the shaft of the femur must slope downward and medially. On the other hand, the head of the femur must face medially to fit into its socket. Therefore, the upper 10 cm or more of the femur are bent medially, forming an angle of 125° to 140° with the shaft. This curved part is known as the **neck.** It terminates above in a spherical articular **head** which faces medialward, upward, and slightly forward, to fit accurately in the acetabulum of the hip bone.¶

¶ The acetabulum and the femoral head both face slightly forward so that the two make their completest 'fit' when the femur is rotated slightly medialward, i.e., when the toes are 'turned in.'

At the site where the femur is bent, two prominences project, one at the convexity on the lateral side, the other at the concavity on the medial side; the two are known as the **greater trochanter** and the **lesser trochanter,** respectively. They have been pulled out, as it were, by the actions of muscles attached to them.

The greater trochanter projects upward and is somewhat rectangular in outline; it receives the insertions of the powerful gluteal muscles that clothe the outer aspect of the blade of the ilium. The lesser trochanter projects medialward and somewhat backward, and is conical; it receives the insertion of the powerful muscle that clothes the inner aspect of the blade of the ilium.

A rough line running obliquely downward and medialward unites the two trochanters across the front of the shaft. It is known as the **intertrochanteric line** and is produced by the strong iliofemoral ligament that forms the very thick front part of the fibrous capsule of the hip joint. A corresponding raised and smooth crest unites the two trochanters behind. It is known as the **intertrochanteric crest.**

Shaft. The cylindrical body, below the neck, is smooth and featureless except for a large ridge running down the middle of its posterior aspect. This is the **linea aspera** (L. = rough line). From it arise large muscles which clothe the femur in their downward course (the vasti) and into it insert a group of muscles that occupy the medial aspect of the thigh (the adductors). The linea aspera divides above into two diverging lines; one passes below the lesser trochanter, the other reaches the base of the greater trochanter.

Lower End. As the knee is approached, the cylindrical shaft becomes expanded sideways and a large, smooth, triangular area exists on the back of the bone between two lower diverging limbs of the linea aspera; this area forms the upper part of the 'floor' of the popliteal region (the region at the back of the knee) and is known as the **popliteal surface of the femur.**

When the bone is viewed from behind, its lower end is seen to be dominated by two articular **condyles** (Gr. = knuckles) which rest on the upper plateau-like surface of the tibia, the larger bone of the leg. These condyles resemble in appearance two wheels placed parallel to each other and each is 2 to 3 cm (about 1″) thick. They project backward beyond the shaft so that a deep **intercondylar notch** or **fossa** separates them. As their circumferences are traced to the front, the condyles blend with the anterior surface of the expanded shaft and their smooth articular surfaces blend also with a trough-like surface situated between and above them, which surface is for the reception of the **patella** or **knee-cap.** It is these wheel-like condyles that slide on the upper surface of the tibia in the movements of flexion and extension of the knee. At the axes of the 'wheels' on the sides of the bone, low **epicondyles** occur and to them are attached the collateral or side ligaments of the knee joint. The *adductor tubercle* is a lump just above the medial epicondyle and is very easily palpated (fig. 105). Further details of the condyles are discussed under 'Knee Joint' on page 101.

Surface Anatomy. The head can be palpated in the groin when the limb is rotated laterally. The neck is not palpable. The greater trochanter lies close to the skin and is the bony prominence felt well laterally some 10 cm (3″ or more) below the crest of the ilium. It is the bone one finds hard to get comfortable when one lies on the side. From a knowledge of the situation of the greater trochanter it is easy to understand that the shaft of the femur descends obliquely through the thigh. It is the presence of the adductor muscles on the medial side that obscures the otherwise unsightly knock-kneed appearance of the human lower limb.

The shaft of the bone is too well clothed with large muscles to be palpable. The condyles are readily felt and their outlines can be traced with the finger by palpating on each side of the patella. The epicondyles are the bony masses felt on each side of the knee within the circumference of the condyles. The adductor tubercle is an easily felt landmark.

TIBIA

The strong and massive bone on the medial side of the leg is the tibia (figs. 107, 108). It alone receives weight transmitted to it by the femur and it conveys that weight to the foot. The salient features of the tibia may readily be made out since the bone lies just deep to the skin where, as the shin, it is palpable from knee to ankle.

Upper End. In the midline below the knee, a bony swelling may be made out. It is the **tuberosity** of the tibia and, when the knee is actively extended, the broad and

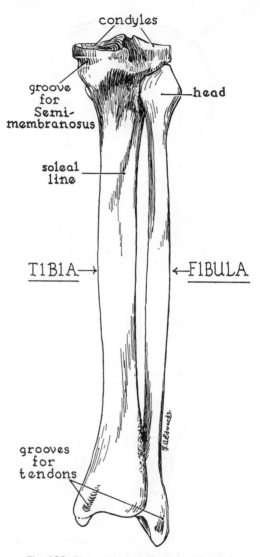

Fig. **108.** Right tibia and fibula from behind

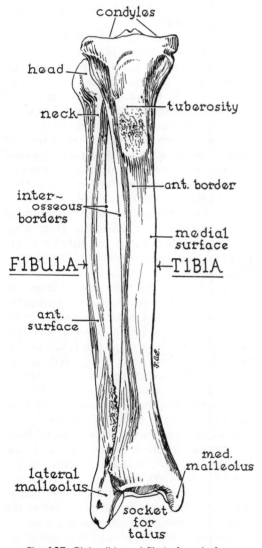

Fig. **107.** Right tibia and fibula from in front

strong tendon of the extensors of the knee can be felt descending to this tuberosity from the patella.

Above the tibial tuberosity, the bone expands into two large **condyles** upon which rest the condyles of the femur. The rounded outlines of the tibial condyles can be traced with the exploring finger except at the back where they lie deep in the popliteal region.

Their upper surfaces present a rather flat plateau for the reception of the femoral condyles (fig. 109). In general, this consists

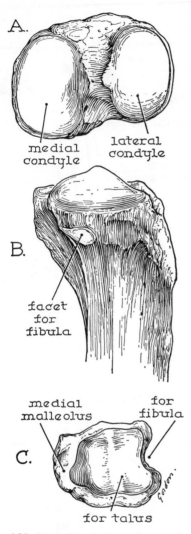

Fig. **109.** Right tibia. A. Upper (proximal) surface.
B. Lateral side of upper end. C. Lower (distal) surface.

of three parts: (1) an oval articular surface (with its long axis from front to back) medially; (2) a circular articular surface laterally; (3) a non-articular, hour-glass-shaped strip between the two which rises to a pair of little eminences in the middle of the plateau.

The long axis of the oval is greater than the diameter of the circle, and the two areas reach to the same transverse level in front; therefore, the medial condyle reaches farther back than does the lateral. The upper part of the shaft curves somewhat backward so that the condyles also project backward beyond the plane of the shaft.

On the under surface of the overhanging lateral condyle lies a small, circular, articular facet for the head of the fibula, the lateral bone of the leg. The fibula at the knee lies, therefore, on a plane behind that of the shaft of the tibia (cf., radius and ulna); it takes no part in the knee joint.

Shaft and Lower End. The sharp, superficial, *anterior border* can be traced in the midline of the leg downward from the tubercle. At its lower end, it swings medially to end as the front edge of the **medial malleolus.** This is a downward-projecting lump of bone that bounds the ankle joint medially and also acts as a pulley for several tendons.

The broad, superficial, *medial surface* of the shaft can be traced in a similar fashion to the same malleolus. On the lateral side of the sharp anterior border lies a large muscle which, above, obscures the *lateral surface* of the shaft. But where this muscle becomes tendinous above the ankle joint, the lateral surface comes to the front and forms the broad area that faces forward just above the ankle joint; it is obscured by the numerous tendons that cross it as they pass to the dorsum of the foot. The *posterior surface* of the shaft is entirely hidden by the large calf muscles, but from what has been made out it is not difficult to appreciate that, on cross-section, the shaft is triangular.

The shaft of the tibia tapers from the expanded upper end until it approaches the ankle when it again expands; the lower end, however, is less expanded than the upper. At the lower end the lump on the medial side has been recognized as the medial malleolus; this projects more than 1 cm as a blunt process beyond the level of the lower end of the shaft and is, therefore, the lowest part of the bone. On the lateral side of the lower end, the expanded bone accommodates the lower end of the fibula. For that reason the shaft presents, on the lateral side of its lower end, a triangular concave surface in which the lower end of the fibula fits snugly and is bound to the tibia by a strong interosseous ligament.

The under surface of the lower end of the tibia (fig. 109) is quadrangular; it rests on the talus and forms the greater part of the socket of the ankle joint. It is articular and the articular surface includes the lateral side of the projecting medial malleolus. The socket which is a mortise is completed laterally by a triangular articular surface on the medial side of the fibular malleolus.

FIBULA

The lateral bone of the leg, the fibula, is like a long thin stick; its shaft is twisted and is molded by the origins and dispositions of the numerous muscles that arise from and are in contact with it (fibula, L. = a skewer).

It has three joints with the tibia and one with the talus, the most proximal of the bones of the foot. Its knob-like upper end possesses a small circular articular surface which fits against the similar facet on the under surface of the overhanging posterior part of the lateral tibial condyle. Its shaft is united throughout its length by an interosseous membrane to that of the tibia.

Its lower end is more pointed than its upper end. The part that projects downward beyond the tibia is known as the **lateral malleolus.** On its medial aspect is the triangular articular area that completes the socket of the ankle joint (see fig. 107).

Two triangles are worthy of note at the lower end. The first is an inverted triangle on the medial side and is the articular area just mentioned. The other is an upright one that stretches upward from the lateral surface of the malleolus; it leads to the anterior border of the fibula. The borders and surfaces of the fibula are rather complex and their details should interest professional anatomists only.

Surface Anatomy. The lateral malleolus and the subcutaneous triangle above it are readily felt; this malleolus descends farther than does the medial one. At the back of the lateral side of the knee the head of the bone is also palpable, and a considerable amount of the shaft can be felt on deep pressure among the muscles on the lateral side of the leg.

The functions of the fibula are threefold: (1) to give origin to muscles; (2) to act as a pulley for tendons passing behind it at the ankle; (3) to act as a lateral 'splint' for the ankle joint. Without the fibula the whole security of that joint is lost; hence this last function is its most important one.

PATELLA

The patella or knee-cap (fig. 110) is a bone developed in a tendon and such bones are called **sesamoid bones.** The patella (L. = little plate) is triangular in outline with its apex down and it is thick but more or less flat. Since it lies just deep to the skin its size and shape are readily made out. It lies in front of the expanded lower end of the femur; hence its posterior surface is articular. It is said to enhance the power of the already powerful extensor muscle of the knee, in whose tendon it is developed, by increasing the leverage for that muscle.

The patella glides in its trough on the femur (fig. 111). As the knee is flexed, the patella maintains a constant relationship with the tibia to whose tubercle the strong tendon of the knee extensor passes from the apex of the patella. Thus, in flexion, the patella is pulled down in its vertical trough to lie in contact with the inferior surface of the femur (fig. 112). However, when one kneels it is the tibial tubercle and the

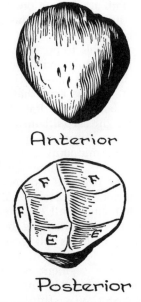

Fig. **110.** Right patella. Different parts of posterior surface are in contact with femoral condyles in full flexion (F) and in full extension (E) of knee joint.

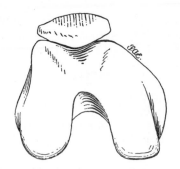

Fig. **111**. The patella in its groove in the (right) femur, viewed from below.

tendon that receive nearly all of the weight, the apex of the patella being only to a slight degree concerned.

Because the powerful extensors of the knee pull obliquely and chiefly from the lateral side of the thigh, they have a tendency to dislocate the patella laterally. This is resisted, in part, by the lateral lip of the trough being projected forward considerably more than the medial one. Sometimes this projection fails to develop and the patella is dislocated laterally with the utmost ease.

BONES OF FOOT

The human foot, in contrast to the hand, has sacrificed all other functions to concentrate on the duties of **weight-bearing** and **locomotion.**

It is not profitable to pursue the subject of homologies (for such exist) between the bones of the hand and those of the foot but, in general, it may be said that there are seven **tarsal bones** and that they correspond to the eight carpal bones.

If one draws an outline of a foot (fig. 113), marks the midpoint of both medial and lateral borders, and joins these points with a line, the result is a division of the foot into anterior and posterior halves. The anterior half includes all the metatarsals and phalanges, i.e., the long bones; the posterior half includes all seven tarsal bones (fig. 114).

If the posterior segment is further divided by an S-shaped line into an anterior third and a posterior two-thirds, one can fit five small tarsals into the smaller anterior compartment and two larger tarsal bones

(talus and calcaneus) into the larger posterior one.

Talus and Calcaneus

Of the tarsal bones, these two are much enlarged and are concerned with receiving the weight of the body above. The talus rests on the calcaneus or 'heel bone' and occupies the socket of the ankle joint.

On the upper surface of the talus, i.e., at the ankle joint, occur only the simple hinge movements of flexion and extension. The calcaneus, on the other hand, accommodates itself to the irregularities of the ground with which it is in contact, and it does so without in any way disturbing the superimposed talus. This mobility of the calcaneus can be demonstrated by grasping the heel when the foot is off the ground and swinging it passively. No movement

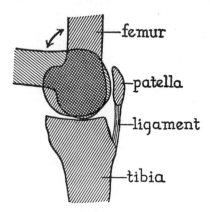

Fig. **112**. Position of patella during flexion and extension of knee.

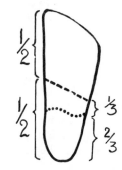

Fig. **113**. Outline of foot. The tarsal bones are confined to the posterior one-half of the foot. The talus and calcaneus occupy the posterior two-thirds of the posterior one-half (cf., fig. 114).

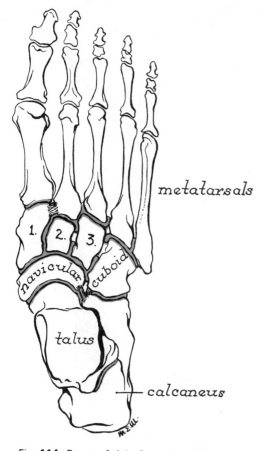

Fig. **114.** Bones of right foot viewed from above (cf., fig. 113). Joints in red; 1, 2, 3 = cuneiforms.

of about 15° to that of the calcaneus. The calcaneus endeavors to provide some support for this overhanging head by projecting a shelf of bone beneath it. This calcaneal shelf is appropriately known as the **sustentaculum tali;** it projects medially beneath only the extreme back part of the head of the talus, so that the fore part of the head has no important *bony* support (fig. 117).

From what has been said, it follows that there are **three articular areas** on the upper surface **of the calcaneus:** (1) the quadrangular area for the body of the talus; it is convex from side to side; (2) a narrow, elongated, oval area on the upper surface of the sustentaculum tali for the neck and back of the head of the talus. This may be continuous with or entirely dissociated from: (3) a small surface on the extreme antero-medial corner of the calcaneus.

Between these articular surfaces on the calcaneus is a non-articular area ($\times$ in fig. 117) occupied by an interosseous ligament; this binds talus to calcaneus and acts as a pivot around which reciprocal movements

takes place at the ankle joint but the foot can, in this way, be turned sole in **(inversion)**, or sole out **(eversion)** (fig. 115). The former is the position of the foot in the newborn infant.

The **calcaneus** is the largest tarsal bone, being 8 cm (3″) long, 5 cm (2″) deep, and nearly 3 cm or more wide.

The **talus** consists of a **body** behind, separated from a **head** in front, by a constriction known as the **neck.** The body of the talus rests on a quadrangular articular surface occupying the middle third of the upper surface of the calcaneus. The head of the talus, however, instead of resting foursquare on the front end of the calcaneus, projects precariously above the medial side of the front end of the calcaneus (see figs. 116, 117). In other words, the long axis of the talus is set at an angle

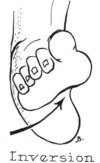

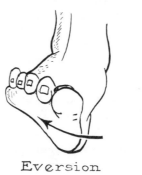

Eversion   Inversion

Fig. **115.** Eversion and inversion of the foot

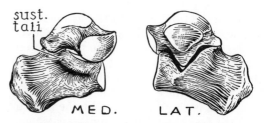

sust. tali

MED.   LAT.

Fig. **116.** Medial and lateral views of left talus and calcaneus in articulation.

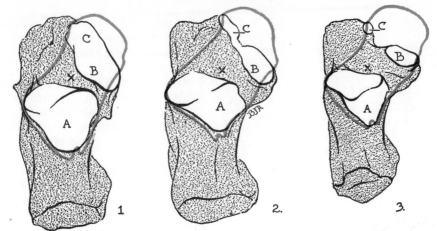

Fig. **117.** Outline of talus (red) superimposed on left calcaneus from above. 1, 2 and 3 show variability of upper surface of calcaneus. A, B and C are three articular surfaces for talus; X is site of attachment of interosseous ligament joining two bones.

take place. These movements give considerable flexibility to the foot.

The posterior third of the calcaneus projects behind the ankle joint as the **heel.** (The official alternative name—*os calcis* —means 'bone of the heel'.) The heel is a lever whereby the powerful calf muscles can extend the ankle and raise the body on tiptoes (fig. 118). Posteriorly, a large tubercle is the only part of the calcaneus to come in contact with the ground through the tough fascia and skin that cover it. The front end of the bone is tilted upward so that the calcaneus is the posterior unit in a longitudinal arch (fig. 119).

Navicular, Cuboid, and Three Cuneiforms

In front of the calcaneus and the talus are five small tarsal bones; these give additional flexibility to the foot. In front of them again lie the five metatarsals which correspond to the metacarpals in the palm (figs. 120, 121).

Since the front end of the calcaneus and of the talus reach to the same forward level, it is convenient to regard the two joints in front of them as a unit, known as the **transverse tarsal joint.** The calcaneus in front carries the cuboid; the round head of the talus is capped in front by the navicular; the two independent joints—calcaneo-cuboid and talo-navicular—constitute the

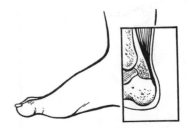

Fig. **118.** The calcaneus is a lever to which is attached the tendon of Achilles.

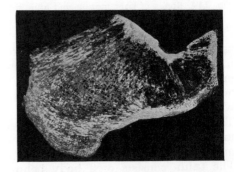

Fig. **119.** Photograph of cut surface of calcaneus showing arrangement of spongy bone into many small struts or columns along lines of weight-transmission.

transverse tarsal joint at which the movements of inversion and eversion are noticeably increased.

The **navicular** (L. = boat-shaped) owes its shape, and thus its name, to the depression on its posterior surface for the head of

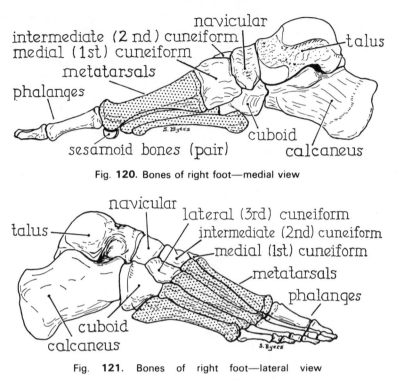

Fig. **120**. Bones of right foot—medial view

Fig. **121**. Bones of right foot—lateral view

the talus. On its anterior surface the three wedge-shaped bones, the **cuneiforms** (medial, intermediate, and lateral), crowd together. In diagrams of the foot from above the bases of the wedges are seen. The apices are less obvious in the sole of the foot. Each carries on its broad anterior surface the base of the corresponding metatarsal (1, 2 and 3).

The **cuboid** occupies all of the area between the anterior surface of the calcaneus and the bases of the fourth and fifth metatarsals which it carries.

Metatarsals and Phalanges

The **metatarsals** are comparable to the metacarpals. They differ in certain features: (1) in general they are longer and more slender; (2) their heads are little if any expanded but since the toes can be hyperextended, the joint surfaces are prolonged onto the backs of the heads; (3) all five are tied together at their heads by **transverse ligaments of the sole** uniting their capsules, so that the great toe has none of the freedom of movement enjoyed

by the thumb; (4) the metatarsal of the great toe is stout and strong but relatively short; its head rests on a pair of pea-sized **sesamoid bones** which transmit the weight to the ground (fig. 120).

The **phalanges** are arranged in similar fashion to those of the hand. In the foot an obvious emphasis is placed on the great toe and the phalanges of the little toe verge on the vestigial. In general, the phalanges of the foot are short and the toes have little functional importance compared to the fingers.

Arches of Foot

Viewed from above, the bones of the foot may be divided longitudinally into two segments: (1) calcaneus, cuboid, metatarsals 4 and 5; (2) talus, navicular, three cuneiforms and metatarsals 1, 2 and 3. There is some functional significance to this arrangement.

The lateral segment of the foot consists of a low longitudinal arch supported chiefly by ligaments (fig. 121). The medial segment consists of a high longitudinal arch supported chiefly by ligaments during standing but also by muscles during locomotion. The first 'line of defense' against flat feet is ligamentous (see p. 110).

Weight-Bearing

In the erect posture, half of the weight of the body is borne by each talus. The talus, in turn, transmits half of its load backward through the calcaneus to the heel (fig. 120) and the other half forward to be distributed equally to six bearing-points at the front. Each of the two sesamoids at the head of metatarsal 1 carries the same weight as each of metatarsals 2, 3, 4, and 5. The foot is not a tripod (as it sometimes is claimed to be) and there is no transverse arch at the heads of the metatarsals during weight-bearing.

In walking, the weight is first borne by the heel and then is spread forward along the lateral edge of the foot quickly to the metatarsal heads. The 'push-off' thrust is concentrated at the head of metatarsal 1, accounting for the large diameter of this bone.

3

Articular System

DEVELOPMENT AND CLASSIFICATION OF JOINTS

The word joint is in such everyday use that there can be no mistake as to what is meant by it; it is a uniting together. When two pieces of wood are to be joined one of a variety of devices may be employed. It may suffice simply to coat the opposed surfaces with glue and press them together. Or, some form of lapped joint such as is seen in the 'siding' of a frame dwelling may be used. By shaping reciprocally the surfaces to be united such secure joints as a tongue-and-groove, mortise-and-tenon, or a dovetail may be fashioned. In all of these joints secure and immobile union is the desired objective (fig. 122).

It is frequently necessary, however, to design a joint that permits movement. The lid of a box, or a door hung on its frame, are examples of joints in which, by the use of metal hinges or some similar device, movement is permitted in a given direction. The machinist goes further and by the use of a metal ball carried on a rod and revolving in a lubricated hollow sphere, produces a joint free to move in any direction and which he consequently calls a 'universal' joint.

Nature is more versatile than man, and in order to understand how similar and better joints come to exist in the human body it will be profitable to trace briefly their development.

It was earlier observed that, in the limbs, the long bones arise out of a core of condensed mesenchyme that occupies the midline axis of the limb bud (fig. 123). In the region between the adjacent ends of two developing bones, this mesenchyme persists to unite them; the subsequent fate of this *primitive joint plate*, as it is called, determines the class of joint later to be found there. Similarly, when bone arises out of membrane, as in the bones of the vault of the skull, it is the fate of the intervening membrane between contiguous bones that engages our attention.

Fibrous Joints. Should the intervening mesenchyme or membrane be little altered it becomes simply a fibrous bond between the bones and so appears in the adult. To this class name **fibrous joint** is applied and it may be defined simply as a joint in which bone is united directly to bone by fibrous tissue. In anatomical language such a joint is known as a *syndesmosis* and examples of it are numerous. It will be recalled that a band or cord of fibrous tissue joining bone to bone is a *ligament* (page 6).

If the amount of connecting fibrous tissue is minimal, it is seen in the adult as a line between contiguous bones, appropri-

75

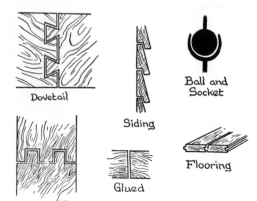

Fig. **122**. Some types of man-made joints

ately known as a **suture,** which may take
various forms depending on the dispositions of the opposed bony edges (fig. 124).
Such forms as overlapping (squamous),
flat (*plane*), or interlocking (*serrate and
dentate*) are common and are found exclusively between the bones of the skull
where movement is neither permissible nor
desirable. Indeed, with increasing years,
the suture lines become obliterated. The
bones themselves fuse together, and the
fibrous joint is replaced by a bony one,
thereafter known as a *synostosis*.

Many of the fibrous joints allow some
movement. The degree of this movement
depends, of course, upon the distance between the bones and the flexibility of the
uniting fibrous tissue. Just above the ankle
joint, the tibia and fibula are united by a
fibrous bond known as the interosseous
tibiofibular ligament; it consists of such
short fibers that the bones are held in the
closest apposition and very little movement is possible between them. A joint
such as this seems to occur where movement is undesirable but a little 'give' is
necessary. The shafts of the radius and
ulna, on the other hand, are united
throughout their lengths by a sheet of
fibrous tissue known as an interosseous
membrane; it is broad enough and sufficiently flexible to allow the considerable
movements of pronation and supination.

Cartilaginous Joints. In their development out of mesenchyme, bones were observed to pass through a cartilaginous

stage. Should the intervening mesenchyme
also become cartilage, but thereafter develop no further, a **cartilaginous joint**
results (fig. 125). In such a case the cartilage is of the same hyaline variety as that
out of which the adjacent bones developed.
Cartilaginous union, however, may result
from a previous fibrous union subsequently
becoming fibrocartilage. Thus, two varieties of cartilaginous joints, *hyaline* and
fibrocartilaginous, exist and, since fibrocartilage is rather less rigid than hyaline,
it allows somewhat freer movement. The
bar of hyaline cartilage uniting the first
rib to the sternum, and the transitory
plates of hyaline cartilage that, in growing
bones, unite (or separate) the shaft and
epiphyses, are examples of the hyaline
variety; intervertebral discs between adjacent vertebral bodies are examples of the
fibrocartilaginous variety. A cartilaginous
joint is called a *synchondrosis*.

Synovial Joints. The third, and func-

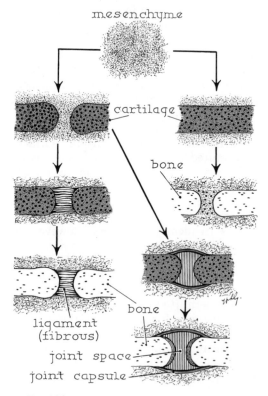

Fig. **123**. Stages in the development of various
types of joints (schematic).

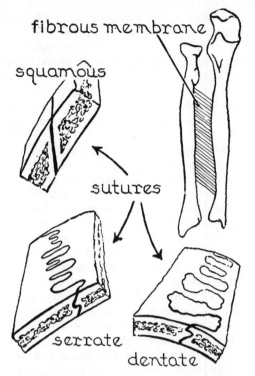

Fig. **124**. Varieties of fibrous joints (see text)

possesses the following *four distinguishing features:* (1) a **potential cavity;** (2) **lubricated articular cartilage;** (3) a **capsule of fibrous tissue,** lined with (4) a **synovial membrane** (see fig. 126).

It is usual to regard the synovial membrane as being continuous with the articular cartilage, i.e., where the cartilage ends the synovial membrane begins. From this attachment to bone, the membrane, as a rule, merely lines the inner surface of the fibrous capsule.

Movements of a joint tend to leave spaces here and there as the bones change their relationship to one another. Nature, abhorring a vacuum, provides *synovial folds and fringes,* which are small tags of synovial membrane. If these tags are filled with fat cells they are called (Haversian) *fat pads.* These slip in and out of potential spaces that are formed or obliterated by the movements of a joint. Sometimes they get pinched, causing pain.

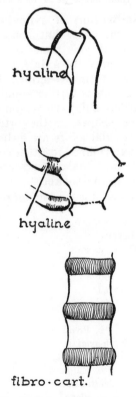

Fig. **125**. Varieties of cartilaginous joints (see text). (In top figure the joint is transitory.)

tionally most important, class of joints is that resulting from the primitive joint plate undergoing special changes. Instead of developing farther, the interior of the plate gradually dissolves or is eaten away and a cavity is formed bounded by a sleeve of fibrous tissue which is all that remains to unite the two bones. The sleeve is known as the **articular capsule.** Friction between the two bones is reduced to a minimum by the original hyaline cartilage, out of which each bone developed, remaining to provide an exceedingly smooth coating known as the **articular cartilage.** Finally, cells of the inner surface of the articular capsule take on the duty of transuding a lubricating fluid which not only further reduced friction, but also serves to nourish the hyaline cartilage which, strange to say, has no blood vessels.

The delicate inner layer of the articular capsule is known as the **synovial membrane** and the name serves to designate the class as the **synovial joints.** A synovial joint may be defined as one that

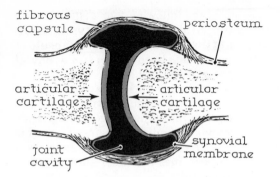

Fig. **126**. Scheme of a synovial joint. (For clarity, two bones are pulled apart and capsule is inflated.)

Diseases of joints, therefore, fall into categories according to whether they involve the bones, the articular cartilage, or the synovial membrane. For example, an excessive production of bone—a common effect of arthritis—may, by its presence, limit movement and cause pain. Again, a softening and erosion of the articular cartilage may result in increased friction and, pain. Or, an over-production of synovial fluid—often the result of a strain or sprain—is the common condition of 'water' on a joint.

The classification of synovial joints depends on the fact that some permit movement in only one plane (**uni-axial**), some in two planes (**bi-axial**), and others in many planes (**multi-axial**). Each of these three groups is further subdivided.

These various subdivisions take cognizance of the shapes of the articulating surfaces. Uni-axial joints are either **hinge** or **pivot;** bi-axial joints are either **condyloid** or **ellipsoid;** multi-axial joints are **plane, saddle,** or **ball-and-socket.**

The following table is given for reference:

Classification of Joints
A. Fibrous
B. Cartilaginous
  1. *Hyaline*
  2. *Fibro-*
C. Synovial
  1. *Uni-axial*
    a. Hinge
    b. Pivot
  2. *Bi-axial*
    a. Condyloid
    b. Ellipsoid
  3. *Multi-axial*
    a. Plane
    b. Saddle
    c. Ball-and-Socket

Examples. The elbow, knee, ankle, and interphalangeal joints are **hinges** (fig. 127). The rotating upper end of the radius in its socket is one example of a **pivot joint;** the dens or odontoid process of the axis revolving in the collar formed by the anterior arch of the atlas and the transverse ligament is another example (fig. 128).

The knuckles or joints at the heads of the metacarpals give their names to the condyloid (Gr. = knuckle-like) type, and provide the best example (fig. 129).

Ellipsoidal joints resemble condyloid joints. In them, one surface is a shallow, convex ellipse, the other is reciprocally shaped; movements occur in the two principal axes of the ellipse. The wrist is the prime example.

Plane joints (fig. 130) are numerous and they are nearly always small. In them, the little surfaces opposed to one another are either flat or nearly so. Gliding movement can occur in almost any plane but it is always slight. The joints are found particularly, though not exclusively, in the bones of the axial skeleton. They occur between

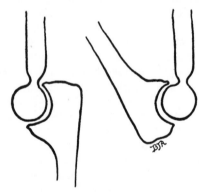

Fig. **127**. Hinge joint—uni-axial (elbow)

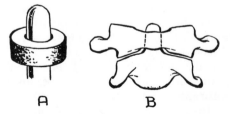

Fig. **128**. Pivot joint—uni-axial. A = schematic; B = atlas and axis (cf. fig. 140).

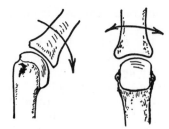

Fig. **129**. Condyloid joint—bi-axial (knuckle)

Fig. **130**. Plane joint—multi-axial (acromioclavicular).

individual vertebral arches (at the articular processes), at the heads and at the tubercles of the ribs, at the front ends of the costal cartilages, at the lateral end of the clavicle where it meets the acromion, and lastly, at the carpus and at the tarsus between the individual bones of the hand and of the foot.

The terms **saddle** and **ball-and-socket** are self-explanatory. They are the freest moving joints and the characteristic feature of them is that they permit not only angular movements but also movements of rotation about the long axis of the moving bone—saddle joints to a limited extent, ball-and-socket joints to a considerable extent. The joint at the base of the thumb is the outstanding example of a saddle joint (fig. 131); shoulder and hip exemplify the ball-and-socket variety (fig. 132).

TYPES OF MOVEMENT

The various movements that occur at synovial joints are sometimes rather complex; it will be profitable therefore to name, illustrate, and define them.

In the familiar exercise of walking, the alternate movements that occur at the hip and at the knee joints are those of **flexion** (bending) and **extension** (straightening). Similar movements of flexion and exten-

sion take place also at the shoulder joints as the arms swing at the sides (fig. 148 on p. 89); the plane in which the limbs move is the **sagittal plane** (see page 19 and fig. 27).

If, in standing in the anatomical position, the arm be raised sideways from the body, the movement is one of **abduction** and the movement that restores the abducted limb to the side is the converse, viz., **adduction** (fig. 148). These movements occur in the **coronal plane** (see page 19 and fig. 28).

Bend your elbow to a right angle and keep it applied to your side. Then swing the forearm and hand medially (fig. 149 on page 89). This movement occurs at the shoulder joint and is **medial rotation;** its converse is **lateral rotation.** The rotation takes place round the long axis of the humerus, and the plane in which it occurs is the **horizontal or transverse plane** (see page 19 and fig. 29).

Flexion and extension, abduction and

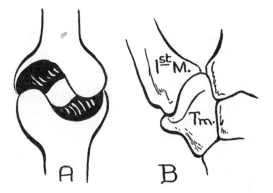

Fig. **131**. Saddle joint—multi-axial. A = schematic; B = base of thumb, showing saddle joint between first metacarpal and trapezium.

Fig. **132**. Ball-and-socket joint—multi-axial (hip).

adduction, medial rotation and lateral rotation, at whatever joint they may occur, involve together three planes in space at right angles to one another. But they may be combined to produce movements in an endless number of intermediate planes. If, for example, the arm is raised forward and lateralward, the movement occurring at the shoulder joint is in a plane intermediate between the sagittal and the coronal, and is a combination of flexion and abduction. Or, if the arm is swung in a circular fashion, the movement, called **circumduction,** is a combination of the four primary movements in which flexion, abduction, extension, and adduction succeed one another. *In describing any movement it is necessary to use these recognized terms and to use them correctly.*

FUNCTIONS OF LIGAMENTS

The varieties of movement that it is possible to perform at any given joint depend on two factors:

1. The shape or configuration of the articulating surfaces.
2. The presence of restraining ligaments.

In some joints the former is the more important factor, in others, the latter. For example, the hook-like shape of the upper end of the ulna practically confines movements at the elbow joint to the sagittal plane, i.e., to flexion and extension. [It even inhibits extension from proceeding beyond the straight line.] At the knee joint, on the other hand, the two rounded femoral condyles roll on the almost flat upper surface of the tibia, so that the shapes of the articulating surfaces can offer no restraint to the movements of the joint. Consequently, the restriction of the movements of the knee joint to practically one plane, wherein the knee resembles the elbow, depends, not on bone, but on the dispositions of the restraining ligaments.

Besides the use to prohibit movement in an undesired plane, ligaments also serve to limit the range or extent of normal movement; this is particularly true in those cases in which the shapes of the bones concerned can be of no assistance. In neither hip nor knee can extension progress beyond a certain point and in both joints it is by the use of strong ligaments that the movement is arrested. It is well to notice these two different uses of ligaments, for when movement in a normal plane is gradually brought to a halt, the ligaments concerned (probably containing many yellow elastic fibers) become progressively taut as the movement proceeds; when abnormal movements are altogether prohibited the ligaments responsible (probably mainly of white non-elastic fibers) at once are tensed, and that often violently (figs. 133, 134). It is in these latter cases that ligaments are so often injured in strains and sprains occurring at joints.

Every synovial joint is equipped with ligaments made of fibrous tissue, a variable number of the fibers being elastic. Sometimes a ligament exists merely as a localized thickening of the fibrous capsule of a joint, and is not an independent structure; it is then spoken of as a **capsular ligament.** At other times it exists as a structural entity situated near a joint but not part of the capsule; then it is called an **accessory ligament.**

Ligaments possess no inherent power of contraction, being static and passive structures. It is incorrect to speak of the actions of ligaments; they have, not actions, but supportive functions. A knowledge of the direction taken by the fibers of a ligament is the key to the movements it can restrain. Finally, several ligaments pass between

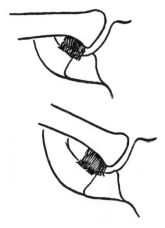

Fig. **133.** Example of a ligament (costoclavicular) gradually taking strain (during elevation of clavicle).

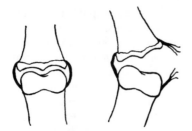

Fig. **134**. A ligament ruptured by abnormal movement (interphalangeal joint).

two points on the same bone and, consequently, can never be affected by any movement. They exist to protect or hold in position some more important structure, usually a nerve. In brief, muscles perform; ligaments either prevent, restrain, or protect.

FUNCTIONS OF DISCS

Fibrocartilaginous discs of varying importance are present in the interior of certain synovial joints, where they may partially or wholly divide the cavity into two compartments. They represent the persistence of a portion of the primitive joint plate, on each side of which a cavity has appeared. Several reasons for their existence have been advanced:

1. The need to lessen shock in a joint, i.e., 'shock absorbers'—questionable.
2. The need to adjust bony articulating surfaces of different shape to one another—possible.
3. The need to permit two types of movement to occur simultaneously —obvious.
4. The need, inside some joints (e.g., sternoclavicular), for a modified restraining ligament—undeniable.
5. The need, in many joints, for a special device to assist in the lubrication of articular surfaces (MacConaill)— gaining acceptance.

HOW TO DESCRIBE A JOINT

In order not to forget any important points about a joint it is generally advisable to consider it under the following headings: (1) type and subtypes; (2) bony parts involved; (3) fibrous and synovial capsule; (4) ligaments and their uses; (5) special features such as discs; (6) movements and the muscles that produce them; (7) special functions; (8) important relationships to other structures; (9) blood supply and nerve supply; (10) surface anatomy.

If, in this book, such a simple plan were followed for each joint, no details would be missed, but the presentation would become heavy-handed and monotonous. Nevertheless, the above organization, thinly veiled, is detectable in any full description of a joint.

JOINTS OF SKULL

The eight bones that bound the cranial cavity and the 14 that form the skeleton of the face are, with the exception of the mandible, united to one another to produce a single complicated unit, the skull. The joints concerned are, almost without exception, fibrous joints possessing no movement. They are recognizable in the adult skull usually as (suture) lines which in later life tend to become obliterated, so that even their situations are sometimes hard to make out.

In the skull there is one pair of synovial joints and one only,* it consists of the two temporomandibular joints whereby the lower jaw is slung from the skull.

Temporomandibular or Jaw Joint

This joint is situated immediately in front of the external ear; it can be felt and its movements can be appreciated by the palpating finger. If the forefinger palpates the joint while the thumb palpates the angle of the jaw, it will be found that, as the mouth is opened, the head of the mandible moves forward and the angle moves backward, and to a like extent. Evidently the center of this movement is near the midpoint of the ramus and not at the joint. An appreciation of this fact explains the anatomical features now to be noted.

The articulating surface on the skull lies immediately in front of the passageway

---

* The anatomist will remind us that the little joints between the ossicles of the ear are synovial joints and are important.

(external auditory meatus) leading from the exterior to the eardrum. This surface is concave in its back part but convex in its forepart, the convex part being called the **articular eminence.**

The **head** (or *condyle*) of the mandible resembles a short, solid cylinder laid on its side, and therefore consists of a convex surface. When the mouth is closed, the convex condyle rests in the concave part of its socket; when the mouth is opened, the condyle rides forward onto the articular eminence, i.e., convex surface meets convex surface. It can be felt to do this by the exploring finger.

The joint space (fig. 135) is divided into

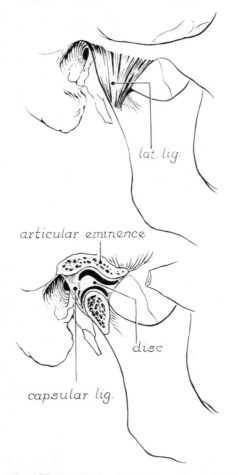

upper and lower compartments by a **disc,** which is concavo-convex on its upper surface and concave on its lower. When the mouth is opened, a hinge-like movement occurs in the lower compartment and a gliding movement in the upper. In the gliding movement, disc and condyle move forward on to the convex articular eminence (fig. 136). One of the muscles of the jaw has partial insertion into the disc, and is capable of pulling it forward when the lower jaw is voluntarily thrust forward or the mouth widely opened.

A 'dislocation' occurs when the condyles slip too far forward—perhaps resulting from too vigorous laughing or yawning—and leaves the victim in the awkward position of being unable to close his gaping mouth (fig. 136).

The jaw can be moved also from side to side, a movement involving a shuttling of the condyle and the disc in the concave part of the socket. This shuttling becomes progressively less as the mouth is opened. Movements in which both axes are involved result in the grinding action of the teeth. The joint capsule is strengthened on its lateral aspect by the presence of a triangular thickening, the **lateral ligament,** whose fibers are gradually tensed as the mouth is opened and aid therefore in preventing dislocation.

JOINTS OF THE VERTEBRAL COLUMN

Two series of joints unite the individual vertebrae of the spinal column: (1) the fibrocartilaginous joints (synchondroses) uniting adjacent bodies; (2) the synovial joints uniting adjacent vertebral arches. Since synchondroses are fundamentally more limited in their movements than synovial joints, it follows that the extent of movement enjoyed by the joints of the vertebral arches is primarily determined by the extent of movement permitted at the discs (see also pages 30 and 31). See figures 137, 138, 139.

It has been observed already (on page 30) that the amount of movement between two adjacent bodies depends on the thickness of the intervening disc. Cervical and lumbar discs are thick, thoracic discs are thin;

Fig. **135**. Mandibular joint. Upper figure: lateral view of capsule. Lower figure: sagittal section of joint to show disc dividing joint into an upper and a lower cavity.

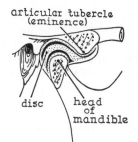

articular tubercle
(eminence)

disc   head
        of
        mandible

a. Mouth closed

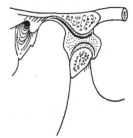

b. Mouth open

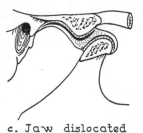

c. Jaw dislocated

Fig. **136**. Relationship of parts of jaw joint with—a, mouth closed, b, mouth open wide, and c, jaw dislocated.

movements, therefore, are freest in the neck and in the lumbar spine, and this must be as true for the joints of the vertebral arches as it is for the bodies.

Joints of Articular Processes

In the neck, the articular surfaces of the joints of the vertebral arches look, in general, upward and downward. As the thorax is approached, a gradual transition occurs until, in the thoracic region, the articular surfaces look forward and backward. At the last thoracic vertebra, an abrupt change occurs whereby, in the lumbar region, the surfaces look medialward and lateralward,

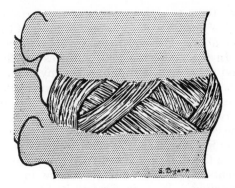

Fig. **137**. Outer surface of anulus fibrosus of an intervertebral disc. Layers of obliquely crossing fibers joint vertebral bodies.

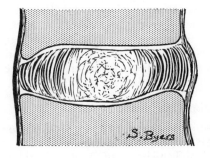

Fig. **138**. Vertical section of an invertebral disc

nucleus   anulus
pulposus   fibrosus

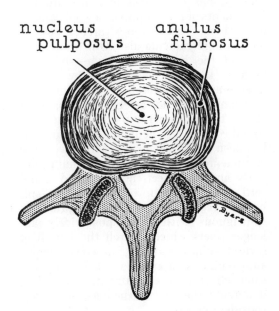

Fig. **139**. Cut surface of an invertebral disc, from above.

i.e., the articular processes of the vertebra below grasp the outer aspect of those of the one above. These changes are significant.

In the neck, the joint surfaces are flat and movement is possible in many planes. The range of movement is quite considerable. Indeed, head movements would be very limited were it not for the flexibility of the cervical vertebrae—witness the restriction in the presence of cervical arthritis.

In the thorax, the same variety of movements is possible but the range in any direction is less, in conformity with the thinness of the discs between the bodies.

In the lumbar region, articular processes grasp each other in such a way that movements in this region are practically confined to two planes: (1) flexion and extension—which, because of the thickness of the intervertebral discs, are much more extensive than in any other region—and (2) sideways bending. Rotation (body twisting) is restricted.

Special Ligaments

The presence of **anterior** and **posterior longitudinal ligaments,** binding the fronts and the backs of the vertebral bodies to one another throughout the length of the column, has been seen to safeguard the movements of the column as a whole and to limit them. The vertebral arches possess restraining ligaments also. One group consists of the **ligamenta flava**—so called because they are rich in yellow elastic fibers—and they stretch between the adjacent laminae of the vertebral arches; being elastic these ligaments tend to restore the spinal column to a neutral position after it has been flexed. They also serve, with the laminae, to cover in the spinal canal posteriorly and so protect the contained spinal cord. A second group unites adjacent spinous processes as **interspinous ligaments.** Contiguous with these posteriorly are longer fibers which stretch the length of several spines and are, in consequence, **supraspinous ligaments.** These have the same effect as the ligamenta flava. Undoubtedly, they relieve the back muscles of considerable work.

In the neck, the supraspinous ligaments are so enlarged as to produce a midline

ligamentous partition separating the thick muscles of one side of the back of the neck from those of the other. It is known as the **ligamentum nuchae** (L. = of neck) and in lower animals serves as an elastic ligament to hold up the head. The horse's mane indicates its position in that animal.

Atlanto-Occipital Joints

These uni-axial, synovial, paired joints whereby the first cervical vertebra, the atlas, supports the head have already been referred to as the joints at which the nodding or 'Yes' movement occurs. Each surface on the lateral mass of the atlas is concave and elliptical.

The anterior and posterior arches of the atlas also are united to the margins of the foramen magnum, above, by **anterior and posterior atlanto-occipital membranes.**

Atlanto-Axial Joints

The atlas and axis are united by three joints, two paired and one medianly placed (fig. 140). The paired **lateral atlanto-axial joints** are plane joints and lie directly below the paired atlanto-occipital joints. The surfaces are nearly flat, circular, and about 1 cm (½″) in diameter. Here, on the axis, gliding movements occur, the principal movement being the one that shakes the head.

When the 'No' movement occurs, the skull and atlas move on the axis as a unit, and third joint, which is at the center of the movement, plays an important part. This

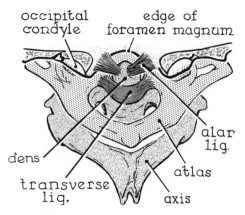

Fig. **140.** Joints of atlas, axis, and occipital condyles of skull (the back of which is cut away).

joint is the **median atlanto-axial joint** and it is an ideal pivotal one in which the dens (Gr. = tooth) or odontoid process, projecting upward from the body of the axis rotates in a collar formed by the anterior arch of the atlas and the **transverse ligament.** A pair of strong, short, 'check' or **alar ligaments** stretches between the tip of the dens and the skull at the margin of the foramen magnum and serves to limit the rotation of the head. They become taut also in head flexion.

The pivot-and-collar mechanism is completely covered and hidden from view by a broad ligamentous band, the *tectorial* (L. = covering) *membrane;* it is the upward and expanded prolongation of the posterior longitudinal ligament already noted as binding together the backs of all the vertebral bodies and discs. This tectorial membrane is attached above to the front margin of the foramen magnum and offers a smooth and continuous sloping surface for the support of the brain stem and spinal cord.

Anterior and **posterior atlanto-axial membranes** have similar dispositions to the atlanto-occipital ones. Movements between the skull and the atlas as well as those between the atlas and the axis are, of course, augmented by the flexibility of the remainder of the cervical vertebrae

JOINTS OF THE RIBS

Synovial joints occur at the heads and tubercles of the ribs and at the sternal ends of the costal cartilages (fig. 141). The joints at the extremities of a rib element, i.e., at the head and at the sternal end of the costal cartilage, indulge in similar movements and are similarly constructed.

The capsule of each is thickened in front and is known as a *radiate ligament* since it fans—indeed, it *must* fan—from the rib head (or costal cartilage) to the adjacent vertebral bodies and disc (or sternum). In the interior of each joint is an **intra-articular ligament** that binds the head (or costal cartilage) to the disc (or sternum) and divides the cavity into upper and lower compartments.

The synovial joint at the rib tubercle occurs where the rib, in its backward sweep, abuts against the transverse process of the vertebra with which it numerically corresponds. Each small joint is reinforced,

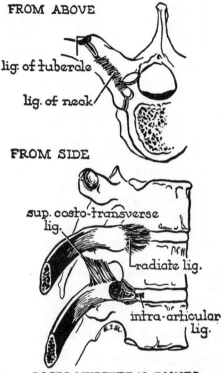

Fig. **141**. Joints and ligaments between ribs and vertebrae.

and the rib more firmly bound to the transverse process, by two ligaments, one on each side of the joint. The strong medial one is simply called the **costotransverse ligament** (*ligament of the neck*); it lies horizontally disposed and it fills the space between the back of the neck of rib and front of the adjacent transverse process of the vertebra. The lateral one is called the **lateral costotransverse ligament** (or *ligament of tubercle*); it is a short but strong cord that passes horizontally lateralward from the tip of the transverse process to the back of the rib just beyond the tubercule. A third and rectangular ligament stretches vertically from the neck of the rib to the transverse process next above. It often produces a flange on the neck of the rib and is known as the **superior costotransverse ligament.**

The following exceptions to the above general plan should be noted:

1. The heads of ribs 1, 11, and 12 reach

the spinal column not at the side of a disc but at the side of vertebral bodies 1, 11, and 12. Hence no double cavities or intra-articular ligaments exist there.

2. The first rib is directly united to the sternum by the first costal cartilage; the union is a synchondrosis.
3. Costal cartilages 8, 9, and 10 reach, not the side of the sternum, but the lower edge of the costal cartilage next above, where synovial joints occur.
4. Costal cartilages 11 and 12 are free at their anterior ends.
5. Ribs 11 and 12 do not articulate with any transverse process; hence they have no articular tubercles.

Movements of Ribs. The student should review the section 'Thoracic Cage and Respiration' on pages 47 to 48.

JOINTS OF THE UPPER LIMB

The only articulation that the upper limb has with the axial skeleton is at the mobile sternoclavicular joint. The clavicle is united to the other bone of the pectoral girdle, the scapula, at two joints. However, since the humerus articulates with the scapula but not the clavicle, forces from the arm are transmitted to the scapula, thence to the clavicle, and thence to the axial skeleton.

JOINTS OF THE CLAVICLE

There are three joints of the pectoral girdle; all involve the clavicle:
1. **Sternoclavicular joint.**
2. **Coracoclavicular joint.**
3. **Acromioclavicular joint.**

Pectoral Girdle

General. The clavicle is the only bone uniting the upper limb to the axial skeleton. It stretches horizontally across the front of the root of the neck where its medial end articulates with the upper lateral corner of the sternum by means of the relatively large and important sternoclavicular joint. The joint can be readily palpated. As the clavicle runs lateralward it also runs somewhat backward, crossing

immediately above the first rib; its lateral end meets the acromion, which, as the free lateral end of the spine of the scapula, is a bony hood over the shoulder joint (fig. 142).

The scapula lies buried amid the muscles that cover the upper part of the back of the thorax and has no bony anchorage to the vertebral column. Because the sternoclavicular joint lies in front, the extremely varied movements that the upper limb is capable of performing are, for the most part, carried out in front of the body. In conformity with this, the scapula lies, not in the coronal plane, but in a plane about midway between the coronal and the sagittal. The bone also projects well beyond the curved upper part of the thoracic wall.

Definitions. The scapula gives attachment to many muscles. It can be braced back in a highly 'military' position, when its vertebral border lies parallel to and approaches the vertebral spines; then the bone is said to be **retracted.** Or, it can be slid forward round the curved chest wall to

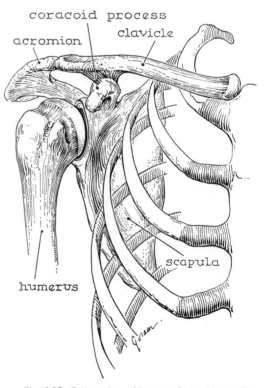

Fig. **142.** Orientation of bones of shoulder region. Note strong coracoclavicular ligament that suspends scapula from clavicle.

an excessively 'round-shouldered' position when its vertebral border lies far distant from the vertebral spines; then the bone is said to be **protracted.** As protraction proceeds, the vertebral border assumes a less and less vertical position since the inferior angle of the bone travels considerably faster than any other part. In other words, the scapula is **rotated** so that the socket of the shoulder joint (glenoid cavity) looks more upward. This combination of rotation with protraction is a desirable feature because it is the position demanded by the hands when they carry out their many duties in front of the body and at, or near, eye level (see page 151).

Besides these movements of retraction, protraction, and rotation, the scapula can be **elevated** (shrugging the shoulders) or **depressed** (drooping shoulders). Elevation is commonly, although by no means invariably, accompanied by rotation of the scapula upward, and is a relatively extensive movement. Depression is exceedingly limited, since, very early in the movement, the clavicle comes in contact with the first rib and cannot permit the scapula to sink farther. See figure 143.

Sternoclavicular Joint

In every one of the above-mentioned scapular movements the sternoclavicular joint (fig. 144) is vitally concerned. It lies at the apex of an imaginary cone. The circum-

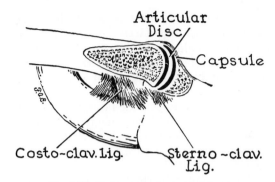

Fig. **144.** Right sternoclavicular joint

ference of the base of this cone is described by the lateral end of the clavicle, which can sweep in a circle. While the lateral end of the bone sweeps at will in the arcs of many circles, the medial end has merely to glide in its socket. The clavicle thus functions to hold the shoulder joint at its proper distance from the midline; it is a strut. A blow on the side of the shoulder or a fall on the outstretched arm imparts its force to the clavicle which transmits it to the axial skeleton at the sternoclavicular joint. With the foregoing facts in mind the following details of the structure of the sternoclavicular joint become meaningful.

The joint surface on the sternum is concave from above downward but slightly convex from front to back. The joint surface on the medial end of the clavicle is not reciprocally shaped; it is rather flat and only slightly convex.

The contact of the otherwise ill-fitting surfaces is improved by the existence of a **disc** which, however, is much more occupied with the duties of absorbing the forces transmitted to the joint along the clavicle from the shoulder region, and preventing the clavicle from being driven out of its socket onto the summit of the sternum. The disc can discharge these duties because it is firmly attached— above, to the clavicle, below, to the first costal cartilage. The rarity of dislocation of the joint testifies to the efficiency of the disc in discharging its duties—even when a hole is worn out in its middle with advanced age.

Excessive protraction and elevation of the clavicle are prevented by the existence

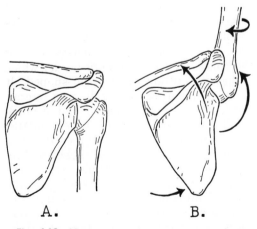

A. B.

Fig. **143.** Movements of scapula, clavicle, and humerus (indicated by arrows) during arm raising overhead (viewed from behind).

of two structures, one a ligament, the other a muscle. Immediately lateral to the sternoclavicular joint, the clavicle is bound to the first costal cartilage by the rhomboidal **costoclavicular ligament** which, being near the center of movement and at first somewhat lax, allows the lateral end of the clavicle considerable excursion in both protraction and elevation before it finally becomes so taut as to bring further movement to a halt. Lateral to the ligament is Subclavius muscle which acts as a dynamic ligament (page 146).

Coracoclavicular Joint

The lateral part of the clavicle crosses immediately above the right-angled bend in the coracoid process. At that site—located in the living at the hollow immediately below the lateral third of the clavicle—the strong **coracoclavicular ligament**† unites the under surface of the clavicle to the bend in the coracoid process (fig. 145). The result is a fibrous joint of considerable strength and, because the ligament has length, it allows some independent movement between the two bones, but the union ensures that in all movements of any considerable extent the clavicle and scapula shall travel together. Further, due to the special oblique direction of its fibers, the ligament transmits forces applied to the scapula at the shoulder region, to the strong and prismatic medial two-thirds of the clavicle.

Acromioclavicular Joint

At the lateral end of the clavicle is found the small and relatively unimportant acromio-clavicular joint (fig. 146); it takes little or no part in the transmission of forces. The site of the joint, 2 to 3 cm (1″) medial to the 'tip of the shoulder,' is often marked by a visible prominence.

SHOULDER JOINT

The shoulder joint is a multi-axial ball-and-socket joint enjoying a remarkable degree of freedom of movement. It is a loose

---

† The ligament has two parts named—according to their shapes—conoid and trapezoid. Their details are unimportant.

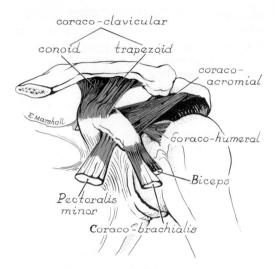

Fig. **145**. Ligaments and tendons of (left) coracoid process from front.

union between two mobile bones and its extreme mobility has been achieved only at the expense of stability and security, dislocation of the joint being a common accident (fig. 147).

Movements of the shoulder *joint* should not be confused with movements of the shoulder *region*. The two very commonly occur together but in any analysis of muscular activities it is helpful to segregate them, since movements of the shoulder region primarily involve the clavicle and scapula, whereas movements of the shoulder joint occur solely between the scapula and the humerus (figs. 147–149).

Articular Surfaces. The glenoid cavity of the scapula would be a vertical oval area, some 4 cm (1½″) in length, but for the concavity on its anterior border which makes its outline rather pear-shaped. It faces as much forward as it does lateralward. It is shallowly concave and of much smaller area than the head of the humerus. The addition of a rim of fibrocartilage to the periphery of the glenoid socket slightly deepens that socket but adds little to the security of the joint; it is known as the **glenoid labrum.** The area of actual contact between the two bones is never very great and a considerable amount of the head of the humerus is at all times in contact with the lax capsule (fig. 150).

Capsule and Ligaments. No other important joint has such a loose capsule. One ligament, however is strong and important, the **coraco-humeral ligament,** which extends from the coracoid process to the greater tuberosity of the humerus. Only its anterior edge is free, the remainder blending with, and thickening, the upper part of the fibrous capsule. The ligament becomes taut on lateral rotation. More important, along with the superior part of the capsule, it is also taut when the arm hangs vertically, forming a 'locking mechanism.' With Supraspinatus muscle, it prevents downward dislocation of the humeral head.

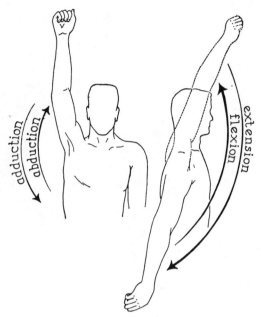

Fig. **148**. Primary movements of shoulder joint defined.

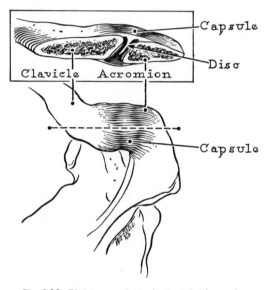

Fig. **146**. Right acromioclavicular joint from above and unopened; inset—coronal section.

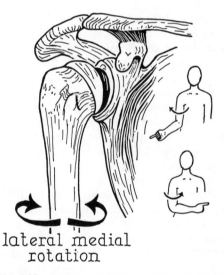

lateral medial
rotation

Fig. **149**. Lateral and medial rotation of shoulder joint defined.

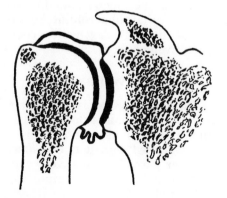

Fig. **147**. Coronal section of shoulder joint

The anterior part of the capsule contains three slight thickenings—the *gleno-humeral ligaments*—which are of no real significance.

Much of the security the joint possesses during its wide excursions is provided by the muscles that surround and move it. Of these, four, as they pass to their insertions

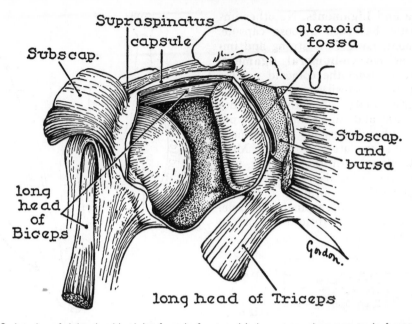

Fig. **150**. Interior of right shoulder joint from in front and below—capsule cut away in front and humerus pulled away from glenoid fossa.

on the two tubercles immediately adjacent to the head of the humerus, blend with the capsule and so reinforce it. By this device the capsule is reinforced in front, above and behind; its weakest site remains below. [The tendon of the long head of the Biceps, one of the muscles of the arm, lies in the groove between the two tubercles having traversed the upper part of the interior of the joint (fig. 150); it may lend additional security by helping to hold the head of the humerus in its socket.]

Movements. The joint may be flexed, extended, abducted, adducted, and rotated (medially or laterally). Any movements that elevate the arm above the level of the shoulder involve extensive scapular and clavicular movements as well. Indeed, it is probable that most movements of the humerus call for some concomitant movement of the glenoid cavity even before the horizontal plane is reached (see page 151 for details).

Coraco-Acromial Arch

The coracoid process in front, and the acromion above and behind, are united to one another by the thin, flat, and triangular **coraco-acromial ligament** (fig. 145). The apex of the ligament is attached to the acromion immediately in front of the acromio-clavicular joint; the base of the ligament runs along the length of the lateral border of the horizontal portion of the coracoid process. Coracoid, ligament, and acromion together have been called the **secondary socket,** and they form a continuous protective arch above the shoulder joint. Immediately under the coraco-acromial arch lies the tendon of supraspinatus muscle blended with the capsule of the shoulder joint. To reduce friction between the tendon and the arch a large bursa—the **subacromial bursa**—is interposed (fig. 151). A general account of bursae is given on page 120.

ELBOW JOINT

The elbow joint proper is the hinge joint between the lower end of the humerus and the upper ends of ulna and radius. But closely associated with it is the **superior radio-ulnar** joint. They share one capsule and their joint spaces are continuous. Indeed, it would be unwise to discuss one without the other.

Bony Parts and Anular Ligament (see pages 55 and 56 for a full description).

The jaws of the deep trochlear notch on the face of the wrench-like upper end of the ulna grasp the trochlea of the lower end of the humerus. This trochlea resembles an hour-glass laid on its side, the medial end of the hour-glass having a greater circumference than the lateral.

'Carrying Angle.' As the ulna swings around the trochlea from flexion to extension the difference in circumference between the two ends of the hour-glass makes itself felt and is responsible for the ulna being forced gradually lateralward and so out of line with the humerus. This difference in alignment between humerus and ulna amounts, when the elbow is fully extended, to about 15° and is spoken of as the **carrying angle.** It is compensated for when the lower end of the radius crosses over to the medial side of the ulna, the angularity of the extended and supinated limb disappearing on pronation.

The radius and the ulna lie side by side but the radius reaches only as high as the coronoid process of the ulna which forms the lower part of the trochlear notch. The circumference of the disc-like head of the radius gives the appearance of having worn for itself a little hollow on the lateral side of the coronoid process where it is constantly revolving. This little hollow is known as the **radial notch of the ulna** and less than one-quarter of the circumference of the radial head can articulate with it at any one time. The remainder of the circumference is accommodated by an incomplete ligamentous ring—incomplete only because its ends are attached to the front and back margins of the radial notch. The ring is known as the **anular‡ ligament** and it is horizontally disposed; it is also of smaller circumference below than above so that, besides completing a collar in which the radial head revolves, it also tends to prevent the radius from being pulled down out of its socket, an accident that sometimes happens, particularly in children, as the result of a too hasty pull on the outstretched hand. The anular ligament is a special development of the fibrous capsule (fig. 152).

‡ Also spelled annular, which is now considered 'incorrect.'

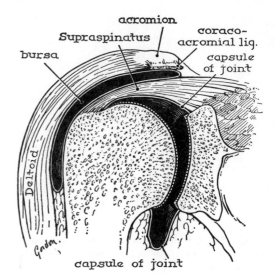

Fig. **151**. Diagram of position and relations of subacromial (or subdeltoid) bursa. Bursa and shoulder joint are shown inflated ; normally these are only potential spaces.

The upper surface of the radial head is circular in outline and slightly concave and, when accompanying the ulna in movements of flexion and extension, moves on the capitulum situated immediately lateral to the trochlea.

Collateral Ligaments and Capsule. Every hinge joint (including elbow) has a thin loose capsule except at its two sides where the capsule is greatly thickened to prevent undesirable side to side movements. Such thickenings are called **collateral** ligaments.

It follows from the above that there are medial and lateral ligaments of the elbow. The **medial or ulnar collateral ligament** is fanshaped. Fastened to the medial epicondyle above, it spreads to the medial edge of the trochlear notch of the ulna. The anterior part is thickened to form a **cord** which plays the biggest part in the prevention of abduction at the elbow (fig. 153). The **lateral radial collateral ligament** runs from lateral epicondyle to the outer surface of the anular ligament and so does not impede supination and pronation. Since the anular ligament grasps the head of the radius firmly in the adult, adduction is effectively prohibited by the radial collateral ligament.

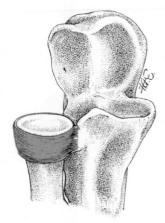

Fig. **152**. The anular ligament holds the head of the radius against the radial notch of the ulna, forming a pivotal joint.

The loose capsule is attached to the lower end of the humerus above the coronoid and olecranon fossae because the corresponding processes of the ulna occupy these fossae in flexion and extension, respectively.

Subcutaneous Bursae. Because of the continuous rubbing of the skin over the olecranon there develop one or more bursae behind the elbow. These are not unlike the structures to be described with tendons (see page 120) and can be compared to empty, cellophane-thin sacs with slippery inner surfaces to reduce friction. When inflamed they balloon up with fluid and are painful (bursitis).

RADIO-ULNAR JOINTS

Pronation and Supination. If the shafts of the articulated ulna and radius are viewed from the side, it is at once obvious that, above, the radius lies on a plane anterior to the ulna (fig. 154). This, of course, is because the head of the radius is carried on the side of the forwardly projecting coronoid process. It is this relationship that makes it possible for the radial shaft to cross in front of that of the ulna while the radial head merely revolves to perform the movement of pronation; the crossing is further aided by the shaft of the radius being, not straight, but bowed a little lateralwards (fig. 155).

Moreover, in pronating the hand it is easy, if not usual, to do so without altering the position of the hand in space. In other words, pronation is accompanied by a shift laterally of the lower end of the ulna. The elbow joint allows this slight ulnar abduction.

Superior Radio-Ulnar Joint

This joint has been discussed with the elbow. It only remains to state that the synovial capsule hangs loosely below the anular ligament as a redundant fold.

Intermediate Radio-Ulnar Joint

The shafts of radius and ulna are united by an **interosseous membrane** (fig. 156) stretching between their adjacent sharp borders; the union is a fibrous joint and is sometimes called the intermediate radio-ulnar joint. The fibers of the membrane take a direction downward and medialward, a fact which suggests that forces, received by the lower end of the radius from the hand, are transmitted, on their passage up the forearm, by the interosseous membrane to the ulna. This function of the membrane has been over-emphasized since the hand is usually in a position of pronation when forces are applied to it, and in pronation the interosseous membrane is rather slack. Moreover, research has shown that the radius shares in the transmission of forces directly to the humerus. The more important use of the membrane would seem to be to increase the area available, both at the front and at the back, for the origins of numerous muscles found in the forearm. See figure 157.

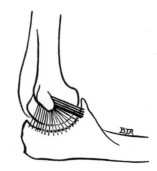

Fig. **153**. Medial aspect of left elbow joint. Note strong anterior cord of fan-shaped ulnar collateral ligament.

Fig. **154**. Direct lateral view of right radius and ulna in articulation.

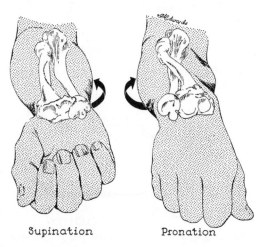

Supination   Pronation

Fig. **155**. Supination and pronation of the forearm.

Inferior Radio-Ulnar Joint

At the lower end of the two long bones of the forearm, it is the ulna that possesses a more or less disc-like head and that gives the appearance of having worn for itself a little hollow on the side of the radius. The hollow is known as the **ulnar notch of the radius.** The ulna is excluded from any participation in the wrist joint by a triangular plate of fibrocartilage lying immediately below its head. This is known as the **articular disc** and it is attached by its apex to the root of the styloid process—the short but stout peg of bone projecting down from the back of the ulnar head. The base of the disc is attached to the radius along the lower margin of the ulnar notch. The apex of the disc is the center of a circle whose arc is described by disc and radius in the movements of pronation and supination (figs. 158–160).

In order to permit the rather extensive movement of pronation, the joint capsule of the inferior radio-ulnar joint must be lax and redundant.

The joint area on the ulna at which supination and pronation occur is quite small (fig. 159); it involves rather less than half the circumference of the head of the ulna.

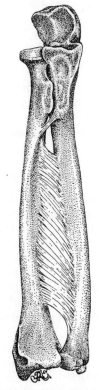

Fig. **156**. Interosseous radio-ulnar membrane

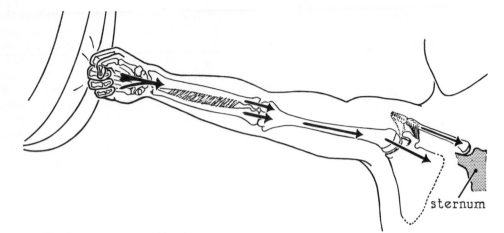

Fig. **157.** Lines of force from knuckles to sternum. Note the important transfer from sternum to clavicle through the strong coraco-clavicular ligament.

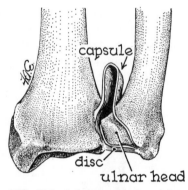

Fig. **158.** Right inferior radio-ulnar joint viewed from front with anterior part of capsule removed to show interior. Note upward redundancy of loose capsule.

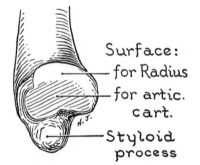

Fig. **159.** Lower end (head) of right ulna seen from lateral side.

WRIST OR RADIOCARPAL JOINT

Articular Surfaces (fig. 160). The lower surface of the triangular disc below the ulnar head continues medially the joint surface at the lower end of the radius. The composite surface—two-thirds radius and one-third disc—has the outline of an ellipse with a transverse diameter of about 5 cm (2″) and an antero-posterior diameter of about 2 cm (¾″). It is slightly concave in both directions.

Below the radius lie the scaphoid laterally and the lunate medially; below the disc lies the triquetrum. The joint between the radius and the disc on the one hand,

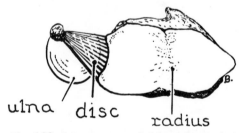

Fig. **160.** Lower aspect of right radius and ulna united by triangular fibrocartilaginous disc. In life, disc completely separates ulna from carpal bones.

and the scaphoid, lunate, and triquetrum on the other hand, is known as the wrist joint. The proximal row of carpal bones is described on page 57.

Movements. The wrist can be flexed or extended, abducted or adducted. In other words, because the joint is ellipsoidal, movements can occur only in the two axes of the ellipse. In abduction the three carpal bones named above shift medially; in adduction they shift laterally.

Capsule and Ligaments. The medial and lateral portions are somewhat thickened and are called the **collateral ligaments** of the wrist; they descend from the styloid processes of radius and ulna to the marginal carpal bones. The capsule is reinforced front and back by fibers that pass obliquely downward and medialward from the lower end of the radius to the carpal bones and force the carpus to move with the radius as a unit during pronation-supination.

It should be pointed out that a common mistake of the beginner is to regard the wrist joint as involving much more than is here described.

JOINTS OF CARPUS AND HAND

Although the joints of the carpus have a continuous single joint cavity, the fact that the proximal and the distal row of bones move on each other as units makes it desirable to give a name to the union between the two rows; it is called the

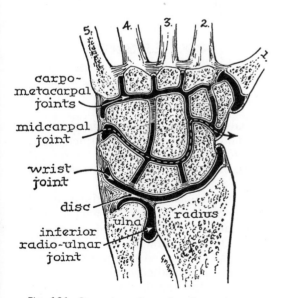

Fig. **161**. Coronal section of wrist and carpal region.

midcarpal joint (fig. 161). It must be realized that this is part of a general joint cavity which extends also between the individual bones of each row and which, proximally, may even communicate with the wrist joint. The common joint space includes the bases of the metacarpals where it is named the *carpometacarpal and intermetacarpal joints*. The insignificant *pisotriquetral joint* and the important **carpometacarpal joint of the thumb** are exceptions in that each is an independent joint.

The **midcarpal joint** acts as a hinge joint and therefore is capable merely of flexion and extension. It moves in conjunction with flexion and extension of the wrist joint to increase the range of these movements for the hand as a whole (see fig. 162). When wrist abduction and adduction occur, however, the midcarpal joint by its sinuosity locks the distal row to the proximal row and so ensures that the hand shall move as a whole (fig. 163).

Carpometacarpal Joints

Of the joints at the bases of metacarpals, only that at the base of the thumb or first metacarpal possesses any considerable range of movement. At that joint, the metacarpal of the thumb articulates with the trapezium by means of the best example of a saddle joint to be found in the body. As has been remarked before, a saddle joint permits angular movements in any plane and a restricted amount of axial rotation. Its mobility is exceeded only by a ball-and-socket joint (see fig. 131, page 79).

Thumb Movements. A cursory inspection will reveal that the thumb metacarpal is 'set' differently from the others. As a result, while the fronts and the backs of the fingers look in the orthodox directions, viz., forward and backward, the corresponding surfaces of the thumb look medialward and lateralward, i.e., at right angles to the fingers. This fact has to be borne in mind in describing the movements of the thumb. For example, the movement that lifts the thumb forward away from the palm is abduction and the movement that carries the thumb laterally away from the fingers is extension (fig. 164). Note that the functional value of the thumb lies in its ability to be 'opposed' to the fingers, and that the movement of opposition occurs at the saddle-shaped joint between trapezium and first metacarpal.

Metacarpal Movements. By passively manipulating the bases of the metacarpals it can be demonstrated that the mild degree of mobility enjoyed by the metacarpals of the little and ring fingers enables one to 'cup' the hand and, perhaps, to improve the grip.

Axial Line; Force Transmission

From all points of view, it soon becomes apparent that the middle finger, the third metacarpal, and the capitate are the axial bones of the hand. To this axis forces are gathered and around it many fine movements of the hand take place (fig. 165). The large capitate supports the base of the third metacarpal and, in part, those of the second and fourth. The second metacarpal (the most fixed and belonging to the 'pointing' finger) is supported by three carpals, viz., capitate, trapezoid, and trapezium. The fourth metacarpal is supported mainly, the fifth entirely, by the hamate which transmits its forces—not to the triquetrum—but to the lunate. Thus the force of a blow struck by any of the knuckles is transmitted to the radius. It is apparent, too, that the second and third metacarpals are in the most direct line of force transmission and are, therefore, the most efficient of the four for the delivery of a blow. Furthermore, the ligaments binding the adjacent carpal bones to the capitate take a direction designed to gather forces to this large, central and axial bone (fig. 165). The importance of recognizing the axial line will be further stressed when the muscles of the hand are discussed.

Metacarpophalangeal Joints

These are the joints between the heads of metacarpals and the bases of the first row of phalanges—the **'knuckle joints.'** Of these joints that of the thumb is the least mobile being similar to interphalangeal joints in structure and function.

When the fist is clenched the 'knuckles' stand out prominently and are the meta-

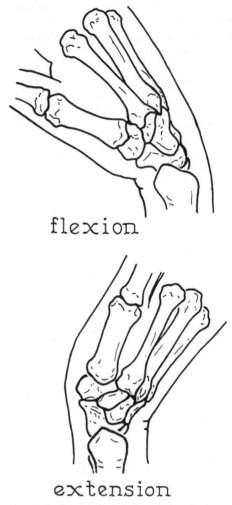

flexion

extension

Fig. **162.** Interplay of bones during flexion and extension of wrist (lateral view; semischematic).

carpal heads; the joints lie just distal to (beyond) these heads and distinctly proximal to the line of the webs between the fingers. They can be felt with the fingernail.

At the knuckle joint, a finger can be flexed until it is at right angles to the metacarpal or it can be extended somewhat beyond a straight line (i.e., slightly hyperextended). The articular cartilage on a metacarpal head needs to clothe the end and the palmar surface of the head, but it needs to extend for only a short distance on to the dorsal surface. If the dried bone is examined, these are the conditions found and, in addition, the articular surface is

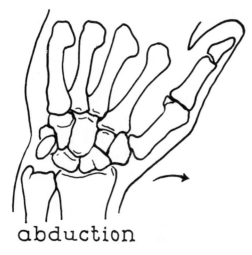

abduction

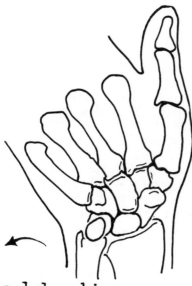

adduction

Fig. **163.** Abduction and adduction of wrist (front view; semischematic).

166, 289). It is attached to the anterior edge of the base of the phalanx, moving in consequence with the phalanx; it has only a loose attachment to the metacarpal. The side of each plate is attached to its neighbor by a strong, square-shaped ligament, each of which is known as a **deep transverse ligament of the palm.** The result is a fibrous band stretching right across the palm and binding the metacarpal heads together (fig. 289, page 169). The palmar

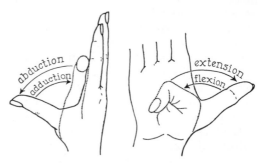

Fig. **164.** Movements of thumb defined

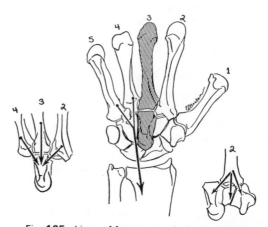

Fig. **165.** Lines of force transmission from knuckles through metacarpal and carpal bones to radius.

seen to be markedly convex from front to back but very much less so from side to side.

The base of a proximal phalanx has a small and slightly concave articular surface which is oval in outline with the long axis transverse.

Palmar Ligaments. Important tendons rub on the front of each joint and so the capsule becomes locally thickened to form a stiff plate of dense fibrous tissue known as a **palmar ligament (or 'plate')** (figs.

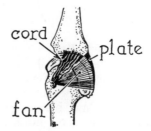

cord

plate

fan

Fig. **166.** A knuckle (metacarpophalangeal) joint from side. 'Cord' becomes taut on flexion.

ligament (plate) of the thumb has no attachment to its neighbor but is free.

Movements; Collateral Ligaments. You will find that when the fingers are extended they can be spread apart, but as they are flexed they are compelled finally to come together. The whole movement is an acquisitive one, the open hand to receive, the closed hand to clutch. Each side of each joint capsule is reinforced by a fan-shaped ligament rather particularly placed (fig. 166). The handle of the fan is attached to the side of the metacarpal head at a site which ensures that the ligament is completely taut on flexion but lax on extension; this is achieved by placing the attachment nearer the distal end than the palmar surface. From this site, the ligament fans out to the side of the base of the phalanx and to the palmar plate; the part to the phalanx is the stronger. These paired **collateral ligaments** prevent abduction and adduction of the flexed metacarpophalangeal joints. It is, in part, the function of these ligaments to bring together the fingers on making a fist. Muscles that control these movements are discussed on page 169.

Interphalangeal Joints

There are nine of these joints and each is a hinge of the purest variety. The head of each phalanx (except the terminal ones of course) is pulley-shaped, the base of the adjacent phalanx possessing a surface to match the two parts of the pulley (fig. 167). The device renders side to side movement impossible, hence the fingers can be flexed or extended but no other movement is possible. The joints are reinforced with little collateral ligaments and equipped with palmar ligaments (plates) similar to those already met except that there can be no transverse ligaments uniting adjacent plates.

The thinnest part of the capsule of all metacarpophalangeal and all interphalangeal joints is at the back. Tendons on the back of the fingers, as they cross each joint in turn, flatten out and, blending with the capsule, afford it additional protection and security.

JOINTS OF THE LOWER LIMB

The joints of the lower limb must be stable because of the great weights they bear, and yet, some of them must be quite mobile as well—especially those lower in the limb. Therefore when we use the often-heard phrase 'limb of stability' we should qualify it with '—relatively.'

SACRO-ILIAC JOINT

The sacrum and ilium are bound together by two joints, one a synovial, the other a fibrous; the dual mechanism is known as the sacro-iliac joint.

The Synovial Joint. It has been remarked that the pelvic girdle, unlike the pectoral girdle, is firmly united to the vertebral column. An examination of the dried hip bone will reveal a large and rough area occupying the posterior third of the inner surface of the iliac blade. The lower (anterior) portion of this area, less rough than the upper, has a fancied resemblance in outline and size to a small ear; it is known as the **auricular surface** of the sacro-iliac synovial joint (see fig. 100, page 62). The corresponding surface on the side of the ala of the sacrum is exactly similar in size and shape (fig. 50, page 29) and so intimate is the contact and so limited are the movements between the two bones, that one is surprised to find a synovial joint here at all. Indeed, the two auricular surfaces—covered with articular cartilage during life—are rather wavy in outline, the total effect being that of a joint in which the applied surfaces are almost interlocking in character (fig. 168).

An appreciable amount of movement can occur at the sacro-iliac joint (enough to

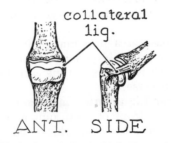

Fig. **167**. An interphalangeal joint is a true hinge

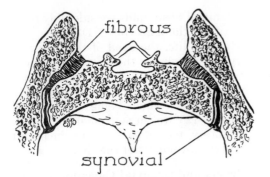

Fig. **168**. Each sacro-iliac joint is synovial in front and fibrous behind. Articular surfaces of synovial joint are normally in close contact but here they are separated to show joint space (dark).

give resilience to the pelvis as it receives weight transmitted to it by the sacrum); the limitation of movement and the physical features noted above encourage changes to occur in the joint whereby, with increasing years, the surfaces tend to adhere and the synovial joint tends to become obliterated, and bony union may occur.

The Fibrous Joint. The upper (posterior) portion of the area on the iliac blade is rough because it is for the fibers of the large and important sacro-iliac fibrous joint. Over this area, the sacrum and ilium are not in direct contact but a cleft exists between them and, behind the cleft, the ilium projects backward beyond the sacrum. The cleft is filled with a mass of fibers known as the **interosseous sacro-iliac ligament** and, behind the cleft, the fibers are known as the **dorsal sacro-iliac ligament.** As if even this were not enough, the summit of the crest of the ilium is united to the tip of the transverse process of the fifth lumbar vertebra by a strong band of fibers known as the **iliolumbar ligament** (fig. 169). Secondary ligaments of the joint are the sacrospinous and sacrotuberous, described on page 64.

All of this ligamentous mass is under constant strain except when the body is recumbent. On it falls the duty of transmitting the weight of the body above to the hip bone. By permitting only a minimal movement, the presence of the ligaments secures the integrity of the synovial joint,

which is helpless to sustain weight by itself, in spite of its interlocking character.

Mechanics of Pelvis (see page 64).

SYMPHYSIS PUBIS

Each pubic bone, where it faces its fellow in the median plane, is coated with a layer of hyaline cartilage and these coatings are united to each other by fibrocartilage (fig. 170). Such a device is known as a **symphysis.** The symphysis pubis is practically an immovable joint.

In sitting up straight, weight falls on the ischial tuberosities which tend to be forced apart; the symphysis pubis in resisting this tendency is a joint under tension. In standing, weight falls on the upper part of the

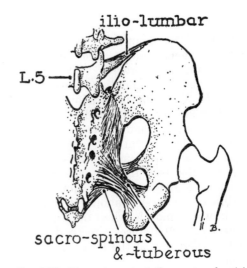

Fig. **169**. Three important ligaments of pelvis (from behind).

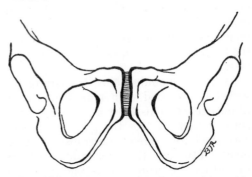

Fig. **170**. Symphysis pubis firmly unites pubic parts of two hip bones in front.

acetabular margin which tends to buckle inward; the symphysis pubis in resisting this tendency now becomes a joint under compression.

In the female, there is rather more movement at both the sacro-iliac joint and at the symphysis pubis than there is in the male. During pregnancy, all of the ligamentous structures of the pelvis become softened and so permit still more movement, a state of affairs of some value in facilitating the passage of the fetal head.

HIP JOINT

The shapes of the bones taking part in the hip joint make it the best example in the body of a ball-and-socket joint; they also make it one of the securest joints in the body.

Articular Surfaces

Socket. The mouth of the large hemispherical **acetabulum** faces downward, lateralward, and forward. Not all of the fossa is articular and lined with hyaline cartilage. A circular area, about 2 to 3 cm (1″) in diameter, at the middle or deepest part of the fossa and a low-ceilinged passageway, about 2 cm (¾″) wide, leading from the circumference to this area, are non-articular (see fig. 98, page 60). The effect is to give to the articular area the outline of a horseshoe about 2 to 3 cm (1″) broad whose mouth, when the hip bone is correctly oriented, faces directly downward and is, therefore, at the lowest part of the joint.

A narrow band of fibers, known as the **transverse ligament of the acetabulum,** joins the two ends of the 'horseshoe.' It bridges the non-articular gap but leaves room under the bridge for vessels to reach the fat pad that occupies the depths of the fossa.

As in the shoulder joint, the rim of the socket is provided with a built-up lip of fibrocartilage. This lip is a complete circle since it, too, bridges the gap and blends with the outer edge of the transverse ligament. Of the two, the lip is much the more important structure and is known as the **acetabular labrum.** It serves to deepen the socket and to offer a smooth and elastic or resilient edge which, firmly embracing

the femoral head, adds considerably to the security of the joint (fig. 171).

Ball. In life, the head of the femur fits very accurately in the acetabular socket (see footnote, page 65). The head is about three-quarters of a sphere and is mounted on a short neck which, being set at an angle of about 125° to the long shaft, translates the rotary movements of the head into angular movements of the shaft.

Fibrous and Synovial Capsule

The **fibrous capsule** of the hip joint is strong, extensive, and important. It is attached to the hip bone a little beyond the acetabular circumference. On the femoral side it reaches in front to the whole length of the trochanteric line and thus encloses the neck as well as the head. This front portion is thickened to form the iliofemoral ligament, one of the thickest and strongest capsular ligaments in the body (fig. 172). The remainder of the capsule, much less strong, is known as the pubofemoral and ischiofemoral ligaments, indicating the bones that are united.

The iliofemoral ligament is triangular in shape but with its sides so thick (sometimes almost 1 cm) as to give it the appearance of an inverted V. The apex of

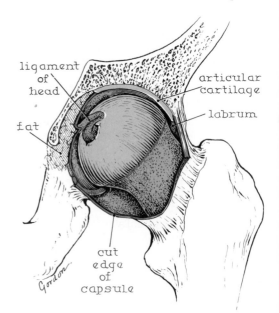

Fig. **171.** Interior of hip joint. Hip bone cut in coronal plane.

the V is attached to the acetabular margin and the anterior inferior iliac spine. The limbs of the V are attached to the two ends of the trochanteric line; between the two limbs the ligament is somewhat less thick though still strong. The iliofemoral ligament is instrumental in preventing the femur from being extended beyond the point at which the lower limb is in line with the trunk; it thus ensures that little or no muscular effort is required to prevent one from rolling over backward at one's hip joint.

Below, behind, and above, the fibrous capsule is much less reinforced and is rather applied than attached to the femoral neck (fig. 173). Figure 172 illustrates a weak area of the capsule in front of the head of the femur. The tendon of Psoas muscle lies in front of this area and gives it support. A perforation often unites the joint space with a bursa deep to the tendon.

Synovial Membrane. Not only does the synovial membrane line the inner surface of the fibrous capsule but it also: (1) lines the neck of the femur between the attachment of the fibrous capsule and the edge of the articular cartilage of the head (fig. 171)—thus the neck is 'inside the joint'; and (2) covers the non-articular area of the acetabulum. From the lower part of the latter a (flattened) tube of synovial membrane runs upward, inside the joint, to a small pit on the femoral head (fig. 171). This synovial tube is called the **ligament of the head of the femur.** In it an (inconstant) artery runs to the head of the femur and gives the ligament its only real importance.

Axiom: The interior of every synovial joint must be smooth, moist, shiny, and slippery. Where there is no articular cartilage there must be synovial lining. The hip joint exemplifies this axiom.

Movements

The hip joint resembles the shoulder joint in that it can be flexed, extended, abducted, adducted, and rotated medialward or lateralward. The movements are much more restricted and the joint, of course, enjoys no movements at all compa-

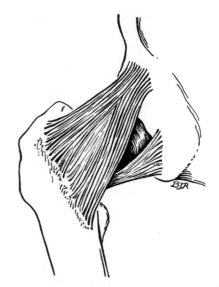

Fig. **172.** Right iliofemoral ligament from in front. It forms the important part of fibrous capsule.

Fig. **173.** Posterior part of capsule of hip is ischiofemoral (and pubofemoral) ligament.

rable to those that the mobile scapula gives to the whole shoulder region.

KNEE JOINT

The knee joint is the largest and the most complex joint in the body. It is a modified hinge joint of peculiar construction.

Consider these facts:

1. If the femoral condyles are viewed from the side, one can be superimposed visually on the other with exactitude, i.e., they have identical out-

lines. But if the lengths of these articulating condyles are measured from front to back, the medial is found to be more than 1 cm longer than the lateral (fig. 174). Therefore, as the tibia swings around on the femoral condyles from complete flexion toward complete extension it 'uses up' all the articular surface available on the lateral side while it still has more than 1 cm to 'use up' on the medial side.

2. The tibia can complete its excursion only by rotating on its long axis. This is a lateral rotation of the tibia (or medial rotation of the femur) and, as it proceeds, a device is needed to bring it finally to a halt. The device is the vertically directed **anterior cruciate ligament.** This ligament also acts as the pivot around which the rotation occurs.

3. One function of a disc in a joint is to adjust surfaces to one another without sacrificing range of movement; the lateral side of the joint can advantageously utilize a disc for this purpose and is provided with one. For it to be effective, it must be mobile; for it to be mobile, the lateral tibial condyle on which it must slide should be flat on top; both conditions exist. The disc is known as the **lateral meniscus** or *lat. semilunar cártilage* (see below).

4. Another function of a disc is to increase the security of a joint by adjusting unmatching surfaces to one another. The medial side of the joint would be more secure if the medial tibial condyle were more concave; its slight concavity is increased by the existence of a disc almost immovably fixed to the top of the medial tibial condyle. This is the **medial meniscus** or *med. semilunar cartilage* (see below) (fig. 175).

5. In such an action as jumping off a chair with the knees bent, the femur has a marked tendency to slide forwards off the plateau-like tibial surface. A strong ligament is needed to resist this unfortunate tendency and it needs to be placed as far back on the

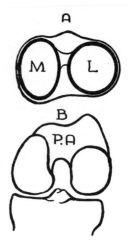

Fig. **174.** A. Upper surface of right tibia = circle—hour-glass—ellipse.' B. Articular surfaces at lower end of left femur (P.A. = patellar area).

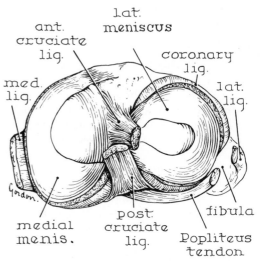

Fig. **175.** Menisci (semilunar cartilages) and ligaments of right knee seen from above.

tibia as possible. This is the horizontally directed **posterior cruciate ligament.**

Bony Parts

For details see pages 66 and 67: 'Lower End of Femur' and 'Upper End of Tibia'; and page 69: 'patella.'

When the knee is fully flexed the relatively flat tibial condyles make but slight contact with the markedly curved and backwardly projecting femoral condyles. As the tibia swings below the femur the area of contact between the two bones increases, since the under surfaces of the femoral condyles are flattened out; as a result, with increasing extension the joint obtains increasing security.

Menisci (or 'Semilunar Cartilages')

The **medial meniscus,** shaped like a moon in its first phase, is fixed to the tibia front and back by its two horns as well as all around its periphery. Since its horns are fixed far apart and since relatively short fibers attach its periphery to the tibia, it can move but little. The medial meniscus serves merely to deepen the socket for the medial femoral condyle.

The **lateral meniscus** (or **semilunar cartilage**) is really not semilunar but circular. Its horns are close together and fixed to the tibia near the center of its upper surface, i.e., near the paired little eminences; its periphery is attached to the tibia by relatively long fibers that do not interfere with its movements (see 'Coronary Ligaments,' below). The meniscus can glide back and forth. As extension approaches its limit, the lateral femoral condyle slips forward as a unit with the lateral meniscus on the upper surface of the lateral tibial condyle. At the same time the rotation (lateral of tibia or medial of femur) occurs which locks the joint in full extension. The final locking movement is usually called the **'screw-home' movement.**

Cruciate Ligaments

The **anterior cruciate ligament** is not only the pivot round which the 'screw-home' movement takes place but it also prohibits any farther movement of either extension or rotation. This ligament lies as near the center of the joint as it can get. It is attached to the tibia just in front of the site where the lateral semilunar cartilage is fixed and, passing upward, it is attached to the femur on the 'inner' (i.e., medial) side of the lateral femoral condyle. Its direction makes it also an efficient mechanism to prevent backward dislocation of the femur (see figs. 176–179).

The **posterior cruciate ligament** is attached to the tibia so far back as to allow

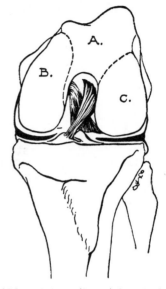

Fig. **176.** Left knee (flexed) from in front. A. = for patella; B. = medial, C = lateral, femoral condyles. Cruciate ligaments and menisci shown in black.

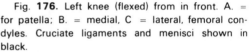

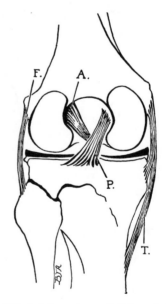

Fig. **177.** Left knee (extended) from behind. F. = cord-like lateral lig.; T. = band-like part of the medial ligament. A. = anterior, P. = posterior, cruciate ligaments.

some of its fibers to arise from the back of the bone below the upper surface. It is the first structure to be seen when the fibrous capsule is opened from the back. Its upper

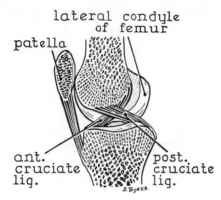

Fig. **178.** Cruciate ligaments of (right) knee joint viewed from medial side (with medial ½ of bones cut away).

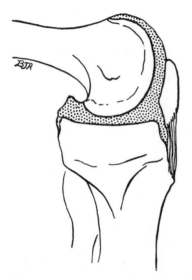

Fig. **179.** Knee in flexion, Stippling indicates synovial joint space.

attachment is to the 'inner' (i.e., lateral) side of the medial femoral condyle. Its fixation to the tibia as far back as possible increases its efficiency to prevent forward dislocation of the femur.

Collateral Ligaments

Several structures prevent displacement sideways, i.e., a lateral or a medial dislocation of either bone. These are: (1) the collateral ligaments; (2) the cruciate ligaments; and (3) the intercondylar eminences.

The thick fibrous capsule of the knee joint, as in all hinge joints, is reinforced on its sides by collateral ligaments. Because a mild degree of rotation is possible in the flexed knee, these ligaments do not become taut until extension is reached. The **lateral ligament** is a stout cord that joins the lateral epicondyle of the femur to the head of the fibula. The **medial ligament** is rather like that of the elbow in that it has two parts—a fan-shaped part which is an intrinsic part of the capsule, and a superficial band which runs from the medial epicondyle down to the medial surface of the tibia some distance below the joint (fig. 177).

The cruciate ligaments act as collateral ligaments, too. Finally, the paired little bony eminences at the center of the upper surface of the tibia fit between the two femoral condyles and help to prevent undue side to side play. At the very beginning of side-play the femur rises up a little on the side of one or other of these eminences, the effect being to tighten at once the collateral and cruciate ligaments.

Fibrous and Synovial Capsule

Fibrous Capsule. The collateral ligaments received separate attention above to emphasize their discreteness, but the remainder of the capsule is equally important.

The aponeurotic (fibrous) insertion of the large extensor muscle of the knee largely provides the fibrous capsule for the front and sides of the joint. In the tendon of this powerful muscle a sesamoid bone developed and became articular. Thus the knee-cap or patella occupies an area on the lower part of the front of the shaft of the femur. Its area is continuous with that for the tibia. Its movements have been noted (page 69) (see figs. 179 and 180).

Behind, in the depths of the popliteal space, the capsule is thickened to form the *meniscofemoral (oblique) ligament* of the knee, which runs upward and lateralward.

Coronary Ligaments. The periphery of each meniscus is attached to the inner aspect of the capsule; the part of the capsule below this attachment is called the coronary ligament. The medial one, in contrast to the lateral, is rather tight and prevents back-and-forth movements of the medial cartilage. This fact is

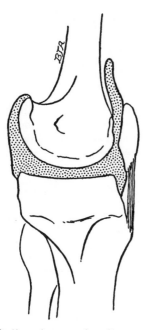

Fig. **180.** Knee in extension. Note upward prolongation of synovial cavity (stippled) above patella.

advanced as a reason for the greater frequency of tears of the medial semilunar cartilage—a rather common athletic injury.

Synovial Membrane. In general, its attachment to all three bones follows closely the edges of the articular cartilage as it does in most joints; but, in front, if the membrane is not to be ruptured when the knee is flexed, it must be lax and redundant when the knee is extended. Therefore, instead of passing directly from the femur to the patella and tibia it rises up for two or three fingers' breadths above the patella before it turns down to reach its patellar and tibial attachments (fig. 181). Further, unlike most joints the synovial membrane cannot merely line the inner surface of the fibrous capsule. It must exclude the cruciate ligaments from the interior of the joint proper. Indeed, the cruciate ligaments develop in a primitive partition which at first divides the joint into two separate spaces each with its own synovial membrane. The partition disappears in front of the cruciate ligaments except for an unimportant thread which runs vertically from the femoral intercondylar notch to just below the patella (infrapetallar ligament). The syno-

vial membrane of the knee joint has many fat-filled free tags that project into the joint; these occasionally get pinched.

Bursae. Several *subcutaneous bursae*, like those at the elbow, are found between the skin and the patella or the patellar ligament (fig. 181). (Housemaid's knee is an inflammation of these bursae.) Besides these superficial bursae which do not communicate with the knee joint space there are several *deep bursae* associated with tendons, which do. ('Bursae' are described on page 120).

TIBIOFIBULAR JOINTS

In order to give the greatest measure of security to the ankle joint, it is desirable that the union between the tibia and the fibula at the lower ends should be as immobile as possible but not rigid. Hence, the **inferior tibiofibular joint** is a fibrous union in which the lower end of the fibula,

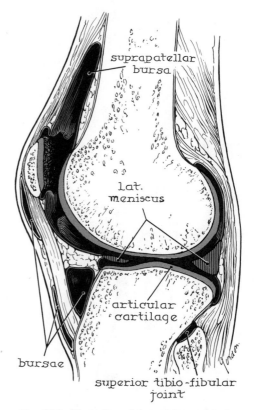

Fig. **181.** Dissection of knee joint; sagittal section through lateral condyles of femur and tibia.

just above its prominent (lateral) malleolus, is received into a hollow on the lateral side of the lower end of the shaft of the tibia. There the two bones are bound together side by side in the closest intimacy by the strong **interosseous tibiofibular ligament** and by strong fibrous bands that here unite the bones front and back; these bands are known as the **anterior** and the **posterior inferior tibiofibular ligaments** and they run obliquely downward and lateralward.

Above this union, the adjacent borders of the two shafts are united to one another by an **interosseous membrane** (fig. 182). Finally, the head of the fibula, taking part in the synovial **superior tibiofibular joint,** possesses a little, flat, oval, articular surface which makes contact with a similar

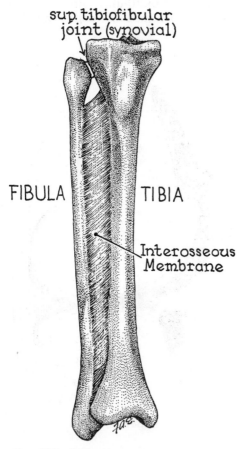

sup. tibiofibular joint (synovial)

FIBULA   **TIBIA**

Interosseous Membrane

Fig. **182.** Right interosseous tibiofibular ligament.

surface under the back part of the overhanging lateral tibial condyle.

Thus, when the two bones are sprung apart by movement at the ankle joint—a movement to be explained in a moment—the upper end of the fibula can slide a little at its synovial joint which may be regarded as an accommodatory joint designed to give a little play to the fibula and to safeguard it against being broken (fig. 183).

ANKLE JOINT

The ankle or **talocrural joint** is a pure hinge; it permits dorsi-flexion and plantar-flexion but any subsidiary movement that may occur, far from being necessary, is undesirable and a sign of weakness in the joint.

Articular Surfaces

The quadrangular surface at the lower end of the tibia is concave from front to back; it rests, and fits, on the convex upper surface of the body of the talus. In the anatomical (neutral) position of the foot the tibia rests, so to speak, on the brow of a hill. When the body leans forward, the tibia slides forward down the hill and the ankle is dorsi-flexed; when the body leans backward, the tibia slides backward down the hill and the ankle is plantar-flexed.

To ensure that no side to side play shall occur, the sides of the body of the talus are grasped by bony flanges, the **medial malleolus** and the **lateral malleolus** (figs. 184, 185).

The superior articular surface on the talus is broader in front than behind, consequently the two flanges are not parallel to one another but are farther apart in front than behind. When the body leans forward, it is this broader part that is engaged between the two malleoli which, as a result, are forced slightly apart, necessitating the slight fibular movement referred to earlier. In this forward position, the talus is more tightly grasped than at any other time and the joint is most secure. When the body leans backward, or better when the 'tip-toe' position is assumed, the narrow part of the talar surface occupies the socket and a little 'rocking' can occur, i.e., the joint is least secure.

Ligaments

As in all hinge joints, the capsule is thickened to form collateral ligaments whose dispositions and functions become meaningful only if the following facts are borne in mind:

1. The line of gravity, when one stands

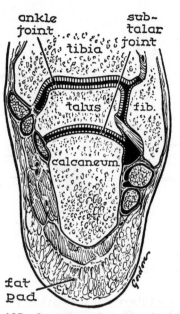

Fig. **185**. Coronal section through the joints above and below the body of the talus, i.e., the ankle and subtalar (talocalcanean) joints. Note the cushion of fat under the heel.

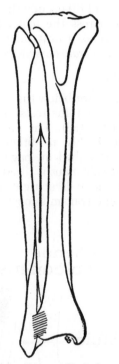

Fig. **183**. Right tibia and fibula from in front. Fibula is firmly bound to tibia but it can be moved slightly in direction of arrow—hence a synovial plane joint above.

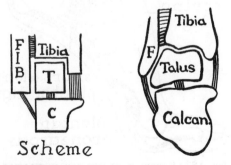

Fig. **184**. Ankle and talocalcanean joints in coronal section (schematic).

erect, falls well in front of the ankle joint. Hence there is a continual tendency for the tibia to slide forward off the talus; the tendency is increased by muscles of the leg that, on their way to the foot, use the backs of the tibia and fibula as pulleys.

2. The medial flange grasps only a shallow area on the upper part of the side of the talus; as a mortise it is much less efficient than the lateral flange which is longer and grasps practically the whole height of the side of the talus.

3. Two bones intervene between the tibia and the ground; ligaments pass from the flanges to both of them (fig. 186).

On the lateral side of the joint, two strong cords efficiently resist the tendency of the leg bones to slide or be forced forward off the talus; both run from the tip of the lateral malleolus. The **posterior talofibular ligament** passes almost horizontally medialward at the back of the joint from a pit just behind the articular surface of the fibular malleolus to the

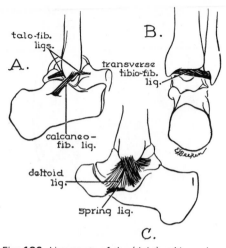

Fig. **186**. Ligaments of the (right) ankle region. A. lateral. B. posterior. C. medial.

projecting tubercle at the back of the talus. The **calcaneofibular ligament** passes distinctly backward and downward to the side of the calcaneus. Since there is little tendency for the leg bones to be displaced backward, only a relatively weak band—the *anterior talofibular ligament*—passes forward from the front of the lateral malleolus to the neck of the talus.

On the medial side of the joint, one single, broad and strong ligament compensates for shortness of the malleolus. It is the **deltoid or medial ligament of the ankle** and it passes from the medial malleolus to the sustentaculum tali of the calcaneus and to the whole length of the side of the talus (fig. 187).

On the back of the joint, the *transverse tibiofibular ligament* joins the malleoli and deepens the socket of the ankle joint.

Twisting of the ankle is resisted by the fibular malleolus and by the deltoid ligament. Of the two, the ligament is the stronger so that a serious 'turning' of the ankle, such as frequently happens in a skiing accident, more commonly results in a fracture of the fibula in association with snapping off of the medial malleolus of the tibia rather than in a tearing of the deltoid ligament. The fibula gives way where it is weakest; this is not at the ankle joint where it is firmly bound to the tibia, but 8 to 10 cm (3-4″) above. The condition is known as Pott's fracture from the London surgeon who, himself, suffered such an accident and gave a classical description of it.

JOINTS OF THE FOOT

Six independent joints exist in the tarsal region and it will be convenient to discuss each in turn. They are:

1. Subtalar
2. Talocalcaneonavicular
3. Calcaneocuboid
4. 'Composite joint'
5. First cuneometatarsal
6. Cubometatarsal

Because (2) and (3) occupy the same transverse plane across the foot and move together almost as a unit they are grouped together as the *transverse tarsal joint*.

Subtalar Joint

This is the joint below the ankle whereby the body of the talus rests on that of the calcaneus (fig. 188). The surfaces of this multi-axial, highly modified plane joint have been described on pages 71 and 72. Because the calcaneus is tilted upward, the plane of the joint is at right angles to the line of weight transmission to the ground (fig. 189).

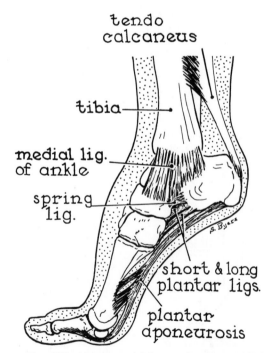

Fig. **187**. Ligaments of foot and ankle—medial view.

The capsule is not noteworthy except that it does not communicate with any other joint. The chief restraining ligaments are those of the ankle which span the talus to attach to the calcaneus. In addition there is the **interosseous talocalcanean**

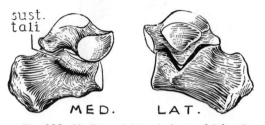

sust.
tali

MED.   LAT.

Fig. **188**. Medial and lateral views of left talus and calcaneus in articulation.

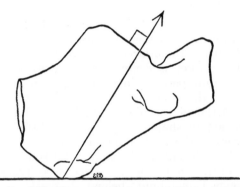

Fig. **189**. Plane of talocalcanean joint is at right angles to the line of weight-transmission to the heel (cf., fig. 121).

ligament. It lies immediately in front of the subtalar joint and directly below the very center of the ankle joint (X in fig. 190). It holds the two bones together and is tightened by any undue movement between them.

Talocalcaneonavicular Joint

This joint involves three bones and three ligaments; of these six structures the navicular, the calcaneus, and the three ligaments combine to form a deep socket for the reception of the more or less spherical head of the talus (fig. 191).

It has already been pointed out that, medially, the head of the talus rests, in part, on the sustentaculum tali. Regardless of this fact, the main mass of the head of the talus projects insecurely in front of the sustentaculum. Here the head is capped by the navicular, between which and the sustentaculum is a triangular interval in which the talus hangs its head and reaches its lowest limit. A very strong ligament fills this interval and on it the head of the talus rests. This ligament, uniting the sustentaculum to the lower edge of the back of the navicular, is known as the **'spring' ligament or plantar calcaneonavicular ligament;** on its strength and integrity depends, in large measure, the security of the whole foot (figs. 187, 192). Should it be stretched and the interval between the

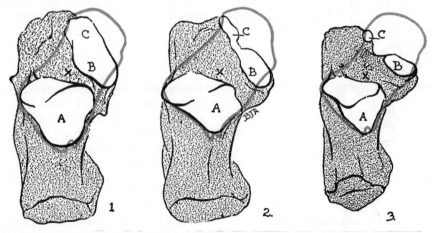

Fig. **190**. Outline of talus (red) superimposed on left calcaneus from above. 1, 2, and 3 show the great variability of the articular surfaces A, B, and C. X is the site of attachment of the interosseous ligament.

sustentaculum tali and the navicular be widened, the head of the talus sinks in this interval and the foot becomes flat.

The spring ligament, at the highest part of the medial longitudinal arch, is at the keystone of that arch. The weight of the body above, transmitted to it by the head of the talus, is distributed backward to the heel and forward to the heads of the metatarsals. The upper articular surface of the ligament is coated with fibrocartilage, and the tendon of a powerful muscle, the Tibialis Posterior, lies immediately below it and reinforces it only during excessive strain.

On the medial side of the joint, the front fibers of the **deltoid (medial) ligament of the ankle** joint reach to the navicular and blend with, and support, the medial edge of the spring ligament; they complete the medial wall of the socket. On the lateral

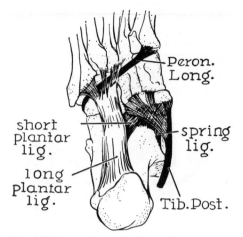

Fig. **192**. Important ligaments and tendons in sole of (right) foot help maintain arches.

side, a much less important ligament unites the calcaneus to the navicular; it is known as the *bifurcate ligament* (fig. 191). The socket is completed above by weak and unimportant fibers joining the navicular to the talus.

Although the joint is somewhat ball-and-socket in shape, its movements little resemble those of a ball-and-socket joint. Gliding occurs at the joint and it is not the ball that moves but the socket (see 'Transverse Tarsal Joint,' below).

Calcaneocuboid Joint

The front end of the calcaneus reaches nearly as far forward as the front end of the talus but is, of course, on a distinctly lower level (fig. 188). The joint surface for the cuboid is not quite flat but rather sinuous. A short beak-like prolongation of the cuboid reaches backward under the calcaneus and is said to give it some support. The movements are chiefly the mild rotary and gliding ones that such a joint suggests.

Just as the strong spring ligament binds the navicular to the calcaneus, so the strong, **short plantar ligament (plantar calcaneocuboid lig.)** binds the cuboid to the calcaneus. Indeed, the two ligaments form an almost continuous structure across this vital region, the one supporting the keystone of the high medial longitudinal arch, the other supporting the keystone of the low lateral longitudinal arch (see fig. 192).

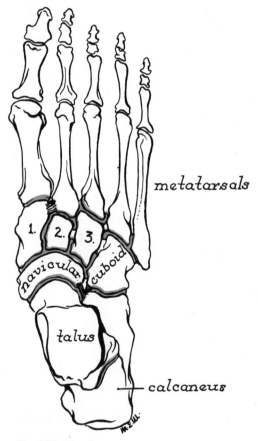

Fig. **191**. The six separate joint spaces of foot (in red); (see text).

Although the joint is capable of some independent movement, it frequently moves in unison with the one for the head of the talus.

Transverse Tarsal Joint

The sinuous line that is anterior to the head of the talus and to the anterior end of the calcaneus in figure 191 represents a plane at which movements of adduction, abduction, inversion, and eversion of the foot occur in co-operation with these movements at the talocalcanean joint. However, the calcaneocuboid and talocalcaneonavicular joints, though acting in unison most of the time, maintain separate joint spaces. In spite of this, they are grouped together as the **transverse tarsal joint.**

'Composite Joint'

No short or convenient expression exists to designate the **'composite joint'** in which the navicular, cuboid, three cuneiforms, and the bases of the second and third metatarsals participate. The custom of considering the parts of the joint as separate entities, while quite proper, is apt to obscure the fact that one common cavity serves them all.

The major portion of the cavity lies in front of the convex anterior surface of the navicular and is between it and the three cuneiforms. Three forward extensions exist between the contiguous surfaces of the cuboid and cuneiforms, the most medial of which, viz., that between the first and the second cuneiform, reaches to the front ends of the second and third cuneiforms and so extends the cavity to the bases of the second and third metatarsals.

The capsule uniting the various bones of this 'joint' is much thicker in the sole than in the dorsum of the foot.

The cuboid and the cuneiforms are also united to their neighbor (or neighbors) by interosseous ligaments. All the movements that occur are of a gliding nature and serve to give flexibility to the region.

Cuneometatarsal Joint of the Great Toe

The first cuneiform projects forward about 1 cm (or ½″) farther than the second and its anterior surface carries the base of the stout first metatarsal. The joint is an independent one and the large, plane, and kidney-shaped contiguous surfaces allow the simple gliding of a plane joint.

Cubometatarsal Joint

The joint surface of the cuboid for the bases of the fourth and fifth metatarsals is flat and relatively small. A large tubercle on the lateral side of the base of the fifth metatarsal projects laterally and posteriorly well beyond the joint. This tubercle is readily palpable in the living and its site indicates on the lateral side of the foot the line of the tarso-metatarsal joints.

Long Plantar Ligament

The strong, important **long plantar ligament** lies on the lateral side of the sole of the foot and extends from the under surface of the calcaneus to the ridge that runs across the cuboid near its front end. It deserves special attention because it helps to maintain the long arches of the foot and because it belongs to several joints. It spans and almost covers the short plantar ligament and it sends forward slips which are attached to the bases of the lateral three metatarsals.

Special Tendons

The tendon of tibialis posterior, a muscle concerned with inversion of the foot, inserts by many slips into most of the bones in the foot after passing beneath the 'spring' ligament and supporting it. Another tendon, that of peroneus longus, crosses the sole from the lateral side to its insertion on the medial edge of the foot. Both tendons contribute to the maintenance of the arches (fig. 192), but only when they are heavily stressed.

Metatarsophalangeal Joints

The anatomical features of these joints correspond, in general, to those of the metacarpophalangeal or knuckle joints; it will be necessary, however, to notice a few important differences.

1. The forward thrust given to the body

by the foot in walking necessitates a position of hyper-extension of the toes; the joint surfaces on the backs of the metatarsal heads are, in consequence, more extensive than those on the backs of the metacarpal heads. This is particularly noticeable in the great toe.

2. It is the great toe that principally imparts the forward thrust to the body, and its metatarsophalangeal joint is equipped with two **sesamoid bones** lying side by side under the metatarsal head; they receive the weight of the body, and so protect the tendon of flexor hallucis longus, which passes below the joint.

3. Because the great toe enjoys none of the free movements so characteristic of the thumb, its plantar plate (cf., palmar plate), in which the sesamoid bones are developed, is united by a **deep transverse ligament of the sole** to its neighbor.

The **interphalangeal joints** of the foot present no noteworthy differences from those of the hand and therefore require no separate description. They too, are pure hinges.

4

Muscular System

VARIETIES OF MUSCLE

The inherent power to move is a property of all living cells but it reaches its highest expression in the muscle cell.

There are three kinds of muscle: (1) voluntary—or striated—or skeletal; (2) involuntary—or unstriated—or smooth; (3) cardiac.

1. **Voluntary muscle** is so named because it is under the control of the will. It is also called **striated muscle** because its cells appear to be striated or striped under the microscope, and many anatomists call it **skeletal muscle** because it is attached chiefly to the skeleton. This type of muscle will be our chief concern in this chapter.

2. **Involuntary muscle** is found everywhere that movement is required without the intervention of the conscious will. This type of muscle chiefly lines hollow organs that must contract automatically—e.g., stomach and intestines, blood vessels, uterus. Because there is no striping of the fibers as in striated muscle, it is also called **smooth muscle** or **unstriated muscle.**

3. Intermediate between (1) and (2) is **cardiac (or heart) muscle** which has many of the characteristics of both. It is not under the control of the will but is striated in appearance (fig. 193).

It should be noted that many of the activities carried on by voluntary muscle become automatic or semi-automatic through constant repetition. This does not constitute involuntary activity since it can be stopped or modified by the will.

Development

The *skeletal muscles* of the trunk develop from the myotomes or myomeres, described on page 13. Those of the limbs develop from the primitive mesenchyme just as the bones and joints of the limb do (fig. 123 on page 76). For a full account, see page 121.

Smooth muscle develops in the walls of hollow organs and vessels from primitive mesodermal cells that are derived from the myotomes (page 121).

Cardiac muscle develops from primitive mesodermal cells which are incorporated very early in the tubular structure that is to become the heart.

STRUCTURE AND ORGANIZATION

Skeletal Muscle

The flesh that clothes the bones is muscle tissue and there are hundreds of muscles of varying importance in the human body. Each is composed of many bundles of individual cells called **muscle fibers**

113

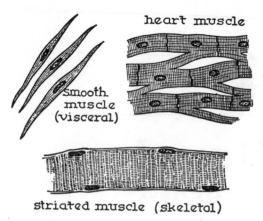

Fig. **193**. Three smooth muscle fibers compared with cardiac (heart) muscle fibers and a short length of a striated muscle fiber (all greatly magnified).

which are so highly elaborate in structure that they are still under intensive investigation by histologists. The specialized protoplasm of the muscle fiber is known as **sarcoplasm.**

A muscle fiber is thread-like, its length being greater than its diameter. An ordinary muscle fiber can be many centimeters long, but its width is about $\frac{1}{10}$ mm. Its ability to produce movement is the result of its power to shorten or contract.

The exceedingly delicate membrane that encloses the sarcoplasm of a muscle fiber is called the **sarcolemma** (sarcos, Gr., = flesh, lemma = sheath) (fig. 194). By its rounded extremities this sarcolemma is attached to the fibrous tissue, called *endomysium* (Gr. = inside muscle), that surrounds the muscle fiber and separates it from its neighbors. A group of muscle fibers is similarly surrounded and separated from other groups by a coarser fibrous tissue envelope called *perimysium* (Gr. = around muscle). The whole muscle, consisting of many groups of fibers, is encased in a still coarser fibrous envelope called *epimysium* (epi, Gr., = upon). This epimysium separates the muscle from its neighbors and, at the same time, permits frictionless movement between them. Finally, groups of muscles are contained in compartments separated from other groups in other compartments by tough fibrous sheets known as **intermuscular septa.** The septa are usually attached to bone and to the invest-

ing deep fascia that surrounds all the muscles like a bandage (fig. 304, page 177). The various fibrous partitions found within and around a muscle act as a framework for its support, and are necessary because muscle tissue is semi-fluid in life.

Motor Unit (fig. 195). Although sarcoplasm is a highly specialized protoplasm able to contract in response to a mechanical, electrical, or chemical stimulus, it normally acts only in response to stimuli from the nervous system. Each muscle is provided with a motor nerve which, on entering the muscle, divides into many branches, bringing all parts of the muscle under its control.

Although the structural unit of striated muscle is the muscle fiber, the functional unit is the **motor unit.** A motor unit begins with a nerve cell (*neuron*) whose *body* is in the central nervous system. The nerve cell has a long fine *fiber* or *process* which runs with hundreds of others in a motor nerve, enters the muscle, and divides into terminal branches varying in number from a few to as many as 2000. Each terminal branch of a nerve fiber ends on a muscle fiber. Thus, the motor unit includes the nerve cell (body and process) and the one-half dozen to many dozen muscle fibers it supplies.

The terminal branches of the nerve fibers each end in a special mechanism known as a **motor end-plate** which has been compared to an electrode. It rests on the muscle fiber; when a nerve impulse reaches it, the end-plate transmits the impulse to the sarcoplasm of the muscle

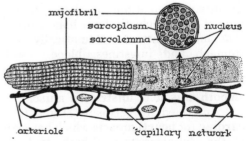

Fig. **194**. A muscle fiber contains many myofibrils bathed in sarcoplasm. Upper figure is a cross-section of the fiber at point indicated by the broken line. Many blood vessels lie between muscle fibers.

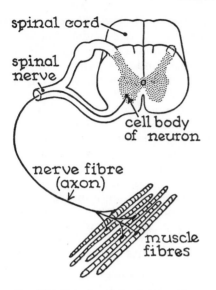

spinal cord

spinal nerve

cell body of neuron

nerve fibre (axon)

muscle fibres

Fig. **195**. Scheme of one motor unit

fiber which gives a brief twitch-like contraction.

All the muscle fibers in a motor unit respond as a unit to an *impulse* or message coming down the nerve fiber and the motor unit as a whole gives only a brief twitch—and then relaxes. The motor units of striated muscle always act in this way (up to 50 times in one second). How then do we get a steady powerful contraction?—by many scattered motor units in a muscle contracting repeatedly and asynchronously (i.e., out of rhythm).

When a muscle fiber contracts, it shortens to two-thirds to one-half its resting length (fig. 196). It follows that the range of movement depends on the length of the muscle fibers or bundles of muscle fibers and that the force exerted depends on the number of motor units thrown into contraction. The mechanical advantage (or handicap) of a given muscle, resulting from its particular situation in respect to the joint it moves, is an important factor influencing range and force. Still other factors such as blood supply, amount of exercise, and temperature affect the responses of muscle.

Smooth Muscle and Cardiac Muscle

The fibers of smooth muscle are smaller and more delicate than those of striated

muscle (fig. 193). They are not attached to bone but are found in greatest amount as the muscular walls of hollow tubes such as the digestive tract, the urinary tract, the blood vessels, and the like. Around a tube they are usually arranged in two sets, an inner one circularly disposed, and an outer one longitudinally disposed. The circular musculature constricts the tube; the longitudinal shortens the tube and tends to widen its lumen. In the digestive tube the co-operative effort of the circular and longitudinal musculature propels the contents along the tube, and the waves of contraction produced by the muscles are spoken of as **peristalsis.**

Heart or *cardiac muscle* is distinctive in at least two respects:

1. Its fibers run together to form a continuous, branching network. Heart muscle, therefore, can and does contract *en masse.*

2. Its fibers are striated but involuntary.

The structural differences exhibited by the three types of muscles reflect functional differences. The contractions of smooth muscle are slow and rhythmic; those of striated muscle are rapid and intermittent; cardiac muscle, contracting rhythmically 60 to 80 times a minute, occupies, functionally, an intermediate position.

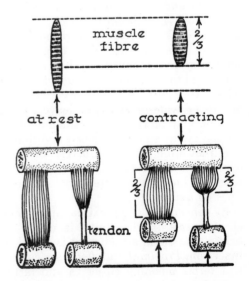

muscle fibre
$\frac{2}{3}$

at rest   contracting

$\frac{2}{3}$ $\frac{2}{3}$

tendon

Fig. **196**. Axiom: the distance that a muscle moves a bone is determined by the length of its fleshy part.

There are even minor differences among individual members of the striated groups. Everyone is familiar with the fact that the muscles of the breast of a chicken are white; those of the lower limb are a reddish-brown. Most human muscles are a mixture of 'white' and 'red.' Indeed, it has been shown that some of the muscles that were traditionally considered 'white' or 'red' are indistinguishable when examined carefully. Further, changing the nerve supply of a muscle or cutting its tendon experimentally can convert red to white and *vice versa*.

Denervation. When a muscle fiber is deprived of its stimulation, it dwindles away (a process known as atrophy) and will eventually die. However, death of the muscle fiber is not as early as it was formerly thought to be. If its nerve supply is restored a year (or even more) later, the muscle will recover its function. It is for this reason that various physical and electrical techniques are employed in cases of paralysis where there is hope for a recovery of the motor nerve.

Attachments of Skeletal Muscle

Most voluntary muscles are attached to bone, but it is not by means of the highly specialized muscle fiber itself that this is accomplished. When a muscle is said to 'arise by fleshy fibers' from a bone the statement is a little misleading. What is meant is either that the delicate sarcolemmae of the muscle fibers are attached to the bone or that as the muscle fibers approach the bone they are gradually replaced by strands of tough fibrous tissue. If, at one end or at both ends of a muscle, the fibrous constituents gather together until a dense closely packed bundle or sheet of fibers results, they constitute a **tendon** or an **aponeurosis;** then the muscle is said to have a fibrous attachment to bone. Attachments 'by fleshy fibers' are apt to be widespread; they leave little if any marking on the bone. Attachments by tendon are at a circumscribed and limited site; they often produce tubercles or tuberosities on a bone. Attachments by aponeuroses are linear; they often produce raised lines on a bone.

By convention, the proximal attachment of a muscle in the limbs is called the **origin,** the distal attachment is called the **insertion.** *Proximal* and *distal* are convenient terms meaning closer to and farther from the axial skeleton respectively. Some persons speak erroneously of the origin as the 'fixed end' and the insertion as the 'moving end.' A consideration of the fact that one end may be moving in some activities and the other end in others makes definition of the 'moving end' impossible. Generally, the origin of a muscle is much more widespread than its insertion which is usually tendinous and near the joint that is moved by the muscle.

Designation of one or the other end of the intrinsic muscles of the head, the neck, and the trunk as origin or insertion is neither easy nor always profitable. Terms such as upper end,' 'medial end,' 'pelvic attachment,' etc., are often preferable.

MECHANICS OF MUSCLE ACTION

A muscle produces movements at joints, and, as a general rule, it can act only on those joints that lie between its origin and its insertion. This is not universally true, but in considering the actions of any muscle, the beginner does well to ask himself first of all what joint or joints the muscle crosses and can influence. Again, in order to avoid confusion, it is well to assume the subjects to be in the anatomical position and to determine what change from this position the muscle effects; it may then be, and often is, desirable to discuss the actions of the muscle in terms of some other assumed position but this should be a secondary consideration.

After the actions of a muscle have been described it is, of course, legitimate to remark on its ability to secure and ensure the integrity of a joint, but let first things come first.

The ability of a muscle to produce move-

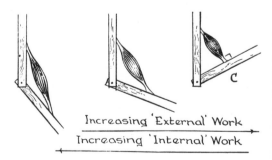

Fig. **197**. Power exerted by a muscle depends in part on its angle of pull.

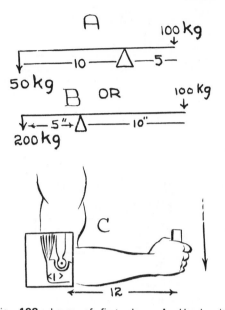

Fig. **198**. Lever of first class. A. Used with mechanical advantage. B and C. Used with mechanical disadvantage. C. Muscle must exert 12 kg force to balance 1 kg resistance.

ment and at the same time to give security to a joint means that the muscle performs *external* and *internal work*. This is an important matter and will be discussed briefly in terms of the mechanics of muscle action.

The muscle that crosses the bend of the elbow from humerus to ulna is known as the Brachialis. When the upper limb is in the anatomical position, the joint is extended, and the line of pull of the Brachialis is almost parallel to the ulna. When the muscle contracts, its power to bend the elbow is at first very limited, most of its energy being expended as it tries, in vain, to pull the ulna straight up. As the elbow bends, the angle of pull of the Brachialis increases and the muscle becomes increasingly efficient. An optimum is reached when the two bones are at rather less than a right angle to one another (fig. 197).

The *optimum angle of pull* for any muscle is a right angle. As the angle of pull departs from a right angle, internal work (holding the joint together) begins to be performed and, to that extent, the power to do external work (bend the elbow) is lost. To the physicist, the whole matter is a

question of the *resolution of forces* and to those who are familiar with that principle the foregoing discussion is unnecessary.

Levers

A voluntary muscle very frequently takes part in a lever system. It is the *force* applied, the bone being the rigid *lever arm* and the joint the *fulcrum*. There are three classes of levers.

1. In a lever of the **first class,** the force is applied at one end of the lever arm, the weight to be moved is at the other, and the fulcrum is at some point intermediate between the two (fig. 198). It is exemplified by a child's see-saw or a balance. Many muscles take part in levers of this class and two features about them are worth noticing: (1) they can be levers of stability; (2) they are often levers of speed. If the force applied is near the fulcrum and the weight to be moved is far from the fulcrum, the speed at which the weight is moved can be very great, but the force applied must be great too. Power is lost but speed is gained and that is often a desirable objective.

2. In a lever of the **second class,** the force is applied at one end of the lever arm, the fulcrum is at the other, and the weight to be moved is at an intermediate point (fig. 199). It is exemplified by a wheelbarrow. Levers of this class always work at a mechanical advantage, in so far as the force required to lift a given weight is always less than would be required were no lever employed at all. Few examples of this class of lever are to be found in the body, perhaps the best example being the one in which muscles are used to open the mouth against resistance. The fact that muscle insertions are so frequently found close to the joints they move precludes many muscles from acting in a lever system of the second class.

3. In a lever of the **third class,** the force is applied at a site intermediate between the fulcrum and the weight to be moved (fig. 200). The muscle involved always operates at a mechanical disadvantage so far as the amount of force required to move a given weight is concerned, but the lever is always one of speed. The class is the

commonest found in the body since it permits a muscle to be inserted close to the joint it moves, and to produce an extensive, rapid, movement with little shortening of muscle fibers.

Gravity

No account of the conditions under which muscles work would be complete were one to ignore the influence of gravity. Some muscles are aided by it, others have to contend against it. For example, when the flexed elbow is gradually extended, it is gravity that operates to control the movement by 'paying out slack' in exactly the same way that a rope does when a bucket is lowered down a well. Only when powerful extension is required, is a muscle called upon to assist gravity. When, on the other hand, the body rises up from a sitting position, muscles are called upon to work against gravity, and it is at least partly due to this fact that the muscles at the back of the hip, at the front of the thigh, and at the back of the leg, are massive and powerful. *Much misunderstanding would be avoided if the influence of gravity in replacing,*

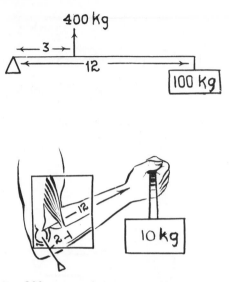

Fig. **200**. Lever of third class. Muscle in lower figure must exert a force of 60 kg to sustain a weight of 10 kg.

aiding, or hindering muscles were constantly borne in mind.

Architecture

Some muscles have most of their fibers running parallel to each other along the long axis of the muscle.* Such a muscle is capable of shortening by about one-third to one-half of the length of its fleshy part because each muscle fiber contracts by this much. The longer the fleshy part of the muscle the greater is the range of its movements.

The total force that a muscle is capable of exerting has been seen to depend on the number of fibers that it possesses. Some muscles sacrifice great range of motion to gain greater power by possessing a special architecture. Short but numerous fibers may run obliquely into a long tendon giving the appearance of a feather (fig. 201). Therefore, muscles possessing this architecture are said to be *pennate* (like a quill). Various types of pennate muscles

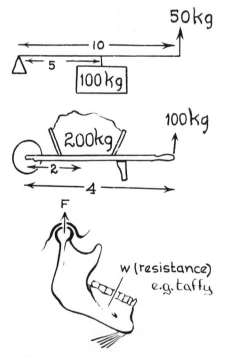

Fig. **199**. Lever of second class (rare in body)

----

*Muscle fibers seldom run the whole length of a muscle because they are only 0.1 to 4.0 cm long. They form bundles of overlapping relays in muscles with long fleshy bellies; more often, however, they form a pennate arrangement which is not obvious on superficial examination.

—unipennate, bipennate, circumpennate, and multipennate—will be met with later.

Actions

The brain knows nothing of individual muscle actions; it is concerned with movements, not muscles. It initiates pictured movements such as bending the elbow, clenching the fist, crossing the knees, or raising the arm. We should therefore choose to analyze a movement rather than to discuss a muscle action, although the latter is the more usual procedure. This is a matter of more than academic importance, for the role of a muscle varies according to the movement in which it participates, and it is desirable that these roles be examined now.

When a muscle is the principle agent in producing a desired movement, it is said to be the **prime mover.** However, muscles often contract to eliminate some undesired movement that would otherwise be produced by the prime mover; then a muscle is said to be the **synergist.** For example, when the fist is clenched, the long flexors of the fingers contract and, if unopposed, would flex the wrist also but, since it is difficult to clench the fist firmly when the wrist is flexed, the extensors of the wrist contract at the same time and so eliminate this undesired wrist flexion by maintaining

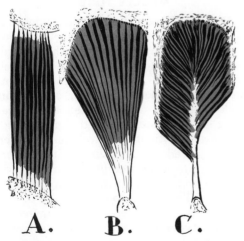

Fig. **201**. Three types of architecture in muscles. A. Parallel fibers end to end. B. Nearly parallel fibers—fan-shaped. C. Pennate (fibers converge on a central tendon).

the joint in a neutral or even an extended position; these wrist extensors act synergically, therefore, with the finger flexors.

The term **antagonist** has been widely misunderstood to mean that prime movers are opposed by other muscles in their normal functions; *they are not.* Only in conditions of spasticity, nervousness, and in performing unskilled movements are antagonist functions of muscles apparent—and they are of dubious value. However, everyone is familiar with the fact that, when some feat requiring considerable muscular effort is to be performed, the body 'gets set' or takes a suitable position to do the work efficiently. A great many muscles that have nothing to do with the actual performance of the specific feat come into play to 'fix' the position of the body as a whole; these are known as **fixators or postural** muscles. In most, if not all, group movements, fixators play their part. When, for example, the elbow is to be flexed, it becomes necessary to fix the shoulder joint in a suitable position, in order to steady the whole elbow region; the shoulder muscles, thus contributing to the efficient working of the elbow flexors, act as fixators.

The performance of even the simplest pictured movement demands the co-operation of a number of muscles. This co-operative effort is spoken of as **co-ordination,** and one portion of the brain (the cerebellum) is largely devoted to it. In the regulation of the activities of prime movers, antagonists, synergists, and fixators, the 'timing' of their activities is vitally important. Muscles get organized to act at a given time in an orderly movement and at no other. They are like players on a stage; they 'have their entrances,' play their parts, and make their exits. When one becomes paralyzed, it disorganizes such a sequence and since, unlike an actor, it has no immediate 'understudy' the ability of other muscles to play their part is seriously impaired. Failure to recognize this timing factor results in the misinterpretation of a relatively extensive disability observed when even one muscle taking part in a sequence is paralyzed (see under Supraspinatus, page 148).

In spite of the foregoing general remarks, we have found that many actions can be and are carried out by isolated actions of single muscles. Indeed, our recent studies show that well-controlled voluntary contractions of single motor units are possible provided that artificial feedback techniques are available.

When a muscle crosses and can influence two joints the chief action of the muscle is to be sought at the more distal of the two joints.

Function. Most muscles are noteworthy for the motion they impart to joints, i.e., their **actions.** The can be stated simply by such expressions as: flexion, extension, abduction, adduction, medial rotation, and lateral rotation. Many muscles have additional important uses or **functions.** A good example is synergistic contractions of the muscles of the wrist when the muscles of the fingers contract. Here there is no actual movement at the wrist, yet the muscles of the wrist are performing a useful function. Another example is when you walk down stairs: the larger calf muscles that control the tendon of Achilles function against gravity and 'let out' the heel in a controlled manner. Wherever they are important such functions will be noted in addition to the actions of a muscle.

Bursae and Sheaths

Devices to eliminate friction are provided wherever a muscle or tendon, in the performance of its movements, is liable to rub on another muscle, tendon, or bone. Such a device is called a bursa (L. = purse) because it is a collapsed bag with cellophane-thin walls whose inner surface is extremely smooth, moist, and slippery (fig. 202). Because the inner surfaces rub together, synovial fluid—just enough for adequate lubrication—is produced by the walls of the bursa.

Sometimes the synovial capsule of a joint can act as a bursa for a muscle in its immediate neighborhood, and sometimes a bursa near a synovial capsule communicates with the cavity of the joint by means of an opening established secondarily.

Where a tendon runs in a tunnel and is subject to friction on all sides, as in the

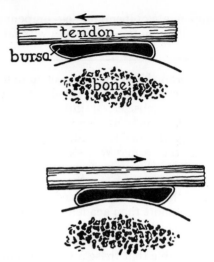

Fig. **202**. A bursa forms wherever a tendon rubs back and forth on a hard structure. Note: normally the bursa is collapsed.

hand and foot, the bursa surrounds the tendon and is known as a tubular bursa or **synovial sheath** (fig. 203). The outer wall of a delicate, elongated, and double-walled synovial sheath lines the tunnel; the inner wall envelops, and is attached to, the tendon. At the entrance to, and exit from, the tunnel the two walls are continuous, so that movements of the contained tendon occur with a frictionless gliding of the inner wall of the synovial sheath on the outer.

MUSCLE TONE

Older textbooks often contain a description of a hypothetical condition of constant activity in a muscle at rest which is referred to as 'tone.' Many unwarranted inferences were drawn from the brilliant work of Sir Charles Sherrington and his colleagues who demonstrated that a cat with its brain stem severed showed automatic activity in its postural muscles when it was placed in a standing position. Tone is the *state of excitability of the nervous system controlling or influencing skeletal muscle.* There are some diseases of the nervous system where tone is increased (spasticity) and others where it is decreased (flaccidity).

Electromyography gives proof that 'tone' defined as 'normal, constant neuromuscular activity at rest' does *not* exist in human

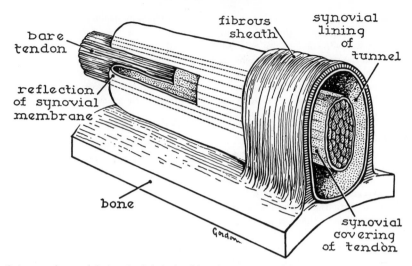

Fig. **203**. Scheme of synovial sheath. A 'window' has been removed in redundant part of sheath (to left) to show how synovial membrane is reflected from tendon to tunnel. Note: normally, tendon fits snugly in its tunnel.

beings. It is true that some persons have difficulty in relaxing but, with care, all skeletal muscles can be completely relaxed.

The difference in the 'feel' of a paralyzed muscle compared with a normal relaxed muscle can be explained by various other changes which take place when a muscle becomes paralyzed. Ironically, muscle fibers that have lost their nerve supply do engage in random invisible contractions at rest—only detectable electromyographically—whereas a resting normal muscle shows no activity whatever.

DEVELOPMENT OF MUSCLES

Myomeres are the blocks of muscle tissue that lie on each side of the developing vertebral column (fig. 16).

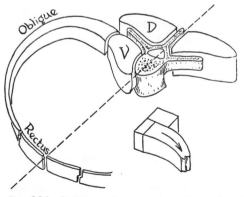

Fig. **204**. Splitting of a myomere. D. = dorsal (posterior); V. = ventral (anterior).

They are the sources of the muscles of the trunk, of the eyes, and of the tongue. Myomeres themselves arise out of a primitive tissue known as mesenchyme. Mesenchyme is a ubiquitous embryonic tissue of high potentiality and is the source of much of the musculature of the body. It is the origin of the (branchiomeric) muscles of the gill or branchial arches and of the muscles of the limbs and their girdles. It gives rise to the smooth muscles of the primitive digestive, respiratory, and genito-urinary systems, and it gives rise to cardiac muscle and the smooth muscle found in the walls of blood vessels.

The muscle derivatives of the myomeres, of the branchiomeres, and of the mesenchyme of the limb buds will be briefly considered in turn.

Myomeric Muscles

When a myomere splits into a dorsal (posterior) and a ventral (anterior) portion, the dorsal portion remains associated with the back of the spinal column (fig. 204). It fuses with similar portions of adjacent myomeres, and so forms the longitudinal musculature of the back. In the adult, this musculature occupies the territory from the back of the sacrum below, to the back of the head above, and from the angles of the ribs on one side to the angles on the other side. It is covered in by a sheet of fascia, the *thoracolumbar fascia*, and this is later overlaid by muscles of the upper limb girdle that have migrated to the vertebral spines.

The ventral portion of the myomere divides into medial and lateral parts. The medial part remains associated with the front of the spinal column where its fate is similar to that of the dorsal portion. The resulting musculature is not an extensive group and is found only in the cervical and lumbar regions.

The lateral part of the ventral portion undertakes the task of providing the muscular walls of the great body cavities. It is known as the oblique musculature and it migrates around the trunk toward the front; in the thorax, it is represented by the intercostal muscles; in the abdomen, by the flat and extensive muscles that result from the fusion of adjacent segments and that constitute the laminated muscles of the abdominal wall. Its representation in the cervical region will be noted in a moment.

As the migrating elements of one side approach those of the other near the midline of the body in front, they do not fuse across the midline, but the 'end' or front of each segment breaks away or splits off and unites with similar portions above and below it to form a single strap-like muscle that lies adjacent to the midline side by side with its fellow of the opposite side. Thus, a series of muscles known as the 'rectus' group comes into being and extends from the chin above to the symphysis pubis below, except that it is absent in the thoracic region where the presence of a rigid bone, the sternum, would preclude it from having any useful function.

The oblique musculature in the neck is represented by a series of three muscles on each side descending from the cervical transverse process to the first and second ribs; the three muscles are called the Scalenes and, together with the cervical portion of the 'rectus' group, they form the (incomplete) walls of a 'cervical cavity' for the housing of the upper parts of the digestive and respiratory systems.

Branchiomeric Muscles

Associated with the cartilaginous 'bows' that are the framework of each branchial arch (see page 43) is a mass of mesenchyme out of which muscles develop. For the most part, their bony attachments reveal their branchiomeric origins although some wander farther afield. Thus the **masticatory muscles** (first arch) are attached to the mandible; the **muscles of expression** (second arch) originally arose from the styloid process but have, for the most part, spread superficially over the face and neck; the principal **muscles of the pharynx, palate, and larynx** (third, fourth, and fifth arches) still have their origins from the hyoid bone and the cartilages of the larynx.

Muscles of Limbs

As the skeleton of the limbs begins to take form, the mesenchyme lying in front of, and behind, the developing bones begins to be converted into muscle tissue. Out of this, by a process of longitudinal cleavage in two planes, individual muscles come into existence arranged in layers and with several muscles in each layer. The value of knowing that muscles are disposed in layers lies in the fact that important nerves and blood vessels are commonly found coursing in the planes thus provided. The mesenchyme at the root of the limb is responsible for the 'girdle' muscles which, in the case of the upper limb especially, migrate toward the axial skeleton. This migration is so extensive that widespread muscles of the upper limb girdle cover the native muscles of the back from head to sacrum, while others cover a large part of the front and sides of the chest.

Classification

The classification of voluntary muscles can be summarized simply as follows.

A. Axial Skeleton
    (1) Vertebral column
    (2) Head and neck
    (3) Thorax
    (4) Abdomen
B. Limbs
    (1) Upper limb
    (2) Lower limb

Each group will be considered in turn. A muscle whose name appears in **bold face** type is important; one whose name appears in *italics* is of lesser importance and the beginner, if he chooses, may safely neglect it.

Aids to Learning

Throughout the section on the muscular system, the student will find various aids to learning and review.

1. Special tables that group together the muscles which produce specific movements of certain joints. These tables are not for memorization; rather, they are for reference and guidance. As an exercise, the student should make such tables himself and use those in the book for corrections.

2. Four large drawings–figures 292 to 295 (pages 170–171)–show the areas of origin and insertion on the bones of the muscles of the limbs. The student should continually refer to these figures.

3. A series of drawings of dissections with the muscles labeled.

4. A series of photographs of living subjects—Plates 1 to 4, preceding page 19. These, too, should be referred to constantly and used as a guide to the observation and palpation of muscles and tendons at rest and during activity.

Advanced students of muscle function are referred also to *Muscles Alive: Their Functions Revealed by Electromyography*, Williams and Wilkins Co., third edition, 1974.

It will be our practice to name the nerve supply when a muscle or group of muscles is described, but to defer an account of the nerves until the nervous system itself is discussed.

Naming of Muscles

Muscles are named according to their various characteristics: shape, size, location, or actions. The full name of any muscle begins with 'Musculus,' always a masculine noun, followed by a masculine adjective and often, in addition, the genitive of a noun referring to a part of the body, e.g., *Musculus Latissimus Dorsi*. [Latissimus, an adjective, means 'broadest'; Dorsi is the genitive of dorsum (which means 'back') and therefore means 'of the back.' The full name means: the broadest muscle of the back.]

Almost always the word 'Musculus' is omitted leaving the rest of the name to stand by itself. For example, Musculus Trapezius and Musculus Brachialis are written simply *Trapezius* and *Brachialis*.

Most Latin names and adjectives that are masculine end in *-us* and to make them plural the *-us* is changed to *-i*. However, as seen in *Brachialis*, above, some adjectives, although masculine, do not end in *-us* and of these students should beware. (To pluralize *-is* change to *-es*—Brachiales.) Another common adjectival ending is *-or* (e.g., Flexor), the plural being *-ores* (although, in the anglicized form, *-ors* is acceptable, e.g., 'the flexors of the elbow').

In textbooks, the names of muscles are given initial capitals, and although this convention is ignored in more advanced works, the student is well advised to observe it.

MUSCLES OF AXIAL SKELETON

Muscles of the Vertebral Column

A. Dorsal or Posterior

When the muscles of the upper limb girdle (fig. 205) are removed from the back, the underlying intrinsic muscles are seen to occupy a pair of broad gutters situated one on each side of the vertebral spines and extending laterally as far as the angles of the ribs, the transverse processes of the cervical vertebrae, and the mastoid process. The territory is limited above by the under, horizontal surface of the occipital bone (at the back of the head); below, it is limited by the back of the sacrum and by the posterior spines of the iliac crests which project somewhat behind the sacrum. The muscles are covered behind by a tough sheet of fascia co-extensive with them, and when this is removed the paired muscular columns are exposed (see figs. 205–209).

Here, there are scores of muscles or muscle bundles to confuse the beginner, but they can be organized into more or less definite groups. First of all, the mass of

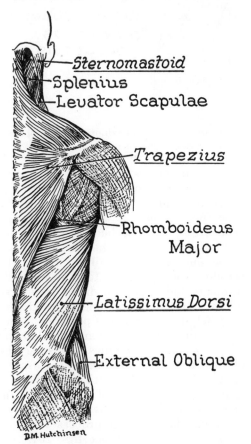

Fig. **205**. These large upper limb muscles (extrinsic muscles of back) cover intrinsic muscles of back.

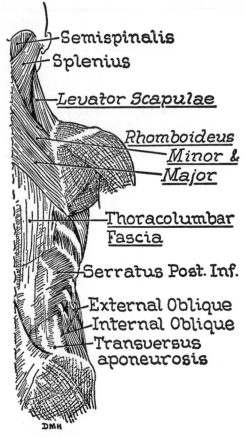

Fig. **206**. Next layer of extrinsic muscles of back. Intrinsic muscles partly revealed.

intrinsic muscles is divided into a superficial and a deep group.

Erector Spinae is the name for the **superficial group.** It arises by a strong aponeurosis from the back of the sacrum and adjacent parts of the iliac crest as a single muscle. As the fibers mount the vertebral column they extend over several segments and in general divide into three columns: (1) a lateral one inserted into (and helping to form) the posterior angles of the ribs and, after many overlapping relays of muscle bundles, reaching the cervical transverse processes—*Iliocostalis;* (2) a middle column inserting into thoracic transverse processes, and after similar relays on cervical transverse processes, reaching the mastoid process—*Longissimus* (L. = longest) (see figs. 207, 208, 210). (3) Confined to the thoracic region is a small column lying medial to the above-

named ones and alongside the tips of the spinous processes to which it attaches—*Spinalis.*

The **deep group,** collectively named the **Transversospinalis,** lies deep to Longissimus along most of the length of the vertebral column. In reality, it consists of a multitude of small muscles in (two or) three layers, whose fibers all run obliquely from the region of transverse processes to the region of spinous processes. Naturally, the muscles of the deepest layer—*Rotatores*—have the shortest span, running from one vertebra to the next. The most superficial layer, **Semispinalis,** is not found below the thoracic region; its bundles have the longest spans (three to six vetebrae), and the upper ones reach the

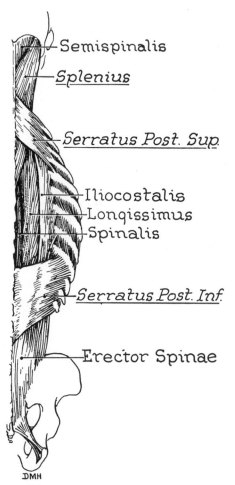

Fig. **207**. Erector Spinae and its immediate relations.

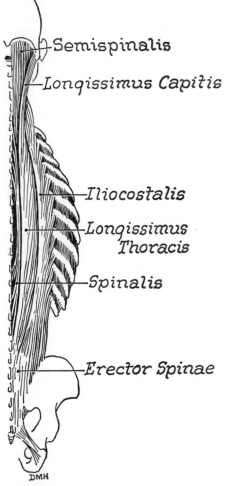

Fig. **208**. Superficial layer of Erector Spinae splitting into its three columns of muscles as it ascends.

the back of the neck where they are separated from their fellows of the opposite side by a midline partition, the **ligamentum nuchae,** already described. The two most important are the **Splenius** (L. = a bandage) (figs. 207, 209) and the **Semispinalis Capitis** (fig. 208). They arise from lower cervical and upper thoracic vertebral arches and are inserted on the skull, the Splenius running obliquely to the mastoid process and adjacent occipital bone, the Semispinalis running vertically to the occipital bone and, with its fellow, covering deeper and smaller muscles now to be noted.

Muscles of Suboccipital Region (fig. 210). On each side, they are:

1. *Obliquus Capitis Inferior (Inferior Oblique),*

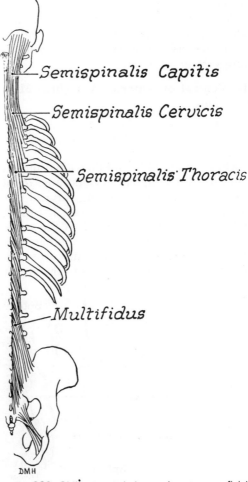

Fig. **209**. Oblique muscle layers deep to superficial layer. (Rotators, the deepest layer, not seen.)

occipital bone. Between these two is a fleshy muscle, **Multifidus,** which begins below from the sacrum (deep to the aponeurosis of Erector Spinae) and, by short relays (two to three vertebrae), it finally reaches the first cervical spinous process (see fig. 209).

Until now we have ignored three superficial muscles each of which partly covers the large mass of muscle described above: (1) **Splenius,** (2) *Serratus Posterior Superior,* and (3) *Serratus Posterior Inferior* (fig. 207). (1) is described below; (2) and (3) are included with the muscles of the thorax where they properly belong (page 137).

Muscles at Back of Neck. Just below the skull, several massive muscles fill in

running almost horizontally lateralward from the spine of the axis to the transverse process of the atlas.

2. *Obliquus Capitis Superior (Superior Oblique)*, running almost horizontally backward (in the sagittal plane) from the transverse process of the atlas to the occipital bone.

3. *Rectus Capitis Posterior Major*, running upward and backward from the spine of the axis to the occipital bone.

4. *Rectus Capitis Posterior Minor*, insignificant, lying deep and medial to 3, and running from the 'spine' (posterior tubercle) of the atlas to the occipital bone.

The first three form a triangle (suboccipital triangle) which is important only for the surgeon.

Actions and Nerve Supply

The actions and functions of all of the muscles of the back are discussed together below.

All of the intrinsic muscles of the back are supplied by posterior branches of the spinal nerves which issue in series from the intervertebral foramina.

B. Ventral or Anterior Vertebral Muscles

The muscles on the fronts of the vertebral bodies are found only in the neck and in the lumbar region.

On each side of the midline a flat muscle known as the **Longus Cervicis** clings to the fronts of the bodies of the cervical and upper three thoracic vertebrae; it runs in relays to the atlas. Lateral to the upper part of the Longus Cervicis another muscle, the **Longus Capitis,** runs from the cervical transverse processes to the occipi-

tal bone in front of the foramen magnum. Uniting the anterior arch of the atlas to the occipital bone directly above, is a pair of short quadrangular muscles on each side, the medial one being the *Rectus Capitis Anterior*, the lateral one, the *Rectus Capitis Lateralis* (fig. 211).

In the lumbar region a powerful muscle, the **Psoas Major** arises from the sides of the lumbar vertebrae and is inserted on the femur. It is a composite muscle much more concerned with femoral movements than with vertebral and, in consequence, its description will be left until the lower limb musculature is dealt with (see page 173).

Nerve Supply. The ventral muscles all are supplied by the anterior branches (rami) of the spinal nerves in their vicinity.

Actions and Functions of Muscles of Back

Actions. It is obvious that the dorsal muscles extend or straighten the column and that their unilateral actions result in bending the column to the same side.

The **Multifidus** and other oblique muscles associated with it are concerned with local movements of the vertebral column, e.g., rotary movements (twisting) of groups of vertebrae. The **Splenius** and **Semispinalis Capitis** extend the head and, if contracting without their fellows of the opposite side, they turn the head, tilting the chin up and to the same side. The ventral muscles of the neck flex both the neck and the head. Their unilateral actions help to turn the chin down and to the other side (see Sternomastoid, page 128).

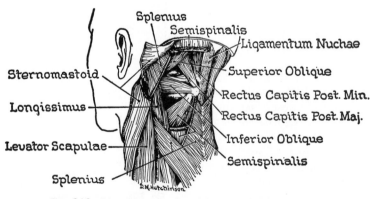

Fig. **210.** Dissection of suboccipital region of the left side

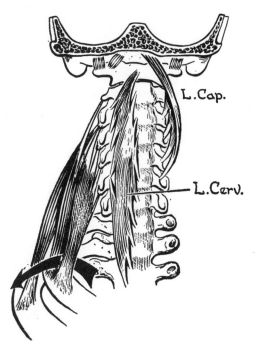

L.Cap.

L.Cerv.

Fig. **211**. Left Longus Capitis; Right Longus Cervicis; and the Right Scalene muscles—Anterior, Medius, and Posterior. Arrow indicates course of subclavian artery between Scalenus Anterior and Medius.

All of the small suboccipital group extend the head, but the Inferior Oblique also turns the face to the same side.

A great many other muscles situated far afield also have important actions on the vertebral column. For example, the whole 'rectus' group flexes the column; unilateral action of the oblique muscles of the abdominal wall bends the trunk to the same side and twists it to the opposite side. In short, almost any muscle one of whose attachments is to the axial skeleton can directly or indirectly influence the vertebral column.

Functions. In walking, the vertical muscle masses on the two sides of the vertebral column contract alternately. Those on the same side as the foot that is leaving the ground are contracting while those on the opposite side may or may not also contract; sometimes the imbalance of activity is the reverse, for no apparent reason.

The *posture of the vertebral column,* including the neck, is regulated by the intrinsic muscles of the back but, contrary to a widely held belief, these muscles are not necessarily all in constant activity. During relaxed standing most of the muscles are quiescent most of the time. Individual groups become active when balance is threatened. When one holds a moderate forward-bending position, the muscles become very active. They relax completely when one bends the back as far forward as possible because the ligaments of the vertebral column (page 84) assume the load. *Muscles are never used where ligaments suffice.*

Muscles of Head and Neck

The intrinsic muscles of the head and neck will be discussed in the following order:

Scalene Muscles
Sternomastoid
Infra-hyoid or 'Strap' Muscles
Muscles of Floor of Mouth
Muscles of Mastication
Facial Muscles
Muscles of Pharynx and Palate
Muscles of Tongue
Muscles of Larynx
Muscles of Eye

For obvious reasons certain muscles of the neck have been considered already with the muscles of the vertebral column (Splenius, Semispinalis, Longissimus, Multifidus, suboccipital group, Longus Capitis, and Longus Cervicis) (see pages 124–126).

Scalene Muscles

There are three Scalenes (Gr. = uneven) on each side, **Scalenus Anterior, Scalenus Medius,** and *Scalenus Posterior* (figs. 183, 184). The **Scalenus Anterior** arises by tendinous slips from the fronts of the transverse processes of the cervical vertebrae (except the highly modified atlas and axis). The slips fuse into a flat muscle which descends obliquely lateralward to be inserted into the upper surface of the first rib.

The **Scalenus Medius,** rather larger than the Anterior, arises from the backs of the same transverse processes plus that of the axis. It then runs parallel to, but

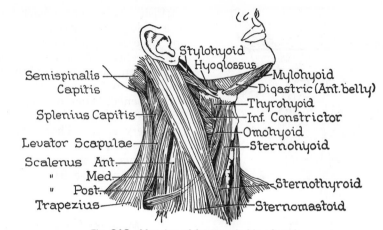

Fig. **212**. Muscles of front and side of neck

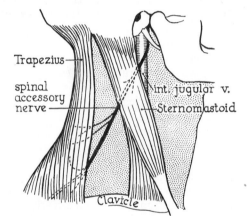

Fig. **213**. (Right) Sternomastoid and some important relations. (Posterior and anterior triangles of neck stippled.)

behind, the Anterior; it also is inserted into the upper surface of the first rib. The *Scalenus Posterior* is but the posterior part of the Medius and its fibers reach the second rib.

The Scalenus Anterior is separated from the Medius: (1) at its origin, by the issuing cervical nerves; (2) at its insertion, by the brachial plexus of nerves and the accompanying subclavian artery crossing the first rib on its way to the upper limb.

Actions and Functions. As regards movements of the neck, the Scalenes pale into insignificance beside the Sternomastoids. Their greatest and most important use is to suspend the thoracic inlet and to maintain its level; they raise the first rib and indirectly the lower ribs during the

inspiratory phase of breathing, being particularly active in forced inspiration.

Nerve Supply. The Scalenes gets twigs from the roots of the plexuses (cervical and brachial) in their vicinity.

Sternomastoid

This large muscle, more properly called **Sternocleidomastoid** because its lower attachment is to the clavicle as well as the sternum, is extremely important and powerful (figs. 212, 213). It runs obliquely down the neck and stands out like a heavy cord when the face is turned to the opposite side. The Sternomastoid lies on the boundary line separating two areas of the neck that have been designated the *posterior triangle* and the *anterior triangle* of the neck. It shelters the great vascular and nerve trunks that course through the neck.

Attachments. The lower attachment of Sternomastoid is to the clavicle and manubrium sterni adjacent to the sternoclavicular joint. The upper attachment is to the mastoid process just behind the ear and to a line passing backward on the skull. Here it meets the origin of Trapezius, an important muscle of the upper limb.

Actions. When right and left Sternomastoids contract simultaneously the head and neck are flexed. This statement can readily be verified by placing the palm of the hand on the forehead in order to resist flexion, when the Sternomastoid can be felt to contract vigorously. Unilateral contraction tilts the chin up and to the

other side—the position assumed in tor-
ticollis (wry-neck), due to shortness or
spasm of one Sternomastoid. The Sternomastoid usually does not initiate the
simple turning of the head to one side; as a
rule it comes into action only during the
later part of that movement.

Nerve Supply. The accessory nerve
pierces and supplies it.

Posterior Triangle of Neck

On the side of the neck is an important
triangular area bounded by the Sternomastoid, the anterior border of Trapezius, and
the clavicle (fig. 212, 213). Filling the
triangle, from below upward, are the Scalene muscles, Levator Scapulae, and
Splenius. Crossing the triangle deep to
fascia is the accessory nerve in its oblique
course downward from the upper part of
the Sternomastoid to the anterior border of
Trapezius. In the lowest part of the triangle, emerging between Scalenus Anterior
and Medius, and resting on the first rib,
are the brachial plexus and the subclavian
artery. It may be said, then, that the
posterior triangle of the neck is the 'root' of
the upper limb. (See also Chapter 16.)

Infra-hyoid or 'Strap' Muscles

Strap-like muscles occupy each side of
the midline of the neck from mandible to
manubrium (figs. 213, 214). Between these
two points lie the hyoid bone and, below it,
the Adam's apple or thyroid cartilage; to
them certain of the muscles gain attachment. The details of their exact attachments are not important and table 1 sufficiently identifies each muscle.

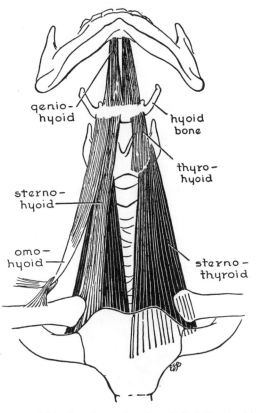

Fig. **214.** 'Strap' muscles of neck. (Left Omohyoid
and Sternohyoid removed as well as muscles of floor
of mouth which normally cover Geniohyoid muscles.)

TABLE 1

Muscle	Origin	Insertion
1. **Geniohyoid**[a]	mandible	hyoid bone
2. **Thyrohyoid**	thyroid cartilage	hyoid bone
3. **Sternothyroid**	back of manubrium	thyroid car-tilage
4. **Sternohyoid**	back of manubrium	hyoid bone
5. *Omohyoid*	{sup. border of scapula {near root of coracoid	hyoid bone

[a] The Geniohyoids have been secondarily taken into the
floor of the mouth by the meeting outside (below) them of the
paired Mylohyoid—the muscular floor of the mouth (see fig.
216).

The hyoid is a U-shaped bone, slung
from the skull by ligament and muscle; it is
strategically situated in the angle where
the front of the neck meets the floor of the
mouth. It gives attachment to the muscle
forming the diaphragm of the mouth
(*Mylohyoid*) and from it the mouth is
forcibly opened by the Mylohyoid and the
Geniohyoid. It is an important anchorage
for the tongue and from it are suspended
the laryngeal cartilages. Because of its
situation between mouth and neck, the
hyoid bone is the most important landmark in the front of the neck.

The muscles protect the organs behind
them (thyroid gland, larynx, trachea and
esophagus) and they stabilize the hyoid
bone and thyroid cartilage. They receive
their nerve supplies from the upper three
cervical nerves.

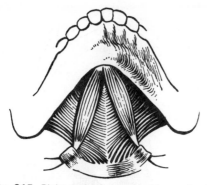

Fig. **215**. Right and left Mylohyoid muscles form floor of mouth. Below them are anterior bellies of right and left Digastrics.

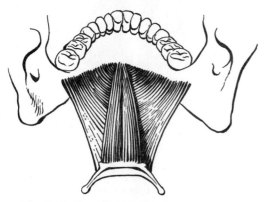

Fig. **216**. Smaller Geniohyoids lie immediately above Mylohyoids (here seen from within mouth).

Muscles of Floor of Mouth

The floor of the mouth is shaped like a shallow trough, and the muscle that forms it is the **Mylohyoid**† (figs. 215, 216). This paired muscle arises from the mylohyoid line on the medial surface of the mandible. The most posterior fibers reach the hyoid bone; farther forward the fibers of the two sides meet and interlace in the midline. An interlacement of fibers is a *raphe* (Gr. = seam).

Actions. When the mandible is immobile, Mylohyoid, Geniohyoid, and the anterior belly of Digastric raise the hyoid bone and *vice versa*.

Nerve Supply—a branch of the mandibular division of the trigeminal nerve (except Geniohyoid which is supplied by the first cervical spinal nerve via the hypoglossal nerve).

† Myle. Gr. = mill, i.e., the lower jaw or grinder.

Digastric. Each Digastric (fig. 217) is a compound muscle consisting of two bellies, an anterior and a posterior, connected by a round intermediate tendon that is fastened to the hyoid bone by a band of fascia. This fascia acts like an inverted sling.

The *posterior belly* arises from a groove on the deep surface of the mastoid process and has running with it a small muscle, the *Stylohyoid*. It is a very important landmark deep to which many important vessels and nerves pass (e.g., carotid arteries, internal jugular vein, cranial nerves VII, X, XI, and XII). Because it develops from facial muscles, it is supplied by the facial (seventh cranial) nerve.

The *anterior belly* develops from, adheres to, has the same actions as, and gets the same nerve supply as the Mylohyoid (see above).

Muscles of Mastication

Four important muscles on each side are devoted to the tasks of biting and chewing. They are named **Temporalis, Masseter, Medial Pterygoid,** and **Lateral Pterygoid.**

The **Temporalis** (figs. 218, 220) is an extensive fan-shaped muscle arising from the side of the cranium and descending to the coronoid process of the mandible where the 'handle of the fan' is inserted. It is a very powerful biting muscle easily seen, when in action, on the side of the head.

The **Masseter** (Gr. = chewer) is a thick quadrangular muscle clothing the outer surface of the ramus of the mandible and

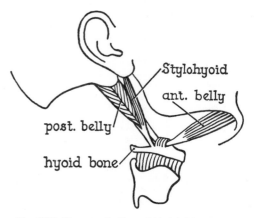

Fig. **217**. The two bellies of (right) Digastric

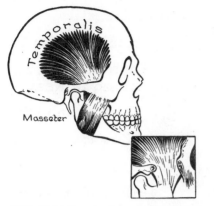

Fig. **218**. Right Temporalis and Masseter. Inset: insertion of Temporalis into coronoid process.

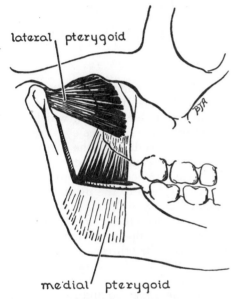

Fig. **219**. The Lateral and Medial Pterygoid muscles both arise from the lateral pterygoid plate.

arising from the zygomatic arch. Its fibers run downward and backward. The exploring finger can trace its outline when the muscle is made to contract by clenching the teeth. The muscle is a powerful biting one. It also protrudes the jaw (figs. 218, 220.)

The **Medial Pterygoid** (fig. 219) lies on the inner side of the ramus of the mandible and has a general shape and direction similar to that of the Masseter. The origin is on the medial surface of the lateral pterygoid plate; its insertion, the ramus in the region of the angle. It has actions

similar to those of the Masseter, but, in addition, it draws the jaw toward the opposite side in grinding movements.

The **Lateral Pterygoid** has an altogether different direction from the other three. It lies, in general, in a horizontal plane which runs backward and lateralward. It arises from the lateral surface of the lateral pterygoid plate (which, therefore, is a 'muscular' plate separating the two pterygoids) and is inserted on the neck of the mandible in close association with the mandibular joint; some of its fibers are inserted on the articular disc. It pulls the condyle and disc forward and so protrudes the jaw; it alternates with its fellow of the opposite side to produce side to side (chewing and grinding) movements.

The **nerve supply** is by the mandibular division of the trigeminal or fifth cranial nerve which division is the nerve of the first pharyngeal arch.

Movements of Mandible. Temporalis, Masseter, and Medial Pterygoid produce the movement of biting. The jaw is protruded by Lateral Pterygoid (with help from Masseter and Medial Pterygoid) and is retracted by the posterior fibers of Temporalis.

In chewing movements, the lower jaw is moved to the right by the left Medial and Lateral Pterygoids (and right Masseter); and to the opposite side by their counterparts. Gravity is sufficient to depress the jaw, but if there is any resistance to overcome, Mylohyoid and Digastric (page 130) come into action.

Facial Muscles of Expression

These are pale and often scattered bundles of fibers having their principal insertions into the deeper layers of the skin of the face. Primarily, they surround the orifices—the eye, ear, nose, and mouth—and are sphincters and dilators. Secondarily, they are the mimetic muscles of the face and scalp capable of expressing a variety of emotions. It will be sufficient to notice some of their more important actions and to name a few of them (figs. 220, 221).

If the forefinger of one hand pulls the outer angle of the eyelid laterally, the

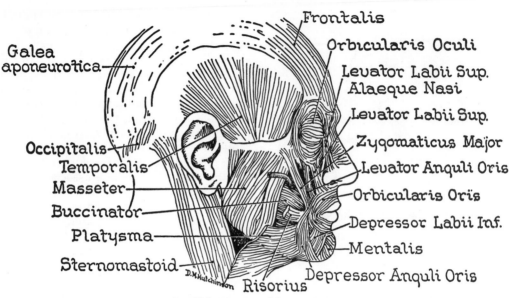

Fig. **220**. Muscles of face and jaw

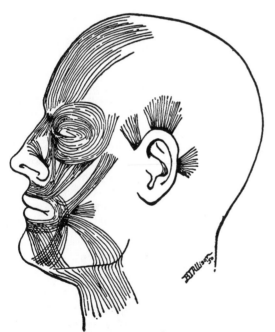

Fig. **221**. Facial muscles of expression control the orifices (see text).

forefinger of the other hand can readily feel a short cord-like ligament at the medial angle of the eye. This is the **medial palpebral ligament** and from it and the adjacent medial orbital margin arise muscle fibers whose loops sweep in concentric circles widely around the orbital margin and in the eyelids. It is the sphincter of the eye, the **Orbicularis Oculi,** which by its contraction 'screws up' the eye, draws the lids medially and gives the eye efficient protection. Its contraction encourages the flow of tears by helping to empty the tear sac situated deep to the medial palpebral ligament (see fig. 619).

An encircling muscle, the **Orbicularis Oris,** lies in the lips, surrounds the mouth and constricts its orifice.

Figures 220 and 221 show well the muscles which run:

1. into the upper lip and raise it;
2. into the lower lip and depress it;
3. into the angle of the mouth;
   a. from above ('smile'),
   b. from the side ('grin'),
   c. from below ('grimace').

A special muscle—**Buccinator** (L. = trumpeter)—lines the cheek and helps to move food in the mouth to between the grinding teeth (fig. 222).

The nostrils can be dilated or compressed although the muscles devoted to these functions are small and insignificant. The external ear also possesses muscles but movements of that organ in man are too insignificant to have any functional value.

An extensive sheet of vertical muscle

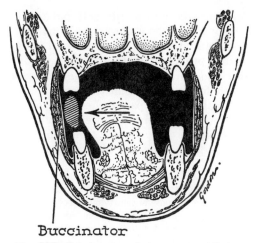

Buccinator

Fig. **222**. Buccinator and the tongue shift lump of food back and forth between molar teeth.

fibers lies immediately beneath the skin and extends from in front of the upper two ribs obliquely upward into the neck. The anterior fibers of this muscle, the **Platysma,** reach, and are inserted on, the margin of the mandible; the posterior fibers cross the mandible and blend with the other facial muscles. It is used to tense the skin of the neck (fig. 220).

The scalp is equipped with a musculo-aponeurotic helmet. The muscle fibers of the **Occipitalis,** arising from the occipital bone at the back of the head, give place to a wide aponeurotic sheet over the vault of the skull; in the forehead, muscle fibers once more appear and are known as the **Frontalis** which, on contraction, elevates the eyebrows and expresses surprise. The fibers of the Frontalis intermingle and blend with those of the Orbicularis Oculi (fig. 220).

Attached to the styloid and to the nearby mastoid process are the *Stylohyoid* and the *posterior belly of the Digastric*, respectively (fig. 223). They run, sheltered by the angle of the jaw, to the hyoid bone and, while their developmental origin and nerve supply assign them to the muscles of expression, they are concerned with suspending and elevating the hyoid bone.

A tiny muscle of this group, the *Stapedius*, is attached to one of the ossicles of the ear and will be referred to with that organ.

The **nerve supply** of all of the foregoing muscles of expression, including Buccina-

tor, is the seventh cranial or facial nerve which is the nerve of the second pharyngeal arch.

Muscles of Pharynx and Palate

The muscles of the pharynx are striated yet it is a familiar observation that once food has reached the pharynx from the mouth no effort of the will can prevent it from being swallowed.

In the wall of the pharynx, the striated muscles are arranged in two layers, an (outer) circular and an (inner) longitudinal. The circular musculature, greater in amount than the longitudinal, consists of three muscles on each side known as the **Superior,** the **Middle,** and the **Inferior Constrictor.** The longitudinal musculature of the pharynx concerns itself also with the soft palate whose muscles will accordingly be discussed with those of the pharynx.

Superior Constrictor (fig. 224). Running from a point just behind the last upper molar tooth to a point just behind the last lower molar tooth, is a delicate 'ligament,' the **pterygomandibular raphe,** which gives origin to two muscles, one running forward in the cheek, the Buccinator, the other running backward in the pharynx, the Superior Constrictor.

The Superior Constrictor spreads like a fan to embrace its own half of the pharyngeal wall. At the back of the pharynx it meets its fellow of the opposite side in a long midline raphe whose upper end

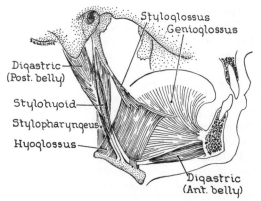

Fig. **223**. Muscles of right styloid process and some related muscles.

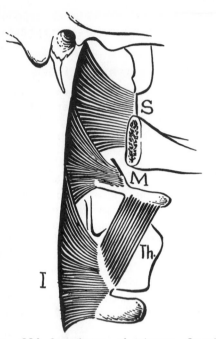

Fig. **224**. Constrictors of pharynx—Superior (S), Middle (M) and Inferior (I) seen on right side; also right Thyrohyoid (Th).

reaches the skull in front of the foramen magnum (a *pharyngeal tubercle* there marks the site). The lower half of the muscle is overlapped behind by the Middle Constrictor.

Middle Constrictor. From its origin on the hyoid bone (fig. 224), the muscle sweeps round the pharynx, fans out and meets its fellow in similar fashion to the Superior Constrictor. The lower half is overlapped behind by the Inferior Constrictor.

Inferior Constrictor. Arising from an oblique line on the thyroid cartilage and from the cricoid cartilage just below, the Inferior Constrictor behaves as do the Superior and the Middle.

Gaps. The Constrictors overlap and provide a complete muscular wall for the pharynx behind, but gaps permit the passage of the auditory tube and *Levator and Tensor Palati*.

Palatal Muscles. These arise from the base of the skull just lateral to the cranial attachment of the pharynx, and descend to the palate. Between them lies the **auditory tube** which, in effect, is the first gill cleft. The *Levator Palati* descends directly to,

and spreads out in, the soft palate. The *Tensor Palati* hooks around a bony pulley (the hamulus) at the lower end of the medial pterygoid plate and then becomes an aponeurosis (*palatine aponeurosis*) which forms the fascial framework of the soft palate. The paired Levators elevate and the paired Tensors tense the soft palate to shut off the mouth from the nasal cavity during the act of swallowing (fig. 225).

Stylopharyngeus and Palatopharyngeus. From its origin on the base of the styloid process (fig. 223), the Stylopharyngeus slips into the pharynx in the interval above the Middle Constrictor. As it does so, it spreads out into a thin and delicate muscular sheet applied to the inner surface of the Middle and Inferior Constrictors; it reaches as low as the thyroid cartilage on whose posterior border the lower fibers of the muscle are inserted.

The 'skeleton' of the soft palate is a fibrous sheet, the palatine aponeurosis, whose front half, because of its direct continuity with the bony hard palate, is immobile. The posterior half, however, enjoys some movement so it is equipped with muscles, two of which, the Levator and the Tensor, have already been seen to descend to the palate from the base of the skull. Two others descend from the palate, the *Palatoglossus* (fig. 226) to the tongue, the *Palatopharyngeus* to the pharynx. These two stand out prominently and, covered with mucous membrane, can be seen behind

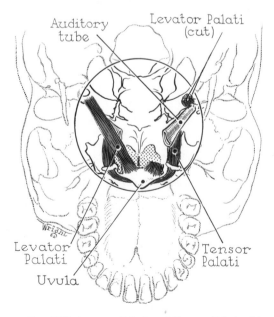

Fig. **225**. Levator Palati and Tensor Palati with related auditory tube.

the back of the mouth where they form the walls of the fossa that lodges the tonsil. The Palatopharyngeus blends with the Stylopharyngeus and so contributes to the longitudinal musculature.

The **nerve supply** of the muscles of the pharynx and palate is from the tenth (vagus) and 11th (accessory) cranial nerves. There are, however, two exceptions: Stylopharyngeus by the ninth (glossopharyngeal nerve) and Tensor Palati by the fifth (mandibular branch of trigeminal nerve).

Muscles of Tongue

The tongue is a muscular and very mobile organ that bulges upward from the floor of the mouth and is covered with mucous membrane. Several groups of intrinsic and extrinsic muscles form the bulk of the tongue. Four pairs of muscles (fig. 226) reach the tongue from origins beyond it and are known as the **extrinsic muscles,** one of which, the Palatoglossus, developmentally belongs to the palate (see above).

The **intrinsic muscles** are arranged in three planes and are named accordingly, the *longitudinal*, the *transverse*, and the *vertical muscles*. They alter the shape of the tongue at will and of them the longitudinal group is the largest and the most important.

The **extrinsic muscles** (excluding the Palatoglossus) arise from three bony sites: (1) from the tip of the styloid process, the **Styloglossus** descends to the side of the tongue; (2) from the lateral part (greater horn) of the hyoid bone, the **Hyoglossus** ascends to the side of the tongue and reaches nearly to the tip; (3) from a little

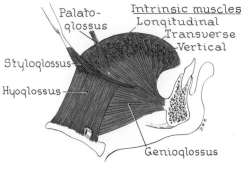

Fig. **226**. Muscles of tongue

bony tubercle situated on the mandible close beside the midline, the **Genioglossus** fans out on entering the under surface of the tongue from base to tip; it is closely applied to its fellow of the opposite side. (Gr., geneion = chin; glossa = tongue.)

Actions. By the combined efforts of the intrinsic and extrinsic muscles, the tongue can, at will, be rendered concave or convex from side to side, and it can be protracted, retracted, elevated or depressed.

Nerve Supply. All of the muscles of the tongue (except Palatoglossus) are supplied by the last (12th) cranial or hypoglossal nerve.

Swallowing

The striated muscles of the pharynx, palate, and tongue are particularly concerned with the act of swallowing which, therefore, will be analyzed now. The series of co-ordinated actions falls into two groups: (1) voluntary movements of the cheeks, mouth, and tongue; (2) 'automatic' movements of the palate, pharynx, and larynx.

The first series of actions is designed to reduce the capacity of the mouth cavity and to utilize the tongue as a chute. Thus, the Buccinators keep the food within the limits of the teeth; the Mylohyoids and Geniohyoids tense and slightly raise the floor of the mouth; the intrinsic muscles of the tongue press the front of that organ against the hard palate; the Styloglossi and Palatoglossi raise the sides of the tongue while the Genioglossi depress it in the midline. The tongue thus becomes a chute and the bolus of food, squeezed between the tongue and the hard palate, slides from the mouth into the pharyngeal entrance.

The second series of actions tenses the soft palate (Tensor), presses it against the posterior wall of the pharynx (Levator)—it is never far away—and so shuts off the nasopharynx. The Palatopharyngeus and Stylopharyngeus draw the pharyngeal wall over the bolus of food as a glove is drawn on a finger. The insertions of these muscles on the posterior border of the thyroid cartilage enable them, at the same time, to raise the larynx so that it ascends behind the hyoid bone which moves forward (Geniohyoids) to make room for it. By this means, the entrance to the larynx is drawn up beneath the posterior part of the tongue. The hyoid bone moves upward too, but it is principally concerned as the bony anchorage for the tongue and the larynx, and its chief movement is a forward one. Finally, the Constrictors and the longitudinal muscles begin the action of peristalsis whereby the food is passed steadily along the length of the digestive tube. Although the muscles of the pharynx—and, for that matter, those of the upper third of the esophagus or gullet into which the food passes—are striated, they hardly qualify for the term 'voluntary' since they act completely automatically.

Muscles of Larynx

An account of these special small muscles will be found on page 228.

Muscles of Eye

An account of the muscles that move the eyeball will be found on page 349.

MUSCLES OF THORAX

The walls of the body cavities are constructed to meet the demands of the contained organs. The brain, not varying in size, is housed in, and protected by, a rigid bony box. The heart and lungs, requiring to expand and contract, are housed in a cage whose walls are partly bone, partly cartilage, and partly muscle. The abdominal organs, requiring a still more flexible cavity, are housed in one whose walls are almost entirely muscular. Thus, many of the muscles enclosing body cavities are more concerned with protection than they are with movements.

The muscles of the thorax are concerned, then, with helping to form a protective cage. However, this is not an ordinary cage for it must be alternately expanded and contracted in size to suck air into the lungs and to force it out again, i.e., to perform respiratory movements. Thus, the muscles of the thorax are also called the **Muscles of Respiration.** They include:

1. **Intercostals—External and Internal**
2. *Levatores Costarum*
3. *Transversus Thoracis*

4. *Serratus Posterior Superior* and *Serratus Posterior Inferior*
5. **Diaphragm.**

Of these, the first and the last are the most important (see figs. 227 and 228).

Intercostals. These are the muscles uniting adjacent ribs and so completing the thoracic wall. They are arranged in two layers, the outer **External Intercostals** and the inner **Internal Intercostals.** The External Intercostals are thick behind, thin and aponeurotic in front; the Internal Intercostals are aponeurotic behind and thicker in front. The Externals run in a direction similar to that taken by the hand placed in the trousers' pocket; the Internals run at right angles to the Externals (fig. 228).

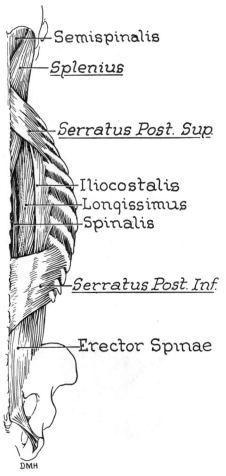

Fig. **227**. Serratus Posterior Superior and Inferior muscles.

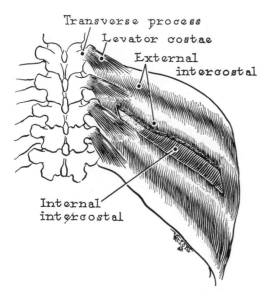

Transverse process
Levator costae
External
intercostal
Internal
intercostal

Fig. **228.** Intercostal muscles and (four) Levatores Costarum muscles.

The actions of the Intercostals have been the subject of much controversy and while it is generally agreed that the Externals raise the ribs and so assist the diaphragm in inspiration, yet it cannot be said that there is like unanimity of opinion with respect to the actions of the Internal Intercostals.

The Intercostals maintain a constant distance between the ribs, and with the Scalenes lifting the first and second ribs upward, they move all the ribs upward and forward in inspiration. They also help to bend the trunk forward and sideways and are postural muscles.

Levatores Costarum. Each Levator Costae—there are 12 pairs—is a triangular little muscle whose apex is attached to the back of a transverse process, and from there the muscle fans out laterally to be inserted on the next rib. The muscle generally is said to raise the back of the rib and so aid in inspiration (fig. 228).

Transversus Thoracis. This thin and unimportant muscle arises from the back of the lower end of the sternum. From there its muscular fibers radiate like the rays of the setting sun, to be inserted on costal cartilages 2 to 7, which they can pull downward and so, perhaps, assist in forced expiration.

Serratus Posterior Superior and *S. P. Inferior* (fig. 227). These are two unimportant muscles of paper-like thinness that have close affinity to the muscles of the thorax even though they can scarcely be said to aid in enclosing any cavity. The Superior spreads

from lower cervical and upper thoracic vertebral spines downward and lateralward to the upper three or four ribs; the Inferior spreads from lower thoracic and upper lumbar spines upward and lateralward to the lower three or four ribs. They belong to the respiratory muscles, and perhaps help to fix the upper and lower parts of the thoracic cage.

Nerve Supply. All of the foregoing muscles of respiration are supplied by intercostal nerves (i.e., anterior rami of thoracic spinal nerves) except the Levatores Costarum whose supply is from posterior rami and the diaphragm (phrenic nerve).

Diaphragm. This important respiratory muscle is discussed more conveniently with the muscles of the abdomen where it is equally important.

Forced respiration. It is obvious that in difficult or forced respiration any muscle having attachment to ribs can be called upon for assistance. If the disposition of such a muscle tends to raise the ribs it is an accessory inspiratory muscle. If its disposition tends to lower the ribs or to compress the costal cage it is an accessory expiratory muscle. For example, the Latissimus Dorsi—one of the widespread muscles of the upper limb girdle that clothes the back of the thoracic cage—contracts vigorously and squeezes the cage in the explosive expiratory act that accompanies a sneeze. The actions of the abdominal muscles in this respect will be noted in a moment.

MUSCLES OF ABDOMEN

General

The anterior abdominal wall, being a muscular one, allows great variation in the size of the cavity it protects; it is built on a simple plan. Its laminated muscles—lying between two fascial layers as do muscles everywhere else—are covered, first with a fatty areolar layer, and then with skin (or peritoneum). In consequence, whether the wall be investigated from without-in or from within-out the same order of layers is met (fig. 229).

The laminated muscles now to be studied are: External Oblique, Internal Oblique, and Transversus. Intimately associated with these three is the important Rectus Abdominis (see fig. 230).

Rectus Abdominis and *Pyramidalis*

A long and powerful strap-like muscle, the Rectus Abdominis, descends on each side of the midline of the abdomen from sternum to pubis. It is 10 cm (about 3–4″) wide above where it arises on the chest from the front of costal cartilages 5, 6, and 7. It narrows somewhat below and is inserted by tendon to the front of the body of the pubis (figs. 230, 231).

The muscle possesses a sheath (which is described below with Internal Oblique) and, within this sheath, a little, triangular, unimportant muscle, the Pyramidalis is found in front of the insertion of Rectus (fig. 231).

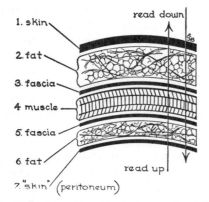

Fig. 229. The order of layers in anterior abdominal wall (see text).

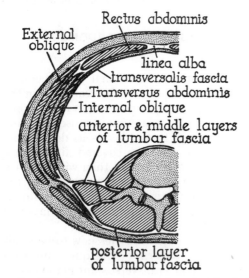

Fig. 230. Schematic cross-section of abdominal wall.

The fleshy part of Rectus Abdominis is interrupted by three transverse fibrous bands, known as tendinous intersections, to which the sheath of the Rectus adheres. Therefore, the muscle does not slide freely in its sheath.

External Oblique

This broad muscle (figs. 232, 233) corresponds in developmental origin and in the direction of its fibers to the External Intercostals; it is the outermost muscle of the abdominal wall.

Origin. It arises a full hand's breadth above the costal margin where it clings by finger-like slips to the outer surfaces of the lower eight ribs.

Insertion. When it descends to the abdomen it quickly becomes an extensive, thin, but tough, aponeurotic sheet. This aponeurosis fuses with its fellow of the opposite side down the midline of the abdominal wall from xiphoid process to symphysis pubis; the line of fusion is

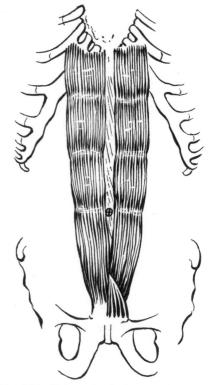

Fig. 231. Right and left Rectus Abdominis and left Pyramidalis.

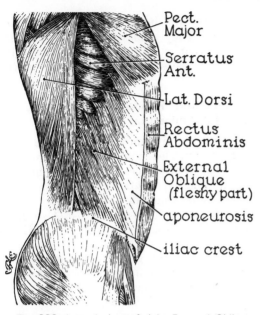

Fig. **232**. Lateral view of right External Oblique and Rectus Abdominis and some important neighbors.

Pect. Major

Serratus Ant.

Lat. Dorsi

Rectus Abdominis

External Oblique (fleshy part)

aponeurosis

iliac crest

known as the **linea alba** or white line. The most posterior fleshy fibers descend vertically from their thoracic origin to insert on the anterior half of the iliac crest (fig. 232).

Below, the aponeurosis has a free, rolled edge which spans across the root of the lower limb at the groin from the pubis to the anterior spine of the iliac crest; this free edge is known as the **inguinal ligament** (of Poupart).

Thus, the muscle has two free edges, a posterior muscular one and a lower aponeurotic one. Just above the medial end of the inguinal ligament, a triangular split in the aponeurosis gives passage to the duct of the testis and its associated structures (the spermatic cord) or, in the female, to a cordlike ligament, the round ligament of the uterus (see fig. 443 on page 245).

Internal Oblique

This broad flat muscle is the intermediate and the thickest of the three muscles. It arises below and is inserted above, so that its fibers run at right angles to those of its companion External Oblique, but in the same direction as those of the External Oblique of the other side (fig. 233).

Origin. It arises from rather more than the lateral one-half of the inguinal ligament and from the anterior two-thirds of the iliac crest.

It is muscular over the lateral half of the abdominal wall and aponeurotic over the medial half.

Insertion. The posterior part of Internal Oblique runs upward and medialward to be inserted on the costal cartilages of the last four ribs. The chief insertion of the muscle is into the midline from xiphoid process to symphysis pubis but when the aponeurosis reaches the lateral border of the Rectus Abdominis (see above) it splits into two layers or lamellae, one passing behind, the other in front of that muscle, thus providing a sheath for it. The lowest quarter of the Rectus however has no such sheath behind, since both lamellae of the Internal Oblique aponeurosis in that region pass in front of the Rectus.

The fibers that arise from the inguinal ligament have a special insertion, described under: 'Weak Spot' and Cremaster muscle, below.

Transversus Abdominis

This is the deepest and thinnest muscle of the abdominal wall and its fibers, as its name implies, run transversely from the periphery of the abdominal wall (fig. 234).

Attachments. Transversus arises behind and runs forward to be inserted in the midline. The origin of the muscle is from the lateral one-third of the inguinal ligament, from the anteromedial two-thirds of the iliac crest, from the lumbar transverse processes by means of a flat aponeurosis, and by short muscle fibers from the whole length of the lower margin of the costal cage. By its costal origin the muscle interdigitates with the costal origin of the diaphragm and it is between the digitations that the lower intercostal nerves escape from the thorax. The aponeurosis of the muscle joins the posterior lamella of that of the Internal Oblique and is similarly disposed (see also 'Weak Spot,' below).

'Weak Spot' of Abdominal Wall

Immediately above the medial half of the inguinal ligament the Internal Oblique and Transversus fail to reach the ligament but have a conjoined, free, arched, lower border. The deficiency in the wall below this border, the **inguinal canal,** is filled by the spermatic cord in the male or by the round ligament of the uterus in the female. In order to provide some strength for this deficient area, the medial end of the arched lower border, called the **conjoint tendon (falx inguinalis),** curves behind the issuing spermatic cord (or round ligament) and gains attachment, behind the inguinal ligament, to the anterior part of the *pecten pubis*, i.e., the anterior part of the pelvic brim. Thus, behind the site of exit of the cord, viz., the **superficial inguinal ring,** the conjoint tendon adds strength where it is most needed. In spite of this, the wall, weakened by the exit of the structures mentioned, often gives way and allows abdominal contents to protrude and so produce a hernia.

Cremaster Muscle. From its free arched border, the Internal Oblique provides elongated muscle fibers that leave it to descend with, and act as a covering for, the spermatic cord. They are, essentially, loops of muscle fibers that descend into the scrotum and then return again to the abdominal wall in the region of the pubic tubercle. They constitute the *Cremaster muscle,* that suspends and can slightly raise the testis (see figs. 444, 445, page 245).

Actions and Functions of Muscles of Anterior Abdominal Wall

Acting like dynamic corsets, these four pairs of muscles hold in the abdominal contents. When necessary, they compress the abdominal contents and so aid in expulsive movements, (e.g., of urine, or the three *Fs*, fetus, feces and flatus). Also they contract vigorously to protect against blows.

They can aid gravity in flexing the trunk—the Rectus being especially useful —but they are seldom called upon. By helping to stabilize the pelvis, they contribute to good posture. Rectus acts most vigorously when a supine person raises his shoulders or raises both lower limbs; this is simply because the pelvis must be stabilized against the leverage applied by the weight of the limbs. Theoretically, although seldom actually, unilateral action of the External and Internal Obliques results in bending the trunk to the side. The

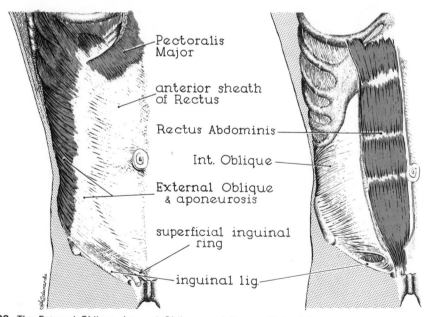

Pectoralis Major

anterior sheath of Rectus

Rectus Abdominis

Int. Oblique

External Oblique & aponeurosis

superficial inguinal ring

inguinal lig.

Fig. **233**. The External Oblique, Internal Oblique, and Rectus Abdominis muscles; the last is exposed by removing its anterior sheath made from the aponeuroses of the other two muscles.

External Oblique of one side with the Internal Oblique of the opposite side produce rotation of the trunk.

Contraction of all the muscles of the abdomen during forced expiration compels the upper abdominal viscera (notably the liver) to thrust the dome of the relaxed diaphragm upward.

Nerves of Anterior Abdominal Wall

——Lower six intercostal nerves and first lumbar nerve.

The Internal Oblique and Transversus are supplied by the main stems of the lower six intercostal nerves, which, when they reach the front ends of their respective intercostal spaces, pass behind the costal cartilages and escape into the abdominal wall. Here they are found deep to the Internal Oblique. Having supplied this muscle and the underlying Transversus, they are conducted by the posterior lamella of the aponeurosis of Internal Oblique to the back of the rectus sheath which they pierce in order to supply the Rectus; their terminations finally become cutaneous near the midline of the abdominal wall.

Lateral branches of the lower six intercostal nerves make their exit from intercostal spaces and supply the External Oblique at the sites of its slips of origin, and continue to the skin to become sensory.

Quadratus Lumborum

This rectangular (quadrate) muscle fills the interval between the 12th rib above and the posterior part of the iliac crest below; it gains additional attachment to the transverse processes of the lumbar vertebrae (see fig. 235). Each muscle helps to bend the vertebral column to its own side. It is supplied by anterior rami of lumbar spinal nerves.

Diaphragm

The diaphragm (fig. 236) is a thin but strong fibromuscular partition separating the thoracic from the abdominal cavity. It is so markedly domed upward that many important abdominal organs in contact with its lower surface enjoy the protection of the lower ribs (fig. 377 on page 215).

Origin. It is attached behind to the vertebral column, at the sides and in front, to the last six costal cartilages and to the xiphoid process of the sternum; it, therefore, is said to have a lumbar and a costal part.

The **lumbar origin** is by two stout muscular 'legs' or **crura** that are fixed to the sides of the first three lumbar vertebrae; these crura straddle the midline aorta. (Singular of *crura* is *crus*.)

Lateral to the crura on each side, two thin fibrous arches, the **medial** and the **lateral arcuate ligaments,** allow the diaphragm to span the muscles of the poste-

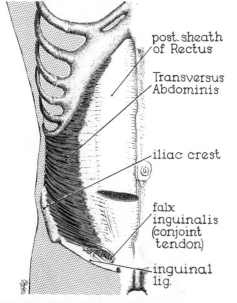

Fig. **234.** Transversus Abdominis and the posterior sheath of Rectus (most of which was removed).

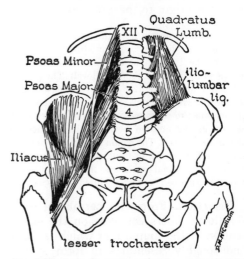

Fig. **235.** Left Quadratus Lumborum and right Ilio-psoas. These are the fleshy muscles of the posterior abdominal wall. Small unimportant Psoas Minor lies in front of the Major (cf., fig. 236).

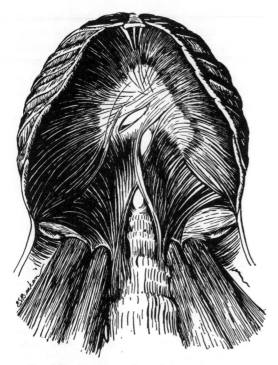

Fig. **236**. Diaphragm from below—its crura and three principal openings: aortic (between crura), esophageal, and caval (in central tendon). Psoas and Quadratus Lumborum seen on both sides of vertebrae.

rior abdominal wall (Psoas and Quadratus Lumborum) and, at the same time, they carry the origin to the 12th rib.

At its **costal origin** from the lower margin of the thoracic cage, the diaphragm interdigitates with the fibers of the Transversus Abdominis, the innermost muscle of the anterior abdominal wall. The muscular deficiency above the 12th rib is the vertebrocostal trigone. It exists because the vertebral and costal origins failed to meet and so close a passageway that existed in the embryo between the thoracic and abdominal cavities. The region is surgically important since the kidney lies in front of it and since, too, a hernia of abdominal organs into the thorax may occur here. The trigone is normally closed by fascia.

Insertion. From the peripheral origin, the fleshy fibers converge on the **central tendon** of the diaphragm which is fibrous and roughly resembles in form a three-leafed clover. When the muscle fibers contract and shorten, they pull on this central tendon and flatten out the dome; thus with each inspiration the thoracic cavity is en-

larged in its vertical diameter at the expense of the abdominal cavity.

Structures Piercing Diaphragm. Besides the **aortic opening,** it possesses an **esophageal opening** and a *foramen for the inferior vena cava* (the great vein returning blood to the heart from the lower half of the body) situated in the central tendon; it is therefore enlarged with each inspiration, the movements, of course, encouraging the flow of venous blood to the heart. Other structures pierce the diaphragm, some passing through one or other of the three main openings noted above, others through smaller openings; they will be noted as occasion demands.

To the upper surface of the central tendon, the pericardial sac, which encloses the heart, is firmly attached.

The **nerve supply** is the phrenic nerve made up of branches from the third, fourth, and fifth cervical nerves; it descends through the thorax from the neck.

Levator Ani

The paired Levatores Ani (fig. 237) together form the funnel-shaped muscular floor of the pelvic cavity and so support the contents of the pelvis.

The **origin** of each is along a line beginning in front from the inner (posterior) surface of the pubic body—these are the most important fibers—and extending along the fascia on the side wall of the true pelvis as far back as the ischial spine.

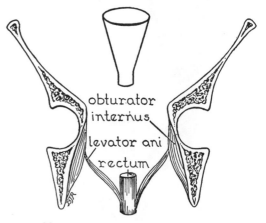

Fig. **237**. Pelvic floor in coronal section is funnel-shaped. Paired Levator Ani muscles form sloping floor which supports organs (cf., fig. 374).

Their **insertion** consists chiefly of a blending of the fibers from the opposite sides near the midline behind, and in front of, the rectum. Some fibers reach the coccyx (see fig. 374 on page 214).

The direction of the fibers is downward, backward, and toward the midline. Anteriorly there is a narrow cleft between the free borders of the two muscles and through this pass the genital and urinary tracts—the vagina in the female being most important.

Actions and Functions. Besides acting as a support for the organs in the pelvis, the Levatores Ani contribute to the voluntary control of the rectum—both in 'shutting it off' and, conversely, in emptying it during defecation. The fibers that help to 'shut off' the rectum form a sling behind the rectum (*puborectal sling*) and by pulling constantly forward cause a kink in the last part of the rectum. During expulsive movements, the Levator Ani contracts vigor-

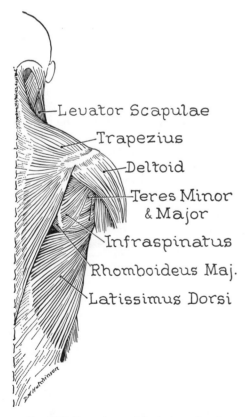

Fig. **238.** Trapezius and Latissimus Dorsi

ously along with the other abdominal muscles.

The **nerve supply** is from the (second), third, and fourth sacral nerves.

Muscles of Perineum

Below the Levatores Ani is a diamond-shaped area surrounding the opening of the rectum—the anus—and the genitalia; it is known as the *perineum*. For a description of the perineum and its special muscles see pages 214 (anal sphincters) and 247 (urogenital diaphragm and muscles of penis).

MUSCLES OF UPPER LIMB

Muscles of Girdle

Since the upper limb girdle as well as the humerus possesses considerable mobility, it is convenient to divide the 18 muscles of the shoulder region into three groups according to their passing: (A) from the axial skeleton to the pectoral girdle only; (B) from the axial skeleton, and spanning the girdle, directly to the humerus, moving both the girdle and the shoulder (glenohumeral) joint; (C) from the pectoral girdle to the humerus, moving the shoulder joint only.

A. AXIAL SKELETON TO GIRDLE

1. Trapezius

This large, flat, triangular‡ muscle (figs. 238, 239) covers the back of the neck and upper half of the trunk.

Origin. The long linear (thus aponeurotic) origin of the Trapezius starts on the skull a short distance behind the mastoid process where it meets the origin of the Sternomastoid. It passes on the occipital bone to the midline *inion* at the back of the head; it is continued down the midline of the neck and back, by the ligamentum nuchae and 12 thoracic vertebral spines.

Insertion. The upper fibers descend to the back of the lateral end of the clavicle; the middle fibers run horizontally to the length of the spine of the scapula; the lower fibers ascend to the tubercle on the spine of the scapula situated two or three fingers'-

‡ the paired muscles have the appearance of a trapezion, Gr. = an irregular four-sided figure.

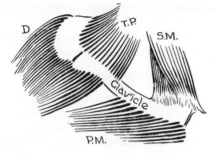

Fig. **239**. Muscles attached to (right) clavicle and acromion: Trapezius (T.P.) and Sternomastoid (S.M.) from above; Deltoid (D). and clavicular head of Pectoralis Major (P.M.) from below.

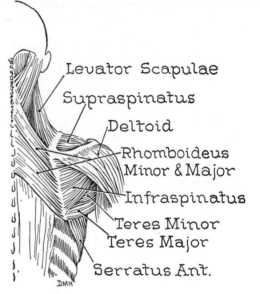

Fig. **240**. Levator Scapulae, Rhomboids, and muscles around scapula.

breadths from the vertebral border. The dispositions of the three groups of fibers on the girdle bones are the clues to the actions.

Actions and Functions. The **upper fibers** shrug the shoulders; they maintain shoulder level but in fatigue they allow the shoulders to droop.

The **middle** or **horizontal fibers** retract the scapula, i.e., they produce the 'standing at attention' position of the shoulders. They steady the scapula at the very beginning of the movement of raising the arm above the head. However, the moment the scapula is required to swing into protraction these horizontal fibers let go, fixation with protraction being then undertaken by the Serratus Anterior (page 145).

The **lower fibers** pull the medial end of the spine of the scapula downward, i.e., they co-operate with the upper fibers to rotate the glenoid cavity upward—a necessary position in raising the arm above the head.

Its **nerve supply** is the same as that of Sternomastoid, i.e., (spinal part of the) accessory or 11th cranial nerve; but, in addition, some direct branches reach it from the cervical plexus.

2. Levator Scapulae

Four tendinous slips arise from the backs of the transverse processes of the first four cervical vertebrae. The resulting long muscle (figs. 210, 212, 238, 240) reaches the upper angle of the scapula where it is inserted. The name proclaims the action. It also aids in tilting the glenoid fossa downward (fig. 243-B). (See table 2.)

TABLE 2

Muscles Acting on Shoulder Girdle (Sliding Movements of Scapula on Thorax)

Elevation (Shrug)	Depression	'Rotation Up'[a]	'Rotation Down'[a]	Protraction (Forward Pull)	Retraction (Pull Toward Vertebrae)
1. Trapezius— upper fibers 2. Lev. Scap. (3. Serr. Ant. —upper fibers)	1. Pect. Major 2. Lat. Dorsi 3. Pect. Minor (4. Subclav.)	1. Trapezius— upper fibers 2. Trapezius— lower fibers 3. Serratus Anterior	1. Lev. Scap. 2. Rhomboids 3. Pect. Major 4. Lat. Dorsi 5. Pect. Minor	1. Serr. Ant. 2. Lev. Scap. 3. Pect. Major 4. Pect. Minor	1. Trapezius (especially middle) 2. Rhomboids 3. Lat. Dorsi

[a] 'Rotation up' results in the glenoid cavity being elevated and made to face upward; 'rotation down' is the reverse.

The **nerve supply** is by branches of C. 3, 4 (and 5) spinal nerves.

3. Rhomboids

These two muscles (figs. 238, 240, 241), Rhomboideus Minor and R. Major, **arise from** the lower end of the ligamentum nuchae (Minor) and from the first four thoracic spines (Major). The fibers run obliquely downward and lateralward to be **inserted on** the vertebral border of the scapula from the level of the spine to the inferior angle; by far the greatest number of fibers congregate at this inferior angle.

The longer the blade of the scapula, the greater the force exerted at the shoulder joint by the muscles using the blade, provided, of course, they take advantage of their opportunity by gaining insertion on the extreme end of the long blade. This is the reason why so many muscles, including the Rhomboids, concentrate their fibers at the inferior angle of the scapula. The primates alone possess a scapula that is longer than it is broad and the primates alone are capable of freely raising the arm above the head.

Actions and Uses. They pull back (retract) the protracted scapula with considerable force and they turn the glenoid cavity downward, thus forcibly lowering the raised arm. They are constantly in use by the woodsman swinging his ax. With

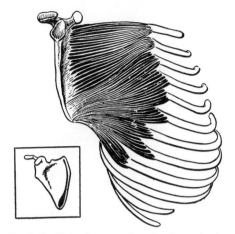

Fig. **242**. Right Serratus Anterior. Inset: its insertion. Note crowding of fibers at anterior angle of scapula. The scapula has been swung away from the chest wall to expose all the muscle.

other muscles they keep the scapula applied to the chest wall (fig. 241).

Their **nerve supply** is by branches of C. 5.

4. Serratus Anterior

One of the most powerful muscles of the girdle, the Serratus Anterior, is perhaps a little difficult to visualize. Halfway around the chest, it **arises** by finger-like slips§ (fig. 242) from the outer surfaces of the upper eight ribs. It quickly becomes a broad sheet of muscle clothing the chest wall behind its origin. It curves backward, in contact with the wall, until it reaches its **insertion** along the whole length of the vertebral border of the scapula; it thus lies between the scapula (with its fleshy clothing of Subscapularis) and the ribs (fig. 241). Its lower and most important fibers approach the vertebral border of the scapula from a direction directly opposite to that of the Rhomboids, its direct antagonists. Like those of the Rhomboids, by far the greatest number of its fibers (five-eighths) are inserted on the inferior angle.

Actions and Uses. It is the most powerful protractor of the scapula. Since it concentrates its efforts on the inferior angle, it also rotates the glenoid cavity upward so that the arm may be raised above

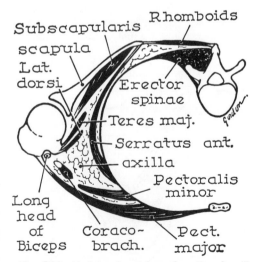

Fig. **241**. Horizontal section of walls of axilla (semischematic). Note how Serratus Anterior (the muscle of medial wall) and Rhomboids hold scapula against thoracic wall.

§ The saw-toothed (or serrated) appearance of the origin gives the muscle its name.

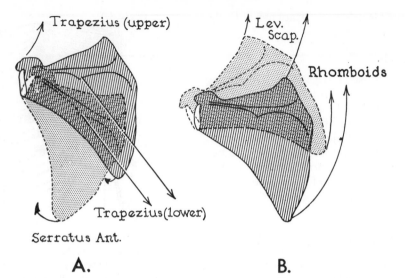

Trapezius (upper)

Lev. Scap.

Rhomboids

Trapezius (lower)

Serratus Ant.

A.   B.

Fig. **243**. Rotation of scapula: A. Turning the glenoid cavity upward as in raising arm. B Downward

the head. By keeping the scapula applied to the chest wall at all times, it acts as an 'anchor' for the bone allowing other muscles to use it as the fixed bone for the movements of the humerus. Without the Serratus Anterior the scapula becomes uselessly mobile. When the arm is raised from the side, the scapula is pulled away from the chest wall in an uncontrollable fashion and the arm can be abducted no farther than to the horizontal position. The resulting jutting backward of the scapula is known as 'winged scapula' and it is a sign of paralysis of the Serratus Anterior.

Nerve Supply. A long nerve from the upper roots (C. 5, 6, and 7) of the brachial plexus runs downward on its lateral surface (near the midaxillary line) and supplies it.

Rotation of Scapula

See figure 243 for a graphic explanation of the forces which move the scapula on the chest wall. See also page 151.

5. Pectoralis Minor

Entirely hidden by the overlying and bigger Pectoralis Major, the Pectoralis Minor (fig. 244) has the shape of a triangle whose base (origin) occupies the front ends of the third, fourth, and fifth ribs just lateral to their cartilages. The apex of the triangle (insertion) is on the medial horizontal border of the coracoid process. The muscle assists in drawing the shoulder

forward and downward; it is used in stretching the arm forward for an object just beyond the reach. Its chief importance lies in the fact that it is a useful landmark for the many structures in the axilla, since, with the coracoid process, it forms a bridge beneath which vessels and nerves must run in order to reach the arm.

Its nerve, the medial pectoral, comes from the medial cord of the brachial plexus.

6. Subclavius

This small, inaccessible muscle (fig. 244) arises from the junction of the first rib and its cartilage. It

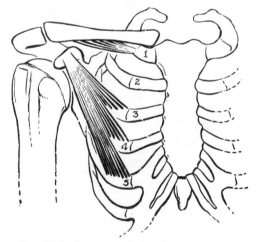

Fig. **244**. Removal of (right) Pectoralis Major (cf., fig. 245) reveals triangular Pectoralis Minor and tiny Subclavius.

runs lateralward and upward, fanning out to occupy the shallow groove along the under surface of the clavicle. The muscle actively restrains excessive elevation and protraction of the clavicle; it is also a useful cushion for the protection of the subclavian vessels as they pass above the first rib and below the clavicle on their way to the limb. Its nerve (C. 5) descends to it from the neck.

B. AXIAL SKELETON TO HUMERUS

1. Pectoralis Major

The **clavicular head,¶** the uppermost quarter of the muscle, **arises from** the front of the medial half of the clavicle and passes downward and lateralward to its linear **insertion**—by a thin but broad aponeurosis—on the lateral lip of the bicipital groove (figs. 245–247).

The **sternal head** is much the larger head. It **arises from** the length of the sternum and from the cartilages of the upper six ribs which it crosses on its way to its **insertion** behind (deep to), and in association with, the clavicular head. The sternal head is responsible for the rounded muscular mass that forms the front wall of the armpit or axilla.

Actions and Functions. The whole muscle is a powerful adductor of the arm and a medial rotator. Each part can act independently. Thus, the clavicular head flexes the shoulder joint (i.e., raises the arm forward) and from this position the sternal head extends the shoulder joint (i.e., carries the arm backward). These facts can be easily confirmed by the student on himself.

Nerves. Coming from the brachial plexus, the lateral pectoral nerve (C. 5, 6, and 7) and part of the medial pectoral nerve (C. 8 and T. 1) enter and supply the Pectoralis Major.

2. Latissimus Dorsi

The name (L. = widest of the back) is an appropriate one since the muscle covers the lower half of the back (fig. 238).

Origin. It arises by means of an aponeurosis from the lower six thoracic, all the lumbar and the upper sacral spines, and

---

¶ Only one part—the **sternal head**—arises from the axial skeleton. However, it would be pedantic to ignore the **clavicular head** at this time. Pectus, L. = breast.

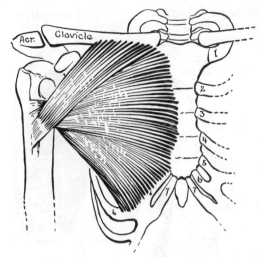

Fig. **245**. Right Pectoralis Major. Clavicular head overlaps sternal head which folds itself into two layers at its insertion.

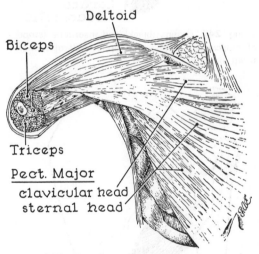

Fig. **246**. The Pectoralis Major and some important neighbors.

from the posterior half of the iliac crest. As the muscle sweeps upward and lateralward to converge on the humerus it gains additional origin, in passing, from the lower three or four ribs and from the inferior angle of the scapula.

Insertion. It approaches the axilla from behind and runs in the posterior fold of the axilla to reach the humerus in the floor of the bicipital groove as a thin, ribbon-like tendon (figs. 248, 257).

Actions and Uses. It extends the arm and rotates it medially (see also Teres Major, below). Because of the above ac-

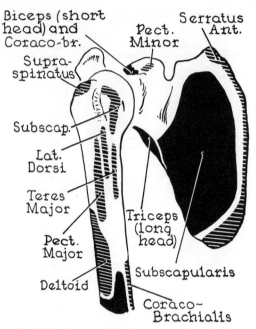

Fig. **247**. Origins (black) and insertions (striped) of muscles of scapula and upper half of humerus (front view).

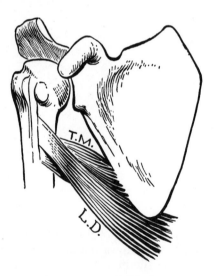

Fig. **248**. Right Teres Major and tendon of Latissimus Dorsi as seen in axilla.

tions and its downward and backward pull on the whole shoulder girdle, Latissimus Dorsi is actively concerned with swimming or paddling a canoe. With Pectoralis Major, it is vigorously used in preventing the shoulder girdle from being driven upward in such situations as ordinary pushing with

the arms or raising the body and suspending it (e.g., on parallel bars).

Its **nerve** (C. 6, 7, and 8) comes directly from the posterior cord of the brachial plexus.

C. GIRDLE TO HUMERUS

1. Clavicular Head of Pectoralis Major

This head has been discussed already with the whole muscle (page 147).

2. Teres Major

Origin. The Teres Major arises from the inferior angle of the scapula.

Insertion. The tendon of Teres Major runs in contact with the back of that of Latissimus and inserts on the medial lip of the bicipital groove (figs. 247, 248).

Its **actions** are mostly similar to those of Latissimus Dorsi, namely, extension and medial rotation of the humerus, but, of course, it does not move the shoulder girdle. It helps to stabilize the upper end of the humerus during abduction.

Its **nerve** is a direct branch (C.5 and 6) from the posterior cord of the brachial plexus.

3. Guardians of the Shoulder Joint or 'Rotator Cuff'

    a. **Subscapularis**
    b. **Supraspinatus**
    c. **Infraspinatus**
    d. *Teres Minor*

It would be profitless to discuss these four deep muscles separately because they are all so closely related to each other, to the shoulder joint capsule, and to a single function—'guarding' the shoulder joint.

    a. **Subscapularis** (figs. 247, 249, 257) This triangular muscle has a multipennate origin from most of the subscapular fossa. Its tendon converges on the lesser tubercle where it is inserted. As the tendon passes across the front of the shoulder joint, a bursa deep to it usually communicates with the cavity of the joint.

    b. **Supraspinatus** (figs. 240, 250, 251). As its name implies, this muscle arises from the fossa (supraspinous) above the spine of the scapula. It passes directly above the shoulder joint to its insertion on

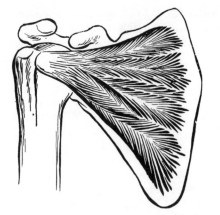

Fig. **249**. Right Subscapularis—guards front of shoulder joint.

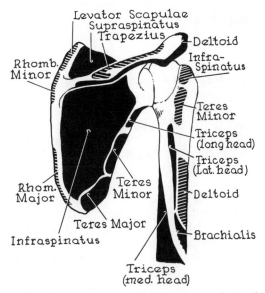

Fig. **250**. Origins (black) and insertions (striped) of muscles of scapula and upper half of humerus (from behind).

the axillary border above the Teres Major, is scarcely a separate entity but receives the name Teres Minor.

The Infraspinatus and Teres Minor pass behind the shoulder joint where their tendons also reinforce the capsule and are **inserted** on the greater tubercle, where the Supraspinatus has the highest insertion, the Teres Minor the lowest (see fig. 253).

Actions and Functions of the 'Guardians.' All four muscles act as 'dynamic ligaments' for the shoulder joint which otherwise is unstable and easily dislocated. They react with contractions of appropriate strength to various forces that tend to pull the head of the humerus out of the very shallow glenoid cavity. Thus the muscles are extremely important even when they are producing no actual movement or action.

Subscapularis is a useful medial rotator

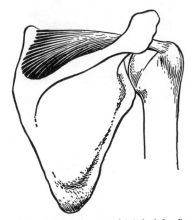

Fig. **251**. Supraspinatus (right) (cf., figs. 252 and 253).

the summit of the greater tubercle. As its tendon crosses the joint it reinforces the capsule by blending with it.

Between the Supraspinatus and the over-lying Deltoid muscle and coraco-acromial arch (page 90) lies one of the largest bursae in the body—the **subacromial bursa** (fig. 254).

c. **Infraspinatus** (fig. 252) and (d) *Teres Minor* (fig. 252). **Origins.** Infraspinatus occupies the fossa (infraspinous) below the spine, where it has a multipennate origin. The lateral part of the muscle, attached to

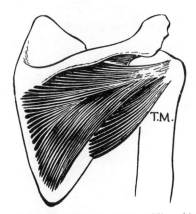

Fig. **252**. Infraspinatus and Teres Minor (right)

of the shoulder joint (along with Pectoralis Major, Teres Major, and Latissimus Dorsi). Both Infraspinatus and Teres Minor extend the arm and they are the chief lateral rotators.

Supraspinatus is essential for normal abduction of the humerus. Its role is to stabilize the head of the humerus by pulling over the top of the head while the Deltoid pulls upward on the midshaft of the humerus. Supraspinatus is especially important at the onset of abduction when Deltoid is pulling directly upward along the length of the humerus. Paralysis of the Supraspinatus leaves the powerful Deltoid

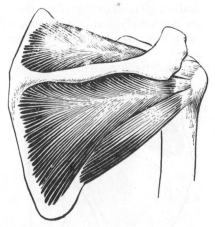

Fig. **253**. Supraspinatus guards shoulder joint above; Infraspinatus and Teres Minor, behind.

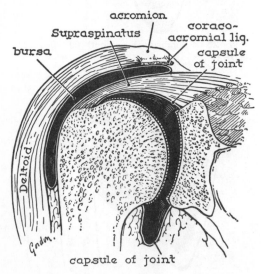

Fig. **254**. Relationships of Deltoid and Supraspinatus.

handicapped to a surprising degree. Indeed, without trick movements or external help Deltoid cannot initiate abduction.

The Infraspinatus and Teres Minor, along with Teres Major, also help to produce smooth abduction at the shoulder joint by pulling downward and stabilizing the head of the humerus. They are said to act as a 'force couple' with the true abductors. A 'force couple' describes two or more muscles pulling in different directions to achieve a single desired movement.‖

Nerve Supply. All four muscles are supplied by C. 5 and 6—the Subscapularis via branches from the posterior cord of the brachial plexus, the Spinati via the suprascapular nerve, and Teres Minor via the axillary nerve.

4. Deltoid

This thick fleshy mass (figs. 254, 255) helps form the roundness of the shoulder. It is a powerful muscle.

It **arises from** a continuous V-shaped bony strip consisting of clavicle, acromion and spine, specifically: (1) the front of the lateral third of the clavicle; (2) the lateral border of the acromion; and (3) the inferior lip of the crest of the spine of the scapula. Except for an intervening and palpable bony strip, the V-shaped origin of the Deltoid embraces the V-shaped insertion of the Trapezius.

The fibers of the Deltoid converge to an **insertion** on a roughness at the lateral aspect of the middle of the shaft of the humerus known as the **deltoid tuberosity.**

Actions and Uses. Deltoid is a powerful abductor of the humerus but cannot initiate that movement since, when the arm is hanging by the side, the line of pull of the muscle is parallel to the shaft of the bone. Supraspinatus provides the correct pull to overcome this difficulty (see above). The Deltoid can carry the arm no higher than to a horizontal position, farther elevation being brought about by tilting the glenoid fossa upwards.

The *anterior fibers* of the muscle co-

‖ Two boys moving a revolving door in a circle are, at any moment, facing and pushing in opposite directions, but they form a 'force couple' that results in the turning of the door.

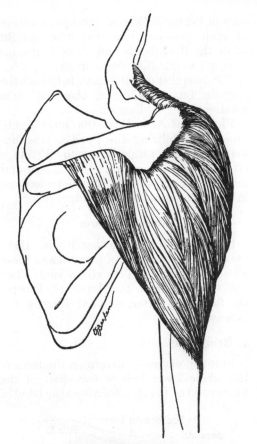

Fig. **255**. Right Deltoid from behind. It covers shoulder joint and related tendons.

operate with the clavicular head of the Pectoralis Major to flex the shoulder, and it is with flexion that abduction is so often combined. The *middle fibers* of the muscle are the most powerful and concern

themselves with abduction alone. The *posterior fibers* are the weakest and take little part in abduction, but, on the contrary, they adduct and extend the abducted arm; this part of the muscle is quite fibrous and of minor importance.

The **nerve supply** of Deltoid is by C.5 and 6 via the axillary (circumflex) nerve, a branch of the brachial plexus.

5. Coracobrachialis

A small muscle, Coracobrachialis (figs. 256, 257) arises from the tip of the coracoid process by means of a tendon which it shares with the short head of the Biceps. It is inserted halfway down the medial border of the shaft of the humerus. It is active when the shoulder joint is flexed and adducted against resistance; but it is of minor importance.

The **nerve supply** is C. 5, 6, and 7 via the musculocutaneous nerve, a branch of the brachial plexus.

Special Movements of the Shoulder Region

Abduction and flexion of the arm not only involve movement at the gleno-humeral or shoulder joint but also movements of the clavicle and scapula. Through the complete excursion of 180°, for each 15° of movement only 10° occurs at the gleno-humeral joint. (See table 3.)

Abduction of Arm. The muscles that abduct the gleno-humeral joint are the Supraspinatus and Deltoid, as explained above. The movements of the scapula consist of: (1) a rather variable pattern of adjustment of its position on the chest wall during the first 30° of abduction, apparently to obtain the best fitting together of

TABLE 3
Muscles Acting on Shoulder (Gleno-Humeral) Joint

Flexion	Extension	Abduction	Adduction	Med. Rot'n.	Lat. Rot'n.
1. Pect. Major —clav. head	1. Lat. Dorsi	1. Deltoid	1. Pect. Major —both heads	1. Pect. Major —both heads	1. Infraspinatis
2. Deltoid— ant. fibers	2. Teres Major	2. Supraspin.	2. Lat. Dorsi	2. Lat. Dorsi	2. Teres Minor
3. Coracobrach.	3. Pect. Major —sternal head		3. Teres Major	3. Teres Major	3. Deltoid— post. fibers
4. Biceps	4. Deltoid— post. fibers		4. Coraco- brach.	4. Subscap.	
(5. Supraspin.)	5. Triceps— long head		5. Triceps— long head	5. Deltoid— ant. fibers	
			(6. Deltoid— post. fibers)		

the joint surfaces (this may actually result in retraction and depression in many persons); (2) 'rotation up' of the scapula (i.e., upward tilt of the glenoid cavity); and (3) protraction of the scapula. Movements (2) and (3) are produced by a force couple illustrated by fig. 243-A. They begin at 30° of abduction of the arm and continue throughout the whole range of 180°.

To allow the whole range of abduction the humerus must be rotated laterally by Infraspinatus and Teres Minor so that the greater tubercle will not impinge on the acromion. As lateral rotation occurs, the anterior fibers of Deltoid with the upper

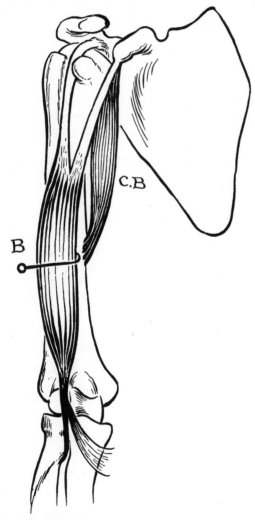

Fig. **256**. Right Biceps Brachii and Coracobrachialis from in front.

ones of Pectoralis Major, taking advantage of their position, hold the arm upright while the Rhomboids and all of Trapezius contract to stabilize the scapula.

The lateral end of the clavicle follows the scapula but in addition there is some rotation along its long axis.

Flexion of Arm. With the 'rotator cuff' muscles stabilizing the head of the humerus, the flexor muscles act to advantage on the humerus (see table 3). The scapular movements are similar (but not identical in detail) to those during abduction.

Muscles of Arm

Excluding the Coracobrachialis, there are only three muscles of the arm and these all act principally on the elbow joint. They are: (1) Brachialis, the flexor; (2) Biceps, the flexor-supinator; and (3) Triceps, the extensor.

1. Brachialis

From an extensive **origin** on the front of the whole lower half of the shaft of the humerus, the Brachialis crosses the bend of

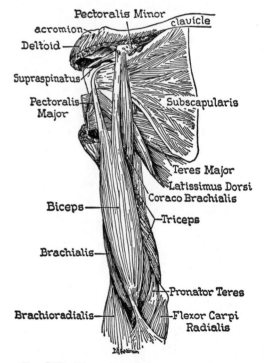

Fig. **257**. Dissection of muscles of right axilla and front of arm.

Fig. **258**. Right Brachialis from in front (Biceps, which covers it, removed).

the elbow as a wide fleshy muscle (figs. 257–259). It is **inserted** by a stout tendon into the tuberosity of 'the ulna situated immediately below the coronoid process.

Actions. The sole concern of the Brachialis is flexion of the elbow joint. The muscle is hidden modestly behind the Biceps which usually gets the credit for the work done by the Brachialis.

Its **nerve** (shared with Biceps and Coracobrachialis) is the musculocutaneous (C. 5, 6, 7), a terminal branch of the brachial plexus.

2. Biceps Brachii

Origin. This, as its name states, is a two-headed muscle of the arm (figs. 256, 257). Its **short head** arises in conjunction with the Coracobrachialis from the tip of the coracoid process. Its **long head** arises

from the root of the coracoid process immediately above the upper end of the glenoid cavity (supraglenoid tubercle). This long head is an elongated, round tendon, which is peculiar in that it traverses the interior of the shoulder joint (fig. 150, page 90). It runs down in the bicipital groove which is converted into a tunnel by fibrous tissue. This necessitates a synovial sheath, and this sheath communicates with the shoulder joint.

Insertion. About halfway down the arm the two bellies fuse and the muscle is inserted by a cord-like tendon into the back of the tuberosity of the radius, which lies below the neck on the anteromedial aspect of the shaft.

Actions. When the hand is pronated the tuberosity of the radius faces backward and the tendon is wrapped round the bone—explaining why a bursa intervenes between the tendon and the front of the tuberosity which is smooth (fig. 265). The contraction of Biceps unwraps its tendon and, by 'whipping' the bone over, supinates the forearm.

Biceps is called into action to help flex

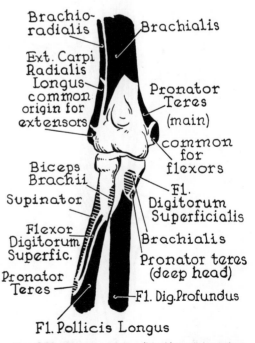

Fig. **259**. Muscle origins (black) and insertions (striped) of right elbow region (from in front).

the elbow strongly against resistance or to supinate the forearm against resistance— assisting Brachialis and Supinator respectively. It is a powerful reserve.

Neither head is of much importance in the actual production of shoulder movements. The short head possibly imitates Coracobrachialis; the long head is of some value in helping to retain the head of the humerus in the glenoid fossa.

Note. A strong *bicipital aponeurosis* passes (as a thin triangular sheet) from the tendon of insertion obliquely medially; it blends with the fascia covering the muscles on the medial side of the forearm. The bicipital aponeurosis shelters the important vessels

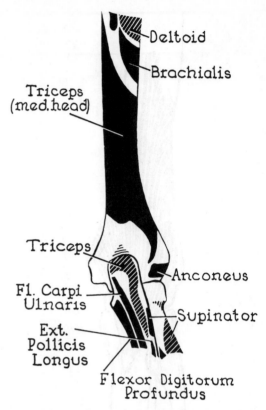

Fig. **261**. Muscular origins (black) and insertions (striped) of right elbow region (from behind).

and nerves at the bend of the elbow and it aids the flexing action of the Biceps.

When the muscle is in action the tendon of insertion juts out in the midline of the elbow and the sharp, curved, upper edge of the aponeurosis can be seen or felt on its medial side.

Nerve Supply. Biceps like Brachialis receives branches from the musculocutaneous branch (C. 5, 6, 7) of the brachial plexus.

3. Triceps Brachii

Origin. Of the three heads of this, the only muscle at the back of the arm, only the long head arises from the scapula while the lateral and medial heads arise from the humerus (fig. 260). The origin of the **long head** is from the infraglenoid tubercle, a rough area immediately below the glenoid fossa. The **lateral head** has a linear origin extending from just below the back of the greater tubercle of the humerus for 5 to 8

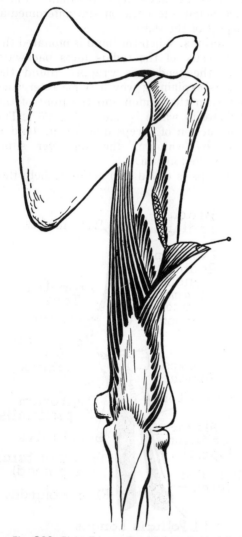

Fig. **260**. Right Triceps Brachii from behind. Lateral head cut to show upper fibers of medial head.

cm (2–3″) down the shaft. The long and lateral heads lie side by side. Together they are comparable to the Biceps in front (fig. 262). Deep to them, lower down on the humerus, lies the **medial head** which has a fleshy origin from the back of the shaft similar to that of Brachialis from the front. Both Brachialis and the medial head of Triceps pick up extra origin from the tough sheets of fibrous tissue which separate them beyond the medial and lateral borders of the humerus—the *medial and lateral intermuscular septa.*

Insertion. The three heads converge, in the lower part of the arm, into a broad aponeurosis which narrows to a stout tendon. This inserts chiefly into the posterior edge of the upper surface of the olecranon process (fig. 261).

Actions and Uses. Triceps is a powerful muscle in spite of the fact that, on most occasions, it enjoys the assistance of gravity. It is the only important extensor muscle of the elbow (see Anconeus, below). Undoubtedly, its large size is chiefly due to its function of holding the elbow steady when the upper limb is used in pushing against resistance. It acts in a lever system of the first class but its lever arm is extremely short (fig. 198, page 117). (See table 4.)

TABLE 4
Muscles Acting in Elbow Joint

Flexion	Extension
1. Brachialis	1. Triceps
2. Biceps	(2. Anconeus)
3. Brachioradialis	
(4. Pronator Teres)	

The long head may help also in extending the shoulder joint.

Nerve Supply. The three heads receive one or two branches each from the radial nerve (C. 6, 7, 8), a terminal branch of the posterior cord of the brachial plexus.

Anconeus, a detached part of the Triceps, is described separately on page 157.

Muscles of Forearm and Hand

Antecubital Fossa. The rather hollow region of the front of the elbow, the (ante)-cubital fossa, contains the tendons of Biceps and Brachialis, the brachial artery and the median nerve, and is, therefore, an important region.

First consider five muscles in the forearm that act either on the elbow joint or on the three radio-ulnar joints where pronation and supination occur. They are: **(1) Brachioradialis, (2) Supinator, (3) Anconeus, (4) Pronator Teres,** and **(5) Pronator Quadratus.**

1. Brachioradialis

This important muscle (figs. 263, 270) is often included with the extensor muscles of the forearm (because of its location and nerve supply), but, in fact, it devotes itself entirely to the elbow.

Attachments. Its name indicates the bones that it joins—the humerus (the bone of the brachium or arm) and the radius. It has its **origin** on the upper two-thirds of the lateral supracondylar ridge which extends from the lateral epicondyle up the shaft of the humerus; it forms the lateral boundary of the antecubital fossa and forms a marked fullness on the lateral side of the forearm. It is **inserted** by a long tendon on the base of the styloid process of the radius.

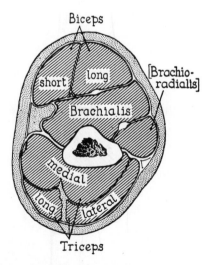

Fig. **262.** Cross-section of arm (semischematic) showing how three heads of Triceps are 'mirror image' of Biceps and Brachialis.

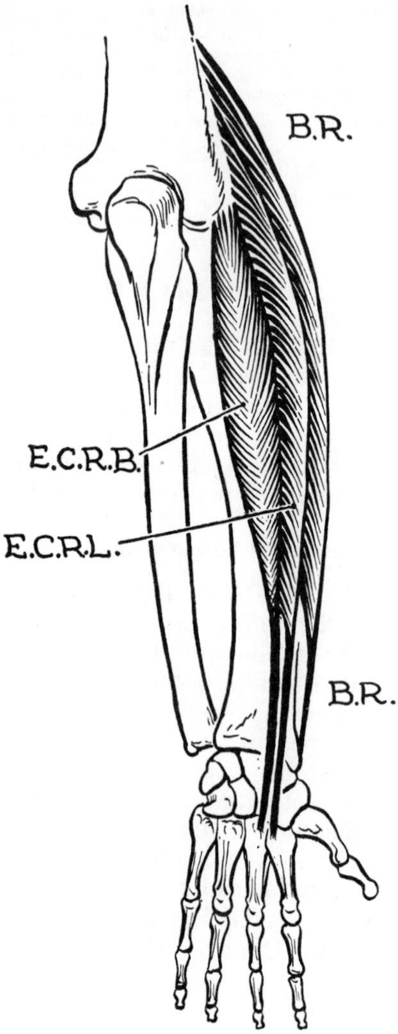

B.R.

E.C.R.B.

E.C.R.L.

B.R.

Fig. **263**. Brachioradialis (B.R.); Extensor Carpi Radialis Longus; E.C.R. Brevis.

Action and Function. Brachioradialis is a strong flexor of the elbow only with heavy loads or during rapid flexion and usually when the forearm is semi-pronated by other agencies. [Not least of these is the normal tendency of the passive structures (bones, joints, and ligaments) to keep the forearm in this semi-prone or 'natural' position; this is the position of rest; it is also the position of greatest advantage for most functions of the upper limb.]

The **nerve supply** is via the radial nerve, a terminal branch of the brachial plexus (C. 5, 6).

2. Supinator

The **origin** is from a crest which descends from the back of the radial notch of the ulna. Its upper fibers gain additional origin from the neighboring parts of the back of the fibrous capsule. From their origin, the fibers run obliquely downward and lateralward behind the radius to wrap around the posterior and lateral aspects of the upper quarter of that bone (figs. 261, 264).

The **insertion** is into most of the area between the oblique lines of the radius; (these lines descend on the anterior and posterior surfaces of the radius from the radial tuberosity to the midpoint of the lateral aspect of the bone).

Actions and Uses. The Supinator supinates the forearm (fig. 265) when no more power is needed than that which is sufficient to overcome the inertia of the joint. Biceps is a reserve which is called upon to help Supinator against resistance. When powerful rotatory forces (e.g., driving a screw) are required, rotation of the shoulder joint accompanies pronation and supination. (See table 5.)

The **nerve supply** comes from the deep branch of the radial nerve which burrows through Supinator on its way to the extensor muscles of the forearm (C. 5, 6).

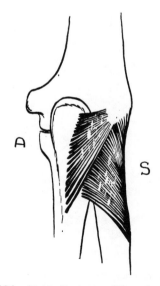

Fig. **264**. Right Supinator (S) and Anconeus (A) from behind.

Fig. **265**. How the two supinators act

TABLE 5
Muscles of Pronation-Supination of Forearm

Pronation	Supination
1. Pronator Quadratus	1. Supinator
2. Pronator Teres	2. Biceps
(3. Flex. Carpi Rad.?)	(3. Abd. Pollicis L.) ⎫
	(4. Ext. Poll. L.) ⎭ ?

3. Anconeus

A small triangular muscle (fig. 264), Anconeus (Gr. = elbow), arises from the back of the lateral epicondyle of the humerus and inserts into a flat triangular area on the ulna behind Supinator.

The muscle is described as a detached part of the Triceps—which it is—having the same action—which it has.

Its **nerve** is a twig of the radial nerve.

4. Pronator Teres

This fusiform muscle (figs. 266, 270) crosses the front of the upper half of the forearm by running obliquely downward from its origin on the medial epicondyle of the humerus to its insertion on the middle of the lateral aspect of the shaft of the radius where the bowing of the radius reaches its maximum.

Actions. The Pronator Teres is specialized for pronation, but because of its position it is said to aid in flexion of the elbow and no doubt it can help when great force is required.

Its **nerve** is a branch of the median nerve.

5. Pronator Quadratus

This muscle, as its name implies, is rectangular. It runs transversely from ulna

to radius across the front of the shafts just above the wrist joint (figs. 266, 270–272). Being deep to many structures, it is difficult to observe or feel.

Its **origin** is an aponeurotic one from a line on the front of the ulna; its **insertion** covers the anterior aspect of the lowest one-quarter of the radius.

Actions. Much larger in cross-section than Pronator Teres and in a more strategic position, Pronator Quadratus is a much more important pronator than is generally recognized. It is active in all ordinary movements of pronation whereas Pronator Teres is called upon to assist when resistance is encountered (cf., Supinator and Biceps).

Its **nerve,** too, is a branch of the median nerve.

DISPOSITION OF MUSCLES IN FOREARM

The front of the elbow joint is crossed by the Biceps and Brachialis and the back is crossed by the Triceps, so that those flexor and extensor muscles of the forearm seeking origin from the humerus are forced to occupy the epicondyles at the sides, the flexors clinging to the medial epicondyle, the extensors to the lateral.

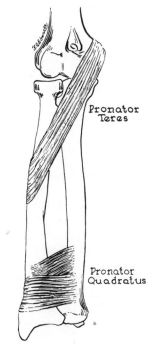

Fig. **266**. Right Pronator Teres and Pronator Quadratus.

Below the elbow joint, there is nothing to prevent other and deeper flexors and extensors from having their origins on the fronts and backs, respectively, of the ulnar and radial shafts (fig. 272). The posterior border of the ulna—subcutaneous and palpable from olecranon to wrist—is the dividing line at the back of the forearm between the extensor and the flexor territory.

All flexor tendons are restricted to the front of the wrist and most of them crowd under the bridge formed by the **flexor retinaculum** (page 58).

FLEXORS IN THE FOREARM

There are six flexors in the forearm arranged in three layers, and the deeper the layer the more distal the insertions of its muscles. In general, the three flexors of the superficial layer act only on the wrist and carpal joints; they are inserted on the bases of the metacarpals. The one flexor of the middle layer reaches the middle phalanges and acts on all intermediate joints in passing. The two flexors of the deep layer reach the terminal phalanges and similarly act on the intermediate joints.

The Three Superficial Flexors—Flexor Carpi Radialis, Palmaris Longus, Flexor Carpi Ulnaris

Origins. These arise (together with the Pronator Teres) from the medial epicondyle of the humerus and their mass bounds the medial side of the antecubital fossa (figs. 267, 270). They are separated from one another by intermuscular fibrous septa from which they gain additional origin. Flexor Carpi Ulnaris arises by a broad aponeurosis from the upper two-thirds of the posterior border of the ulna.

The Palmaris Longus is sometimes missing but never missed. When present, its somewhat delicate tendon stands forward prominently in the exact midline when the wrist is flexed against resistance.

Insertions. The stout tendon of Flexor Carpi Radialis can be felt a little lateral to the midline of the wrist. It crosses the distal skin crease of the wrist in contact with the tubercle of the scaphoid and is inserted on the bases of metacarpals 2 and 3.

The Palmaris Longus is inserted into a triangular fibrous sheet, the **palmar apo-**

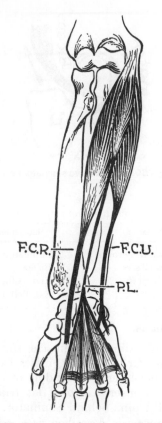

Fig. **267**. The three (right) superficial flexors— Flexor Carpi Radialis, Palmaris Longus, Flexor Carpi Ulnaris.

neurosis (fig. 268), which occupies the palm of the hand and is there firmly bound to the deeper layers of the skin. The aponeurosis is an important protection for underlying structures and it accounts for the fact that the skin of the palm cannot be readily pinched up. The aponeurosis is present even when the muscle is absent.

The tendon of the Flexor Carpi Ulnaris can be felt in front of the ulnar border of the wrist where it is inserted at once on the pisiform bone—which also can be felt at the distal skin crease. The pisiform, in turn, is anchored distally by two strong, short ligaments, one going to the hamate and the other to the base of metacarpal 5; thus the pisiform is regarded as a sesamoid bone developed in the tendon.

Actions and Functions. All three flex the wrist, although Palmaris is not very powerful. The Flexor Carpi Radialis is not an important abductor of the wrist (for that matter, the wrist can be abducted very little from 'neutral'), but the Flexor Carpi Ulnaris is an important adductor, acting in concert with the Extensor Carpi Ulnaris. (See Table 6.)

All the muscles of the wrist are quite important for their synergistic use in stabilizing the wrist so that the fingers and fist work to best advantage.

Nerve Supply. Flexor Carpi Radialis and Palmaris Longus receive branches from the median nerve; the Flexor Carpi Ulnaris, from the ulnar nerve.

The Intermediate Flexor—Flexor Digitorum Superficialis (F.D. Sublimis)

The **origin** of this muscle (figs. 269, 271, 272) is from a curved *oblique line* that reaches from the medial epicondyle of the humerus to the middle of the shaft of the radius. Where the origin crosses the interval between the ulna and the radius, it consists of a fibrous arch or span which

allows the passage beneath it of the important median nerve and ulnar artery.

Course and Structure. From the origin, muscle fibers descend vertically, and above, they occupy the width of the forearm. As the wrist is approached the muscle splits and becomes four tendons which lie in the midline of the wrist and are arranged two in front (for the middle and the ring finger) and two behind (for the index and

Fig. **269**. Flexor Digitorum Superficialis (Sublimis) (cf., fig. 274).

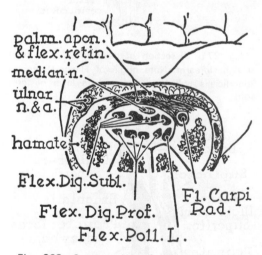

Fig. **268**. Cross-section of right carpal tunnel (semischematic).

TABLE 6

Muscles Acting on Wrist (Radiocarpal) Joint and Midcarpal Joint

Flexion	Extension	Abduction	Adduction
1. Fl. Carpi Rad.	1. Ext. Carpi Rad. L.	1. Ext. Carpi Rad. L.	1. Ext. Carpi Uln.
2. Fl. Carpi Uln.	2. Ext. Carpi Rad. Br.	2. Ext. Carpi Rad. Br.	2. Fl. Carpi Uln.
3. Palm. Longus	3. Ext. Carpi Uln.	3. Abd. Poll. L.	
(4. Abd. Poll. Longus)		4. Fl. Carpi R.	

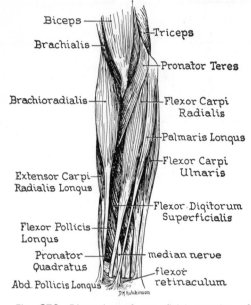

Fig. **270.** Dissection of superficial muscles of front of right arm.

the little finger); the radial head of the muscle belongs to the middle finger tendon.

The four tendons pass deep to the flexor retinaculum (fig. 268), traverse the palm, and diverge to the roots of the four fingers (figs. 269, 283, 284); each is accompanied behind by a tendon of the Flexor Digitorum Profundus, to be described in a moment. In front of the first phalanx each Superficialis tendon splits to permit the passage through it of the Profundus tendon (fig. 274).

The tendons run within osseo-fibrous tunnels in the fingers (*fibrous flexor sheaths*) to prevent bow-stringing and they are provided with *synovial sheaths* to reduce friction in the tunnels (fig. 273).

Insertion. The details of the tendon insertion are shown in figure 274 (see also fig. 284). The arrangement is such that it ensures free movement of the Profundus tendon through the 'perforation' of the Superficialis tendon.

Actions and Uses. The chief joint that Superficialis flexes is the proximal interphalangeal joint. The Superficialis and Profundus act to clench the fist, but since powerful clenching is possible only if the wrist is not flexed, the two muscles are

prevented from flexing the wrist by synergistic action of the wrist extensors.

The **nerve supply** is from the median nerve—which lies in naked contact with

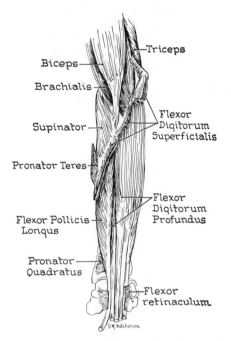

Fig. **271.** Dissection of deep muscles of front of right forearm. Note origin of Flexor Digitorum Superficialis (which is intermediate layer of forearm).

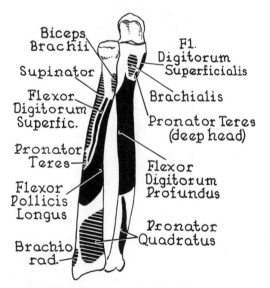

Fig. **272.** Muscular origins (black) and insertions (striped) on front of right radius and ulna.

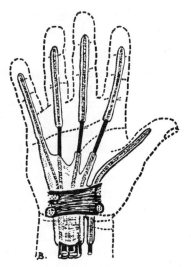

Fig. **273**. Scheme of synovial sheaths for flexor tendons in fingers and running deep to flexor retinaculum at wrist.

the muscle after passing deep to its fibrous arch.

The Two Deep Flexors—Flexor Digitorum Profundus and Flexor Pollicis Longus

Origins. Between the origin of the Flexor Digitorum Superficialis above and the Pronator Quadratus below, there remains an extensive smooth surface composed of the middle two quarters of radius, ulna, and interosseous membrane. From the lateral half (i.e., radius and adjoining membrane) arises the long flexor muscle of the thumb (L. = pollex); the remaining surface is for the deep flexor of the fingers (fig. 272). These origins are fleshy and leave no markings on the bones.

Courses and Insertions. The two muscle-bellies narrow down to five tendons which pass, side by side, as the deepest structures in the *carpal tunnel* (behind the flexor retinaculum) into the hand and spread to the five digits (figs. 275, 283, 284). They accompany and then 'pierce' the Superficialis tendons as described above. Their insertions are into the bases of the distal phalanges of the four fingers.

The long tendon of Flexor Pollicis Longus runs unaccompanied to the base of the terminal phalanx of the thumb and has a

fibrous tunnel and a synovial sheath similar to those of the fingers.

Actions. Both muscles chiefly flex distal interphalangeal joints (fig. 276).

Nerve Supply. The median nerve supplies all of the Flexor Pollicis Longus and the adjoining half of the Flexor Digitorum Profundus; the medial half of Profundus is supplied by the ulnar nerve.

EXTENSOR MUSCLES IN THE FOREARM

Dispositions. Only two layers of extensor muscles exist, a superficial and a deep. Even this statement needs qualifying for, halfway down the forearm, the superficial layer is split into a medial and a lateral group by the deep layer out-cropping between them to become superficial also.

The superficial extensors have a common origin from the lateral epicondyle of the humerus and consist of: **Extensor Carpi Radialis Longus** and **Extensor Carpi Radialis Brevis** which lower down become the lateral group (fig. 279); and **Extensor Digitorum,** *Extensor Digiti Minimi,* and **Extensor Carpi Ulnaris** which lower down become the medial group (figs. 279, 280, 282). The deep extensors come to the surface between the Ex-

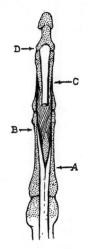

Fig. **274**. Insertions of Flexor Digitorum Profundus (at D) and Flexor Digitorum Superficialis (at C). Latter tendon splits (at A) and some of its fibers cross (at B) to opposite side.

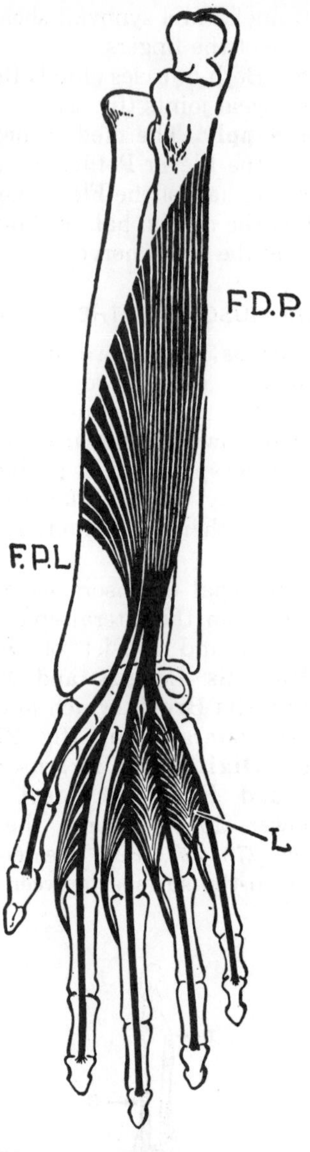

FD.P

F.P.L

L

Fig. **275**. Flexor Digitorum Profundus (F.D.P.) and Lumbricals (L.) arising from its tendons are in same deep plane as Flexor Pollicis Longus (F.P.L.).

tensor Carpi Radialis Brevis and the Extensor Digitorum.

Extensor Retinaculum. The extensor tendons occupy the lateral side as well as the back of the wrist and, lying in bony grooves, are held in place by a transverse band of fibrous tissue called the **extensor retinaculum.** From the deep surface of the extensor retinaculum a series of fibrous partitions pass to the margins of the bony grooves, and through six com-

partments thus produced run the 12 extensor tendons (fig. 277).

Superficial Extensors

Origins. The **Extensor Carpi Radialis Brevis, Extensor Digitorum,** and *Extensor Digiti Minimi* confine their origins to the common tendon attached to the lateral epicondyle. The **Extensor Carpi Radialis Longus** arises not only from the lateral epicondyle of the humerus but also above the epicondyle, from the lower third of the supracondylar ridge; it is this fact that distinguishes the muscle as the 'longus.' [This muscle is closely associated with the Brachioradialis which, it will be recalled, arises from the upper two-thirds of the supracondylar ridge.]

Extensor Carpi Ulnaris spreads its origin down from the common tendon to include an aponeurosis from the upper two-thirds

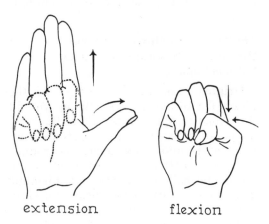

extension   flexion

Fig. **276**. Flexion and extension of thumb and fingers defined.

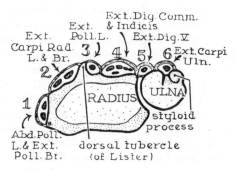

Fig. **277**. Six tunnels enclose the extensor tendons at the wrist.

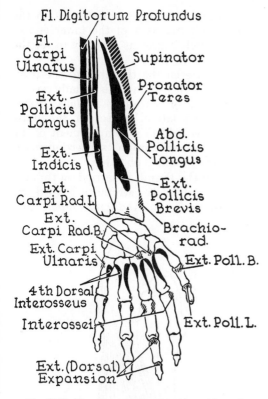

Fl. Digitorum Profundus
Fl. Carpi Ulnarus
Supinator
Pronator Teres
Ext. Pollicis Longus
Abd. Pollicis Longus
Ext. Indicis
Ext. Pollicis Brevis
Ext. Carpi Rad. L.
Ext. Carpi Rad. B.
Brachio-rad.
Ext. Carpi Ulnaris
Ext. Poll. B.
4th Dorsal Interosseus
Interossei
Ext. Poll. L.
Ext. (Dorsal) Expansion

Fig. **278**. Muscular origins (black) and insertions (striped) of back of forearm and hand.

of the posterior border of the ulna (cf., Flexor Carpi Ulnaris).

Insertions. The extensors of the wrist imitate the flexors, inserting on the bases of the same metacarpals. The Extensor Carpi Radialis Longus inserts on the base of metacarpal 2, the Brevis on 3, and Extensor Carpi Ulnaris on 5 (fig. 278).

The **Extensor Digitorum** (fig. 280) splits into four round tendons a little above the wrist. These run distally down the back of the hand to the bases of the fingers, where they flatten and spread out into what are known as the **expansions of the extensor tendons.** Since these expansions are joined by the tendons of important little muscles of the hand, they will be more fully considered later. It is enough to note now that the expansions are the sole tendinous structures on the backs of the fingers and are, therefore, responsible for finger extension (see also page 169 and figs. 290, 291).

At the backs of the metacarpals, oblique fibrous bands unite adjacent extensor tendons. The tendon of the index finger is freest of these restricting bands and consequently enjoys the most complete individual movement.

The *Extensor Digiti Minimi* is merely a detached medial slip of the Extensor Digitorum (fig. 282). It provides a second and more or less separate tendon for the little finger, although it is hard to see why the little finger needs this double equipment. At the base of the finger, the tendon merely contributes to the formation of the extensor expansion and so loses its identity.

Actions. Since the contributions that Extensor Digitorum itself makes to the extensor expansions can be traced no farther than the bases of the middle phalanges, the muscle can have no power to extend the terminal phalanges. Indeed, its principal action is to extend the metacarpophalangeal joints and for them it is the sole extensor.

The Extensores Carpi Ulnaris, Radialis Brevis, and Radialis Longus extend the wrist and acting with their companion flexors produce side to side motion at the wrist (adduction and abduction).

The important *function* of the extensors as synergistic stabilizers of the wrist has been mentioned earlier.

Deep Extensors

Three deep out-cropping extensor muscles are almost exclusively devoted to what may properly be called 'supination of the thumb.' The movement is the contrary one to that performed when the thumb is 'pronated' in order to oppose it to the other digits in such an everyday action as, say, picking up a pencil. A fourth and lowest muscle is present which joins the tendon of Extensor Digitorum of the index finger; it can, of course, produce only extension of the index.

In a series from above downward the out-cropping muscles are: **Abductor Pollicis Longus,** *Extensor Pollicis Brevis,* **Extensor Pollicis Longus,** and *Extensor Indicis* (figs. 281, 282).

Their adjoining **origins** are from the

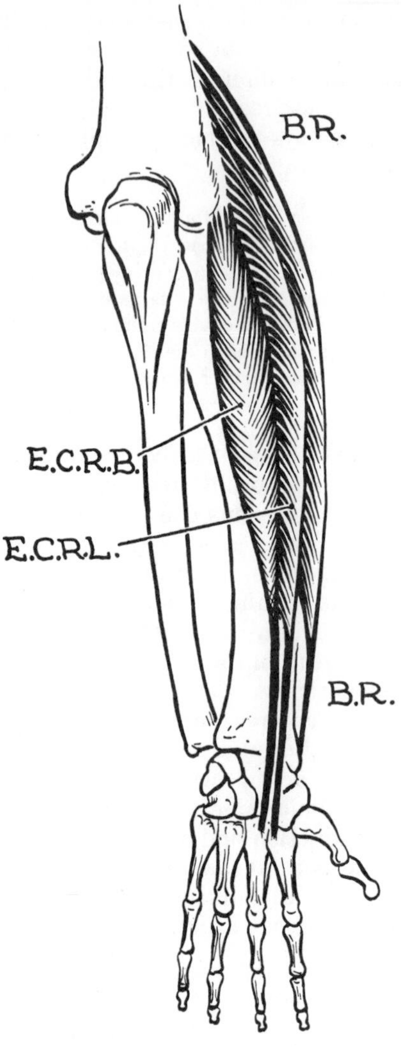

B.R.

E.C.R.B.

E.C.R.L.

B.R.

Fig. **279**. Brachioradialis, Extensor Carpi Radialis Longus and E.C.R. Brevis are closely related.

back of the interosseous membrane (and adjacent bones) below the level of Supinator. (The exact origins are unimportant.)

Insertions and Actions. The actions of these muscles depend on the sites of insertion of their tendons and the direction of pull; these are easily investigated by the student on his own hand. Hold the wrist constantly straight with the palm facing the floor and draw the thumb laterally and backward to point at the ceiling. Now the tendon of *Extensor Pollicis Longus* will be seen running to its insertion on the dorsum of the base of the distal phalanx of the

thumb and it can be traced back to the tubercle on the back of the lower end of radius which it uses as a pulley to set the direction of its pull.

Next, stretch the thumb into the position of its widest span. This usually reveals the tendon of *Extensor Pollicis Brevis* crossing the lateral aspect of the wrist (where it is held in a compartment of the extensor retinaculum) to its insertion on the base of the proximal phalanx.

Now, if the thumb is carried on through an arc until its tip stretches toward the floor, the tendon of *Abductor Pollicis Lon-*

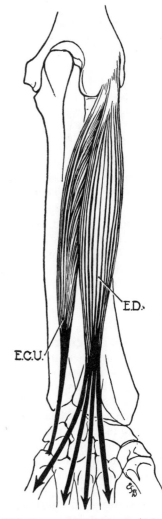

E.D.

E.C.U.

Fig. **280**. Extensor Carpi Ulnaris (E.C.U.) and Extensor Digitorum (Communis) (E.D.) (see also fig. 290).

gus can be felt and seen as the most lateral tendon on the anterior aspect of the wrist. Its insertion is on the base of the first metacarpal. At the wrist, its tendon is in a compartment of the extensor retinaculum which is far enough forward for the muscle to abduct the thumb (see Thumb Movements, page 95), to flex the wrist, and to assist the extensors of the wrist and thumb in their stabilizing functions, one of the most significant being that of preventing the hand from being forced ulnarward (as in holding a heavy jug by its handle).

Between the tendons of the Extensor Pollicis Longus and the closely related

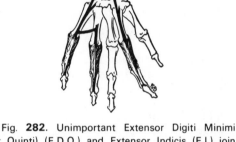

Fig. **282.** Unimportant Extensor Digiti Minimi (or Quinti) (E.D.Q.) and Extensor Indicis (E.I.) join common extensor tendons to little and index fingers.

Abductor Longus and Extensor Brevis, there will appear a hollow, called the **anatomical snuff box.**

Nerve Supply of the Extensor Muscles of the Forearm. All of the extensor musculature of the forearm, including the Abductor Pollicis Longus, is supplied by the radial nerve or its large deep terminal branch, the posterior interosseous nerve.

MUSCLES IN THE HAND

The 18 small but important muscles in the hand fall into several convenient categories or groups. These groups will be discussed in the order in which they are met during a dissection and their nerves will be discussed at the end (page 168). Finally, a summary of the effects of the complex, aponeurotic, extensor expansions of the fingers will be given.

Thenar and Hypothenar Muscles**

General. The **flexor retinaculum** was identified as a rather square, tough, and

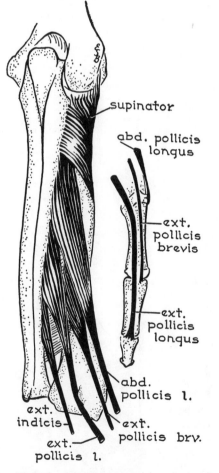

Fig. **281.** The deep extensors (out-cropping muscles) on back of right forearm. They lie in same deep plane as Supinator. Insertion of three to thumb shown in smaller figure.

_____

** Thenar, Gr. = hand, came to refer to the ball of the thumb; hypo, Gr. = under.

fibrous ligament that, by its attachments to the four bony prominences of the marginal carpal bones, formed a bridge deep to which long tendons ran to enter the palm. This retinaculum and these prominences give origin to six muscles, three of the ball of the thumb—the **thenar muscles,** and three of the ball of the little finger—the **hypothenar muscles.** Owing to the mobility of the thumb metacarpal, the thenar muscles are much the more important and the hypothenar muscles are their mirror-image (figs. 283–285).

The **thenar muscles** are named: **Abductor Pollicis Brevis, Opponens Pollicis,** and *Flexor Pollicis Brevis*. The corresponding **hypothenar muscles** are named *Abductor Digiti Minimi, Opponens Digiti Minimi,* and *Flexor Digiti Minimi Brevis* (figs. 283–285).

The strap-like **Abductors** lie on the borders of the hand and **arise from** the proximal part of the flexor retinaculum and the proximal marginal prominences. Each is **inserted** on the 'abducting' side of the base of the first phalanx of its own digit. The one lifts the thumb forward (abduction) (fig. 286); the other pulls the little finger away from the ring finger.

The two **Opponens** muscles, largely hidden by the Abductors, **arise from** the distal parts of the flexor retinaculum and from the distal marginal prominences.

They are **inserted** along the whole lengths of their respective metacarpal shafts. Because the Opponens Pollicis rolls the metacarpal of the thumb toward the midline of the hand and opposes it to the fingers, it is very important.

The *short flexors* of the thumb and the little finger arise from the distal margin of the flexor retinaculum and each is inserted on the palmar aspect of the base of the proximal phalanx of its own digit. Although both are small, the Flexor Pollicis Brevis is important.

The Four Lumbricals

Arising from the radial (lateral) side of each of the four tendons of Flexor Digitorum Profundus in the palm—and from the ulnar (medial) side of a neighboring Profundus tendon where possible (figs. 275,

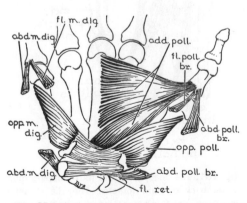

Fig. **284.** Deeper dissection of muscles of palm.

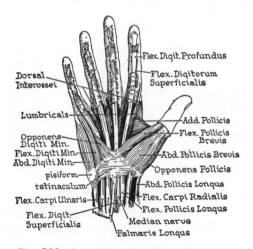

Fig. **283**. Superficial dissection of muscles of palm.

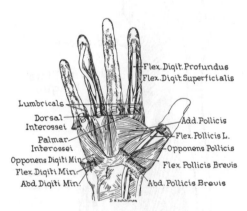

Fig. **285.** The three thenar muscles, three hypothenar muscles, and Adductor Pollicis in right palm.

TABLE 7
Muscles Acting on Fingers

Metacarpophalangeal Joints				Proximal Interphalangeal Joints		Distal Interphalangeal Joints	
Flexion	Extension	Abduction	Adduction	Flexion	Extension	Flexion	Extension
1. Interossei 2. Lumbricals (3. Long flexors)	Long extensors	Dorsal Inteross. Abd. Dig. V	Palmar Inteross.	1. Flex. Dig. Superfic. (2. Flex. Dig. Prof.)	1. Interossei 2. Lumbricals (3. Long extensors)	Flex. Dig. Prof.	1. Interossei 2. Lumbricals

TABLE 8
Muscles Acting on Thumb

Carpometacarpal Joint					Metacarpophalangeal Joint		Interphalangeal Joint	
Flexion	Extension	Abduction	Adduction	Opposition	Flexion	Extension	Flexion	Extension
Flexor Poll. Br.	1. Abd. Poll. (2. Ext. Poll. Br.) (3. Ext. Poll. L.)	1. Abd. Poll. L. 2.. Abd. Poll. Br.	1. Add. Poll. (2. 1st Dorsal Inteross.)	Opponens Poll.	Flex. Poll. L. Flex. Poll. Br.	1. Ext. Poll. Br. (2. Ext. Poll. L.)	Fl. Poll. L.	Ext. Poll. L.

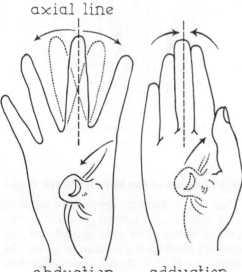

axial line

abduction   adduction

Fig. **286**. Abduction and adduction of thumb and fingers defined.

283)—are these well-named, worm-like muscles.

The delicate tendon of **insertion** of each one runs distally to join the expansion of the extensor tendon on the back of the finger.

Actions. The lumbrical muscles are said to place the hand in the 'writing' or 'billard cue' position (metacarpophalangeal joints flexed, interphalangeal joints extended). (See table 7.)

Adductor Pollicis

Deep to the long flexor tendons and Lumbricals and before the 'floor' of the palm is reached, there stretches between the third metacarpal and the thumb a fleshy triangle of muscle known as the Adductor Pollicis. It arises by its base from the length of the shaft of the third (middle or 'axial') metacarpal bone (and from the carpals adjacent to its base) and it is inserted by its apex on the medial side of the base of the first phalanx of the thumb—an insertion which it shares with the little flexor (figs. 284, 285).

Actions and Uses. As its name indicates, it draws the thumb toward the palm (see Thumb Movements, page 95 and fig. 286). It gives the power to the grasp which is sorely missed when this muscle is paralyzed. (See table 8.)

The Seven Interossei

Since the *axis of the hand* is the third metacarpal and the third (middle) finger, it is convenient to define all movements toward this axis as adduction and all movements away from this axis as abduction. In conformity with this convention it is to be observed that abduction of the middle finger itself can be to either side, adduction simply restoring that finger to the axial line.

For the performance of the movements of abduction and adduction it is apparent that each of the five digits needs to be equipped with two muscles—a total of ten. Since the thumb has its own special abductor and adductor and the little finger its own abductor, only seven muscles remain to be provided. Between the metacarpals and arising from their shafts lie these seven muscles appropriately known as the **Interossei.**††

A little observation and thought will reveal that four of these seven are required for abduction (index 1, middle 2, ring 1) and three are required for adduction (index 1, ring 1, little finger 1).

Because the four abducting Interossei **arise from** the opposite surfaces of adjoining metacarpals (fig. 287), they are best seen in the back of the hand where they fill the intervals between the metacarpals; they are known as the **Dorsal Interossei.** The three adducting ones are known as the **Palmar Interossei;** they arise from the anterior borders of metacarpals 2, 4, and 5 (figs. 284, 288).

All seven tendons of **insertion** join the extensor expansions more proximally than do the Lumbricals (figs. 289–291).

Actions. Besides their principal actions to spread and bring together the four fingers, they also have the same actions as the smaller Lumbricals, i.e., flexion of the metacarpophalangeal (knuckle) joints and extension of the interphalangeal joints.

---

†† Singular = Interosseus; the adjectival form = interosseous.

Fig. **287**. The four Dorsal Interossei spread fingers 'apart' (away from axial line).

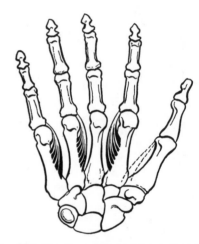

Fig. **288**. The three Palmar Interossei move fingers 'together' (toward axial line).

Nerve Supply of the Muscles in the Hand

The thenar muscles are usually supplied by a special (recurrent) branch of the median nerve. The only muscles of the hand supplied by the median nerve are the lateral two (i.e., first and second) Lumbricals. All of the remaining muscles (hypothenars, Interossei, and Adductor Pollicis) are always supplied by the ulnar nerve. Not infrequently the ulnar nerve extends its supply to all the Lumbricals and even part of the thenar muscles. Indeed, in a small percentage of hands *all* of the muscles are supplied by the ulnar nerve.

Actions of Extensor Expansions of Fingers

On the backs of the fingers are the complex aponeurotic extensor (or dorsal) expansions composed of the tendons of the Interossei and Lumbricals (figs. 290, 291).

When the elongated tendons of the Ex-

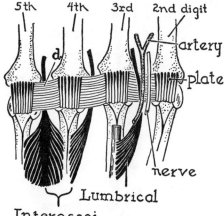

Fig. **289.** Scheme of relationships of tendons, arteries and nerves to deep transverse ligaments (d) of palm.

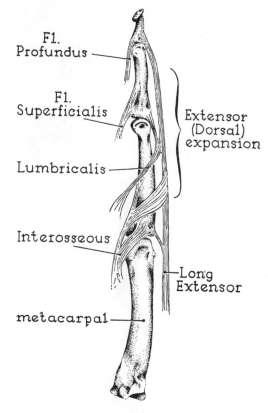

Fig. **291.** Insertions of the tendons of a finger

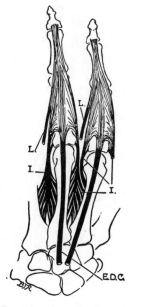

Fig. **290.** Posterior view of right middle and ring fingers to show composition of (dorsal) expansions of extensor tendons (E.D.C.). I. = Interossei; L. = Lumbricals.

tensor Digitorum (and Extensores Indicis et Digiti Minimi) enter the fingers and flatten out, they occupy the midline of the fingers and can be traced to the bases of the second phalanges but no farther. It has been pointed out that the long extensors are principally concerned with the knuckle row of joints for which they are the sole extensors, and that, even though they may be traced just beyond the first row of interphalangeal joints, they have only some weak extending action on these joints and no action on the distal joints of the fingers. It is the side portions of the expansions, made up entirely of Lumbrical and Interosseous tendons, that—proceeding onward to the bases of the terminal phalanges—are responsible for extension of proximal and distal interphalangeal joints. Both joints are extended synchronously.

There are many other details which are only of importance to the plastic surgeon.

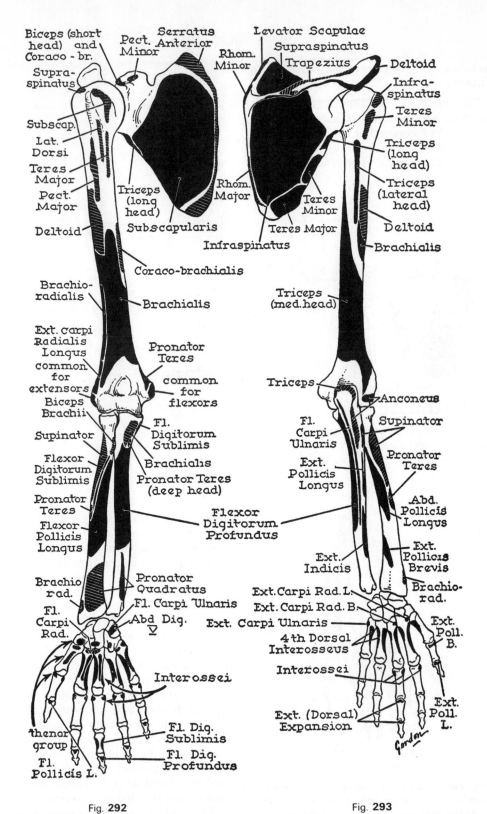

Fig. **292**

Fig. **293**

Fig. **292.** Anterior view of origins (solid black) and insertions (hatched) of muscles of upper limb.
Fig. **293.** Posterior view of origins (solid black) and insertions (hatched) of muscles of upper limb.

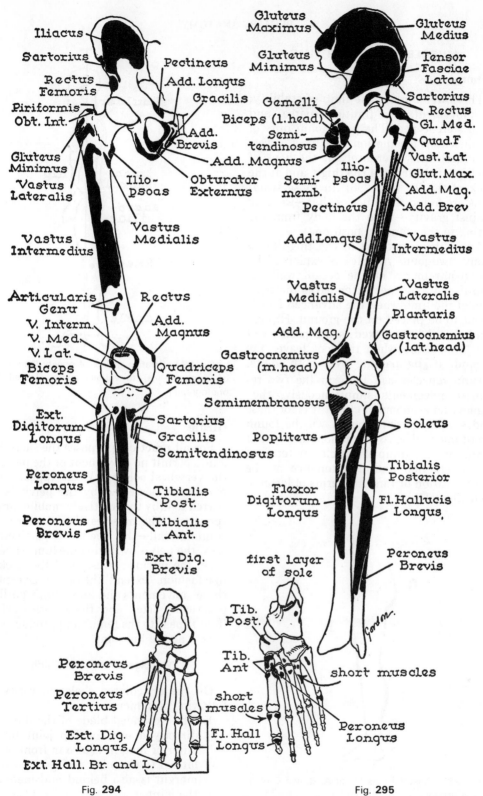

Fig. 294

Fig. 295

Fig. 294. Anterior view of origins (solid black) and insertions (hatched) of muscles of lower limb.

Fig. 295. Posterior view of origins (solid black) and insertions (hatched) of muscles of lower limb.

MUSCLES OF LOWER LIMB

General

The lower limb is the limb of stability; its movements are limited, coarse, and often stereotyped. The paramount duty of its muscles is locomotion. The most powerful muscles lie alternately at the back of the hip, at the front of the thigh, and at the back of the leg (fig. 296). In those situations are found the muscles that—working against gravity—raise the body from the sitting to the standing position.

Again, the existence in the thigh of so many 'two-joint' muscles is explained by the requirements of the act of walking. Hip flexed and knee extended is the position of the limb when the advanced foot is about to be placed on the ground. Hip extended and knee flexed is the position of the 'hind' limb when the foot leaves the ground. If, by passing over both joints, certain muscles can perform the two required movements simultaneously, they achieve an economy of effort. It is for these and similar reasons that it will be found profitable to discuss the actions and uses of a muscle, or a group of muscles, in terms of the production and maintenance of the erect position or (and) in terms of locomotion (fig. 297).

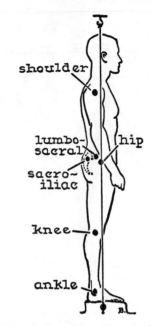

Fig. **297**. Relation of line of gravity (here shown by a plumb-line) to various joints. (Traced from photograph.)

Girdle

For all practical purposes the sacro-iliac joints permit no movement of the pelvis on the vertebral column. What appears to be movement possessed by the pelvis is imparted to it by the relatively mobile lumbar vertebrae. The muscles of the anterior abdominal wall—the Rectus in particular—'flex' the pelvis by flexing the lumbar vertebrae; the intrinsic muscles of the back, in like fashion, 'extend' the pelvis. Except for these movements the lower limb girdle is practically fixed and the muscles arising from it pass to the femur and produce their effects at the hip joint.

Muscles of Hip and Thigh

On the articulated skeleton, verify the following statements.

1. The expanded blade of the ilium lies directly above the hip joint but its anterior margin is also in front of and lateral to the joint, while its posterior margin is also behind and medial to the joint.

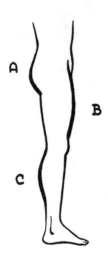

Fig. **296**. Powerful muscles at A, B, and C work against gravity.

2. The pubic portion of the hip bone lies medial to, and in front of the joint.

3. The ischial portion of the hip bone lies below, medial to, and behind the joint.

4. The greater trochanter of the femur stands farther laterally than any part of the hip bone, so that all muscles inserted on the greater trochanter approach it from a medial direction. By pulling the trochanter upward, the thigh is abducted; by rolling the trochanter forward, the thigh is rolled or rotated medialward; by pulling the trochanter backward, the thigh is rolled or rotated lateralward.

The muscles of the hip and thigh will be described in the following groups.

   A. Muscles crossing the front of the hip joint.

   B. The three gluteal muscles (and the Tensor Fasciae Latae).

   C. The six lateral rotators.

   D. Muscles of the front of the thigh.

   E. Muscles of the medial side of the thigh.

   F. Muscles of the back of the thigh.

A. MUSCLES CROSSING THE FRONT OF THE HIP JOINT

Psoas Major and Iliacus

The two muscles lie side by side and, because they share a common tendon of insertion and have a common action, are often regarded as one—the **Iliopsoas.**

Origins. The Iliacus occupies, and arises from, the hollow of the false pelvis, i.e., the inner surface of the expanded iliac blade; the Psoas clings to the sides of the bodies of the lumbar vertebrae, their discs, and transverse processes. A few fibers reach up to the 12th thoracic vertebra within the thorax. The Psoas is an elongated fleshy muscle corresponding to the tenderloin of beef (fig. 298).

The two muscles come together, fuse, and cross the front of the hip joint immediately in front of the femoral head.

Insertion. The stout tendon of insertion passes downward, backward, and slightly

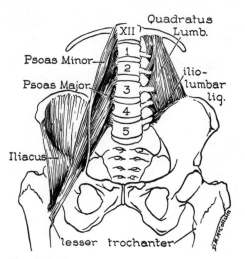

Fig. **298.** Iliacus and Psoas are one muscle—Iliopsoas.

lateralward to be inserted on the lesser trochanter of the femur.

Actions and Functions. Both parts of the Iliopsoas flex, abduct and laterally rotate the hip joint. Probably this muscle contributes to the stability of the hip during walking and standing upright. The Psoas alone may act upon the lumbar vertebrae but what movement it produces there is not known clearly.

The student may enjoy, and possibly profit by, joining in the guesswork on this and other secondary actions of Psoas. He must be warned, however, that authorities often hold opposite views based on the same observations. For example, some claim that Psoas is a medial rotator while others claim that it is a lateral rotator of the thigh. Electromyography proves the latter to be true.

The **nerve supply** of Iliopsoas is by the lumbar plexus, Psoas directly from L. 2 and 3, Iliacus from the femoral nerve (L. 2, 3, and 4).

Psoas Minor, a small muscle (often absent in man), lies in front of Psoas Major (fig. 257). Its long tendon blends with the fascia of the Major and attaches to the pelvic brim.

B. GLUTEAL MUSCLES

The three gluteal (Gr. = rump) muscles are all large and powerful and the qualifying adjectives **Maximus, Medius,** and **Minimus** merely indicate their relative sizes.

1. Gluteus Maximus

One of the largest, thickest, and coarsest-grained muscles in the body, the Gluteus Maximus is placed entirely behind the hip joint (figs. 299, 300). It is shaped like a parallelogram whose long sides run downward and lateralward.

Its **origin** is from the mass of ligaments that binds the backs of the sacrum and ilium together, and from the sacrotuberous ligament. The muscle has also a small area of origin from the extreme posterior part of the iliac blade.

Its **insertion** is extensive along a continuous line from just below the level of the hip joint to just below the knee joint. About one-quarter of the insertion is into the back

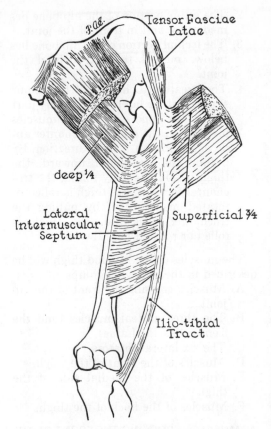

Fig. **300**. Three-quarters of Gluteus Maximus runs into iliotibial tract. Many of these fibers then run to linea aspera via lateral intermuscular septum.

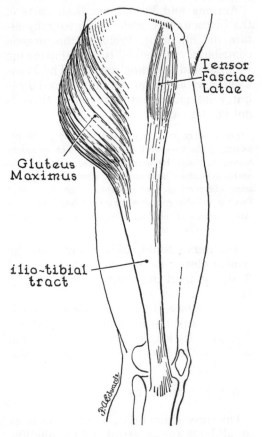

Fig. **299**. Right Gluteus Maximus and Tensor Fasciae Latae.

of the shaft of the femur along a prominent roughness that starts immediately below the greater trochanter (gluteal tuberosity). The remaining three-quarters run into a tough, wide vertical strip of fascia which is part of the *fascia lata* (see below) and is known as the **iliotibial band** or **tract.** This band attaches below to the lateral tibial condyle and can be felt on the lateral aspect of the extended knee. It also is attached to the length of the linea aspera by many of its fibers 'dipping deep' along the *lateral intermuscular septum.*

Actions and Uses. Gluteus Maximus is an extensor of the hip joint but it is used only when the joint has to be extended with power, e.g., against the weight of the body (gravity). It is used, therefore, in rising from a sitting or stooped position, climbing a hill, or going upstairs; it is used in running but not in walking on the level. The muscle is also a powerful lateral rotator of the extended thigh although it loses this power if the thigh is flexed. The adducting action ascribed to the muscle is more theoretical than real.

When the iliotibial tract is made taut by Gluteus Maximus and Tensor Fasciae Latae (see below), it becomes a 'splint' for the knee and helps to make the lower limb a rigid column.

Nerve Supply. Entering the middle of its deep surface is the inferior gluteal nerve (L. 5, S. 1, 2), a branch of the sacral plexus.

Fascia Lata. This strong, thick, deep fascia of the thigh has mostly circular fibers which enclose the muscles of the thigh like a bandage. On the lateral side, however, many vertical fibers form the distinct iliotibial tract, described above.

The transversely disposed fold of skin that limits the prominence of the buttock below is called the **natal fold (fold of the buttock).** It is not produced by the lower border of the Gluteus Maximus since that border crosses the fold and runs obliquely downward and lateralward beyond it. The fold results from a binding of the superficial to the deep fascia whereby the fat of the buttock is confined to that region.

2. Tensor Fasciae Latae

It **arises** from the outer surface of the ilium just behind the anterior superior spine and runs vertically downward. With Gluteus Maximus, it is **inserted** into the iliotibial tract of the fascia lata (figs. 299–301, 305).

Actions and Functions. Because of its position—anterolateral to the hip joint—it can help to abduct, medially rotate, and perhaps flex the hip joint. [Recent investigations seem to show it is an extensor of the knee joint also.]

Actually, it probably is little concerned with any of these actions in the normal person. Its important use is to brace the knee so that, in walking, that joint can take the weight without 'buckling' while the other foot is off the ground and the body is swinging forward; the same function is demonstrable if one merely stands on one leg. This explains why the muscle is inserted into the iliotibial band.

Its **nerve supply** is by the superior gluteal nerve (L. 4, 5, S. 1), a branch of the sacral plexus.

3. Gluteus Medius

This large, powerful, important muscle (figs. 302, 303, 316) has an extensive area of **origin** on the outer surface of the ilium but much farther forward than the origin of the Gluteus Maximus (fig. 301). In general, the Gluteus Medius lies in the same coronal plane as the hip joint, rather more in front of it than behind it.

The muscle is **inserted** on the greater trochanter of the femur (figs. 301, 314).

Actions and Uses. It not only abducts the thigh, but the anterior part of the muscle also rotates the thigh medialward.

If, when standing erect, the support of one limb is suddenly removed, the Gluteus Medius of the other side at once springs into action in order to prevent the pelvis from falling to the unsupported side. It acts as a dynamic ligament, and the

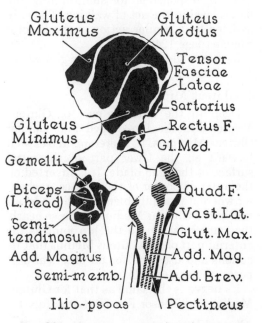

Fig. **301.** Muscular origins (black) and insertions (striped) on right hip bone and upper part of femur, from behind.

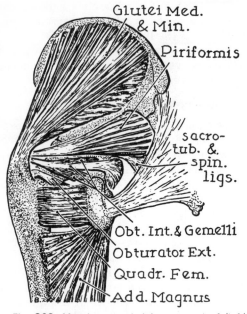

Fig. **302**. Muscles revealed by removal of (left) Gluteus Maximus and hamstring muscles are posterior part of Gluteus Medius and Minimus; six small lateral rotators and upper part of Adductor Magnus.

alternate action of the muscles of the two sides is responsible for maintaining the pelvis level in the act of walking. They are assisted by a similar action of the Erector Spinae (page 124).

Its **nerve** also is the superior gluteal (L. 4, 5, S. 1).

4. Gluteus Minimus

Deep to the anterior part of Gluteus Medius lies the fan-shaped and smaller Gluteus Minimus. It, therefore, **arises** well forward on the lower part of the outer surface of the iliac blade. It is **inserted** on the anterior border of the greater trochanter where it partially blends with the insertion of the Gluteus Medius. It has so nearly the same relations to the hip joint as the anterior part of the Gluteus Medius that its actions and uses are, for all practical purposes, the same.

Its **nerve** is the same as that for Gluteus Medius and Tensor Fasciae Latae (q.v.).

Gluteal Bursae

As each gluteal muscle approaches its insertion it is protected against friction on

the trochanter by an interposed bursa. That deep to the Gluteus Maximus is an extensive one.

C. THE SIX LATERAL ROTATORS

Behind and below the Gluteus Minimus, the back of the hip bone is crossed by a series of small muscles and tendons which are hidden from view by the Gluteus Maximus. They all pass behind the hip joint and their tendons—except the one to be noted—crowd to be inserted into the medial surface of the greater trochanter (figs. 302, 303, 317).

They are described briefly in order from above downward.

1. Piriformis, arises inside the pelvis from the front of the sacrum and passes out the greater sciatic foramen which it almost fills.

2. Obturator Internus, arises inside the pelvis from the membrane that fills the obturator foramen. Its tendon makes an angle of 90° to the muscle as it passes out the lesser sciatic foramen where it uses the smooth bone below the ischial spine as a pulley. It is protected against friction by a bursa.

3, 4. The two twins or *Gemellus Superior* and *Gemellus Inferior* arise above and below the pulley for, and are inserted into, the tendon of Obturator Internus (figs. 302, 317).

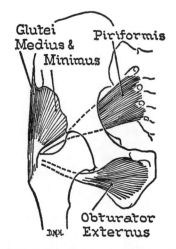

Fig. **303**. Right Obturator Externus, Piriformis, and Glutei from the front (semischematic).

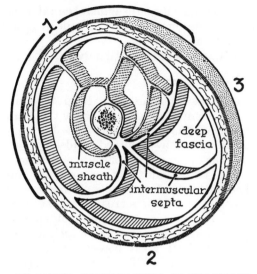

THE THREE COMPARTMENTS OF THE THIGH

Fig. **304**. Three compartments of thigh—anterior (1), posterior (2), and medial (3).

5. Quadratus Femoris, a fleshy little rectangular muscle running almost horizontally lateralward from the ischial tuberosity to the quadrate tubercle—a low smooth eminence at the middle of the trochanteric crest (figs. 302, 317).

6. Obturator Externus, arises outside the pelvis from the obturator membrane. Deeper than the other five, Obturator Externus is hidden from view by Quadratus Femoris. Its tendon runs obliquely upward across the back of the neck of the femur to a pit in the angle between the greater trochanter and the neck (figs. 302, 303, 317).

Actions. All six are lateral rotators of the hip, and probably do little else.

Nerve Supplies. Each has its own small branch from the sacral plexus nearby.

ORGANIZATION OF MUSCLES OF THE THIGH

The elongated muscles of the thigh are disposed in three groups: (1) **Extensors** (or **Muscles of Front of Thigh**); (2) **Adductors** (or **Muscles of Medial Side**); (3) **Flexors** (or **Muscles of Back of Thigh**). The terms 'extensors' and 'flexors' refer to actions on the knee joint. To avoid ambiguity, the three alternative terms that indicate position should be used.

Figure 304 is a schematic cross-section of the thigh and it indicates the three compartments that enclose the groups.

D. MUSCLES OF FRONT OF THIGH

The shaft of the femur, descending obliquely through the thigh, is clothed with muscles, one on its medial aspect, one on its lateral aspect, and one in front which is overlapped on its sides by the other two. These are the three **Vasti** and they are not entirely separated from one another or from the **Rectus Femoris** which descends in front of them from the hip bone to the knee; the three Vasti and the Rectus are grouped together under the name **Quadriceps Femoris.** Superficial to the Quadriceps there descends a very long and obliquely running muscles, the **Sartorius.**

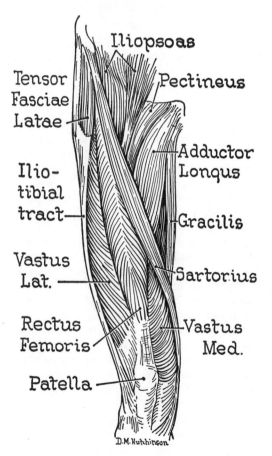

Fig. **305**. Dissection of muscles of front of (right) thigh.

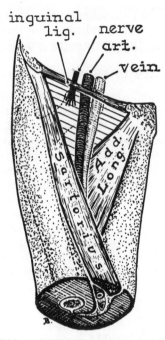

Fig. **306**. Boundaries and three chief contents of (right) femoral triangle. Also note position of femoral artery and vein deep to Sartorius in the cross-section of the midthigh.

1. Sartorius

This, the longest muscle in the body (fig. 305), has the longest fibers. It **arises** from the anterior superior spine of the ilium and, running obliquely, makes its way, as the most superficial muscle of the thigh, to the medial side of the knee where it is **inserted,** by a thin, flat, expanded tendon, on the upper part of the shaft of the tibia.

Actions and Functions. Sartorius is not a powerful muscle in man. Grasp the lower limb and bring the insertion of Sartorius as close as possible to the origin: the hip will be flexed, abducted, and laterally rotated; the knee flexed and medially rotated. Are these, then, the actions of Sartorius? Actually, Sartorius by itself cannot produce any of these movements except very weakly.‡‡ It possibly plays a stabilizing role during walking.

Sartorius is most important as a land-

---

‡‡ Sartor, L. = tailor. The muscle was believed to bring the lower limb into the traditional cross-legged sitting position of a tailor.

mark because the **subsartorial canal,** containing the femoral vessels, is deep to its middle one-third (fig. 306).

Its **nerve** is the femoral nerve (L. 2, 3, 4).

2. Rectus Femoris

It **arises** (a) by a straight head (tendon), from the anterior inferior spine of the ilium which is situated just above the acetabulum; and (b) by a curved head (tendon), from the upper margin of the acetabulum.

The two tendons of origin join and thence the fusiform belly of the muscle runs straight (L. = rectus) down the front of the thigh where it is easily palpated. It has a special type of pennate architecture which gives it great power over a short range (figs. 305, 307). The belly narrows below to a tendon that is **inserted** on the

Fig. **307**. Right Rectus Femoris extends knee or flexes hip.

upper border of the patella but, since this patella is held at its apex to the tibia by the short (but stout and strong) patellar ligament, the Rectus, of course, pulls on the tibia (figs. 308, 311).

Actions and Uses. Rectus Femoris is the only member of the Quadriceps group that crosses the hip joint and that can, therefore, flex the hip and (or) extend the knee, a combination of movements which advances the limb in walking; when used vigorously it helps to deliver a kick.

Its **nerve,** too, is the femoral nerve.

3. The Three Vasti

Origins. The **Vastus Medialis** *arises from* the linear aspera and its medial 'branches.' It reaches, above, as high as the lower end of the trochanteric line; below, it

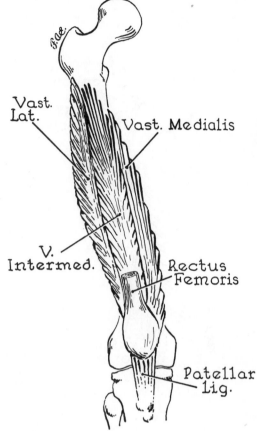

Fig. **309.** Vasti from in front with Rectus Femoris removed.

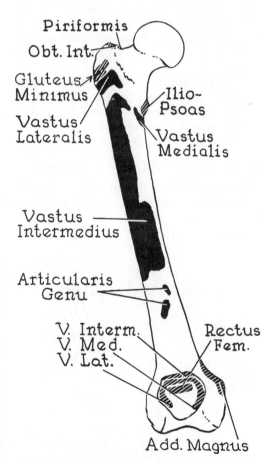

Fig. **308.** Muscular origins (black) and insertions (striped) on right femur and patella.

occupies only the upper third of the medial supracondylar line (figs. 309, 315).

Similarly, the **Vastus Lateralis** *arises from* the linea aspera and its lateral 'branches.' It reaches, above, as high as the base of the greater trochanter; below, it occupies only the upper third of the lateral supracondylar line. The muscle arises also from the tough lateral intermuscular septum that separates it from the hamstring muscles (figs. 305, 310, 314, 315).

The **Vastus Intermedius** *arises from* the front of the femoral shaft. It reaches as high as the trochanteric line and it is hidden from view by the other two Vasti which wrap themselves round the femur and meet one another in front (figs. 309, 314, 315).

Insertions. Vastus Medialis and Later-

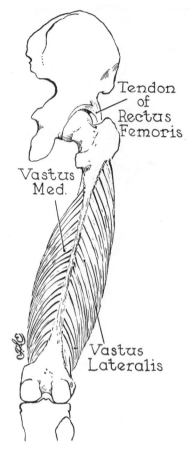

Fig. **310**. The Vastus Medialis and V. Lateralis arise from the linea aspera on back of femur. Note tendon of Rectus Femoris arising from ilium.

does not remain in constant contraction to maintain extension of the knee—even in the erect posture. The locking of the knee (page 102) with, perhaps, the contraction of Tensor Fasciae Latae, is sufficient. This is demonstrated by: (1) the practical joke of striking a sharp blow to the back of the knees of a standing victim; and (2) the fact that a patient with paralysis of both Quadriceps muscles can maintain a standing position.

Because Vastus Medialis is muscular right down to the patella and is inserted well down along the border of that bone it helps to correct the tendency of the patella to be dislocated laterally.

A few scattered fibers of the deep part of the Vastus Intermedius are attached to the synovial capsule of the knee joint where it 'bags' upward for three finger's breadths above the patella. These fibers are known as the *Articularis Genu* and they serve to pull the capsule up when the knee joint is extended.

Nerve Supply of Vasti. Each Vastus receives one or two branches (upper and lower) from the femoral nerve (L. 2, 3, 4). These nerves generally enter the anterior borders of the muscles.

alis are inserted into the sides of the Rectus tendon and into the patella (fig. 311) to be described below; the Vastus Intermedius is inserted into the back of the Rectus tendon and, with it, into the base (upper border) of the patella. Sheets of fibrous tissue on both sides of the patella (known as expansions of the Vasti) reach also to the tibia and to the fibula, where they form the most important reinforcement of the fibrous capsule of the knee joint (fig. 305).

Actions and Functions of Vasti. The Vasti along with Rectus Femoris are the powerful extensor complex of the knee. No muscle can extend the knee if Quadriceps is paralyzed—obviously a serious disability. (See tables 9, 10.)

On the other hand, the Quadriceps group

E. MUSCLES OF MEDIAL SIDE OF THIGH

General. When the articulated skeleton is viewed from in front, it is apparent that

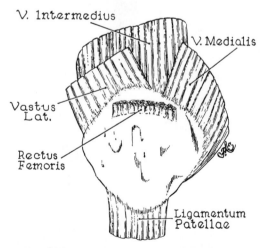

Fig. **311**. The insertions of the four heads of quadriceps femoris on the right patella.

TABLE 9
Muscles Acting on Hip Joint

Flexion	Extension	Abduction	Adduction	Med. Rotation	Lat. Rotation
1. Iliopsoas	1. Gluteus	1. Gluteus	1. Add. Mag-	1. Gluteus	1. Six small lat-
2. Sartorius	Maximus	Medius	nus	Medius	eral rota-
3. Pectineus	2. Semitend.	2. Gluteus	2. Add. Longus	2. Gluteus	tors:
4. Rectus Fem.	3. Semimemb.	Minimus	3. Add. Brevis	Minimus	⎛ Piriformis
(5. Tensor F. L.)	4. Biceps	3. Tensor F. L.	(4. Gracilis	3. Tensor F. L.	⎪ Obt. Int.
(6. Add. Long.)	Fem.	4. Iliopsoas	(5. Pectineus	4. Add. Mag.—	⎨ 2 Gemelli
(7. Add. Brev.)	5. Add. Mag.	(5. Piriformis)	(6. Obt. Int.)	horiz.	⎪ Obt. Ext.
(8. Add. Mag.—	—post	(6. Sartorius)	(7. Gemelli—2)	fibers	⎝ Quad. Fem.
ant. fibers)	fibers		(8. Obt. Ext.)	5. Add. L.	2. Gluteus Maxi-
			(9. Quad. Fem.)	6. Add. Br.	mus
				(7. Pectineus)	Iliopsoas

TABLE 10
Muscles Acting on Knee Joint

Flexion	Extension	Med. Rotation of Tibia	Lat. Rotation of Tibia
1. Hamstrings:	1. Quadriceps	1. Popliteus—'unlocks'	Biceps Fem.—
⎛ Semimemb.	Femoris	knee	when knee bent
⎨ Semitend.	⎛ Rectus Fem.	2. Sartorius ⎞	
⎝ Biceps F.	⎪ Vastus Lat.	3. Gracilis ⎪ when	
2. Gastrocnem.	⎨ Vastus Interm.	4. Semimemb. ⎬ knee	
3. Sartorius	⎝ Vastus Med.	5. Semitend. ⎠ bent	
4. Gracilis	2. Tensor F. L.		
(5. Popliteus)			
(6. Plantaris)			

the angular interval seen on the medial side of the obliquely running femur must be filled in, if the lower limb is not to have an unsightly knock-kneed appearance. This is accomplished in the living by a strap-like muscle, the **Gracilis** (fig. 312), descending vertically from near the midline symphysis pubis to the medial side of the tibia just beyond the knee, and by a group of muscles filling in the V-shaped interval between the Gracilis and the femoral shaft. The muscles are known as the **Adductors.**

The names of the Adductors are: **Gracilis, Pectineus, Adductor Longus, Adductor Brevis,** and **Adductor Magnus;** they will be considered together.

Origins and Positions. They arise from the front of the pubis and they run downward and lateralward to pass behind the origin of the Vastus Medialis. With the exception of the Gracilis, they are all triangular in outline and are arranged in layers like the pages of a book. The highest is the **Pectineus** (figs. 305, 312) and it arises from the superior ramus of the pubis as far back as the pectineal line, i.e., the front part of the pelvic brim; it lies adjacent to the Psoas of which it is probably the divorced medial portion. [Pectineus, unlike the other adductors, is supplied by the femoral nerve which also supplies the Psoas.] The **Adductor Longus, Adductor Brevis,** and the **Gracilis** are crowded together on the front of the body of the pubis. The **Adductor Magnus** (figs. 313, 314, 316, 317) arises from the ischiopubic (inferior) ramus and extends its origin as far back as the ischial tuberosity.

Insertions. From the limited origin, the muscles fan out to insert along a vertical line beginning just below the lesser trochanter, continuing along the entire length of the linea aspera and medial supracondy-

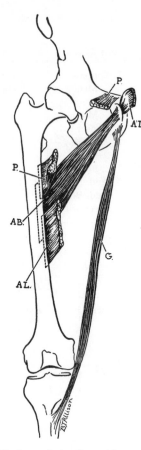

Fig. **312**. Four of the five adductors of thigh—
Pectineus (P); Adductor Brevis (AB); Adductor
Longus (AL, AL); and Gracilis (G) from in front
(right side).

lar line to the adductor tubercle (fig. 315).
The lowest attachment (that of Gracilis) is
to the upper part of the tibia.

On the linea aspera, the Adductor Longus occupies
the middle two-quarters, the Pectineus is inserted
above it, and the Adductor Brevis lies half behind the
Pectineus and half behind the Longus. The massive
Adductor Magnus occupies the length of the linea
aspera from a little below the quadrate tubercle to the
adductor tubercle. [The Quadratus Femoris is its
divorced uppermost portion.] The Gracilis reaches the
medial side of the shaft of the tibia a little below the
knee, where its insertion is closely associated with
that of the Sartorius and the Semitendinosus (q.v.).

Actions and Uses. As a group, the adductors can produce powerful adduction.
Although they are obviously not needed for

maintaining posture, they are quite active
during locomotion. Adduction alone does
not account for this activity; flexion and
medial rotation of the thigh are also involved.

Because they are on a plane in front of
the hip joint, most of the adductors can
flex as well as adduct. Because they pass
anterior to the femoral head, they assist in
medial rotation as proved by electromyography. Further, since the fibers of the
Adductor Magnus that reach the adductor
tubercle are the ones that arise from the
ischial tuberosity, they lie on a plane behind the joint and can, therefore, assist in
extension of the hip.

Nerve Supply. The main nerve to the
adductors is the obturator nerve (L. 2, 3, 4)
which enters them from above after passing through the upper part of the obturator

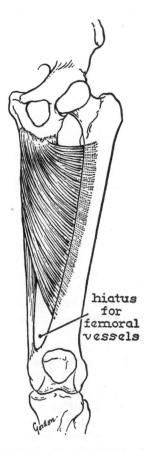

Fig. **313**. Adductor Magnus (left) lies behind the
other adductors (cf., fig. 312).

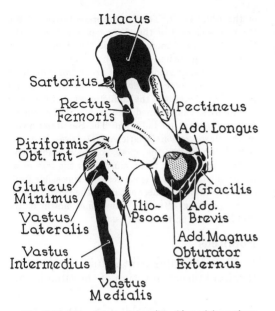

Fig. **314.** Muscular origins (black) and insertions (striped) on right hip bone and upper part of femur, front view.

medial side of the knee to the front and is inserted into the tibial shaft in close association with the Sartorius and the Gracilis.

Surface Anatomy. When the knee is flexed against resistance, the tendons of insertion of the hamstrings are easily identified bounding the depression behind the knee known as the **popliteal fossa.** The tendon of Biceps Femoris can be traced to the head of the fibula. It should not be confused with the iliotibial tract which is on the lateral aspect of the knee joint. The tendons of the 'medial hamstrings' are close together but the Semitendinosus can be made to stand out more sharply on the lateral side of the Semimembranosus.

Actions and Uses. They extend the hip and (or) flex the knee. In walking, as the foot leaves the ground to take a step forward, the hamstrings contract momentarily in order to take the weight of the partially flexed leg; as soon as hip flexion

foramen. [Pectineus, however, gets most of its supply from the femoral nerve.]

F. MUSCLES OF BACK OF THIGH

Hamstring Muscles

The group of muscles lying behind the Adductor Magnus is known as the **hamstrings.** These are very long and extend from a common origin on the ischial tuberosity to beyond the back of the knee joint. They are: **Biceps Femoris, Semimembranosus,** and **Semitendinosus.**

The **Biceps Femoris** (figs. 316–318), as its name implies, has two heads. In addition to its *long head*, it has a fleshy *short head* of origin from the middle of the linea aspera and upper part of the lateral supracondylar line. The muscles is inserted on the head of the fibula which, it will be recalled, articulates with the back of the lateral tibial condyle. The **Semimembranosus** (figs. 316, 317, 319), whose upper one-third is membranous (i.e., aponeurotic), is inserted on a comparable site on the back of the medial tibial condyle. The **Semitendinosus** (figs. 317, 320) has a long rounded tendon that curves round the

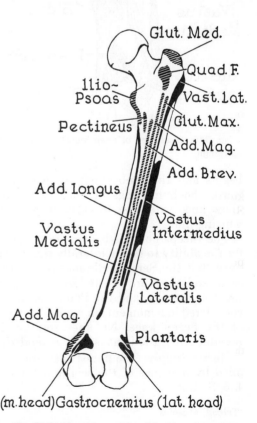

Fig. **315.** Muscular origins (black) and insertions (striped) on right femur; viewed from behind.

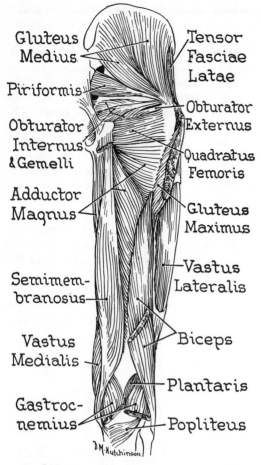

Fig. **316**. Dissection of deep muscles of back of right thigh.

begins and the limb starts to swing forward, the hamstrings at once relax and allow knee extension to occur in the advancing limb.

The Biceps Femoris is alone responsible for the ability to rotate laterally the flexed knee, but the Semimembranosus and the Semitendinosus are assisted by the Sartorius, the Gracilis, and the Popliteus (to be considered in a moment) in medial rotation of the flexed knee. No knee rotation is possible when the joint is fully extended.

Nerve Supply. The hamstrings are supplied by branches of the sciatic nerve (L. 4, 5, S. 1, 2, 3).

'Tripod' Muscles

We have encountered three strap-like muscles of the thigh that are inserted together on the medial side

of the tibial shaft. They are representatives of the three regions and of the three motor nerves of the thigh; they arise, moreover, from the three outlying points of the pelvis. Thus, the **Sartorius** of the extensor region (femoral nerve) arises from the anterior superior spine of the ilium—the most lateral point. The **Gracilis** of the adductor region (obturator nerve) arises from immediately beside the symphysis pubis—the most medial point. The **Semitendinosus** of the flexor region (sciatic nerve) arises from the ischial tuberosity—the most posterior point. This arrangement of the three muscles as the legs of an inverted tripod (fig. 321) suggests their possible function as a stabilizing mechanism for the pelvis capable of co-operating with the 'splint' mechanism of the iliotibial tract on the lateral side.

Popliteus

When the knee joint is fully extended, a terminal lateral rotation of the tibia is said

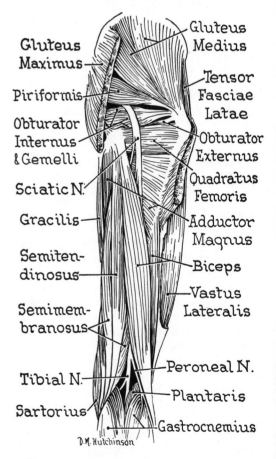

Fig. **317**. Superficial dissection of back of right thigh with most of Gluteus Maximus removed. Note sciatic nerve.

Fig. **318**. Right Biceps Femoris—the lateral hamstring.

Fig. **320**. Semitendinosus justifies its name

Fig. **319**. Right Semimembranosus and Popliteus.

Fig. **321**. 'Tripod' muscles (see text)

to 'lock' the joint. The key that unlocks the joint is the triangular **Popliteus** (fig. 319) extending obliquely across the lower part of the back of the knee.

Attachments. Popliteus arises in the horizontal groove on the lateral femoral condyle by a rounded tendon that rubs against the back of the lateral semilunar cartilage and superior tibiofibular joint. [It has a bursa which may communicate with this joint and, near the lateral semilunar cartilage, usually communicates with the knee joint.] Its insertion is fleshy and broad and occupies the area above an oblique line crossing the upper part of the back of the tibia (fig. 295).

Actions. It is concerned with the almost 'involuntary' rotation that has to accompany the beginning of flexion. When the *flexed* knee is voluntarily rotated the hamstrings, not the Popliteus, are the principal muscles concerned.

Its **nerve** is a branch of the medial popliteal nerve, the tibial division of the sciatic nerve.

Muscles of Leg

General. By palpation in the living, it is easy to establish that tendons cross the ankle joint in four situations: (1) in front of the ankle; (2) behind the lateral malleolus; (3) behind the ankle; (4) behind the medial malleolus. It is, therefore, convenient to divide the muscles of the leg into four groups and to discuss them in the following order:

A. Extensors, in front of the ankle.

B. Peronei, behind the lateral malleolus.

C. Superficial Flexors, behind the ankle.

D. Deep Flexors, behind the medial malleolus.

Figure 322 is a schematic cross-section of the leg. It shows the compartments for the four groups of muscles and the way they are separated from one another by fibrous intermuscular septa.

A. EXTENSORS (DORSIFLEXORS)

The Extensor muscles lie on the lateral side of the tibia and in front of the fibula. The word 'Extensor' as applied to these muscles refers to the fact that they straighten or turn the toes up, i.e., extend them. However, it is confusing to refer (correctly) to turning up of the foot at the ankle as 'extension' when obviously the ankle is being bent rather than straight-

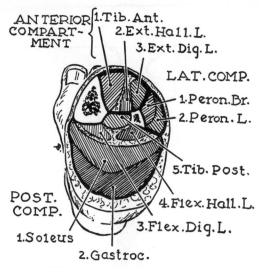

Fig. **322.** Cross-section at midcalf (semischematic).

ened. Indeed, some anatomists insist on calling the above movement 'flexion.' The difficulty can be avoided by using the terms **dorsiflexion** and **plantarflexion.**

The extensor muscles (dorsiflexors) are: **Tibialis Anterior, Extensor Digitorum Longus,** and **Extensor Hallucis Longus.** The lower and lateral part of the Extensor Digitorum Longus has received a separate name, the **Peroneus Tertius,** and a little muscle entirely confined to the dorsum (upper surface) of the foot, the **Extensor Digitorum Brevis,** properly belongs to the extensor group. The five muscles will be described in turn; their nerve supplies will be discussed together at the end.

1. Tibialis Anterior

This muscle (figs. 323, 328) is the most important, the most powerful, and the most medial of the group. It **arises from,** and is closely applied to, the lateral surface of the tibia. It also picks up origin from the neighboring interosseous membrane and from the tough deep fascia which encloses the muscle.

Insertion. When the muscle is made to act, its tendon can be traced to the base of the first (great toe) metatarsal and to the adjoining medial cuneiform (fig. 323).

Actions and Uses. This is unquestionably the chief muscle that (a) dorsiflexes the

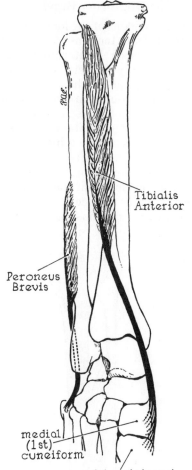

Fig. **323**. Right Tibialis Anterior and Peroneus Brevis from in front.

inserted in a fashion quite similar to that observed for their counterparts in the hand (Extensor Digitorum).

Actions. It is the extensor of the lateral four toes and helps to dorsiflex the ankle weakly. The toes are capable of considerably more hyper-extension than are the fingers.

3. Peroneus Tertius

A functionally important characteristic of the human Extensor Digitorum Longus is that most of its lateral half is in the process of splitting away from the remainder to form a new muscle, the **Peroneus Tertius,** i.e., a third Peroneus muscle (figs. 325, 328). Hence a tendon is found that is usually inserted on the shaft of the fifth metatarsal. Man needs to place his foot foursquare on the ground. No other primate has achieved this power and, in man, this special muscle designed to help raise the lateral border of the foot has not yet attained complete structural independence

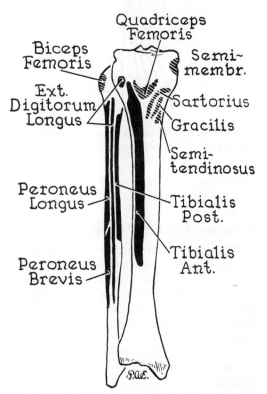

Fig. **324**. Muscular origins (black) and insertions (striped) on front of right tibia and fibula.

ankle; and (b) inverts the foot. It is used in walking to bend the foot up and so prevent stubbing of the toes as the advancing limb swings forward. Its paralysis results in 'foot-drop' and a high-stepping gait.

2. Extensor Digitorum Longus

This muscle can be felt with ease lateral to the Tibialis Anterior. Its principal **origin** is from the upper two-thirds of the narrow anterior surface of the fibula (figs. 324, 325, 328).

In front of the ankle joint the muscle divides into four tendons which diverge on the dorsum of the foot to the four lateral toes. The tendons are reinforced and are

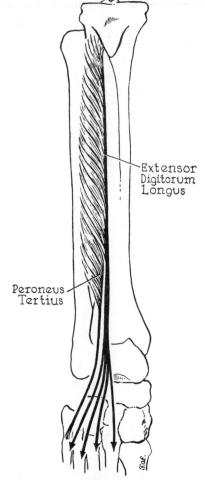

Fig. **325.** Right Extensor Digitorum Longus and, splitting from it, the Peroneus Tertius.

of the other two evertors, Peroneus Longus and Peroneus Brevis.

4. Extensor Hallucis Longus

This special extensor for the great toe **arises** from the middle two quarters of the front of the fibula. Its tendon appears at the ankle close beside that of the Tibialis Anterior and can be seen in action if the great toe (hallux) is extended vigorously (figs. 326–328).

The muscle is **inserted** on the base of the distal phalanx of the great toe and its **actions** are extension and hyper-extension of the great toe. It assists in dorsiflexion of the ankle.

5. Extensor Digitorum Brevis

With the tendons of the long extensor are the more delicate tendons of a short extensor. This Extensor Digitorum Brevis (fig. 329) is the only muscle arising on the dorsum of the foot and has no counterpart in the hand. If you examine your own foot, you will find a soft muscle mass immediately in front of the lateral malleolus. A person who has injured his ankle commonly mistakes the little muscle—which he has never observed before—for a bruise. The muscle **arises** from the extreme front end of the upper surface of the calcaneus.

Four slender tendons pass to the four

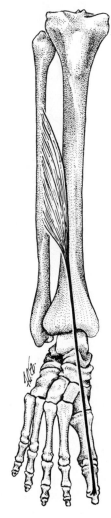

Fig. **326.** Right Extensor Hallucis Longus

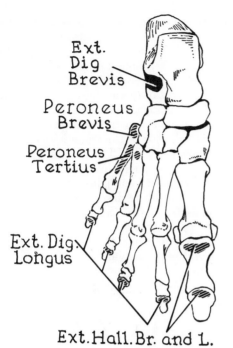

Fig. **327**. Muscular origins (black) and insertions (striped) on dorsum of right foot.

medial toes where, except in the great toe, they join the long extensor tendons; in the great toe the tendon remains independent and, as the **Extensor Hallucis Brevis,** is inserted on the base of the proximal phalanx (fig. 327).

Nerve Supply of Extensor Muscles

The deep peroneal nerve (L. 4, 5, S. 1), enters the area above and runs down on the interosseous membrane to supply direct branches to Tibialis Anterior, Extensor Digitorum Longus and Peroneus Tertius, Extensor Hallucis Longus, and—running to the dorsum of the foot—Extensor Digitorum Brevis.

Retinacula and Synovial Sheaths

Four tendons enter the foot on the front of the ankle (Tibialis Anterior, Extensor Hallucis Longus, Extensor Digitorum Longus, and Peroneus Tertius); two enter behind the lower end of the fibula (Peroneus Longus and Peroneus Brevis, discussed below), and three behind the lower end of the tibia (Flexor Hallucis Longus, Flexor

Digitorum Longus, and Tibialis Posterior). All of these tendons are kept from bowstringing or from slipping out of place by fibrous bands or loops known as retinacula

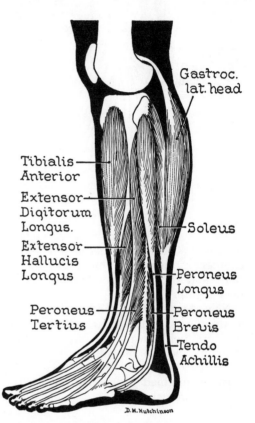

Fig. **328**. Muscles of (left) leg from lateral side (semischematic).

Fig. **329**. Extensor Digitorum Brevis, the only muscle arising from dorsum of foot.

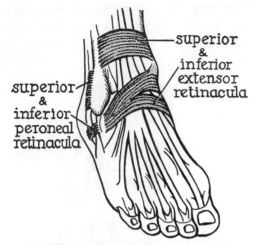

Fig. **330**. Extensor and peroneal retinacula

Insertions. Peroneus Brevis is inserted into the base of the fifth metatarsal (just in front of the prominent and easily felt tuberosity). Peroneus Longus is finally inserted into the lateral side of the base of the first metatarsal (great toe) and into the medial cuneiform behind it. This site, it will be recalled, is near the insertion of the Tibialis Anterior to the same bones.

Actions and Uses. The peronei evert the foot. Peroneus Longus is a plantarflexor of the transverse tarsal joint (as is Tibialis Posterior, yet to be described).

The usefulness of the evertors (and invertors) of the foot becomes apparent when one walks over rough ground; the foot is set down and assumes various awkward posi-

(singular = retinaculum). They are: the *superior and inferior extensor retinacula* in front (fig. 330), the *superior and inferior peroneal retinacula* on the lateral side (fig. 330), and the *flexor retinaculum* on the medial side.

Now, because the tendons around the ankle must slide freely up and down where they are 'held down,' they are enclosed in synovial sheaths or tubular bursae like those in the hand. Often these bursae become inflamed or swollen (*synovitis*, a type of bursitis).

B. THE PERONEI (FIBULARES)

Origins and Courses. The **Peroneus Longus** arises from the upper two-thirds (figs. 322, 331, 334), the **Peroneus Brevis** (figs. 322, 323, 328) from the lower two-thirds, of the lateral side of the fibula.§§ Their tendons can be felt behind the lateral (fibular) malleolus which they use for a pulley. The Peroneus Brevis can be readily traced forward to its insertion. The tendon of Longus can be felt no further because it plunges deeply into the sole where it lies in a groove on the under surface of the cuboid and crosses the sole from lateral to medial side.

_____

§§ Peroneus, Gr. = Fibularis (L.). These muscles are also properly called Fibularis Longus and Fibularis Brevis. Perone, Gr. = fibula (L.) = a long pin or skewer.

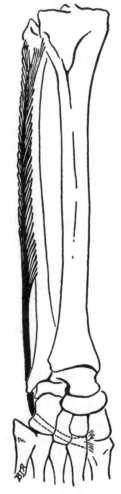

Fig. **331**. Right Peroneus Longus from in front

tions which are maintained—at least briefly—by the muscles. When the muscles are 'caught off guard' a sprained ligament is the result.

The **nerve supply** is by the superficial peroneal branch of the lateral popliteal nerve (L. 4, 5, and S. 1).

Peroneus Tertius

This part of the Extensor Digitorum Longus has been discussed with the extensors (page 187).

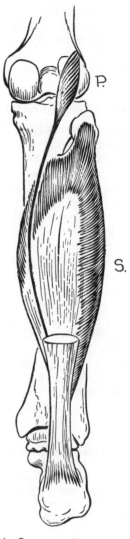

Fig. **333**. Right Gastrocnemius removed to show Soleus (S.) and Plantaris (P.).

C. SUPERFICIAL PLANTARFLEXORS

Gastrocnemius

Soleus

Plantaris

The two large powerful muscles responsible for the calf of the leg are named Gastrocnemius (Gr. = belly of the leg) and Soleus (L. = sole, the flat fish whose shape Soleus imitates). Together these two are sometimes referred to as *Triceps Surae*. Associated with them is the interesting but

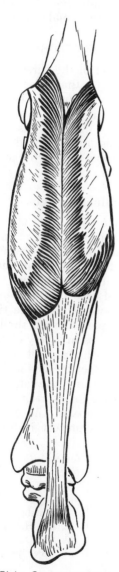

Fig. **332**. Right Gastrocnemius and tendo calcaneus from behind.

unimportant Plantaris (planta, L. = foot or sole of foot). Each muscle will receive special consideration first (origins and characteristics); then insertion, actions and functions, and nerve supply of the group will be considered.

Origins and General Characteristics

The **Gastrocnemius** (figs. 322, 328, 332) is the most superficial and is a muscle that crosses two joints and that consists of two bulging, parallel bellies. The medial and lateral heads arise by fleshy fibers and tendon from the back of the femur just above the respective femoral condyles. The heads quickly unite to form a single muscle, which, less than halfway down the calf, becomes a rather wide and flat tendon.

The **Soleus** (figs. 322, 328, 333, 334) has a continuous horseshoe-shaped origin from the back of the tibia and of the fibula. On the tibia, it can reach no higher than an oblique line above which lies the Popliteus. Between the two bones, a fibrous arch allows the muscle to bridge the vessels and nerves that descend to the leg from the back of the knee.

The *Plantaris* shares an origin with the lateral head of the Gastrocnemius. The muscle is about the size of a finger but has a long string-like tendon (fig. 333).

The little Plantaris seems fated to get into trouble. Its delicate tendon can be ruptured by excessive muscular action and the resulting condition is a very painful type of 'charley horse.'

Tendo Calcaneus

The three muscles share a single tendon of insertion called the tendo calcaneus or *tendo Achillis*. This is stout and strong; it stands out prominently at the back of the ankle joint and is inserted into the back of the calcaneus (figs. 328, 333).

Actions and Functions

The fibers of the Gastrocnemius are so short that the muscle is unable to flex the knee and plantarflex the ankle at the same time; it cannot take up enough slack. Therefore, the muscle can do one or other of these actions but never both together. The Soleus is the great plantarflexor of the

ankle; when the knee is flexed, the Gastrocnemius can be of no assitance. (See table 11.)

Both muscles are active during the 'take-off' phase of the foot in walking and running and are sorely missed when paralyzed. Contrary to popular belief, they need not be—and, in many persons, are not—active during ordinary standing.

Nerve Supply

The three muscles are supplied by branches from the tibial nerve.

D. DEEP PLANTARFLEXORS

Under shelter of the three calf muscles, are three muscles that arise: one from the back of the fibula; one from the back of the tibia; one from the back of both bones and their uniting interosseous membrane. Their tendons pass behind the medial (tibial) malleolus (or nearby) and enter the sole of the foot.

1. Flexor Hallucis Longus

This muscle (figs. 332, 334, 335) arises from the lower two-thirds of the posterior aspect of the fibula. It is destined for the great toe and for that reason demands and secures a position of priority.

Course and Insertion. Its tendon passes behind the lower end of the tibial shaft, where it occupies a broad shallow groove. Thereafter, it lies in a very distinct groove on the back of the talus; it finally reaches the sole of the foot by passing beneath the sustentaculum tali of the calcaneus. As the tendon passes to its insertion on the base of the distal phalanx of the great toe, two sesamoid bones protect it from pressure

TABLE 11
Muscles Acting on Ankle Joint

Plantarflexion	Dorsiflexion
1. Soleus	1. Tibialis Ant.
2. Gastrocnemius	2. Ext. Hall. Long.
3. Peroneus Long.	(3. Ext. Dig. Long. and
(4. Plantaris)	Peroneus Tertius)
(5. Peroneus Br.)	
(6. Tibialis Post.)	

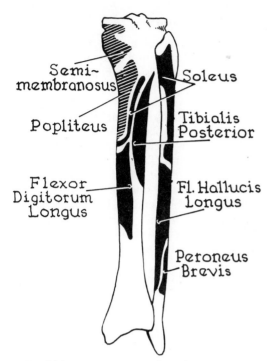

Fig. **334**. Muscular origins (black) and insertions (striped) on back of right tibia and fibula.

malleolus. It then passes alongside the free medial border of the sustentaculum tali—not below. In the sole, the somewhat oblique direction of its tendons is in part corrected by the pull of an accessory muscle, the **Quadratus Plantae** or **Flexor Accessorius,** arising from the medial side and inferior aspect of the calcaneus (fig. 340).

As the tendon of the Digitorum approaches the toes, it divides into four

where it crosses the head of the first metatarsal.

These sesamoids are relatively large. During walking they receive the whole weight of the body just before the next step is taken.

Action and Uses. It flexes the great toe and assists the calf muscles in plantarflexion of the ankle, although alone the deep flexors are unable to lift the body on tip-toes.

2. Flexor Digitorum Longus

This is a smaller muscle that Flexor Hallucis Longus although it provides tendons for all four lateral toes (fig. 336).

Origin. It arises from a strip of bone on the posterior surface of the tibia (fig. 334). Both flexor muscles of the toes pick up extra origin from the neighboring surfaces of fibrous tissue, e.g., fascia covering the Tibialis Posterior. Their fleshy origins overlap, and almost completely hide, that of Tibialis Posterior.

Course and Insertions. Its tendon runs downward and passes behind the medial

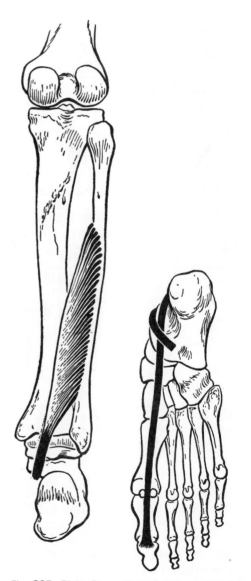

Fig. **335**. Right Flexor Hallucis Longus from behind and in sole (crossed by Fl. Digitorum Longus).

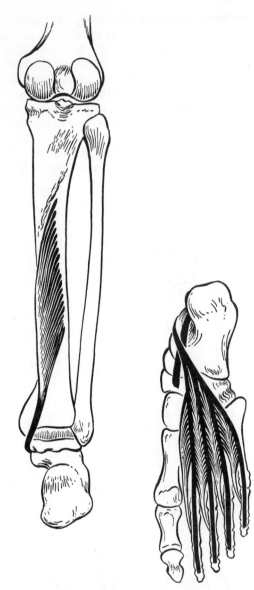

Fig. **336**. Right Flexor Digitorum Longus from behind and in sole where tendons give origin to four Lumbricals. Flexor Accessorius omitted—see fig. 340.

3. Tibialis Posterior

This, the deepest muscle of the leg, is large, important, and lies in the same transverse plane as the bones of the leg (figs. 322, 337, 338).

Origin. A compartment (open only below) is formed by the interosseous membrane, the opposite surfaces of tibia and fibula, and an enclosing fascial membrane that lies behind Tibialis Posterior. From the inner surface of this osseo-fibrous pocket the muscle fibers arise.

Course and Insertion. Its tendon passes behind the medial malleolus and runs forward on the side of the deltoid ligament of the ankle. It spreads out into several slips the most important of which is inserted into the tuberosity of the navicular.

Actions and Functions. Like Tibialis Anterior, the Posterior is an invertor of the foot. Like Peroneus Longus, it plantarflexes the transverse tarsal joint ('forefoot flexion') and has only slight plantarflexor effect on the ankle joint. (See table 12.)

Nerve Supply of Deep Flexors

The (posterior) tibial nerve (L. 5, S. 1, 2) runs between three muscles and supplies

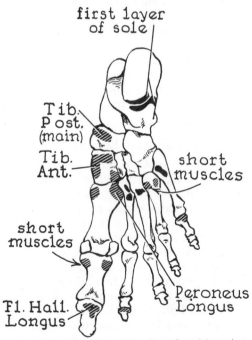

Fig. **337**. Muscular origins (black) and insertions (striped) in sole of right foot.

smaller tendons that behave in exactly the same manner as the tendons of the Flexor Digitorum Profundus of the hand, viz., they perforate the tendons of the Flexor Digitorum Brevis (see page 197) which corresponds, in the foot, to the Flexor Digitorum Superficialis in the hand.

Actions. Flexion of the small toes is its only confirmed important action.

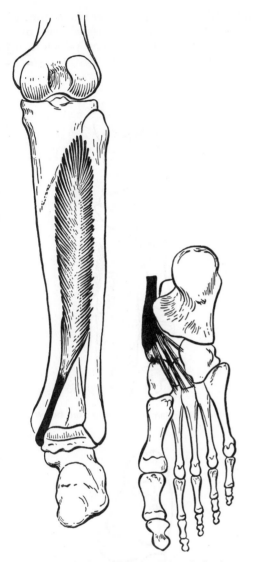

Fig. **338**. Right Tibialis Posterior. It is deepest muscle in calf.

them with branches and then continues into the sole of the foot where one of its branches supplies Accessorius.

Muscles of Foot

The architecture of the foot is similar to that of the hand and those interested in doing so can work out an impressive series of homologies. But the duties of the hand are multifarious, those of the foot are very specialized. The specialization that has characterized the development of the human foot has robbed it of many functional possibilities that the very primitiveness of the human hand has allowed that member to retain. The fact that several muscles in the foot possess names implying functions they are totally unable to perform serves as a reminder that the modifications of the bones, joints, ligaments, and muscles of the foot, whereby, much more than the hand, it has departed from the basic mammalian pattern, have equipped the foot for its two outstanding duties of bipedal support and locomotion—quite unique among mammals.

There are but two essential origins for the intrinsic muscles of the foot. One is at the back of the under surface of the calcaneus where, in the sole, it possesses a considerable tuberosity; the other is near the bases and along the shafts of the metatarsals.

Muscle Layers. There are four layers of muscles and tendons in the sole. The first layer consists of three 'tie-beam' muscles; the second consists of the two tendons of the long flexors of the toes, the Quadratus Plantae muscle, and four Lumbricals; the third consists of three short muscles associated with the great and little toes; the fourth consists of seven Interossei and the tendons of Tibialis Posterior and Peroneus Longus; it is on the 'skeletal plane.'

Plantar Aponeurosis. The central part of the deep fascia of the sole of the foot is

TABLE 12
Muscles Acting on Intertarsal Joints

Inversion	Eversion	Plantarflexion
1. Tibialis Ant.	1. Peroneus Long.	1. Peroneus L.
2. Tibialis Post.	2. Peroneus Br.	2. Tibialis Post.
	3. Peroneus Tertius	3. Abductor Hall.
		4. Abd. Dig. Min.
		5. Flexor Dig. Br.
		(6. Peron. Br.)
		(7. Long flexor tendons to toes)
		(8. Lumbricals and Quad. Plantae)

Fig. **339**. First layer of muscles of sole

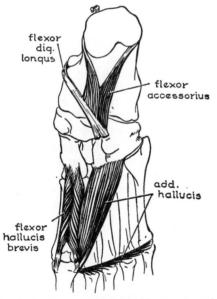

Fig. **340**. Flexor Accessorius (Quad. Plantae) of second layer. Flexor Hallucis Brevis, Adductor Hallucis Obliquus, and Transversus are in third layer.

greatly thickened, and acts as a tie-beam for the arches of the foot. It is firmly attached to the tuberosity of the calcaneus behind, and to ligamentous structures near the metatarsal heads in front. In some animals it is continuous with the Plantaris tendon (cf., palmar aponeurosis).

First Layer

Origins. The large weight-bearing tuberosity at the back of the calcaneus is a

pillar that is common to both longitudinal arches. From it, three muscles arise and, like tie-beams for the arches, pass forward to the toes (figs. 337, 339).

Fig. **341**. Plantar Interossei from below

Fig. **342**. Dorsal Interossei from above. Axis of foot is second toe, not as in hand.

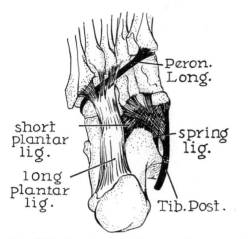

Fig. **343**. Insertions of Peroneus Longus and Tibialis Posterior in sole of right foot.

From medial to lateral side these muscles are: (1) Abductor Hallucis; (2) Flexor Digitorum Brevis; and (3) Abductor Digiti Minimi.

Abductor Hallucis is **inserted** into the base of the first phalanx of the great toe by a tendon which, near its insertion, adheres to the surface of the medial sesamoid of the great toe.

Abductor Digiti Minimi is **inserted** into the base of the proximal phalanx of the little toe.

Flexor Digitorum Brevis lies deep to the plantar aponeurosis and is **inserted** into the second phalanges of the four lateral toes.

Actions and Uses. Their names give no indication of the function of the muscles of the first layer. They may help to maintain the arches of the foot during locomotion along with the ligaments in the sole of the foot.

Second Layer

The **tendons** of Flexor Hallucis Longus (fig. 335) and Flexor Digitorum Longus with the associated *Quadratus Plantae* (fig. 340) have already been described. The *four Lumbricals* in this layer are similar in most details to those of the hand (fig. 289).

Third Layer

The **Flexor Hallucis Brevis** (fig. 340) is a fleshy little muscle with two parallel bellies, arising from the ligaments in the region of the base of the metatarsal of the great toe and inserting into the base of the proximal phalanx. Its two tendons adhere to the two sesamoid bones (which are often described as being 'in' the tendons). Also adhering to the medial sesamoid is Abductor Hallucis, already described.

To the lateral sesamoid there also adheres the tendon of insertion of **Adductor Hallucis,** a two-headed muscle; the larger *oblique head* arises near the base of the metatarsals, the tiny *transverse head* arises from the capsules of metatarsophalangeal joints.

Flexor Digiti Minimi Brevis is an insignificant little imitator of Flexor Hallucis Brevis.

Actions and Uses. These three small muscles act on their metatarsophalangeal joints; those that are attached to the sesamoids control their excursions.

Fourth Layer

The **tendons** of Peroneus Longus (fig. 343) and Tibialis Posterior (figs. 338, 343) have been described. The *Interossei* (figs. 341, 342) are small but not unlike those of the hand in general plan. [The axis of the foot is along the second toe, however, and this is evident in the arrangement of the four Dorsal (abductor) and the three Plantar (adductor) Interossei.] Their actions and functions do not appear to be important.

Nerve Supply of Intrinsic Muscles of Foot

The muscles of the foot are supplied by medial plantar and lateral plantar nerves, the terminal branches of the (posterior) tibial nerve. In general, the medial plantar nerve corresponds to the median nerve in the hand while the lateral corresponds to the ulnar nerve. The plantar nerves enter the sole with the long flexor tendons deep to Abductor Hallucis.

5

The Skin

The skin or common integument is the outermost, waterproof, and protective covering of the body. It is capable of withstanding wear and tear of a lifetime and, at the end, still offering a presentable appearance. Without it, the fluid environment of the cells would escape and life would be impossible. But it is more than a covering. It is an excretory organ, a tactile organ, a heat-regulating organ, a warehouse, and a factory.

As a waterproof covering, skin possesses **sebaceous** (oil) **glands** which grease and preserve the surface; the glands are absent in the palm and sole where such greasing would be a nuisance. Skin pigment protects against the rays of the sun. As an excretory organ, skin possesses **sweat glands** capable of eliminating a considerable part of excess quantities of water and salts; it is therefore an important aid to the lungs and kidneys. As a tactile organ, skin possesses **hairs**—the most sensitive touch end-organs in the body—while between and within its surface cells naked nerve fibers serve as (touch?) temperature and pain end-organs. As a heat-regulating organ, skin possesses a rich blood supply and its arterioles, by constriction, can conserve heat and, by dilatation, can dissipate it. As a warehouse, it stores chlorides; and as a factory, it utilizes the ultraviolet rays of the sun to manufacture **vitamin D** the anti-rickets vitamin essential to the growth and maintenance of bone and teeth. The fingernails are modified appendages that develop from skin as do the mammary glands (see page 253).

Structure

The skin consists of two layers, a deeper fibrous one and a superficial cellular one. The deeper fibrous layer is called the **corium** (L. = leather), or **dermis** (fig. 344). The superficial cellular layer is called the **epidermis.** The fibers that constitute the dermis are both white (non-elastic) and yellow (elastic); the latter give the skin its elasticity and enable it to resist deformity. Loss of elastic fibers explains the wrinkled skin of old age. Adjacent to the epidermis, the dermis is corrugated—each corrugation consisting of a series of elevations known as **papillae.** The surface of the epidermis conforms in outline to the corrugations of the dermis and the resulting ridges are especially noticeable in the hands where they increase the grasping efficiency and are used for identification in fingerprinting. The fibers are less densely packed in the deeper parts of the dermis and no clear line of demarcation separates the dermis from the loose and fatty areolar tissue which constitutes the underlying **superficial fascia.**

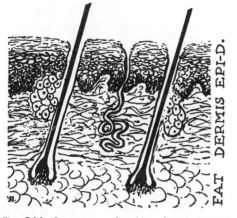

Fig. **344**. Structure of skin (semischematic, greatly enlarged cross-section). Note two hair follicles with associated oil glands and a coiled tubular sweat gland.

The cells of the epidermis are several layers thick; the deepest are columnar and are known as the **germinative layer.** They become progressively flatter as the surface is approached where dead cells are constantly shed or are rubbed off as they are replaced from the germinative cells.

Nails

The nails, like the claws and hooves of lower animals, develop from the skin (not the bone) on the backs of the distal phalanges. Each grows from an area of epidermis called the *nail root* that lies deep to the proximal margin. This edge is overlapped by a fold of skin which tapers off distally along the lateral margins of the nail. The site of active growth at the root corresponds roughly to the opaque, white, semilunar part of the nail called the *lunula*. The rest of the nail is translucent and, because the *nail bed* is very well-supplied with blood vessels, it is pink in good health. Pressure on it causes blanching beneath as the blood vessels are occluded. It must be remembered that a nail itself is tough dead tissue similar to the dead outermost protective layer of the epidermis growing from deeper lying living cells (fig. 345).

Hair

Hairs are modified skin; they begin as tubular downgrowths of all the layers of the skin. These downgrowths frequently reach into the superficial fascia and are known as **hair follicles** (fig. 344). From this region the hair grows up inside the follicle; the part filling the follicle is the **root of the hair;** the part projecting from the surface is the **shaft of the hair.** Contrary to popular belief hairs are not hollow but are solid. The **bulb of the hair** at the bottom of the follicle receives the ends of sensory nerve fibers. Smooth muscle fibers pass downward from the epidermis to the deep part of a hair follicle. Their contraction, resulting from cold or fright, produces 'goose flesh' and causes the hairs to 'stand on end.'

The **distribution of hair** has a general and an individual pattern. Hairs of different types (ranging from very fine and almost invisible ones to heavy coarse ones) are everywhere on the skin with the exception of the palms, soles, and the moist surfaces of the lips and genitalia.

The sexual pattern appears with puberty. Coarse hair begins to grow in the axillae of both sexes, and both surrounding and on the genitalia. The pubic hair of the male usually extends up to the umbilicus in contrast with that of the female which is confined chiefly to the pubic area. Coarse hair also appears on the chest of the growing man and often becomes quite profuse over most of the torso, limbs, and face.

Women occasionally grow sparse hair in the area around the nipples and on the upper lip as well as coarse hair over the arms and legs; this is not related to the

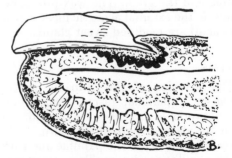

Fig. **345**. Fingertip in semischematic longitudinal section to show nail bed. Note fibrous bands mooring 'pulp' of finger to phalanx.

functional disturbances but appears to be determined by heredity. Heredity is an important factor in both the pattern of hair growth as well as baldness.

Coarse hairs become noticeable in the ear canal and nostrils in the late 30s and slowly grow more profuse with advancing age, particularly in men.

The color of a hair is determined by the amount of a pigment in its core; it is the same pigment, *melanin*, which determines the color of skin. Melanin is found within the living cells of the epidermis in all races. In light-skinned races, these cells respond to sunlight by increasing their store of pigment. However, tanning is coincidental to the production and release of vitamin D by the epidermis. Indeed, the very increase in tan obstructs the passage of ultraviolet rays and so reduces the rate of vitamin-production.

Glands

It is into the sides of the follicles above the smooth muscle that the **sebaceous glands** open; the glands lie in the dermis and provide the natural oil for the hair. Minor infections ('pimples') commonly begin at these sites.

Sweat glands are highly coiled delicate little tubes; the coil is more deeply situated than a sebaceous gland and may lie in the superficial fascia. Its long delicate duct runs a wavy vertical course to the surface. In the palms and soles the ducts open on the summits of the ridges. Sweat glands are found everwhere in the skin. They are particularly large and numerous in the axillae and the groin. The wax-producing glands of the external acoustic meatus (the ear canal) are modified sweat glands.

Vessels

The **arteries of the skin** lie in the subjacent superficial fascia. From them two networks of arterioles are formed: one is in the deepest part of the dermis; the other is just beneath the papillae and from it the papillae receive capillary networks.

Many **lymphatic vessels** (page 292) form a profuse network in the corium and

Fig. **346**. Semischematic enlarged drawing of lymphatic capillary network in a tiny piece of skin with an area of about one square mm. (After Kampmeir, greatly modified.)

drain into the larger subcutaneous lymphatics (fig. 346). The **cutaneous nerves** (page 299) end in both the epidermis and corium, being very numerous around hair follicles. Both fine naked branches and special little oval bodies (end-organs) act as sensory receptors (page 299).

Fingerprints. Everywhere, the surface of the skin has minute ridges and furrows easily visible through a hand lens. On the palms and soles these are thrown into characteristic patterns. They are well-marked on the palmar aspect of the fingers, providing the well-known fingerprints used for personal identification.

Subcutaneous tissue is also known as superficial fascia. It is loose areolar tissue with a widely varying content of fat cells. When its fat cells are numerous and heavily laden with fat, fascia is known as adipose tissue. The subcutaneous layer of yellow fat, which is almost the rule among well-nourished persons, is a storage depot. This store is quickly mobilized when the food intake does not meet the needs of the individual.

Generally, the subcutaneous fascia is loose and not made of tough threads, but in certain areas of the body it is converted into dense cords or bands that moor the skin to the underlying bone or deep fascia. In the palmar aspects of the fingers, the palm itself, and the sole of the foot, these bands are particularly significant for they prevent the skin from sliding around on the deeper structures.

6

Digestive System

Alimentary Canal

The alimentary canal is a tube that extends throughout the length of the body. It is also referred to commonly as the **digestive tract** or **gastro-intestinal tract** (although these terms are open to certain criticisms). In the embryo, the tube runs along the ventral or anterior surface of the notochord just as the neural tube runs along the dorsal or posterior surface; it is, therefore, at first relatively straight and uncomplicated. As development proceeds, the tube becomes markedly coiled in the abdominal region as a result of a rapid increase in its length, and it becomes of varying caliber as a result of differing rates of growth of its walls. It finally attains a length of some four times an adult's stature.

General Structure (fig. 347). The wall of the tube consists of a secreting and absorbing **mucous membrane (or mucosa)** backed by areolar tissue (known as submucosa) and covered, outside the areolar tissue, with two **muscular coats**—an outer longitudinal one and an inner circular one. Throughout the abdominal and pelvic cavities, the tube possesses still another coat since it is closely invested by peritoneum—the serous membrane lining the walls of those cavities and reflected on to the walls of the tube; this is known as the

serous coat (**serosa**). Throughout the neck and thorax, the tube has no serous investment; other local differences in the structure of the wall will be noted as they are met.

MOUTH AND PHARYNX

Pharynx

The entrance to the alimentary tube is the pharynx. The upper end of the pharynx reaches the base of the skull behind the posterior nasal apertures and opens forward into the nose. Between palate and tongue it communicates with the mouth or oral cavity. Well below the back of the tongue it divides into the first parts of the air passage (larynx, or voice-box) and of the food passage (esophagus or gullet). The term "throat" is hardly explicit enough when one is referring to the pharynx.

It is protected behind by the bodies of the upper six cervical vertebrae from which it is separated by loose areolar tissue that permits movement of the tube on them (fig. 348).

At its upper end of the pharynx is about 5 cm (2") wide and 2–3 cm (1") from front to back. It narrows as it passes downward until, at the back of the larynx or voice-box where the pharynx becomes the esophagus, the front and back walls are in contact with

201

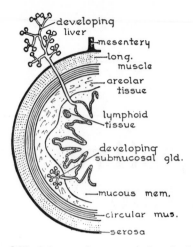

Fig. **347**. Scheme of structure of wall of intestine in partial cross-section.

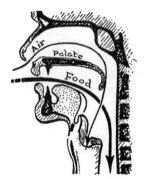

Fig. **348**. Air and food passages cross.

one another separating only to permit the passage of food. A little lower, the almost circular tube has an overall diameter of about 2 to 3 cm (1″).

The front wall of the pharynx is deficient since it opens forward into the nose, the mouth, and the larynx. The nose and mouth are separated from one another by the **hard** (bony) **palate.** This is continuous behind with the **soft** (fibromuscular) **palate** which, projecting backward, reaches almost to the posterior wall of the pharynx. The elevated soft palate is capable of shutting off the nasopharynx during the passage of food from the mouth.

The entrance to the larynx lies below and behind the back or vertical part of the tongue and projects freely into the interior of the pharynx from in front; this makes the outline of the pharynx in a cross-sec-

tion at this level definitely U-shaped (figs. 348, 349).

From what has been said it is apparent that the pharynx is divisible into three **parts:** the **nasal pharynx,** also known as the **nasopharynx,** the **oral pharynx,** and the **laryngeal pharynx.** It is further apparent that the pharynx is a tube common to the respiratory and the digestive systems.

The nasopharynx, being entirely respiratory, is made non-collapsible by the attachment of its walls to bone (fig. 350). The pathway for inspired air must cross that for food and the oral pharynx is, of course, both respiratory and digestive (fig. 348).

Special Features of Nasal and Oral Parts of Pharynx. On each side wall of the nasopharynx near the posterior nasal apertures (and so hidden by the soft palate when one looks into the mouth) is the opening of a passage, the **'pharyngotympanic'** or **auditory** (Eustachian) **tube** (fig. 539 on page 302). This tube leads upward and lateralward to the middle ear cavity. It serves to equalize atmospheric pressure on the two surfaces of the tympanic membrane thus enabling that structure to vibrate in response to sound waves (fig. 351). A swelling of the pharyngeal mucous membrane caused by hay fever or a severe cold in the head, may block the passage and result in a temporary impairment of hearing. [Behind the opening of each tube there is a slit-like lateral cul-de-sac of the pharynx, the *pharyngeal recess.*]

Lymphoid Tissue. The entrance from mouth to pharynx, the **fauces,** is guarded by lymphoid tissue—most active during that period of life when there is an irresisti-

Fig. **349**. The larynx indents the pharynx from in front (schematic horizontal cross-section).

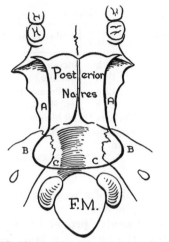

Fig. **350**. Walls of pharynx are attached to medial pterygoid plates (A) that bound posterior nasal openings (nares) and to a semicircle that crosses petrous bones (B) and occipital bone (C) in front of foramen magnum (F.M.).

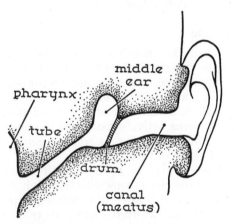

Fig. **351**. Auditory tube brings air from pharynx to middle ear which is completely separated from meatus by the eardrum or tympanic membrane (schematic).

ble impulse to put everything in the mouth. Lymphoid tissue guards against the entrance of germs into the body. The largest collection of this tissue forms a pair of firm, oval, almond-shaped masses—the **(palatine) tonsils.** These lie in a pair of recesses or fossae on the side walls of the oral pharynx (fig. 352). Each tonsillar fossa is bounded by a pair of prominent folds of mucous membrane underlying which is a pair of muscles—Palatoglossus in front and Palatopharyngeus behind. The folds are

very apparent on looking in the open mouth.

The tonsils are continuous with more irregular masses of lymphoid tissue existing on the back of the tongue **(lingual tonsil)** and on the under surface of the palate. Thus, a complete ring of lymphoid tissue surrounds the entrance from mouth to pharynx. Another collection exists in the midline at the summit of the nasopharynx, which, if enlarged, hangs like a bunch of grapes from the pharyngeal roof; the collection is known as the **pharyngeal tonsil** or **'adenoids'** and it may partially block the entrance to the auditory tube and so require removal.

Fascial and Muscular Coats. Lying between the pharyngeal mucous membrane and the muscular layer is a thin but relatively strong layer of fascia known as the **pharyngobasilar fascia;** it is this fascia which, acting as a framework for the pharynx, is attached, above, to the base of the skull and, in front, to the medial pterygoid plates; the plates bound the entrance from the nose to the pharynx. Outside the pharyngobasilar fascia, the order of the muscular coats differs from that occurring throughout the rest of the tube for, within, lies the thin and ill-defined longitudinal musculature and, without, lies the circular musculature known as the constrictors of the pharynx; these have already been described (page 133; see also fig. 491). The muscles are covered outside by a thin and delicate areolar sheet known as the **buccopharyngeal fascia,** and this is continuous at the pterygomandibular raphe with the fascia covering the outside of the Buccinator—the muscle forming the side wall of the mouth. It is the buccopharyngeal fascia that lies in front of the cervical vertebrae.

Mouth and Teeth

The space bounded by the lips and cheeks on the one hand, and by the upper

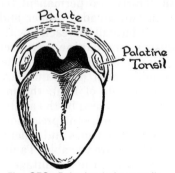

Fig. **352**. Paired palatine tonsils

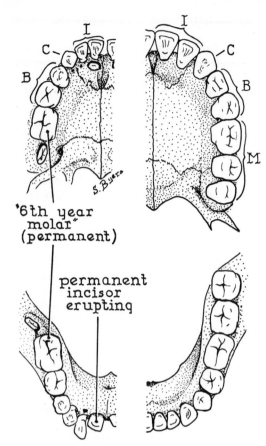

'6th year
molar'
(permanent)

permanent
incisor
erupting

Fig. **353**. Child's (left) and adult's (right) upper and lower dental arcades compared. (Child aged 7.)

TABLE 1

Deciduous	Time (in months)	Permanent	Time (in years)
Incisors (8)	6th	Incisors (8)	6th
Canines (4)	18th	Canines (4)	9th
1st Molars (4)	12th	1st Premolars (4)	12th
2nd Molars (4)	24th	2nd Premolars (4)	12th
Note: lateral incisors may be considerably delayed.		1st Molars (4)	6th
		2nd Molars (4)	12th
		3rd Molars (4)	18th

The structure of a tooth in its socket is shown in figure 354.

The **oval cavity proper** lies within the confines of the teeth. It is bounded above by the hard palate and below by mucous membrane covering salivary glands and certain muscles of the tongue. The muscular floor of the mouth (oral diaphragm) has been described (page 130).

Tongue

When the mouth is closed the tongue almost fills the oral cavity. The tongue is a highly mobile, muscular, and tactile organ, and it plays an important part in articulate speech. It is also accessory to the digestive tract, being actively concerned in mastication and swallowing as well as being the chief organ of taste.

The tongue is covered with mucous membrane continuous with that lining the oral cavity. It consists of a mobile, horizontal, anterior two-thirds and a more or less fixed, vertical, posterior one-third. This latter helps to form the anterior wall of the pharynx, and its surface has an entirely different appearance from that of the anterior two-thirds.

Papillae of Tongue. The roughness of the mucosa of the upper surface of the tongue is due to many little projections called papillae (L. = nipples). There are innumerable **filiform** (L. = thread-like) **papillae** which are arranged in parallel rows and give the tongue a rasping function. This rasping function in man is of minor importance. Scattered among the filiform papillae are the larger **fungiform** (L. = mushroom-like) **papillae** and these are equipped with taste buds (see below).

and lower gums and teeth in apposition on the other, is called the **vestibule of the mouth.** It is lined with mucous membrane which covers the muscles of the lips and cheeks and it is reflected on to the gums. The mucous membrane contains many tiny mucous glands.

Teeth. Thirty-two permanent teeth are arrayed in two arcades, upper and lower (fig. 353). Bilaterally in each jaw they are: two **incisors,** one **canine,** two **premolars (bicuspids),** and three **molars.** The deciduous or milk teeth of infancy are 20 in number (four sets of five), the molars of infancy being replaced by the adult premolars. There are no infant precursors of the adult molars. The times and order of eruption of the teeth are variable. Table 1 gives an average; it has the advantage of being easily remembered.

Where the horizontal part of the tongue meets the vertical, there is a V-shaped row of very large **vallate papillae** (fig. 355). These are like miniature turrets and each is surrounded by a moat, hence the name. The walls of the moat are generously equipped with **taste buds** which are highly specialized nerve endings. A perfectly dry tongue cannot taste and the sense itself is limited to but four discriminations—*bitter*, *sweet*, *salty*, and *sour*. Many of the finer sensations attributed to taste are actually received by the organ of smell.

Behind the vallate papillae, the surface of the tongue is irregularly smooth since

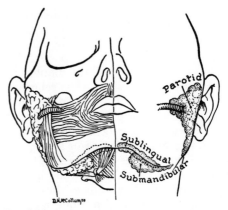

Fig. **356**. The three paired salivary glands form a continuous series from ear to ear. The parotid duct pierces the Buccinator muscle.

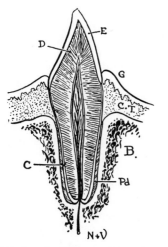

Fig. **354**. Longitudinal section of an incisor tooth in its socket. B = bone of jaw; C = cement; CT = connective tissue of gum; D = dentine; E = enamel; G = gum; Pd = peridontal membrane; N and V = nerve and vessels in root canal.

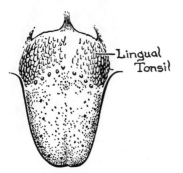

Fig. **355**. The lingual tonsil on the back of the tongue is limited in front by a row of vallate papillae.

here the mucous membrane covers a collection of lymphoid tissue, the **lingual tonsil**.

Muscles of Tongue (see page 135).

Salivary Glands of Mouth

There are three pairs of glands that are associated with the mandible (fig. 356) and that secrete saliva into the mouth through ducts. These salivary glands are: (1) the parotid; (2) the submandibular (submaxillary); (3) the sublingual.

The parotid gland, as its name proclaims, lies beside the external ear. It is packed in the space between the mastoid and styloid processes behind, and the ramus of the mandible in front. It reaches as far medialward as the side wall of the pharynx. The gland, in effect, is wrapped around the upper part of the ramus of the mandible and the two muscles that clothe, one its outer, the other its inner surface. From the superficial portion of the gland, which in part covers the Masseter, issues the duct of the gland and this tubular structure runs forward to the anterior border of the Masseter immediately below the zygomatic arch. The duct turns deeply here and pierces the Buccinator to open into the mouth beside the second upper molar tooth; here, by careful inspection in a good light, its mouth may be seen as a slit-like whitish orifice. Many vessels and the facial nerve traverse the gland. Mumps is an infection of the parotid by a specific virus.

The submandibular gland lies in front of the lower part of the parotid gland and is under shelter of the body of the mandible. It is wrapped around the posterior free border of the Mylohyoid, the diaphragm of the mouth. Because of this relation its duct, which issues from its deep part, is enabled to enter the oral cavity. The duct runs forward, below the tongue, on the Hyoglossus and Genioglossus and opens by piercing the mucous membrane of the floor of the mouth near the midline. This pinpoint orifice lies close beside the frenulum—the fold of mucous membrane running in the midline from the under surface of the tongue to the floor of the mouth.

The sublingual gland, the smallest of the salivary glands, lies well forward beneath the mucous membrane of the floor of the mouth on the lateral side of the submandibular duct and lingual nerve. Its secretion is discharged into the mouth by a series of minute apertures located on the summit of the sublingual fold at the side of the frenulum.

The six salivary glands secrete well over a liter of saliva a day. The most important use of this abundant flow is the moistening and preparation of food for swallowing. Saliva also cleanses and refreshes the mouth. Its role in the digestion of starch—very real in a test tube—seems to be unimportant in a living person.

ESOPHAGUS

The pharynx ends at the level of (and in contact with) the sixth cervical vertebra to become the collapsed, muscular tube known as the esophagus* or gullet. The larynx becomes the trachea (windpipe) at this same level and the latter tube lies in bare contact with the front of the esophagus. Because the junction of larynx and trachea, the cricoid cartilage, is palpable—a short distance (2–3 cm) below the 'Adam's apple'—its level is easily found.

The total length of the esophagus is about 25 cm (10″) Its upper one-fifth lies in the neck; its lower four-fifths lie in the

thorax (fig. 357). In general, it occupies a midline position and its upper half is in direct contact with the vertebral bodies behind, and with the trachea in front. Since the trachea is less than half of its total length, the lower half of the esophagus passes beyond it and is next found in contact with the back of the heart. It has come forward a little so that, behind the esophagus, the aorta can traverse the lower part of the thorax in contact with the vetebral bodies. The lower end of the esophagus swings in a gentle curve to the left and pierces the diaphragm, below which, and therefore in the abdominal cavity, its last 2 to 3 cm may be seen as it opens into the stomach.

With the beginning of the esophagus, the muscular wall of the digestive tube assumes the characteristic arrangement of an outer longitudinal coat and an inner circular one. The longitudinal muscle takes origin from the back of the cricoid cartilage, the lowest or 'foundation' cartilage of the larynx. It is in the upper part of the esophagus that a gradual transition occurs from striated or voluntary muscle to smooth or involuntary muscle.

Automatic peristaltic contraction begins

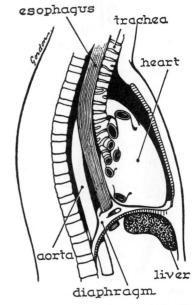

Fig. **357**. Some important relationships of esophagus (semischematic).

---

* Also spelled oesophagus; pronounced with accent on second syllable; derived from Gr. = I carry food.

in the esophagus so that food does not fall down by gravity but is propelled by muscle action. That is why it is possible to drink a glass of water standing on one's head.

STOMACH, INTESTINES, AND ASSOCIATED ORGANS

The abdominal part of the alimentary canal consists of the following consecutive parts:

1. Stomach
2. Small intestine
   a. duodenum
   b. jejunum
   c. ileum
3. Large intestine
   a. cecum and appendix
   b. ascending colon
   c. transverse colon
   d. descending colon
   e. pelvic (or sigmoid) colon
   f. rectum and anal canal (ending at anus).

The associated organs are:

1. Liver and gall bladder
2. Pancreas
3. Spleen.

These structures will be discussed in turn. Finally, a short section will be devoted to peritoneum. A discussion of blood supplies and nerve supplies will be reserved for the circulatory and nervous systems.

STOMACH

The stomach† is the most dilated portion of the digestive tract and has an average capacity of about one liter or quart. It is subject to considerable variation in shape and size, but an average stomach is J-shaped in general outline and has a maximum length of about 25 cm (10″) and a maximum breadth of about 14 cm.

Parts of Stomach

The borders and openings of the stomach are labeled in fig. 358.

The **lesser curvature** carries downward

_____

† Stomach is *not* synonymous with the belly (abdomen), contrary to common polite usage. The Greek word for stomach is *gaster*; from it comes the adjective *gastric*.

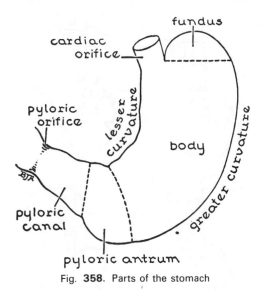

Fig. **358**. Parts of the stomach

the line of the right border of the esophagus and throughout most of its extent is nearly vertical. The **greater curvature** is subject to considerable variation in length and position depending on the condition of the stomach at the time of examination. A small amount of gas—probably swallowed air—is usually found in the **fundus,** which is the upward-ballooned recess lying above the level of the **cardiac orifice.** The stomach gets progressively narrower below, ending as the **pyloric canal** that leads to the **pylorus,** the exit of the stomach.

Position and Relationships

Almost the whole of the stomach lies to the left, its fundus being the main contact of the left part of the diaphragm which separates the stomach from the heart and left lung. Other contacts of the stomach are liver (to the right), spleen and left kidney (posteriorly), coils of small intestine (below), and anterior abdominal wall. A large part of the stomach enjoys the shelter of the left lower ribs. The great variability in the shape and position of the stomach is illustrated in figure 359.

The **cardiac orifice** has a true sphincter of variable competence; also, the passage of the esophagus through the diaphragm may enable that muscle to offer some sphincteric action. The **pyloric orifice,** lying 3 cm (1″) to the right of the midline

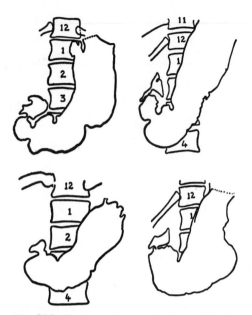

Fig. **359**. Tracings of x-ray shadows of barium-filled stomachs of four different persons to show variations in shape and position.

and about 5 cm (2″) below the tip of the sternum, is guarded by a very powerful sphincteric muscle, the result of a marked increase in the normal circular musculature of the tube; the orifice, slightly open most of the time, permits fluids to be squirted out, but prevents escape of solids.

Structure

The **wall** of the stomach is very muscular since one important duty of the organ is to break up the food; it is commonly forced to undertake duties more properly the concern of the teeth. Besides coats of **longitudinal** and **circular musculature,** the organ possesses a third, innermost and incomplete coat of **oblique fibers.** These are particularly evident in the region of the fundus.

The **mucous membrane** is thick and velvety, and has the appearance of a honeycomb. In the body and the pyloric end, it is thrown into folds or ridges which run parallel to the neighboring curvatures. Besides the **mucous glands** (found in any mucous membrane) the **gastric glands** consist of two types of cells, the **chief cells**

which secrete pepsin (and other enzymes) and the **parietal cells** which secrete hydrochloric acid. These types of glands are abundant in the body and fundus but are absent from the pyloric region where many mucous glands are found.

The *greater* and *lesser omenta*, sheets of peritoneum associated with the stomach, are described on page 222.

SMALL INTESTINE

Various parts of the small and large intestine have names which end in -*um*. The reason is that the noun *intestinum* is omitted and words like duodenum, cecum, and rectum were originally adjectives describing intestinum, e.g., intestinum duodenum (see figs. 360, 361).

DUODENUM

The first and shortest part of the small intestine, the duodenum‡ has a horseshoe-shaped course only 25 cm (10″) long. [The esophagus, stomach, and duodenum are all about the same length.] It starts 2 to 3 cm to the right of the midline and ends the same distance to the left, the two ends being but 5 cm (2″) apart.

Four parts are recognized. The **first** or **superior part,** about 5 cm (2″) long, runs upward and backward; it is movable and comes in contact with the liver. The **second** or **descending part,** about 8 cm (3″) long, turns sharply to run vertically downward along the right side of the second and third lumbar vertebral bodies and in front of the length of the medial border of the right kidney. The **third** or **horizontal (inferior) part,** about 10 cm (4″) long, runs horizontally to the left across the body of the third lumbar vertebra. The **fourth** or **ascending part,** about 2–3 cm (1″) long, rises so that, at its end, it is only a little below the level of the pyloric orifice. All but the first part are bound down to the posterior abdominal wall (retroperitoneal).

---

‡From duodenarius, L. = containing 12. Herophilus (c. 344 B.C.) is said to have described the length of the duodenum as 12 fingers' breadths, but the term duodenum is of more recent origin.

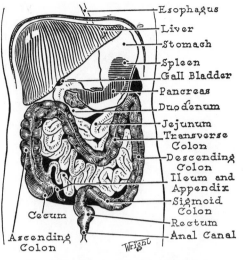

Esophagus
Liver
Stomach
Spleen
Gall Bladder
Pancreas
Duodenum
Jejunum
Transverse
Colon
Descending
Colon
Ileum and
Appendix
Sigmoid
Colon
Rectum
Anal Canal
Cecum
Ascending
Colon

Fig. 360. Scheme of alimentary tract

The head of the pancreas fills and over-flows the 'horseshoe.'

The duodenum is relatively thick-walled and its mucous membrane possesses large **duodenal digestive glands** and is thrown into numerous folds. In the interior of its second part, along the concave side, can be found a small orifice common to the bile duct (from the liver) and the pancreatic duct so that it is provided with additional digestive enzymes (see figs. 362, 383).

JEJUNUM AND ILEUM

About 6 meters (20 feet) of mobile small intestine begin at the duodenojejunal junction.

The upper two-fifths of the mobile small intestine are known as the **jejunum;** the

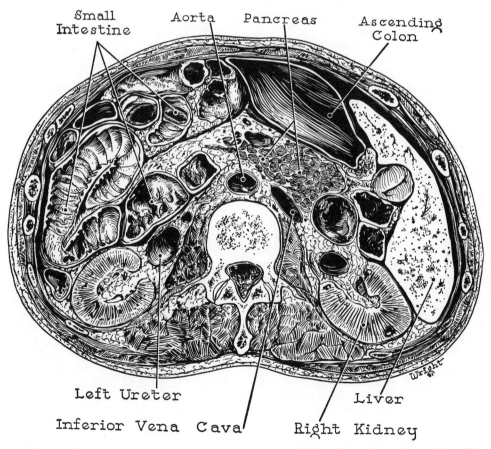

Small Intestine   Aorta   Pancreas   Ascending Colon

Left Ureter

Inferior Vena Cava   Right Kidney   Liver

Fig. 361. Transverse section through abdomen at about level of L. 2 or 3 vertebra to show relationship

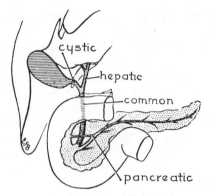

Fig. **362**. The horseshoe-shaped duodenum surrounds head of pancreas and receives bile and pancreatic ducts along inner curve.

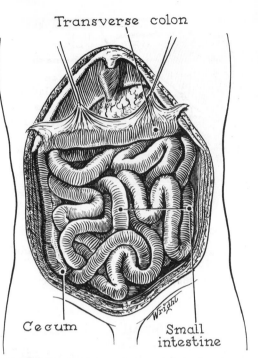

Fig. **363**. Small intestines revealed when transverse colon and greater omentum are reflected upward.

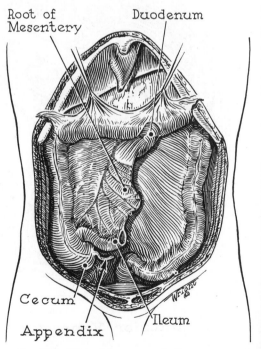

Fig. **364**. With jejunum and ileum removed, root of mesentery is revealed.

lower three-fifths are known as the **ileum.**§ The caliber of the tube decreases gradually from a diameter of nearly 5 cm (2″) at the duodenum to 2 to 3 cm (1″) at the end of the ileum; it varies, of course, with muscular activity (see fig. 363).

Mesentery. Each end—**duodenojejunal junction** and **ileocecal junction**—is moored to the posterior abdominal wall, but the intervening intestines are mobile and slip around into ever-changing coils and loops. Because a blood supply and a nerve supply are so vital, the mobile gut has a free double sheet of peritoneal (i.e., serous) membrane, known as **mesentery,** attaching it to the posterior body wall and transmitting vessels and nerves to the gut. The attachment or *root* of the mesentery to the body wall is along the short oblique line (about 20 cm or 8″ long) joining duodenojejunal junction to ileocecal junction (see figs. 364 and 277 on page 502). Because, on the other hand, the attachment to the small gut is so very long, it is hardly necessary to point out that the mesentery must be fan-shaped. See also the section on peritoneum, page 220.

Mucosa (fig. 365). The mucous membrane of the small intestine is equipped with villi. Each **villus** is a microscopic finger-like process that projects into the lumen of the gut. Villi exist from pylorus to

§ Note the spelling—ileum for gut; ilium for bone. Jejunum, L. = empty (cf., English word jejune). At autopsy, the jejunum is often empty.

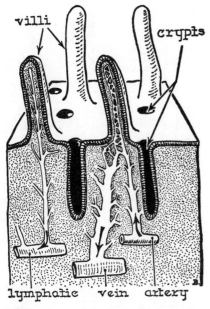

Fig. **365**. Three-dimensional diagram of a microscopic piece of small intestine. The villi are drained by both veins and lymphatics.

ileocolic valve, i.e., throughout the length of the small intestine. Their enormous numbers—like pile on a carpet—give a velvety feeling to the exploring finger. They markedly increase the absorbing area and each villus, containing minute blood vessels (arteriole and venule) and a lacteal (lymph vessel), is to be regarded as a microscopic organ devoted to the absorption of digested material. At the bases of the villi are the mouths of the **intestinal glands.** Because these glands project down into the gut wall they are called **crypts** (of Lieberkühn); they secrete the intestinal digestive enzymes.

In the large intestine glands are present but villi are absent.

It is helpful to observe that digestive activity becomes progressively less from upper to lower end. This gradual functional change is reflected in gradual anatomical changes which are listed below:

1. A larger caliber and a thicker muscular wall above, where peristalsis is strongest.
2. A thicker mucous membrane thrown into many more folds (fig. 361) and possessing many more villi above,

where digestion and absorption are more active.
3. A richer blood supply above.
4. More fat stored between the peritoneal layers below, where less activity is present.
5. More lymphoid tissue in the intestinal wall below, where debris and rejected material accumulate.

LARGE INTESTINE

Ileocecal Valve, Cecum, and Appendix

The ileum ends by 'pouting' into the medial side of a reservoir—the cecum (L. = blind) or first part of the large intestine. At this orifice is the **ileocecal valve** (fig. 366). The orifice is a transverse slit between two 'lips' which pout into the cecum and allow no regurgitation of cecal contents into the ileum.

Next to the stomach, the **cecum** is the largest calibered part of the digestive tract. It occupies the right iliac fossa and is closely related to the lateral half of the inguinal ligament and the adjacent anterior abdominal wall. Besides receiving the ileum, it possesses a diverticulum, the **vermiform** (L. = worm-like) **appendix** (figs. 364, 366).

The **appendix** is the narrowest part of the digestive tract having failed to keep pace in growth with the rest of the cecum. It varies in length from 5 to 15 cm, or even 20 cm or more, and it has about the same caliber as that of a large worm. It is a cul-de-sac or 'dead-end' which opens into the cecum a little below the ileocecal

Fig. **366**. Upper 'lip' or flap of ileocecal valve directs stream downward into cecum.

orifice. It is attached to the last section of
the ileum by a triangular fold of peritoneum, the **mesoappendix** (fig. 367). Its
position is quite variable. In its rich equipment with lymphoid tissue the appendix
resembles the tonsils—indeed it has been
called the 'tonsil of the abdomen.' Like the
tonsils, the appendix is frequently infected
and then becomes a menace.

The cecum is usually retroperitoneal and
relatively fixed, though it may possess a
'mesentery' and, in consequence, be unduly mobile. It is continuous above with
the ascending colon.

The Colon

The ascending, transverse, descending,
and pelvic parts of the colon form the sides
of a 'picture frame' which is deficient in the
right lower corner (fig. 368). The contained
'picture' is the small intestine lying in
closely packed loops which are in constant
activity (fig. 363). Parts of the 'frame'
(transverse colon and pelvic colon) are
mobile, too (fig. 369).

The **ascending colon** is the shortest part
of the colon and ascends retroperitoneally
in the right flank. It reaches to the under
surface of the liver where it makes a
right-angled turn to become the transverse
colon. The bend is known as the **right colic
flexure** and it lies in front of the lower pole
of the right kidney, and in contact with the
duodenum.

The **transverse colon** is mobile and,
being twice as long as the distance across
the abdominal wall, it hangs downward at
its middle—as far as the navel (umbilicus)

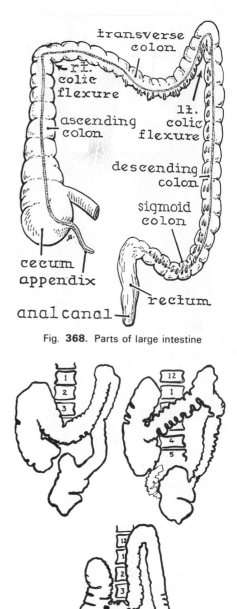

Fig. **368**. Parts of large intestine

Fig. **369**. Tracings of x-ray shadows of the large
intestines (filled with barium) in three different
persons to show variability of shape and position.

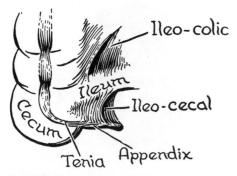

Fig. **367**. Cecum and appendix from in front,
and two related peritoneal fossae (ileocolic and
ileocecal) of interest to surgeons.

when the person stands erect. It hangs from
a mesentery and is attached to the posterior surface of an apron-like peritoneal
structure, the greater omentum, which is

discussed under peritoneum (see fig. 370 and compare with 363 and 364). The transverse colon lies below the stomach and in front of coils of small intestine. It rises at its left end and reaches the spleen where it makes an acute-angled bend—the **left colic flexure**—to become the descending colon.

The left colic flexure lies in front of the lower half of the left kidney. [Here a triangular fold of peritoneum passes laterally from colon to diaphragm; it is known as the *phrenicocolic ligament* and it serves as a shelf for the support of the spleen.]

The **descending colon** is the narrowest portion of the large intestine. It descends retroperitoneally in the left flank to the pelvic brim where it becomes the **sigmoid colon.** The sigmoid colon has an inverted V-shaped mesentery—the **sigmoid mesocolon**—whose root runs at first upward along the pelvic brim and then downward to the middle of the sacrum. This mobile sigmoid colon varies in length from 15 to 40 cm (6″ to 16″) or more; it is succeeded by the **rectum** (described below).

Structure and Function of Colon

One of the distinctive features of the colon is that its longitudinal coat does not surround the wall but consists merely of three muscular bands. These are known as the **teniae coli.** They begin at the base of the appendix and, being shorter than the colon itself, they pucker the wall of the colon so that it consists of a series of **sacculations** or **haustra;** if they are cut the sacculations disappear (see figs. 360, 364, 367, and 368). The rectum does possess a more or less complete longitudinal coat but it, too, is shorter than the rectal wall and is responsible for the sinuosity of the rectum.

Little fat bodies, contained in tag-like redundancies of the peritoneum, hang from the wall of the colon—especially the descending colon—and these, too, give the large bowel a distinctive appearance; they are known as **epiploic appendages** (fig. 368).

The content of the cecum and ascending colon is watery; that of the sigmoid colon is solid, so that considerable absorption of fluid occurs in the colon. The mucosa possesses test-tube-like crypts but not villi.

Rectum

The rectum (L. = straight) occupies the midline. It is about 15 cm (6″) long and lies in front of the lower half of the sacrum whose curvature it follows. It also takes a somewhat sinuous course and therefore belies its name (fig. 371, 372).

It passes downward beyond the sacrum and coccyx and, continuing the forward curvature of the lower end of those bones, reaches, in the male, to the back of the

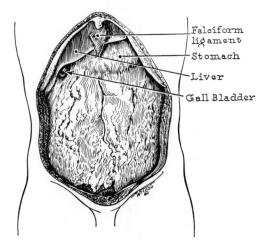

Fig. **370.** Greater omentum hangs like an apron from greater curvature of stomach.

Fig. **371.** Rectum (from left side). Last part, the anal canal, runs backward (cf., fig. 308).

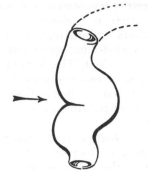

Fig. **372**. Rectum from in front is sinuous.

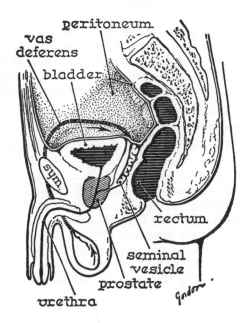

Fig. **373**. Median sagittal section of male pelvis illustrating relationships of rectum. (Rectum, being sinuous from side to side, is 'opened into' several times by the midline section.)

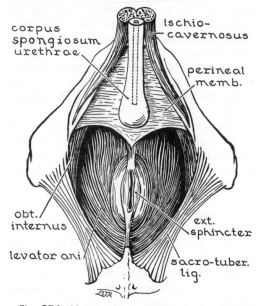

Fig. **374**. Male perineum. Each of paired ischio-rectal fossae is a (normally fat-filled) space below Levator Ani on side of rectum. Anterior half of perineum belongs to urogenital system.

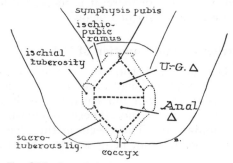

Fig. **375**. Boundaries and two subdivisions of perineal region (scheme).

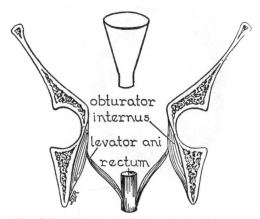

Fig. **376**. Ischiorectal fossa is the space on each side of rectum below the Levator Ani muscle.

prostate gland (fig. 373); in this part of its course it is as much horizontal as it is vertical. In the female it is in contact with the back of the vagina.

It next turns sharply downward and backward, pierces the Levatores Ani (the pelvic diaphragm) and becomes the **anal canal** which, after a course of about 4 cm (1½″) opens to the exterior at the anal orifice **(anus).**

Anal Sphincters (fig. 374). Of the two, the upper and less important **internal**

sphincter is merely a thickening of the circular muscle coat and is involuntary; the lower or **external sphincter** is voluntary striated musculature derived from the muscles that originally surround the cloaca (page 15). When the internal sphincter relaxes—an effect of the autonomic nervous system—the onus for the retention of rectal contents falls on the external sphincter—a muscle under voluntary control which is continuously active except during defecation. The Levatores Ani blend with the longitudinal musculature of the anal wall and draw it upward in defecation.

Below the Levatores Ani is the diamond-shaped **perineal region** or **perineum** (figs. 374–376). The front half of the diamond is bounded by the symphysis pubis and the diverging ischiopubic rami which lead backward to the ischial tuberosities. The back half is bounded by the sacrotuberous ligaments (each overlaid by a Gluteus Maximus) and their meeting behind at the coccyx.

The triangular posterior half of the perineum is known as the **anal region** or **triangle.** Here a quantity of fat fills the wedge-shaped space that lies below the Levator Ani and between it and the side pelvic wall (fig. 376). The space is known as the **ischiorectal fossa** and it allows the anal canal to expand during defecation.

The triangular anterior half of the perineum is known as the **urogenital region** or **triangle** (page 247).

LIVER, GALL BLADDER, PANCREAS, AND SPLEEN

The liver and the pancreas are derivatives of the digestive tract, having grown out from its wall to which they are still connected by their ducts. The spleen has no duct and is not a digestive organ; but, in position and blood supply, it is closely associated with the digestive system.

Liver

The liver (or *hepar*, Gr.) is the largest organ in the body, weighing about 1.5 kg (50 oz) or about one-fiftieth of the total body weight. It is a solid, reddish-brown, pliant organ situated mainly on the right side of the body and in contact with the under surface of the dome of the diaphragm, most of the organ thus enjoying the protection of the lower ribs (fig. 377).

Surfaces and Borders. Its *diaphragmatic surface* conforms to the shape of the diaphragm with which it is in contact. It has right, anterior, superior, and posterior parts which are not demarcated. The posterior part presents a wide deep notch where the organ straddles the vertebral column.

The *visceral surface* is separated from the diaphragmatic one in front by a sharp edge and is molded by the organs with which it is in contact below.

The liver reaches to the left to 2 to 3 cm (1") below the left nipple (fig. 378); here the organ tapers to an apex. Figure 378 shows the relations of its right and anterior aspects to the chest wall and it will be observed, too, that only a small area (at the subcostal angle) descends below the level of the costal cage. Figure 379 shows the areas of contact of neighboring organs which produce depressions on the visceral surface; it also shows the small caudate

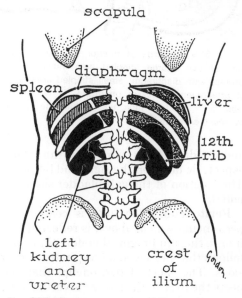

Fig. **377.** Surface anatomy of liver, spleen, and kidneys from behind.

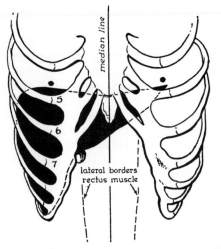

Fig. **378**. Liver is protected by lower right ribs. Gall bladder peeks below lower edge of liver and costal margin at lateral edge of right Rectus.

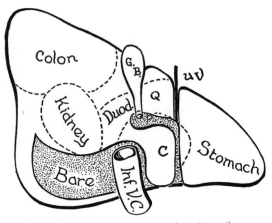

Fig. **379**. Contacts of inferior (or visceral) surface of liver. Large stippled area indicates bare contact with diaphragm; small stippled area is line of attachment of lesser omentum. C. = caudate lobe; Q. = quadrate lobe; uv. = obliterated umbilical vein (i.e., round lig.).

and quadrate lobes, which, on the inferior aspect, lie between the right and left lobes. The location of the gall bladder should be noted.

Peritoneum. The liver is coated with peritoneum which, above, is reflected from it onto the diaphragm; therefore that organ follows that muscle in its respiratory excursions. The sheet of peritoneum that connects the liver to the lesser curvature of the stomach (described on page 222) is the

lesser omentum; it and the **falciform ligament**—the sickle-shaped fold whose lower edge carries the remains of the umbilical vein (the ligamentum teres or round ligament)—indicate, by their attachments to the liver, the surface boundaries of the lobes.

Porta Hepatis. A deep transverse slit on the visceral surface (between caudate and quadrate lobes) is the 'door' or porta of the liver. To it runs the lesser omentum carrying the important vessels and ducts of the liver. These are the **hepatic artery** and the **portal vein** which enter the liver, and the **right and left hepatic ducts** which leave the liver and immediately join to form the **common hepatic duct.**

Functions. The liver is a complex factory and warehouse. The blood returning in the veins from the digestive tract is laden with **glucose** (and other sugars) derived from the food. It is conveyed to the liver whose cells are largely concerned in converting the glucose into **glycogen.** When bodily activity demands it, the stored glycogen, reconverted to glucose, is delivered to the general circulation. Thus the liver maintains the amount of glucose present in the body at a constant level.

The liver makes and stores **vitamin A**—a vitamin essential to the well-being of mucous membranes and other epithelial cells. It also stores iron, plays an important role in metabolism of protein and fat, and acts as a detoxifying agent.

From the hemoglobin resulting from the death of red blood cells (page 255) the liver makes **bile** which, on delivery to the duodenum, attacks and emulsifies ingested fats and is largely responsible for fat absorption. The liver, too, elaborates **heparin,** a vital substance in the prevention of blood clotting, and, like the spleen, its possession of 'scavenger' cells, (phagocytes) which attack germs makes it an important organ in increasing resistance to disease.

It can now be appreciated why the liver is provided not only with a blood vessel for the nourishment of its own cells (the **hepatic artery),** but also with a blood vessel (the **portal vein)** that conveys to it all the blood from the digestive tract.

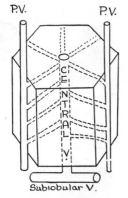

Fig. **380**. Traditional diagram of a lobule of liver. P.V. = branches of portal vein. Central veins drain lobules to sublobular veins, the tributaries of the hepatic veins. (Duct system and arteries omitted for clarity.)

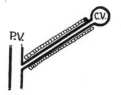

Fig. **381**. Blood from branches of portal veins passes via spaces between cells of liver to central vein which drains lobule of liver (very schematic diagram).

Structure (figs. 380, 381). Deep to its peritoneal coat, the liver is encased in a thin, fibrous capsule from which fibrous partitions pass into the interior to divide the organ into units known as **liver lobules.** If the liver is sectioned, these lobules, each a millimeter or two in diameter, may be seen with the naked eye. Each lobule consists of cords of liver cells¶ radiating from a center. Between adjacent lobules lie three structures—a branch of the hepatic artery, of the portal vein, and of the bile duct. At the center of the lobule lies a solitary tributary of the **hepatic vein.**

¶ Recently, these have been shown to be plates or sheets of cells rather than cords. Further, the traditional concept of the lobule had been modified; in the newer view, the 'interlobular' branches of the portal vein lie in the middle of a unit called an *acinus*. Around the periphery of an acinus are the tributaries of the hepatic veins (the 'central' veins of the traditional lobule) which drain the liver.

As the blood traverses the lobule from periphery to center, i.e., from portal to hepatic vein, it runs the gamut of the vital liver cells whose biochemical processes have been briefly noted.

Hepatic Veins. The tributaries of the hepatic veins unite with one another and finally the large and main hepatic veins empty into the inferior vena cava where that vessel lies buried in the back of the liver. The inferior vena cava delivers its blood to the heart.

Gall Bladder and Bile Ducts

When the **gall bladder** is full of bile its size and shape is that of an elongated pear (fig. 382). It lies in a depression at the edge of the visceral surface of the liver with which it is in bare contact. Its rounded extremity or **fundus** projects beyond the liver (fig. 370) and is in contact with the anterior abdominal wall at the point where the border of the right Rectus Abdominis crosses the ninth costal cartilage. The gall bladder is a reservoir for bile with a capacity of about 50 cc and is not vital to life. Some animals possess a gall bladder, others do not.

In direct contact with its thin muscular coat, the relatively thick mucous membrane of the gall bladder presents a honeycomb appearance and is concerned with the concentration of bile. But an over-concentration leads to a precipitation of bile salts in the gall bladder with a consequent

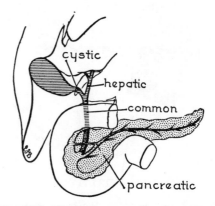

Fig. **382**. Bile and pancreatic ducts open in common into duodenum. An accessory pancreatic duct is often present above main one.

formation of a gall stone or stones; then the individual is better without his gall bladder. The duct of the gall bladder—the **cystic duct**—joins the **common hepatic duct** which descends from the liver; the resultant duct is the **(common) bile duct** (fig. 383).

The **bile duct** (*ductus choledochus*) lies in the free right margin of the lesser omentum (page 222) and is accompanied by the portal vein and the hepatic artery. Just before it enters the duodenum, the bile duct is joined by the duct of the pancreas. Guarding the common entrance is a sphincter (of Oddi) which, except during a meal, is closed. Thus bile, descending continuously down the hepatic duct, is forced to take the by-path up the cystic duct to the gall bladder. During a meal the gall

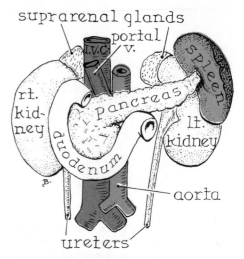

Fig. **384**. Relationships of important structures lying on posterior wall of abdomen.

bladder contracts, the sphincter relaxes, and bile is squirted into the duodenum.

Pancreas

The pancreas (Gr. = all flesh; figs. 360, 361, 382, 384) is a soft, pliable, solid, and fragile organ lying entirely behind the posterior parietal peritoneum. In outline it has the shape of a chemist's retort but it is flattened from front to back and is about 15 cm (6″) long. Its *head* occupies and overflows the concavity of the duodenum and lies in front of the bodies of the first and second lumbar vertebrae. Its elongated *body*, about 2 to 3 cm (1″) wide and 12 to 15 cm (6″) long, is continuous with the upper part of its head and crosses almost horizontally to the left in front of the middle of the left kidney to reach the spleen; this terminal part of the organ is called its *tail* (figs. 360, 384).

The pancreas manufactures a potent digestive enzyme which its **duct,** traversing the length of the interior of the organ, delivers to the duodenum. The entrance into the second part of the duodenum on its medial aspect is shared with the common bile duct.

Besides producing a digestive enzyme, certain special cells of the pancreas manufacture **insulin,** an active hormonal agent in sugar metabolism.

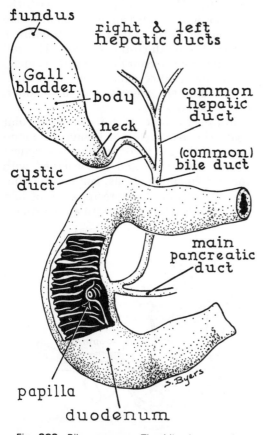

Fig. **383**. Bile passages. The bile duct receives main pancreatic duct and opens at a papilla in the second (descending) part of the duodenum.

Spleen

A person's spleen is about the size of his fist and lies in the upper left part of the abdominal cavity in contact with the diaphragm (figs. 360, 377, 384); the diaphragm separates it from the left pleural sac (page 236) and the ninth, tenth, and 11th left ribs whose protection the spleen enjoys. It is well around toward the back of the body and is invested with peritoneum. Its **hilus** (where its blood vessels enter) faces medially (fig. 385) and it is connected to the greater curvature of the stomach and to the front of the left kidney: these peritoneal 'ligaments,' the **gastrolienal (gastrosplenic)** and the **lienorenal**, are described on page 222. Their dispositions allow the spleen considerable mobility.

Contacts (fig. 385). Contact with the diaphragm gives the spleen a rounded, smooth, lateral or diaphragmatic surface which is separated by a circumferential border (usually notched anteriorly) from three medial or **visceral surfaces** which, from their contacts, are named **gastric, renal,** and **colic.** At the hilus of the spleen, the tail of the pancreas reaches and is in contact with it.

Structure. Deep to its serous (peritoneal) coat the spleen has an elastic fibrous covering containing smooth muscle fibers. From this investment elastic partitions, reaching toward the hilus, divide the organ into compartments. In these compart-

ments is a sponge-work of specialized cells separated by many blood channels called sinusoids. The elastic nature of the framework allows the spleen considerable variation in size.

Functions. The spleen is intimately associated functionally with the blood. It produces one kind of white blood cells **(lymphocytes)** and is a principal residence of the 'scavenger' (reticulo-endothelial) cells of the body. It removes the debris of red blood cells that have disintegrated in the circulation, destroys worn out red blood cells, and filters out bacteria.

By its power to retain red blood cells in its interstices, it helps to regulate the number in circulation. On demand, it adds blood to the general circulation by contracting. It is a built-in blood tranfusion service, but it is not as important for this function in man as it is in other mammals.

PERITONEUM

On entering the abdominal cavity, the digestive tract becomes highly mobile. It does not return to the vertebral column until it enters the pelvis, and the mobility that it enjoys in the abdomen is made possible by its investment with peritoneum, an investment that must be considered now.

The muscles of the abdominal wall, like muscles everywhere else, are invested with fascia. The parts of this fascia lining the muscular surfaces that face the abdominal cavity are continuous from muscle to muscle, so that there exists a complete investment in the form of a fascial bag, and it is important to observe that, outside this bag lie the muscles and nerves, while inside it lie the vessels and organs.

But the vessels and organs do not lie free in the abdominal cavity. They, too, have a covering or investing membrane and this, too, forms a sac-like lining and is known as the peritoneum (see Mesothelial Tissues, page 4). The peritoneal sac may be described as a closed serous sac or membrane between which the fascial investment of the muscles lies a (thick) layer of fatty areolar tissue; in the areolar tissue develop and lie blood vessels and organs. The

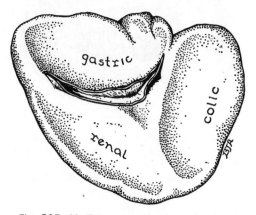

Fig. **385**. Medial aspect of spleen showing contacts of stomach, left kidney, and colon, and the centrally placed hilus.

interior of the serous sac is called the **peritoneal cavity** which actually is only a potential cavity and contains nothing but a little lubricating fluid.

The organs and vessels develop and grow in the posterior part of the abdominal cavity and therefore behind the peritoneum. As the organs increase in size (and length) they obtain more room for themselves by pushing the peritoneum ahead of them and so reducing the volume of its cavity. Some organs do no more; they are then covered in front with peritoneum and are said to remain retroperitoneal (retro, L. = behind) (fig. 386).

Mesenteries. Other organs—requiring greater mobility—progress much farther and 'break away' from the posterior wall. However, this 'breaking away' cannot be complete because the organs must still receive their many vessels and nerves. These vessels and nerves pass to the 'free' organs between the two layers of peritoneum (forming a single sheet) which the organ has 'pulled' off the posterior abdominal wall. Such a double-layered sheet of peritoneum that joins an organ to the body wall is known as a **mesentery** (fig. 386).

There are four named mesenteries in the adult. They are:

1. the *mesentery* (of the small intestine),
2. the *transverse mesocolon* or mesentery of the transverse colon,
3. the *sigmoid mesocolon* or mesentery of the sigmoid colon,
4. the *mesoappendix* or mesentery of the appendix.

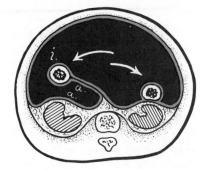

Fig. **387**. Where the intestine—chiefly ascending and descending colon—'falls against' the posterior wall, the mesentery is lost by adherence of layers a and a, to give the final result shown on the right side.

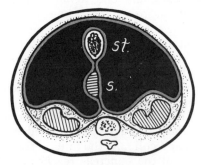

Fig. **388**. The spleen (s.) develops in the dorsal mesentery of the stomach (st.).

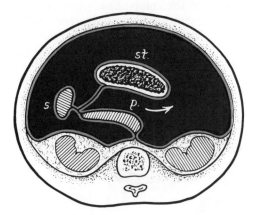

Fig. **389**. The spleen and the pancreas (p.), which also develops in the dorsal mesentery, fall to the left, the stomach (st.) to the right. The double sheet of peritoneum joining stomach to spleen is the gastrolienal ligament.

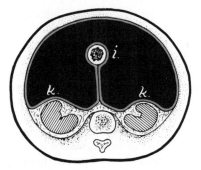

Fig. **386**. The kidneys are retroperitoneal and immobile. The small intestine achieves mobility by means of a mesentery. Continuous red line = peritoneum.

The peritoneum of the body wall (or parietes) is known as **parietal peritoneum;** that which covers the organs (or viscera) is **visceral peritoneum.**

Fig. **390**. The peritoneum behind the pancreas is resorbed and the pancreas becomes retroperitoneal. The arrow leads out of mouth of lesser sac into greater sac.

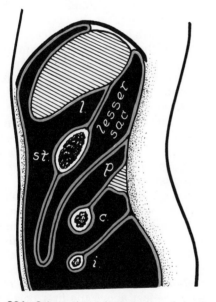

Fig. **391**. Schematic sagittal section of developing peritoneal sheets. The lesser omentum joins liver (l.) to stomach (st.). The dorsal mesentery of the stomach 'billows' downward in front of pancreas (p.), the transverse colon (c.) and the mesentery of the latter.

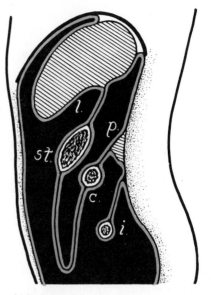

Fig. **392**. Scheme of final appearance of greater omentum seen in sagittal section joining lower edge (greater curvature) of stomach (st.) to transverse colon (c.). It hangs like an apron in front of small intestines (i.). See fig. 370.

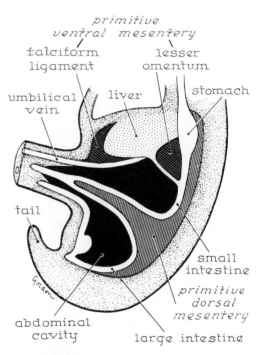

Fig. **393**. Small intestine bulges for a time outside abdominal cavity of embryo at umbilicus. Umbilical vein runs to liver in free edge of a developing peritoneal fold, the falciform ligament. From lower end of gut another blind sac (the urachus) also extends a short distance into umbilical cord.

Loss of Mobility. Certain parts of the gut which earlier had become mobile become adherent again to the posterior abdominal wall and their mesentery becomes completely absorbed. They now give the appearance of never having been free (fig. 387). The parts which are retroperitoneal as a result of such a process are:

1. most of the duodenum and pancreas,

2. the cecum,

3. the ascending colon,

4. the descending colon.

'Rotation' of Large Intestine. Complications also appear as the result of an elaborate migration of the various parts of the colon, which, by elongating and twisting in a counterclockwise spiral, achieves its adult form (the 'picture frame').

The **mesentery of the small intestine** remains relatively primitive and its attachment extends from duodenum to cecum which are both retroperitoneal. It has been described with the small intestine (page 210).

Lesser Omentum and Falciform Ligament. The upper part of the digestive tube—the part that is chiefly destined to become the stomach—has, in the embryo, not only the common posterior mesentery (dorsal mesogastrium) but also a mesentery joining it to the anterior body wall as far down as the umbilicus (ventral mesogastrium). This does not remain simple for long. Along its free edge a duct grows out from the primitive gut and forms an expanding organ, the **liver,** between the two layers of the ventral mesogastrium (fig. 393). This leaves a double-layer sheet of peritoneum joining the liver to the stomach and another joining the liver to the anterior abdominal wall. The former is the **lesser omentum** and has the bile passages in its free edge; the latter is the **falciform** (L. = sickle-shaped) **ligament** and has the umbilical vein in its free edge (fig. 393). The umbilical vein carries blood from the placenta, via the umbilical cord, to the liver of the fetus. After birth it becomes a fibrous cord, the *round ligament* (*ligamentum teres*) of the liver. A large area—the so-called bare area—of the liver comes into naked contact with the diaphragm behind (see also fig. 379 on page 216).

Gastrolienal Ligament and Lienorenal Ligament. The dorsal mesentery of the stomach goes through a series of changes that are interesting though complicated. In the upper part of the dorsal mesentery, the spleen develops and then migrates to the left. That part of the double layer of peritoneum that joins stomach to spleen is the adult **gastrolienal** (or **gastrosplenic) ligament;** the part from the spleen to body wall (actually the left kidney *on* the body wall) is the **lienorenal ligament** (lien and splen, both L., = spleen) (figs. 388–390).

Greater Omentum. The lower part of the dorsal mesogastrium, on the other hand, balloons downward and adheres to the mesentery of the transverse colon and to the transverse colon itself (figs. 391, 392). It hangs like an apron beyond the transverse colon and is known as the greater omentum (see fig. 370 on page 213).

The greater omentum hangs down in front of the 'picture' (small intestines) in the 'picture frame' of the colon. It is often thickened with enclosed fat. Because it will migrate to the site of any inflammation in the peritoneal cavity and wrap itself around such a site, the greater omentum is commonly referred to as the 'policeman' of the peritoneal cavity.

It should be apparent, then, that **omenta** are double-layered sheets of peritoneum joining one organ to another. They are:

1. the greater omentum,

2. the lesser omentum,

3. the gastrosplenic (-lienal) ligament.

'**Lesser Sac.**' One part of the peritoneal sac is almost closed off from the remainder—it is the **bursa omentalis** (*lesser sac*). It lies behind the stomach and lesser omentum and has a small opening— the *epiploic foramen* or *mouth of the lesser sac*—just above the first part of the duodenum. The bursa omentalis is limited to the left by the spleen and its peritoneal connections; and below by the transverse colon and its mesentery, and by the greater omentum (fig. 392).

Numerous small *peritoneal folds* occur in various parts of the abdominal cavity. The little pockets behind these folds are known as *peritoneal fossae* (fig. 367 on page 212).

7

Respiratory System

For the performance of its many functions every living cell of the body must be provided with oxygen and must eliminate carbon dioxide. The blood circulates within tiny vessels in the neighborhood of every body cell and it is the red cells in the blood that bring the tissues oxygen and take away carbon dioxide. In the lungs, the red blood cells discharge their carbon dioxide to the air and from it receive new cargoes of oxygen. The lung is the essential organ of respiration and the nose, pharynx, larynx, trachea, and bronchi are the names of the parts of a continuous and open passage leading to the lung from the exterior; they are spoken of as the **respiratory passages** and will be discussed in turn.

NOSE

The nose is a bilateral passage leading from the exterior to the nasopharynx which lies behind the nose. The openings in front are known as the (**anterior**) **nares** (**nostrils**), and the openings behind are known as the **choanae** (*posterior nares*).

The nostrils are oval, small, guarded by hairs that act as filters, dilatable, and bounded by cartilage. The skeleton of the external nose is chiefly cartilage, attached above to the nasal bone and supported in the midline from behind by the nasal septum (described below). The fleshy *ala*

(or wing) on the side of each nostril is fibro-fatty tissue (fig. 394).

The choanae are rectangular, large, rigid and bounded by bone covered with mucous membrane (fig. 350). The floor of each nasal cavity is narrow and horizontal; it extends from the nostrils to the choanae and is the upper surface of the bony palate (fig. 395).

The **nasal septum** is a vertical midline partition dividing the nose into two cavities. The partition is bony above and behind, but cartilaginous below and in front; it is usually deflected (i.e., 'buckled') somewhat to one side (fig. 396).

The roof of each cavity is so narrow as to be slit-like. The thin bony roof is pierced by many minute foramina and, in consequence is known as the **cribriform plate** (of the ethmoid). Through the foramina pass the many tiny olfactory nerves which ascend from a restricted area of the upper portion of the nasal mucous membrane devoted to the sense of smell. This area is on the septum and adjacent lateral walls and, in man, is no larger than 2 sq. cm.

Conchae and Meatuses

The lateral wall of each nasal cavity is complicated by the fact that from it project medialward three curved, thin, scroll-like shelves of bone known as the **conchae or**

223

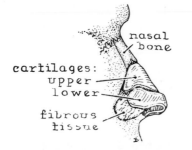

Fig. **394**. Nasal cartilages (from right side). Note that the fleshy part of the 'wing' of the nose is not cartilage.

turbinates. The lower two of these, the inferior and the middle concha, extend from front to back along almost the whole length of the lateral wall, and below each is a space known as the **inferior** and the **middle meatus,** respectively; the highest concha, the superior, is rudimentary and below it lies the **superior meatus** (L. = passage). These three meatuses open medially into the **nasopharyngeal** (or **common) meatus,** hence its name (see figs. 395–398).

Paranasal Air Sinuses

The several important bones surrounding the nasal cavities are hollow; their cavities are known as the **paranasal air sinuses** and these open by restricted apertures into the nasal cavities from which they developed as outgrowths (figs. 399–401). They are lined with mucous membrane continuous with that of the nasal cavity. Absent or very small at birth, they reach their maximum size with maturity. These bilateral paranasal sinuses are:

1. **The maxillary air sinus,** situated in the maxilla below the orbital cavity. In shape, it resembles a three-sided pyramid (tetrahedron) laid on its side with its base (medial wall of sinus, lateral wall of nose) toward the nasal cavity and its apex near or in the zygomatic or cheek bone. The roof of the sinus is the floor of the orbit (fig. 395). Of the remaining two walls, one faces the cheek below the eye, the other is the anterior limit of the infratemporal region and is inaccessible.

The maxillary is the largest of the air sinuses and it opens into the middle meatus by an aperture which, unfortunately, is high up in the medial wall of the sinus (fig. 401). That is why, in order to drain the sinus, the surgeon finds it necessary to make a new opening lower down. The roots of the upper teeth sometimes project into the sinus.

2. **The ethmoidal air cells (or sinuses)** —a collection of eight or ten cells— situated between the lateral nasal wall and the medial wall of the orbit. They are essentially a rectangular box made up of individual cells and are co-extensive with the whole medial orbital wall; they are separated from the orbital cavity only by a thin paper-like lamina of bone and they open into the middle and superior meatuses (see fig. 401).

3. **The frontal air sinus** (fig. 397)—in the frontal bone above the superior orbital margin, i.e., deep to the eyebrow. It may be small or so extensive as to reach right across the whole length of the superior orbital margin. It opens downward into the middle meatus by a relatively long and narrow curved passage (figs. 400, 401).

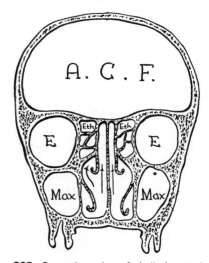

Fig. **395**. Coronal section of skull. Arrows in air passages. A. C. F. = anterior cranial fossa; E = orbital cavity (for eye); Eth = ethmoidal air sinuses; Max = maxillary air sinus.

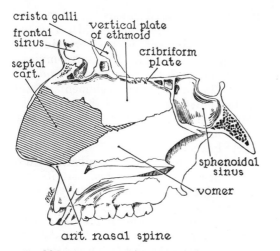

Fig. **396**. Nasal septum (viewed from the left side)

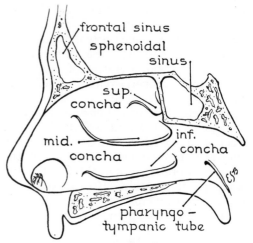

Fig. **397**. Lateral wall of nose

4. **The sphenoidal air sinus,** situated in the body of the sphenoid behind the back of the upper part of the nasal cavity. It, too, is paired and opens into a recess, behind and above the superior concha, known as the *spheno-ethmoidal recess*. It has important relations to the nerve of vision and to many other structures entering the orbit or lying behind the orbital entrance (figs. 397, 398, 400, 401).

Nasolacrimal Duct

The lacrimal or tear sac is situated in the medial corner of the orbital cavity behind the medial palpebral ligament. Its duct runs downward between the bones and opens into the inferior meatus of the nose.

This is the pathway for the drainage of the tears produced in normal or slightly increased quantities. In crying, however, the duct is overwhelmed and the tears spill out over the lower lid.

Nasal Mucosa

The bones of the nasal cavities and their extensions are lined with a thick mucous membrane which, because it lines bone, is known as **mucoperiosteum.** It is richly supplied with blood vessels, particularly large venous channels, and it is capable of considerable swelling. That is the reason for the heavy and stuffed feeling that

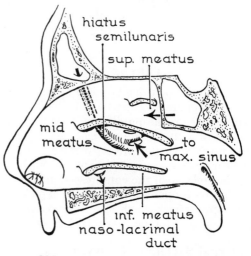

Fig. **398**. Conchae amputated to expose underlying meatuses into which various holes open (cf., fig. 330). (The opening of the sphenoidal sinus, indicated by the horizontal arrow, is unusually low in this specimen.)

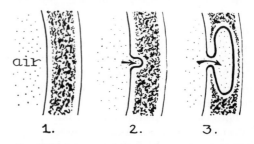

Fig. **399**. Scheme of development of an air sinus (pneumatic bone).

accompanies a cold in the head. The sinuses are readily infected from the nose
because their mucoperiosteal linings are
directly continuous with that of the nose.

Because of the richness of its blood
supply, the mucoperiosteum is efficient to
warm the inspired air; because of the
whirling effect produced on the current of
inspired air by the conchae, the warming
effect is enhanced while dust particles are
efficiently filtered out and cling to the
lining membrane; both of these effects are
lost in mouth breathing. Because the olfactory mucous membrane is high up, smell
appreciation is heightened by sniffing.

PHARYNX

The filtered and warmed air reaches the
nasopharynx which has been described
(page 202). Passing down the oral pharynx
it enters the larynx (see fig. 539 on page
302).

LARYNX

The larynx or voice-box is a specialized
piece of apparatus capable of utilizing the
expired air for the production of voice. Its
ability to shut off the air passages allows
voluntary muscular effort to raise the air
pressure in the respiratory tract below. The
sudden release of the compressed air when
the larynx is opened results in an explosive
expiratory effect which clears the passages
of mucus and foreign particles—coughing.

'Internal Skeleton' of Larynx

Below the larynx and continuous with it
is the trachea or windpipe. Where the
larynx joins the trachea, it is encircled by a
ring of cartilage and fibrous membrane
which together form the basic 'internal
skeleton' of the larynx. This 'skeleton' is, in
effect, the upper part of a tube whose lower
part is the trachea. The upper end of the
tube appears to be thrust from below into
the lower anterior part of the pharynx.
[Actually, the respiratory tract is an outgrowth from the pharynx.]

The upper specialized section of the
respiratory tube is modified to serve as a
foundation on which the rest of the larynx
is built (see fig. 402). Behind and below,

the section is cartilaginous (C); above and
in front, it is membranous (M). The cartilaginous element is the **cricoid** (Gr.
= ring-like) **cartilage** and it resembles a
signet ring with the wide part behind (fig.
403). The membranous element is the
conus elasticus, a tent-shaped elastic
membrane attached around its lower edge
to the anterior half of the cricoid cartilage.
Each lateral part of the conus elasticus
may be referred to as the **cricothyroid** or
cricovocal membrane because its free
upper edge is the fibrous basis of the **vocal
cords** (a, a in fig. 402).

Sitting side by side on the back of the
cricoid ring are two little pyramidal **arytenoid cartilages.** Each has a triangular
base and three triangular sides so each is
actually tetrahedral in shape. The anterior
angle of the base (b) is called the **vocal
process** because to it is attached the posterior end of the vocal cord. The lateral angle
(c in fig. 402) is called the **muscular
process** because it receives the insertions
of muscles.

'External Skeleton' of Larynx

The anterior ends of the vocal cords are
closer to one another than the posterior and

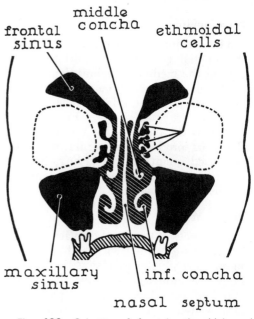

Fig. **400**. Scheme of frontal, ethmoidal, and
maxillary air sinuses.

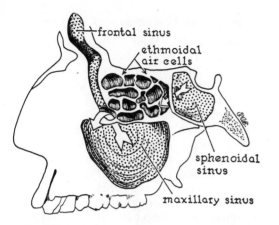

Fig. **401**. The paranasal air sinuses (of right side) opened up from inside the nasal cavity. White arrows pass out of mouths of maxillary and sphenoidal sinuses.

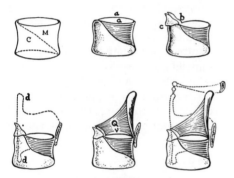

Fig. **402**. Construction of the larynx (see text)

each is fixed near the angle of union of two large and protective plates of cartilage which together constitute the single **thyroid cartilage** (fig. 404). In figure 402 the plate of one side is indicated by dotted lines. The midline prominence formed by the union of the two plates is the familiar Adam's apple. Each plate has a free and somewhat thickened posterior border prolonged upward and downward into horns (d, d).

On the medial surface of the tip of each inferior horn is a circular joint surface which articulates with a raised and similar joint surface on the side of the cricoid cartilage. By these horns the cricoid cartilage is gripped and at these cricothyroid joints the thyroid cartilage can swing like the vizor of a medieval knight. When the

'vizor' is lowered the cords are tensed; when the 'vizor' is raised the cords are relaxed (fig. 405).

The longer superior horn lies vertically below the greater horn of the hyoid bone to which it is attached by the *lateral thyrohyoid ligament*. This ligament is the posterior edge of the **thyrohyoid membrane** which stretches between the upper border of the thyroid cartilage and the **hyoid bone;** thus, the whole larynx is suspended from the hyoid bone and can (and does) move up behind the hyoid during the act of swallowing. In turn, the hyoid is suspended by muscles of the floor of the mouth.

Parts of Larynx Above Vocal Cords

Rising directly upward into the interior of the pharynx from the angle between the two thyroid plates is yet another cartilage (figs. 402, 406). It is the soft **epiglottic cartilage**—soft because, unlike the others, it is an elastic cartilage. It is shaped like an elm leaf or tennis racket fixed by its very short handle. It reaches so far up that its upper rounded end rises above the level of the hyoid bone and faces the back of the tongue to which it is attached. Recently, new evidence has revived the old concept

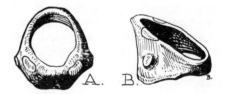

Fig. **403**. Cricoid cartilage. A. From above. B. From right side. The size and shape are those of a signet ring.

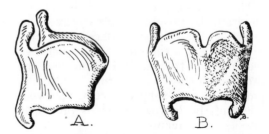

Fig. **404**. Thyroid cartilage. A. From right side. B. From in front. The paired laminae (plates) and upper and lower horns are obvious.

Fig. **405**. Raising 'vizor' relaxes vocal cords (broken line); lowering tenses them (see fig. 409).

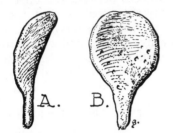

Fig. **406**. Epiglottic cartilage. A. From right side. B. From behind.

that the epiglottis—at least in part—acts like the lid of a box for the larynx during swallowing.

Stretching between the lateral border of the epiglottic and that of the arytenoid cartilage in fig. 402 is the (paired) **quadrangular membrane** (Q). Each has a free upper and a free lower border. The upper border (**ary-epiglottic fold**) bounds the laryngeal entrance and slopes backward; the lower horizontal border is the **vestibular fold** (V) or **false vocal cord** and it lies a little above the (true) vocal cord. Thus the arytenoid cartilages behind, the quadrangular membranes at the sides, and the epiglottic cartilage in front, bound a little chamber which, lying above the vocal cords, acts as a resonating chamber for the voice (fig. 407).

The outer surfaces of the cricovocal and quadrangular membranes are covered with muscles. Their inner surfaces are lined with mucous membrane which, between the false and true cords, i.e., between the two membranes, also forms a laterally projected canoe-shaped recess, the **ventricle of the larynx,** which is prolonged into a small **saccule.***

Figure 408 summarizes the essential architecture of the larynx.

. * In the gorilla the saccule is enormous, reaching the front of the chest and armpit.

Muscles and Movements of Larynx

The muscles, their actions, and their functions are depicted in figures 409 to 415.

Besides the movements at the paired cricothyroid synovial joints whereby the cords are tensed or relaxed, important movements occur at the crico-arytenoid joints which also are synovial. Here rotary and gliding movements of the arytenoid cartilages occur whereby the vocal cords are separated or brought together. Thus the space between the cords, known as the **rima** (L. = a crack) **glottidis,** may be: (1) widened, as in deep breathing or singing a low note; (2) narrowed, as in singing a high note; or (3) entirely occluded, as in the preparatory stage of coughing.

The pitch of the voice depends on the simple physical principles of sound production. The female vocal cords are generally shorter, more taut, and closer together. This results in a higher pitched voice.

The muscles clothing the outer or lateral surface of the quadrangular membrane may be regarded as a sphincteric group effective (by regulating the size of the resonating chamber) in modifying the tone

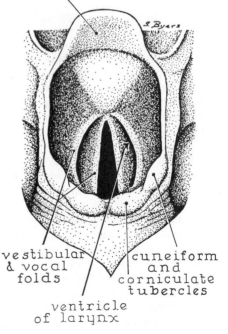

Fig. **407**. Larynx viewed from above and behind

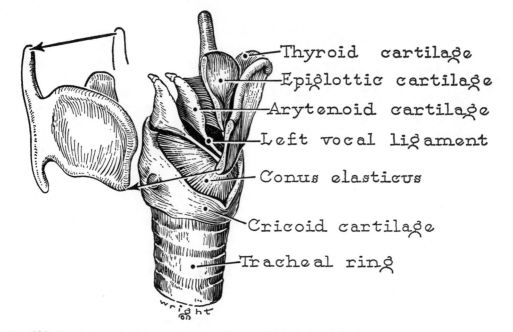

Thyroid cartilage
Epiglottic cartilage
Arytenoid cartilage
Left vocal ligament
Conus elasticus
Cricoid cartilage
Tracheal ring

Fig. **408**. Semischematic right anterior view of larynx. Right half of thyroid and epiglottic cartilages removed to reveal deeper structures.

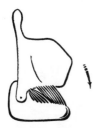

Fig. **409**. Cricothyroid muscle tenses cords

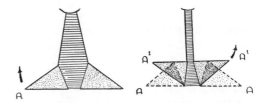

Fig. **410**. Rotation of arytenoid cartilages (A to A') approximates cords (high notes) (see fig. 414).

and quality of the voice, and in 'shutting off' the inlet of the larynx during swallowing.

Nerve Supply of Larynx

The **recurrent** (or *'inferior'*) **laryngeal nerve** rises from below and enters the larynx behind the cricothyroid joint; it is the motor nerve to the muscles. The **superior laryngeal nerve** descends from above. One large branch (the *internal laryngeal*) enters the larynx by piercing the thyrohyoid membrane; it is the sensory nerve to the mucous membrane. Another (the *external laryngeal*) supplies one muscle, Cricothyroid, the important tensor of the cords.

The laryngeal nerves are branches of the tenth cranial or vagus nerve. The recurrent and external nerves (both motor) are in contact with the thyroid gland (page 361) just before they enter the larynx. If they are injured during an operation for goiter, various serious disturbances of the voice may result.

TRACHEA

Continuous below with the larynx is the windpipe or trachea.† It is an elastic tube 10 to 12 cm (4–5″) long and with a diame-

---

† Tracheia (Gr.), meaning 'rough,' was applied as an adjective to 'artery' which meant 'windpipe.' Until Harvey's time (1628 A.D.), the arteries were supposed to carry air, and were referred to as the 'smooth arteries,' while the trachea was the 'rough artery.'

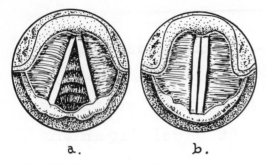

a.   b.

Fig. **411**. Larynx as seen through laryngoscope:
a, vocal folds abducted. b, adducted.

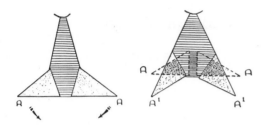

Fig. **412**. Rotation of arytenoids from A to A′
separates cords (see fig. 413).

ter equal to that of a forefinger. The
fibroelastic tissue which comprises its
wall is reinforced—except behind—by
about 20 horseshoe-shaped **rings of cartilage** (fig. 416).

Behind, where the trachea is in contact
with the esophagus, its wall contains involuntary muscle fibers. Because of its elasticity, the trachea can be stretched, as it is
when the head is 'thrown back.'

Course and Relations

Half of the trachea lies in the neck, half
in the thorax, and it ends at the level of the
sternal angle by dividing into a right and a
left bronchus. Lying on each side of the
upper part of the trachea and lower part of
the larynx is a lobe of the **thyroid
gland**—an important gland of internal
secretion. The two lobes are connected
across the front of the upper two or three
tracheal rings by an isthmus; the lobes
themselves reach up to the sides of the
thyroid cartilage (fig. 635 on page 361).

On the sides of the trachea and esophagus ascend and descend the great vessels of
the head and neck. The front of the cervical

portion of the trachea is protected by the
infrahyoid ('strap') muscles.

Behind the manubrium, the front of the
trachea is obscured by the **thymus gland**
(especially large in infants) and, behind
the gland, by the great veins and the great
arteries proceeding to and from the heart.
As the **aorta**—the great artery leaving the
heart—passes from front to back, it arches
over the left bronchus in contact with the
left side of the tracheal termination (fig.
417); a smaller structure—the **azygos vein**
which drains the chest wall—arches forward in similar fashion over the right
bronchus.

BRONCHI AND LUNGS

Main or Principal Bronchi

The two bronchi do not branch quite
symmetrically from the trachea and, while
both are rather more than half the caliber
of the trachea, yet the right one is wider,
shorter, and in more direct line with the
trachea than the left. It is wider because
the right lung is larger than the left; it is

Fig. **413**. Two muscles at back of larynx. a.
Posterior Crico-arytenoids (paired) separate cords; b.
Arytenoideus (unpaired) approximates arytenoids
(and thus cords).

Fig. **414**. Lateral Crico-arytenoid, an approximator
of cords.

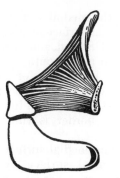

Fig. **415**. Muscles covering outer aspect of quadrangular membrane are constrictors.

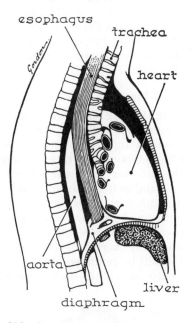

Fig. **416**. Some important relationships of the trachea (semischematic).

trapulmonary), the walls of the bronchi are supported all round with spirals, rings, and plaques of cartilage; between these and the mucous membrane lies a more or less complete coat of smooth muscle.

Lungs

The lungs are the two gross anatomical organs whose essential structures are bronchi, bronchioles, and alveoli. They are housed in the side portions of the thoracic cavity, i.e., the parts mainly enclosed by the ribs, and each is shaped like half a cone (split longitudinally) (fig. 418).

Surfaces and Borders. Each lung has: (1) a rounded **apex,** reaching above the first rib into the root of the neck; (2) a concave **base or diaphragmatic surface,** resting on and fitting the upper surface of the dome of the diaphragm; (3) a **medial surface,** molded by the structures in contact with it (fig. 419) and separated from its fellow of the opposite side by a wide

shorter and more vertical because the arch of the aorta thrusts the lower part of the trachea a little to the right. Foreign bodies falling down the trachea are more likely to enter the right bronchus.

The bronchi lie behind the pulmonary vessels (pulmo, L. = lung) that leave or enter the heart on their way to or from the lungs. These extrapulmonary parts (i.e., the parts outside the lung) are no more than 5 cm (2″) long and are accompanied by the pulmonary vessels. The structure of the extrapulmonary parts is the same as that of the trachea. Within the lung (in-

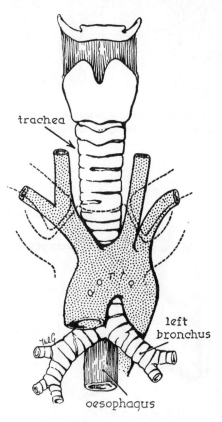

Fig. **417**. Trachea and bronchi

'space'—the **mediastinum**—composed of many important structures; thus, most of the medial surface is known as its *mediastinal part;* (4) a rounded **costal surface,** conforming to the shape of the costal cage (figs. 425, 427).

In the central part of the mediastinal part of the medial surface is a large area where bronchi and pulmonary vessels plunge into the lung. The area is known as the **hilus** (or *hilum*) of the lung. The vessels and bronchi and associated structures together constitute the **root** of the lung.

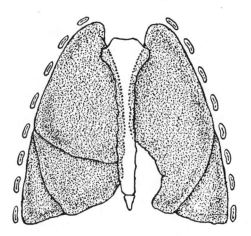

Fig. **418**. Lungs from in front. Right has upper, middle, and lower lobes; left, upper and lower lobes only.

In front, the medial surface is separated from the costal surface by a sharp **anterior border** (fig. 418) and, since each lung tries, as it were, to make contact in front of the heart with its fellow of the opposite side—an effort in which it is not quite successful—the portion of lung adjacent to the anterior border is thin and wedge-shaped.

Behind, the medial surface is separated from the costal surface by a full and rounded '*posterior border*,' more correctly called the *vertebral part of the medial surface*. Thus the lung is much more massive and voluminous behind than it is in front. The concave diaphragmatic surface is separated from the costal and mediastinal surfaces by a sharp circumferential or **lower border.**

Lobes and Fissures

The **right lung** is shorter but rather more voluminous than the left; it is partially divided into lobes by two fissures. The **oblique fissure** separates the **superior** from the **inferior lobe;** the superior lobe lies above and in front, the inferior lobe lies behind and below. If the hands be clasped behind the head and the elbows be brought as near together as possible at the sides of the head, the vertebral borders of the scapulae will be rendered oblique and will indicate the lines of the underlying **oblique**

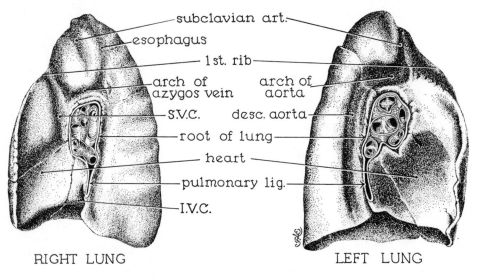

subclavian art.
esophagus
1st. rib
arch of   arch of
azygos vein   aorta
S.V.C.   desc. aorta
root of lung
heart
pulmonary lig.
I.V.C.

RIGHT LUNG   LEFT LUNG

Fig. **419**. Impressions made on the medial surfaces of the lungs

fissures of the lungs. A **horizontal fissure** marks off a small **middle lobe** in front; it conforms in direction to the overlying fourth rib.

The **left lung** resembles, in general appearance, the right except that it possesses no horizontal fissure and, therefore, no middle lobe.

Divisions of Bronchi and of Lobes

The **right lung** consists of three lobes and the right bronchus provides three secondary or **lobar bronchi**—one for each lobe. That for the upper lobe is given off before the bronchus enters the lung substance.

From the three secondary bronchi there issue ten tertiary or **segmental bronchi**— three for the upper lobe, two for the middle lobe, five for the lower lobe. Each segmental bronchus supplies one of the ten **bronchopulmonary segments,** each of which is a more or less individual entity (figs. 420, 421).

The **left lung** consists of two lobes and the left bronchus provides two secondary bronchi—one for each lobe. The main bronchus does not divide until after it has entered the lung substance. The secondary bronchus to the upper lobe furnishes four tertiary bronchi (corresponding to those for the upper and middle lobes on the right). The secondary bronchus to the lower lobe furnishes four tertiary bronchi (instead of five); the left lung thus comprises eight segments.

With the widespread recognition of these lung segments as more or less discrete lung entities, the surgery of the lung has become somewhat modified and attempts are made to remove a diseased segment without undue interference with the rest of the lobe.

Structure of Lung

As the bronchi divide they become progressively smaller until they verge on the microscopic, when they are known as **bronchioles.** Until this stage is reached, the walls of the bronchi continue to possess cartilaginous support, which, however, becomes progressively less in amount. When the bronchioles are reached, the flakes of cartilage that contribute to the patency of

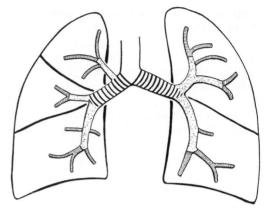

Fig. **420**. Scheme of main, lobar (secondary) and segmental (tertiary) bronchi (see text).

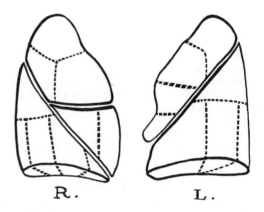

R.   L.

Fig. **421**. Right lung has ten segments, left has eight (see text).

the smaller bronchi entirely disappear. In general it may be said that as the cartilage decreases in amount muscle tissue in the bronchial wall increases in relative amount. Asthma is a condition in which spasm of the bronchial muscles is associated with difficulty in respiration.

Repeated divisions of the bronchioles finally result in terminal bronchioles where active gaseous interchanges begin. A terminal bronchiole with its dependent structures is known as a **respiratory unit** of lung tissue. About 50 to 100 of these make a *lung lobule* of which there are many hundreds.

Figure 422 depicts a *terminal* or *respiratory bronchiole* opening out into *alveolar ductules* which open into a variable number of *alveolar saccules*. The wall of each saccule is made up of a number of ultimate units known as *alveoli* (L. = small hollows). Diagrams of this general arrangement always convey the impression of bunches of grapes, but it must be remembered that the alveoli are spaces filled with air.

Surrounding each alveolus is a network of capillaries and, at the single-celled wall

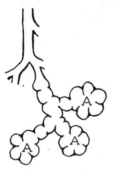

Fig. **422**. A respiratory unit. Fifty to 100 of these make a tiny lobule. Bronchioles divide into alveolar ductules which open into alveolar saccules and alveoli (A).

of the alveolus where the red cells roll along in the adjacent capillaries, interchange of oxygen and carbon dioxide takes place.

Each lung has a framework of **fibro-elastic tissue** for the support of its respiratory structures and each, of course, is pervaded with blood vessels whereby the red blood cells reach the alveoli. It is the very important elastic tissue of the lung framework that permits the lung to deflate passively following an active and voluntary inspiratory effort.

On the surface of the lung, the bases of individual pyramidal **lobules** may be made out as little polygonal areas each about a square cm in size; they are outlined by pigment lying in the fibro-elastic septa supporting the lobules. The pigment is carbon and the mottled appearance that it gives is characteristic of the lung of a city dweller. The lung of the country dweller retains its original healthy and uniformly pink color. Each lobule—incompletely separated from its neighbor by fibro-elastic septa—is furnished with a bronchiole which gives off numerous terminal or respiratory bronchioles. Some authorities name the respiratory unit the **primary lobule** and the larger

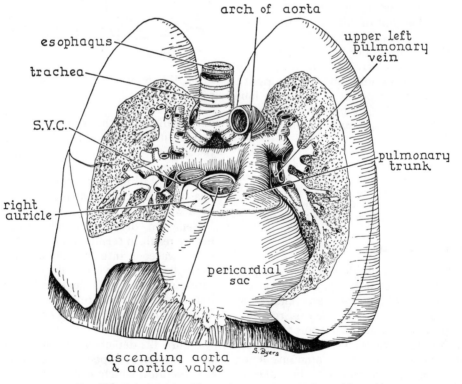

Fig. **423**. Relationship of lungs, heart, great vessels, and bronchi

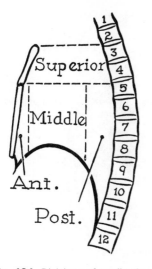

Fig. **424**. Divisions of mediastinum

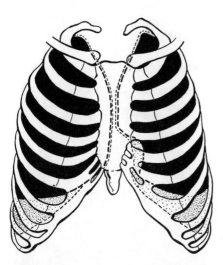

Fig. **425**. The two lungs (solid black) fail to fill lower parts of two pleural sacs (stippled).

unit separated from its neighbors by fibro-elastic septa, the **secondary lobule.**

Blood Supply and Nerve Supply of Lungs

The blood supply of the lungs is described on page 265. The lungs are supplied with branches of the sympathetic nervous system and of the vagus nerve which is parasympathetic. These control the caliber of the air passages and the amount of secretion by the mucosa. They also form the sensory pathway for reflexes such as the cough reflex.

Respiratory Movements

These have been discussed on page 48.

MEDIASTINUM

The thoracic cage houses many other important structures besides the lungs, and for their accommodation there exists, between the lungs, a wide oblong and central region known as the mediastinum; it extends from the inlet above to the diaphragm below, and from the sternum in front to the vertebral column behind; it is bounded on each side by the mediastinal part of the medial surface of a lung and a pleural cavity (to be discussed in a moment) (fig. 423).

For descriptive purposes, the mediastinum is subdivided, by an imaginary horizontal plane that passes through the sternal angle, into a **superior** and an **'inferior' mediastinum;** behind, this plane cuts the disc between the fourth and fifth thoracic vertebral bodies (fig. 424).

The superior mediastinum contains the thymus gland, the great vessels above the heart, the lower half of the trachea, and

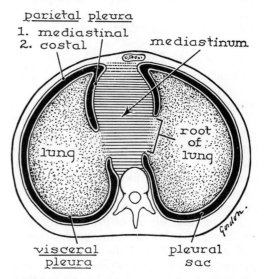

Fig. **426**. Horizontal section of thorax (semischematic). The right and left pleural sacs, each enclosing a lung, are completely separated by the mediastinum.

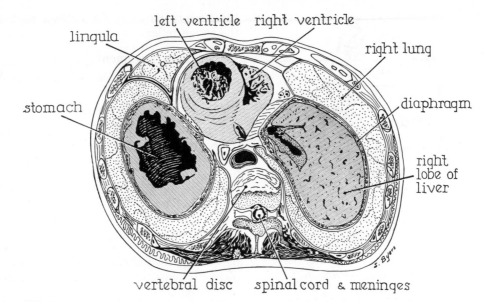

Fig. 427. A horizontal section through disc T.7 reveals relationship of lungs to mediastinal structures and to liver and stomach (which occupy the concavities of the domes of the diaphragm).

part of the esophagus. The 'inferior mediastinum' is further subdivided by the pericardium containing the heart into: (1) the unimportant *anterior mediastinum*, a narrow region in front of the heart; (2) the **middle mediastinum** containing the heart; (3) the **posterior mediastinum** behind the heart, and containing part of the esophagus and the great vessels lying in front of the lower thoracic vertebral bodies.

PLEURA

Each lung is enclosed in a **pleural 'sac'** or **cavity** which is a moist, smooth-walled, potential space.

The **pleura** is a thin, delicate, transparent serous membrane identical in structure to peritoneum. The pleura that lines the interior of the 'half-thorax' which contains the lung is **parietal pleura;** the adherent coat on the lung is **visceral** or **pulmonary pleura** (figs. 425–428).

The parietal pleura of each cavity lines rib cage, diaphragm, and mediastinum, and it bulges like a dome into the root of the neck behind the clavicle. Therefore, it has **costal, diaphragmatic, and mediastinal parts,** and a **cupola.**

The parietal pleura is continuous with the visceral pleura at the root of the lung,

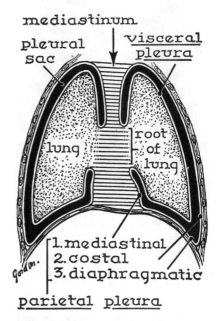

Fig. 428. Schematic coronal section of thorax to show parts of pleura and separation of right and left pleural sacs by mediastinum.

where a sleeve of pleura encloses the bronchi and pulmonary vessels. The visceral pleura lines the lobes of the lung where they come in contact and form fissures (but it does not 'dip in' to outline the eight or ten bronchopulmonary segments of the lung).

The pleural sac is a friction-eliminating device whereby, with every respiration, two moist and smooth pleural surfaces (pulmonary and parietal) glide on one another, just as do the walls of a synovial sheath or a bursa. Should the pleura become rough, dry, and inflamed (as in pleurisy), the resulting friction accounts for the pain that accompanies every inspiration.

Pleural Recesses. The lung does not completely fill the available potential space in the pleural cavity and so in some areas parietal pleura rubs on parietal pleura. These areas of the pleural sac that are unfilled by lung are known as pleural recesses (figs. 425, 428). They are located in three places: (1) and (2) where the periphery of the diaphragm is close to the costal wall (**right** and **left costodiaphragmatic recesses**); (3) on the left side where the pericardium is close to the sternal wall (left costomediastinal recess).

Nerve Supply. Pulmonary pleura is insensitive. The parietal pleura has a liberal sensory supply—the costal part from intercostal nerves, the diaphragmatic part from the phrenic nerve.

Pneumothorax. A lung has so much elasticity that it will collapse to a small size when exposed to the atmosphere.

8

Urinary
System

A special system is devoted to the task of maintaining constant the alkalinity and chemical composition of the blood; this it does by removing from the blood waste products and excesses of water and salts. The organs discharging these duties are the **kidneys;** the waste products (water and salts) they remove constitute the **urine,** and this is conveyed to the **urinary bladder** by paired ducts known as the **ureters.** The urine gradually accumulates in the bladder. By the contraction of its musculature, the bladder empties itself through a midline duct, the **urethra,** which discharges to the exterior. The act of *micturition* is under learned voluntary control and is discussed with the male urethra on page 247.

KIDNEYS

The kidneys (L., renes) are paired, bean-shaped organs that lie, buried in a mass of fat, in the upper part of the abdominal cavity and behind the peritoneum; they are placed higher than is popularly supposed, their lower poles reaching no lower than the level of the umbilicus. The dimensions of a kidney are about $2.5 \times 5 \times 10$ cm ($1'' \times 2'' \times 4''$).

Relationships

Posterior (fig. 429). The upper half of each kidney rests behind on the vertebral part of the diaphragm; it enjoys the bony protection of the last two ribs (fig. 430). The lower half rests on the adjacent muscles of the posterior abdominal wall (Psoas, Quadratus Lumborum, and Transversus Abdominis). The two kidneys are separated from one another by lumbar vertebral bodies, and the posterior surfaces of the kidneys are dimpled by the transverse processes of these vertebrae.

Anterior Relationships (fig. 431). The important anterior contacts differ for each kidney, as will be briefly noted.

The upper half of the **right kidney** is covered in front by the large right lobe of the liver with which its upper pole is in bare contact; here the kidney produces a distinct and rounded impression on the liver. In front of a medial strip of the right kidney, the second part of the duodenum (with the pancreas in its concavity) descends vertically; while immediately medial to the right kidney runs the inferior vena cava, the largest blood vessel in the body. The angle between the rounded upper pole of the right kidney and the inferior vena cava is occupied by the triangular right suprarenal gland—an important gland of internal secretion. In front of the lower half of the kidney lies the hepatic flexure of the colon and coils of small intestine.

The **left kidney** is usually placed a little

238

higher than the right. The spleen lies in front of its upper lateral part, the crescentic left suprarenal gland in front of its upper medial part, the tail of the pancreas in front of its middle or hilus. Between these three organs the left kidney is separated by peritoneum from the posterior surface of the stomach. The lower half of the left kidney is in contact with the (left)

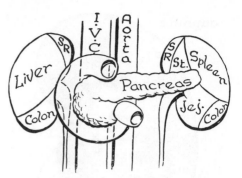

Fig. **431**. Anterior contacts of kidneys. SR. = suprarenal gland; St. = stomach; Jej. = jejunum. (I.V.C. = inferior vena cava.)

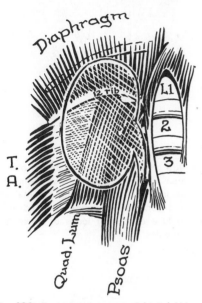

Fig. **429**. Posterior contacts of (right) kidney

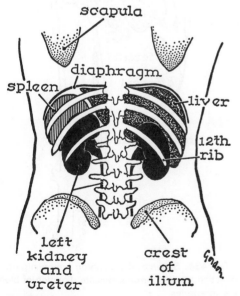

Fig. **430**. Surface anatomy of liver, spleen, and kidneys from behind.

splenic flexure of the colon and with some jejunal coils of small intestine.

'Cavity' of Kidney

The lateral border of each kidney is full and rounded; the medial border presents a deep concavity known as the **hilus** (or, sometimes, *hilum*). This hilus (L. = the point of attachment of a seed) leads to the **sinus of the kidney** which is occupied by the expanded and funnel-shaped upper end of the ureter and its branchings; the expansion is known as the **pelvis** of the ureter or kidney and the branchings are known as **calices** (plural of **calix**, Gr. = cup). *Calices* is also spelled *calyces* (fig. 432).

Structure of Kidney

The cut surface of a kidney (fig. 432) is not of uniform color. The outer, pale area, which is about 2 cm thick in most areas, is known as the **cortex** (L. = bark); the inner darker area is known as the **medulla** (L. = marrow).

The medulla consists of about half a dozen isolated triangular areas each with its base adjacent to the cortex and its rounded apex approaching those of its fellows at the sinus. These **renal pyramids** have a longitudinally striped appearance, whereas the cortex has a granular one. Between adjacent pyramids 'cortical' substance projects as **renal columns.** The meaning of these differences is understood only when the microscopic structure of the kidney is studied.

Under the microscope, the kidney (Gr.,

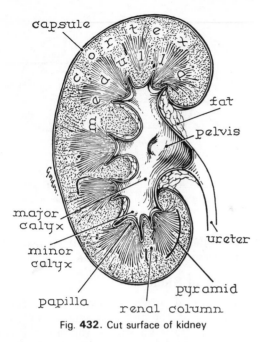

Fig. **432**. Cut surface of kidney

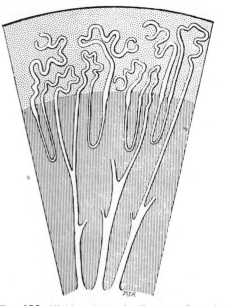

Fig. **433**. Highly schematic diagram of a microscopic area of kidney showing four nephrons. Stippled area indicates 'cortical' tissue containing glomeruli and convoluted tubules. Hatched area indicated medulla containing loops of Henle and collecting tubules.

nephros) is seen to contain enormous numbers of individual tubular units known as **nephrons,** the cells of whose walls are in the most intimate relationship with the blood circulating in many special capillaries. The parts of a nephron, in order , are: *glomerular capsule, first (proximal) convoluted tubule, loop of Henle,* * *second (distal) convoluted tubule,* and *collecting tubule.* These are represented in figures 433 and 434. The walls of the nephron are always only one cell thick but the cells of the various parts have characteristic differences in appearance.

The blood first comes into relationship with a nephron by means of a ball of capillaries known as a **glomerulus** (L. = a small ball of twine). This rests in a cup-like depression (a glomerular capsule) formed by an invagination of the dilated and blind end of a nephron. The glomerulus is fed by a tiny artery (afferent arteriole) and drained by a much smaller one (efferent arteriole) (figs. 434, 435). The compound structure of capsule and glomerulus is known as a renal **corpuscle** and its function is to filter the circulating blood by

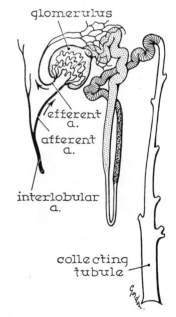

Fig. **434**. Scheme of a nephron (emptying into a collecting tubule). Arrows indicate direction of blood flow in arteries.

* Prominent German anatomist of 19th century.

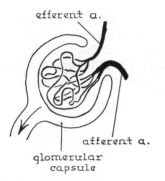

efferent a.

afferent a.

glomerular capsule

Fig. **435**. Scheme of microscopic glomerulus (see text).

removing from it water, glucose (a sugar), and salts.

Having passed through the glomerulus, what is left of the blood passes to a capillary field surrounding the first convoluted tubule, the loop of Henle, and the second convoluted tubule, where most of the water, all of the glucose, and all of the sodium chloride are, under normal conditions, resorbed by the blood; other salts are, in some cases largely, in other cases entirely, retained in the urine. The tubule carrying away rejected products ends by joining others, until, finally, a relatively large collecting tubule discharges at the apex of a medullary pyramid into the pelvis of the ureter.

In the cortex and renal columns lie the kidney corpuscles and the convoluted tubules; here they give to the kidney a granular appearance.

In the pyramids (in part, in the cortex) lie the loops of Henle and the collecting tubules; it is the presence of collecting tubules converging on the apex of a pyramid that gives to that structure a striated appearance. About a dozen collecting tubules open at the papilla on the apex of a pyramid and several apices are embraced by a calix.

When diseased, the kidneys may permit deleterious products to circulate in the blood stream and the resulting auto-intoxication, or uremic poisoning as it is called, accounts for a train of extremely serious symptoms. An inflammation of the kidney is a nephritis and it may be acute or chronic. A kidney is a vital organ although one may be removed without any ill effects, provided the other is healthy. The urinary tract presents a wide variety of congenital anomalies (fig. 436).

URETERS

The funnel-shaped pelvis of the ureter leaves the kidney at the hilus (fig. 432); it tapers rapidly to a narrow tube with a relatively thick muscular wall and a small lumen. The ureter descends vertically behind the peritoneum and on the muscles of the posterior abdominal wall. It enters the pelvic cavity toward the back by crossing the brim of the pelvis about 5 cm (2″) from the midline, where it lies in front of the lower end of the common iliac artery, one of the two terminal branches of the aorta (page 267). Dipping deeply into the pelvic cavity, the ureter next swings forward in a gentle curve to the back of the urinary bladder whose upper posterolateral angle it enters (fig. 437). The ureter is about 25 cm (10″) long, about half of it is in the pelvic cavity.

Structure. The ureters possess a definite muscular coat and a mucosa which is continuous with (but unlike) that of the bladder below and of the renal tubules above.

URINARY BLADDER AND URETHRA

Urinary Bladder

The urinary bladder is a hollow, thick-walled, muscular organ situated in the

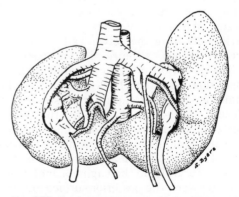

Fig. **436**. An important anomaly: horseshoe kidney (kidneys fused across midline). This occurs once in several hundred persons.

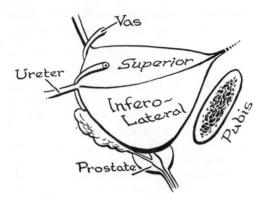

Fig. **437**. Surfaces of urinary bladder and some important relations. Seminal vesicle in contact with (unseen) posterior surface is not labeled (see also fig. 447).

forepart of the true pelvic cavity, and to the shape of this forepart of the cavity it conforms. It is, of course, capable of considerable change in size and shape but, if hardened *in situ* when it is empty, it possesses four surfaces each of which has the shape of an equilateral triangle whose sides are about 7 cm (3″) in length (see figs. 437, 447).

The triangular superior surface, the only one covered with peritoneum, has its base behind and its apex in front; this apex lies behind the symphysis pubis. The surface is horizontal and on it rest coils of intestine; in the female the body of the uterus overhangs this surface from behind.

The paired lateral (or inferolateral) surfaces face the side walls of the pelvis from which they are separated by loose areolar tissue. When the bladder is pulled away artificially from the pelvic wall, a space is created known as the *retropubic space.*

The posterior (or inferoposterior) surface faces downward and backward and, in the male, has parts of the internal sex organs closely applied to it. Behind these, the posterior surface rests, in the male, on the forwardly inclined lower part of the rectum (fig. 449). In the female, the neck of the uterus (cervix) and the vaginal canal intervene between the posterior surface and the rectum (fig. 458 on page 251).

The paired ureters open into the posterior wall of the interior of the bladder 2 to 3

cm (1″) apart after having burrowed through the bladder wall from the posterolateral 'corners.' Both openings are only 2 to 3 cm (1″) from the exit (i.e., the urethra) at the lowest part of the bladder. The triangular area of the bladder wall between the three openings is known as the *trigone* (fig. 438).

Structure. The bladder wall consists chiefly of interlacing bundles of involuntary muscle lined on the inner surface by a specialized mucosa capable of great stretching (except in the area of the trigone).

Urethra

The beginning of the urethra in both sexes is advantageously situated at the inferior angle of the bladder like the mouth of a funnel. However, the rest of the urethra is very different in the male and the female.

The **female urethra** is 2 to 3 cm (1″) long and is closely applied to the front wall of the vagina into the vestibule of which it opens (fig. 458, page 251).

The **male urethra** and the **control of micturition** are described on pages 247 and 248 (see also fig. 438).

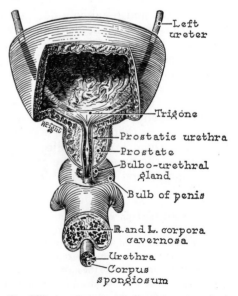

Fig. **438**. Interior of male bladder and urethra

9

Generative or Reproductive System

Certain special cells are set apart in the body of the male and in that of the female for the purpose of perpetuating the life of the species; the male cells are known as **spermatozoa** (Gr. = seed-animals), the female cells as **ova** (L. = eggs).

The essential male organs of generation are the paired **testes,** and the number of spermatozoa produced by them in the course of an average lifetime has been roughly estimated at 400,000,000,000; up to about 200,000,000 are emitted in a single ejaculation. The essential female organs of generation are the paired **ovaries,** and the number of immature ova present in each ovary at birth is about 200,000. In the course of 30 years of functional activity between puberty and the menopause, only a single ovum matures and breaks free from one or other ovary each month, so that a total of about 400 mature ova are produced; the remainder are absorbed without reaching full development.

A spermatozoon is shaped like a tadpole and has a total length of about 0.05 mm. The head of a spermatozoon is about one-tenth of its total length or about .005 mm long, i.e., considerably less than the diameter of a red blood cell; the remainder consists of a neck, body, and a long thread-like flagellum or tail which gives motility to the cell (fig. 439).

An ovum, however, is not motile, but it is relatively large, a human ovum being just visible to the naked eye. It is a round cell with a diameter of about 0.1 to 0.2 mm,

i.e., about four times the length of a spermatozoon or 40 times the length of its head (fig. 440).

Fertilization occurs when a spermatozoon meets an ovum in the uterine tube (described below) after it has traveled about 12 cm (5″) of the female generative tract in about two hours. Should it be successful in meeting an ovum, its head is engulfed by or burrows through the surface of the ovum to gain the interior when the life of a new individual may be said to have begun; in most species, and certainly in man, no more than a single spermatozoon penetrates an ovum which, once impregnated, is resistant to further attack. Should the spermatozoon fail to meet an ovum it is capable of surviving for a short but variable period in the female generative tract, perhaps several days. Certainly, spermatozoa can be kept alive, under ideal conditions outside the body, for many days.

MALE REPRODUCTIVE ORGANS

Scrotum and Development of Testis

The scrotum is a midline pendant sac of loose skin and subcutaneous fascia evaginated from the lower part of the anterior abdominal wall; in it reside the paired testes. Involuntary muscle fibers, the **tunica dartos,** lie in the subcutaneous fascia of the scrotal wall and these are responsible for the shrinkage and wrinkling of the scrotal wall in the presence of cold.

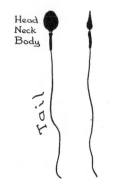

Head
Neck
Body

Tail

Fig. **439**. A spermatozoon—two views (greatly magnified.

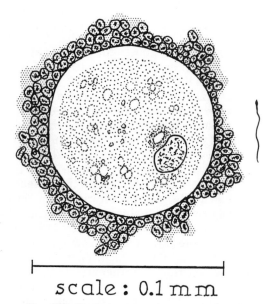

scale: 0.1 mm

Fig. **440**. Ovum is enormous compared to spermatozoon (shown to its right).

The interior of the scrotum consists of two compartments each of which is completely filled by a testis and its associated structures. Since only the skin and its subcutaneous fascia are evaginated to form the scrotal wall, the remaining layers of the abdominal wall must be penetrated by the testis, its vessels, nerves, and duct as they journey from the abdominal cavity to occupy the interior of the scrotum. The weakening of the anterior abdominal wall consequent upon this journey explains the frequency of **inguinal hernia** or **rupture** (see figs. 441–445).

The penetration is not a simple one, for the testis drags into the scrotum with it representatives of each of the layers it passes through; these tubular prolongations of fascia and elongated loops of muscle fibers are called the **coverings of the spermatic cord** in contradistinction to the vessels, nerves, and duct which are known as the **constituents of the spermatic cord.**

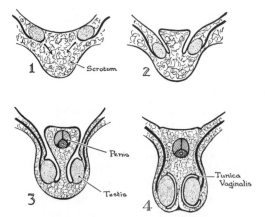

Fig. **441**. Descent of the testis into the scrotum (see text).

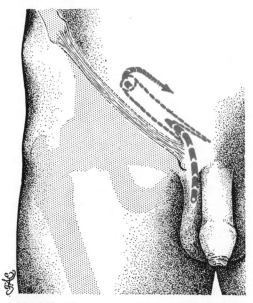

Fig. **442**. Diagram of right spermatic cord running through inguinal canal, arrowhead indicating direction of ductus deferens running to back of prostate in pelvis.

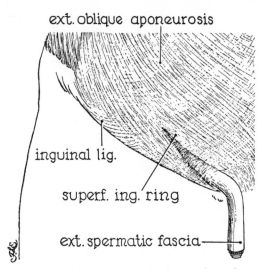

Fig. **443**. Spermatic cord emerges just above pubic bone through a triangular weak spot in External Oblique aponeurosis called the superficial (or external) inguinal ring.

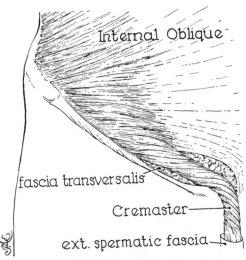

Fig. **444**. Muscle fibers (Cremaster), derived from Internal Oblique, cover spermatic cord.

Each testis develops near the kidney and lies behind the peritoneum to the back of which it is attached. As a result of this attachment, it drags a tubular prolongation of peritoneum with it, so that it still has a peritoneal investment in the scrotal sac. This peritoneal evagination finally lines its own half of the scrotal sac and covers most of the testis, just as a pleural sac covers most of a lung. The continuity between scrotal 'peritoneum' and abdominal peritoneum is usually lost; should it persist, an open peritoneal passage unites the abdominal and scrotal cavities and a congential hernia results. The peritoneal investment of the testis is known as the **tunica vaginalis.**

Testis and Epididymis

The testis is a solid ovoid organ about 4 cm (1 ½″) long. When its tunica vaginalis is removed, its thick fibrous coat, known from its whitish appearance as the **tunica albuginea,** is exposed. From this fibrous coat septa pass into the interior of the organ to divide it into about 250 *lobules.* The septa converge on the middle of the rounded posterior border of the testis where the vessels, nerves, and duct have their entrance or exit. This region is known as the *mediastinum testis.*

Each of the lobules contains from one to three coiled tubules known as **seminiferous tubules;** these are the essential structures of the testis, for the cells lining the tubules are converted into spermatozoa which break free and are wafted along the tubules toward the mediastinum. Each

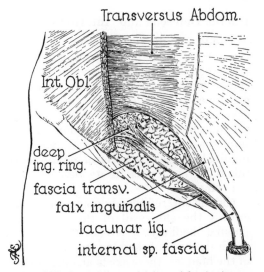

Fig. **445**. Spermatic cord 'pierces' fascia deep to Transversus (transversalis fascia) at deep inguinal ring that lies below the free edge of the arching muscle. Medially, the tendons of Transversus and Internal Oblique form the conjoint tendon which inserts into the pubic bone.

tubule when unraveled is about 60 cm (2 ft) long (fig. 446).

In the mediastinum, the tubules meet to form a network known as the *rete* (L. = network) *testis*. From the upper end of the rete, 15 to 20 *efferent ductules* leave the testis. Each efferent ductule, finer than the lead of a pencil, is highly coiled and, when unraveled, about 20 cm (8″) long; each opens into the duct of the epididymis.

The testis is also an endocrine organ. After puberty, certain of its cells liberate into the blood stream the male hormone, *testosterone*.

The epididymis (Gr. = upon the twin), the first part of the duct of the testis, is remarkable for the length and tortuosity of its course. In it are stored the spermatozoa. The epididymis begins at the upper pole of the testis and ends at the lower pole; between these two points—no more than 4 cm (1½″)—it is so elaborately coiled that, if unraveled, it would be seen to be as long as the digestive tract. The epididymis occupies the back of the testis to which it is loosely attached by fibrous tissue; at its lower pole it finally becomes a more or less regular tube known as the ductus (or vas) deferens.

Ductus Deferens and 'Internal Organs'

The vas or **ductus deferens** ascends on the medial side of the epididymis and, accompanied by vessels and nerves (the constituents of the spermatic cord), is enveloped in muscle fibers and fascia (the coverings of the spermatic cord); it pierces the muscles of the anterior abdominal wall immediately above the medial end of the inguinal ligament thus retracing the course taken by the testis in its descent (figs. 441–445).

Having entered the abdomen, the ductus crosses the side wall of the pelvis and, passing medially, it crosses the ureter to reach the back of the bladder. On the back of the bladder, there exists an outgrowth or diverticulum of the ductus deferens, the **seminal vesicle** (fig. 447). This also lies applied to the back of the bladder and is on the lateral side of the ductus deferens. The seminal vesicle is a somewhat coiled and branched outgrowth providing a sticky

secretion which is discharged with the spermatozoa. The short duct common to the seminal vesicle and the vas deferens is known as the **ejaculatory duct** which, traversing the prostate gland, opens into the upper (prostatic) portion of the urethra (figs. 447–450).

The prostate gland lies below the bladder and surrounds the first 3 cm of the urethra. It is a solid organ about the size of a chestnut and it provides a watery secretion which is added to those of the testes and the seminal vesicles. The prostate contains many smooth muscle fibers which, on contraction, empty the prostate (as though it were a sponge) through minute openings into the urethra. Enlargement of the prostste may lead to obstruction of the urethra and require removal; it is also a frequent site of cancer in advancing years.

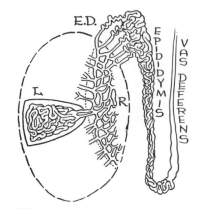

Fig. **446**. Scheme of structure of testis and its duct. L. = lobule (compartment); R. = rete or network; E. D. = efferent ductules.

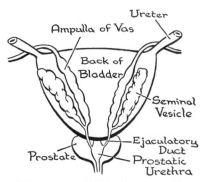

Fig. **447**. Male internal reproductive organs associated with bladder (posterior view).

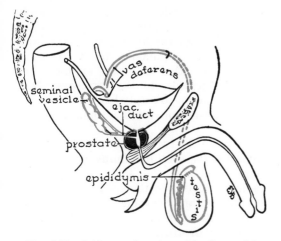

Fig. **448**. Scheme of male pelvis (from right side). Reproductive organs in red. Note three parts of male urethra—prostatic, membranous, penile.

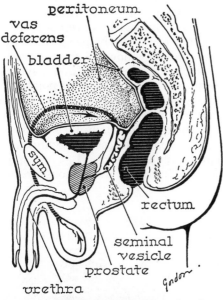

Fig. **449**. Median sagittal section of male genitalia to show relationships.

The prostatic urethra, about 2 to 3 cm (1″) long, is large, fusiform, and dilatable (figs. 447–450).

Urogenital Triangle; Control of Micturition

It will be recalled that the region below the Levator Ani is known as the **perineum;** its triangular posterior half (**anal region** or **triangle**) has been described (page 214).

The triangular anterior half is known as the **urogenital region** or **triangle.** A stout fibrous **perineal membrane** fills the interval below the symphysis pubis and between the diverging inferior pubic rami. Above the membrane is a similarly disposed thin, transverse sheet of striated muscle known as the **urogenital diaphragm.** Above the muscle is yet another (but delicate) fibrous membrane. Thus the very short section of urethra that immediately succeeds the prostatic portion pierces the two membranes (and the intervening muscle) and is therefore known as the **membranous urethra.** The central part of the transverse muscle surrounds the membranous urethra and acts as a voluntary sphincter for it (figs. 448–452).

It is now generally agreed that there is no true internal (involuntary) sphincter at the exit of the bladder. In some unexplained manner the smooth muscle of the bladder around the exit does help to prevent the leakage of urine. When the bladder distends, messages (impulses) pass to the spinal cord along sensory nerves and set up an 'involuntary reflex.' The muscles of the bladder are stimulated by their motor nerves to contract and expel the stored urine. The onus for retention of urine then

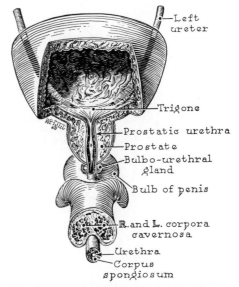

Fig. **450**. Interior of prostatic, membranous, and spongy urethra. Note bulbo-urethral gland (ducts not shown).

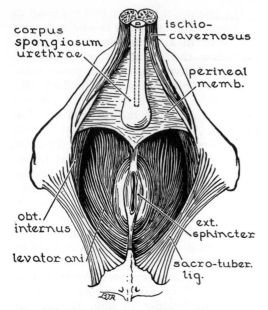

Fig. **451**. Male perineum. Anterior half—the uro-genital triangle—has perineal membrane stretching across. To this membrane corpus spongiosum of penis is attached centrally. Urethra enters penis here by piercing membrane. Paired cavernous bodies of penis (corpora cavernosa) attach to bone on each side and are covered with special ischiocavernosus muscle. (Bulbospongiosus—not shown—covers corpus spongiosum, which is bulbous here.) Note cross-section of penis.

falls on the voluntary sphincter which, with increasing insistence, calls for relief. In paralysis of the lower half of the body (paraplegia), commonly due to a spinal cord injury, the voluntary sphincter no longer functions, so that the paraplegic must be taught to empty his bladder before the distension produces automatic empty-ing.

Between the two membranes lie also the paired *bulbo-urethral glands*, each smaller than a pea (figs. 450, 452); they empty by delicate ducts into the succeeding portion of the urethra. Beyond the membranous portion, the urethra is known as the **penile** or **spongy urethra.**

Penis

The spongy urethra (figs. 452–454) tra-verses the length of the spongy body or corpus spongiosum, the central member of the three cylinders of cavernous erectile tissue that comprise the penis (L. = tail).

Two paired cylinders of erectile tissue— the **corpora cavernosa** (L. = cavernous bodies)—are fixed by their pointed, fi-brous, posterior extremities to the two inferior pubic rami. As they pass forward they converge and finally fuse; below and between them, the central cylinder—the **corpus   spongiosum** (L. = spongy body)—is fixed by its bulbous posterior

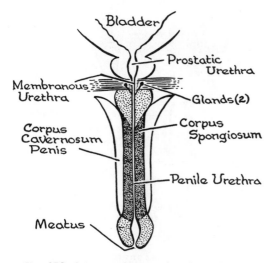

Fig. **452**. Scheme of parts of male urethra and three cylinders of erectile tissue that form penis.

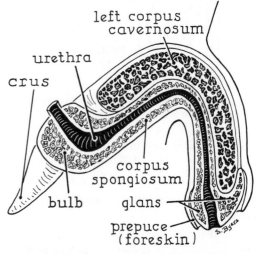

Fig. **453**. Structure of penis cut sagittally.

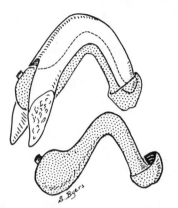

Fig. **454**. Corpora of the penis. The corpus spongiosum (stippled) fits under the united right and left corpora cavernosa.

extremity to the under surface of the perineal membrane. Its bulbous anterior extremity is known as the **glans** (L. = acorn) (see figs. 448–454).

Each cylinder of erectile tissue is covered by a muscle at its attachment to bone or membrane (fig. 451). These muscles are the (paired) Ischiocavernosus and the (paired but midline) Bulbospongiosus. Their contraction aids in obstructing the return of venous blood from the erectile tissue and so may help in maintaining erection.

The **prepuce** (or foreskin) is a movable cuff-like covering of skin for the glans (fig. 453). Its surgical removal is known as circumcision (L. = a cutting around).

FEMALE REPRODUCTIVE ORGANS

Ovaries and Uterine Tubes

An ovary, about half the volume of a testis, develops beside the kidney and, like the testis, is adherent to the back of the peritoneum; it migrates no farther than the side wall of the pelvis where it takes up its permanent position. It lies almost free in the peritoneal cavity, hanging only by a mesentery known as the *mesovarium*. The surface of the ovary is unlike that of any other abdominal organ because it is covered by a layer of its own specialized cells rather than by peritoneum. This layer is called the *germinal layer*. Deep to it is the

stroma, which is a mass of connective tissue containing *ova* in various stages of maturity.

When an ovum matures it ruptures through the surface of the ovary into the peritoneal cavity. Fortunately it usually rolls, falls, or is attracted into the **uterine tube** whose trumpet-shaped and fimbriated (fringed) mouth is in the immediate vicinity since it is attached to the ovary by one of its fimbriae (fig. 455).

The uterine tube, whose fimbriated end partially enwraps the ovary, lies in the upper part of a transverse fold of peritoneum known as the **broad ligament**—to be presently described. The tube runs for about 10 cm (4″) and more or less horizontally medialward, to the upper part of the side of the midline **uterus** or womb. It pierces the thick muscular wall and conducts the ova to the uterus.

Should a fertilized ovum fail to reach the uterus, its normal destination, it may continue to develop for a time in the uterine tube. Such a condition is an *ectopic* (Gr. = out of place) *pregnancy*. It can lead only to a bursting of the tube, and, therefore, it requires surgical intervention.

Uterus and Its Attachments

The uterus is a thick-walled, hollow, muscular organ about the size and shape of an 8 cm (3″) long pear and with the stalk down. It is interposed between the urinary bladder in front and the rectum behind; it is both tilted forward and bent forward so as to overhang the bladder from behind and above. The portion of the uterus above the level of the entrance of the uterine tubes is known as the **fundus** of the uterus; this is succeeded by the **body** of the uterus. The lower constricted portion (the stalk of the pear) is the **cervix** or **neck** of the uterus and it bulges downward into the vault of the vaginal canal whose walls embrace and surround it. See figures 455–459.

Attached to the uterus immediately below the point of entrance of the uterine tube is the **round ligament of the uterus.** It corresponds in position (though not in origin) to the male ductus deferens. Like the ductus, it traverses the anterior ab-

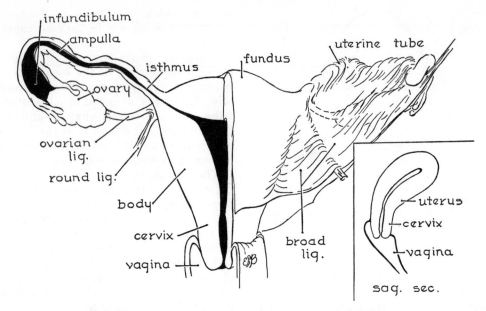

Fig. **455**. Female internal reproductive organs seen from in front. On right of page, structures are intact, showing broad ligament with tube in its upper border and ovary behind; on left side, broad ligament is removed, and uterus and tube opened to show interiors and walls. Inset: sagittal section viewed from right side to show forward tilt of body of uterus.

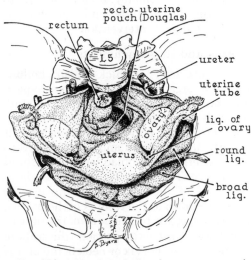

Fig. **456**. Internal genitalia of young woman *in situ*.

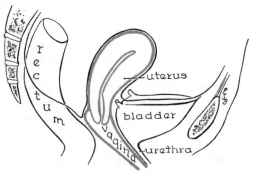

Fig. **457**. Position and relations of uterus and vagina (in sagittal section). Note covering of peritoneum (solid black line) reflected to upper surface of bladder and to vault of vagina. Resulting peritoneal fossa between rectum and uterus is recto-uterine pouch of Douglas.

dominal wall and is lost in the tissues of the labium majus—the female homologue of the scrotum. The round ligament helps to maintain the forward tilt and bend of the uterus. As is the case with other pelvic organs, the point of exit of the uterus (cervix) is its most fixed point and the one from which the organ can expand.

The uterus projects freely into the peritoneal cavity and, therefore, has a coat of peritoneum. Stretching from each side of the uterus to the side wall of the pelvis is a double sheet of peritoneum in whose upper free border lies the uterine tube. This sheet or 'fold' of peritoneum is the **broad ligament** (figs. 455, 459). From the back of the broad ligament the ovary hangs by its mesovarium.

The peritoneum on the back of the uterus reaches the upper end of the vaginal canal; from there it passes backward onto the front and sides of the rectum. The deep peritoneal pocket which thus lies between uterus and rectum is known as the **recto-uterine pouch.** It is the lowest point of the peritoneal cavity. [The surgeon makes use of this fact in draining the peritoneal cavity of foreign matter (e.g., blood).]

Cavity and Structure of Uterus. The cavity of the uterus is extremely small. The walls are almost a centimeter thick (fig. 459) and composed chiefly of involuntary (smooth) muscle. In pregnancy the cavity expands enormously and there is considerable growth of the walls. The mucosa is thick and velvety and, under the microscope, it is seen to be liberally supplied with blood vessels and many simple tubular glands.

When an ovum is fertilized it reaches the cavity of the uterus (in a day or so) and becomes embedded in the mucosa which has become greatly thickened in anticipation. Thereafter, changes occur that have been described under 'Embryology' (page 8). On the other hand, if the ovum has not met any sperm, the thickened mucosa sloughs off—in a process known as **menstruation**—about two weeks after the ovum was released from the ovary (ovulation). In other words, ovulation occurs approximately midway between successive menstrual periods, or *vice versa*. The sloughing off of the mucosa is naturally accompanied by bleeding that lasts several days. Then the mucosa repairs itself and begins building up again only to undergo the same destruction if, once more, fertilization does not occur.

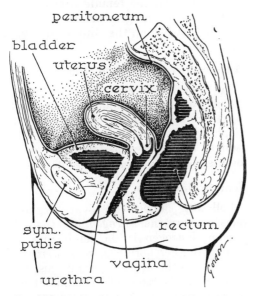

Fig. **458**. Median sagittal section of female genitalia to show relationships.

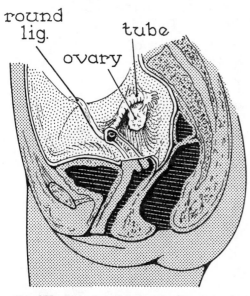

Fig. **459**. Right broad ligament after uterus has been removed. The uterine tube is in its superior margin, the round ligament on its front, and ovary lies behind on the side wall of the pelvis minor.

Vagina and External Genitalia

The vagina (L. = sheath; pronounced va-jĭ′ nah) is the tubular canal (about 10 cm or 4″ long) leading downward and forward from the cervix to the exterior, where its orifice is surrounded by the external genitalia (figs. 457–460). Normally its walls are collapsed but it is capable of enormous stretching during the passage of the infant in childbirth. The external genitalia occupy the anterior part of the perineum (urogenital triangle). In the virgin, the lower end of the vagina is partially obstructed by a thin membrane (the *hymen*) which has a central irregular perforation.

The lower end of the vagina and the hymen are covered by the paired **labia minora** (labium, sing., L. = lip). Outside of these, the much larger **labia majora** lie in contact in the midline and cover the deeper structures. The labia majora are continuous with a mound of skin and

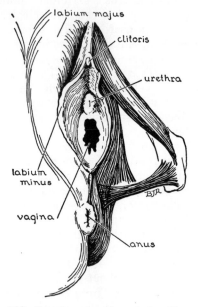

Fig. **460**. Female external genitalia. Dissection on one side exposes special muscles comparable to male (cf., fig. 451).

subcutaneous fat in front of the symphysis pubis, known as the *mons pubis*.

Hidden by the labia majora, the anterior ends of the labia minora meet at a small projecting lump known as the **clitoris**. Although the clitoris is, developmentally and structurally, the homologue of the penis (fig. 461), the **female urethra** does not traverse it but opens between the labia minora, just behind the clitoris and below the hymen.

Muscles of Female Perineum

The muscles are homologous to those in the male with special modifications resulting from the piercing of the urogenital diaphragm (muscle) by the vagina. They tend to be disposed as sphincters of the lower end of the vagina (see fig. 460).

Female Hormones

Following puberty the ovary begins to release a hormone, *estrogen*, into the blood stream. Estrogen is the feminizing hormone. Another hormone, *progesterone*, is secreted by the ovary especially just after each ovum is released. The function of progesterone is to prepare the lining mu-

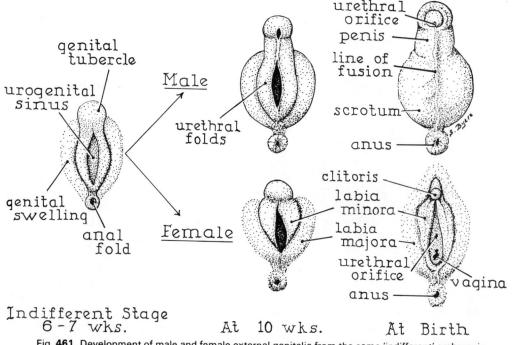

Fig. **461**. Development of male and female external genitalia from the same 'indifferent' embryonic structures.

cosa of the uterus for the embedding of a fertilized ovum. The output of these sex hormones is controlled by other hormones, especially those of the hypophysis or pituitary gland at the base of the brain.

MAMMARY GLAND

Because the female breast starts to become functionally active in the later stages of pregnancy, it is properly regarded as an organ accessory to the reproductive system.

The mammary gland lies in front of the Pectoralis Major in the deeper part of the skin where it developed as a modified sweat gland (fig. 462). The size and contour of the gland depend on the quantity of fat

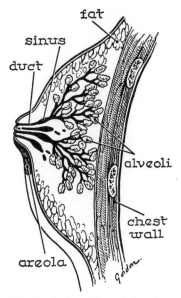

Fig. **463**. Vertical section of female mammary gland (semischematic) showing only two of the glandular lobules.

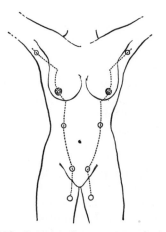

Fig. **462**. Occasionally accessory nipples occur along the 'nipple line.' In mammals that have litters the multiple nipples appear along this same line.

lying in the meshes of the connective tissue framework of the organ. Between the fat lobules lie about 20 compound alveolar glands arranged like grapes on a stalk (see fig. 7 on page 5, and fig. 463). Their ducts (*lactiferous ductules*) converge and open on the summit of the **nipple**—a central elevation containing involuntary muscle and capable of erection. Beneath the nipple, which is surrounded by a circular pigmented area known as the **areola,** each duct possesses a little dilatation (*lactiferous sinus*) following which it is constricted as it enters the nipple.

10

Circulatory System

GENERAL

The first nine months of life for the individual are lived in a water environment, and it is altogether probable that life itself began in the water. Certainly for long ages it existed only in the water, and when living forms ventured upon a terrestrial existence they maintained life (and continue to do so) only because they carried a water environment about with them.

The water or fluid environment that surrounds every living cell of the body is known as **tissue fluid;** it is derived from the blood, renewed from the blood, and returned to the blood. In this tissue fluid the cells of the body find all the materials they require, not only for their continued existence and well-being but also for the efficient performance of their specific functions.

It is the function of the circulatory system to ensure that blood reaches all parts of the body in order that every cell may receive nourishment; so efficiently is this duty discharged, that even the most trivial cut on any part of the surface of the body is inevitably accompanied by bleeding.

Because tissue fluid derived from the blood bathes every cell, the circulatory system has been likened to an irrigation system. But blood is not uniformly distributed. Tissues in which cells predominate require and receive a greater proportion of blood than do those in which intercellular substances predominate. In other words, it is the active living cell, not the less vital product of its activities, that needs blood and one can gauge with reasonable accuracy the vascularity of a tissue by estimating the proportion of its cellular to its non-cellular composition.

Circulation. In 1628 William Harvey, an English physician, published a book in which he set forth his epoch-making discovery that the blood circulates, i.e., it travels in a circle. He pointed out that the blood leaves the heart via the arteries and returns to the heart via the veins. Because he lived before the invention of the microscope, the discovery of the connection between the arteries and the veins was denied to him, but he knew and insisted that such must exist. This connection is now known to be a network of minute channels known as **capillaries** and, because it is at the capillaries that the interchange between blood and tissue fluid takes place, it is the capillaries that are the vital parts of the circulatory system.

Before Harvey's day, the blood was thought to ebb and flow in the veins, an erroneous deduction from the observation that after death blood is found in the veins but the arteries are empty. The arteries were considered to be the carriers of the 'vital spirits.' If

254

they bled when cut, the phenomenon was explained on the assumption that with the escape of 'vital spirits' blood rushed in to take its place.

BLOOD

In an average adult male there are about 5 to 6 liters (5 quarts) of blood. This is made up of fluid known as **plasma** (about 55 per cent) and **cells** (about 45 per cent). Roughly 90 per cent of the plasma is water; the remaining 10 per cent being dissolved material necessary to the nourishing and functioning of the tissues of the body to which the blood is conveyed. The blood cells are known as (1) red blood cells (or erythrocytes, or red corpuscles); and (2) white blood cells (or leucocytes, or white corpuscles). In addition to these, there are many smaller bodies—said to be pieces of cells—known as platelets (fig. 464, g).

Cells

Erythrocytes (Gr. = red cells; fig. 464, a). These are formed in bone marrow—particularly in the cubical or short bones, the flat and irregular bones, and the ends of long bones—where they pass through several stages before reaching maturity. A mature red cell has lost its nucleus and has become completely filled with hemoglobin; it is released into the circulating blood as a disc about seven microns in diameter (one micron = $\frac{1}{1000}$ mm). Viewed in profile the disc is concave on its two surfaces and has a rounded thick rim (fig. 464). This shape is said to be ideal for bringing all parts of the cell close to its surface.

Hemoglobin is a wonderful, highly specialized compound of protein and iron. It is an ideal oxygen carrier since it combines avidly with oxygen but surrenders it with readiness to the cells of the body. If the red cells are low in hemoglobin or if the number of red cells in circulation is reduced below normal, a condition of anemia results.

Red blood cells exist in such enormous numbers that one cubic mm of blood contains between five and six million of them. An adult's total blood contains about 30 million millions. Each red cell wears out in about four months and, dying on duty, disintegrates in the blood stream. The spleen gathers the remains and, at the

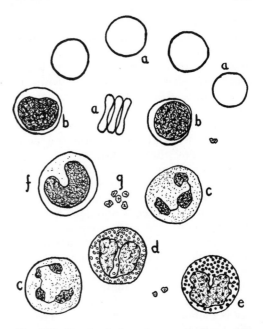

Fig. **464**. Blood cells (greatly magnified). a = red blood cells; b = lymphocytes; c = neutrophile (polymorphonuclear); d = eosinophile; e = basophile; f = monocyte; g = platelets.

same time, ruthlessly destroys any whole red cell whose usefulness is impaired. Since the life of a red cell is but four months, new cells must be produced at an incredible rate and even in these days of mass production the ability of bone marrow to 'turn out' about four million red cells every second of every minute remains an impressive accomplishment.

Leucocytes (Gr. = white cells; fig. 464, b, c, d, e, f). A well-organized state possesses a standing army for the protection and defence of its citizens. So, too, the community of cells that is the human body possesses its army of specialized warrior cells. These are of two types: (1) **mobile cells;** (2) **fixed cells.**

1. The **mobile cells** are the leucocytes and, because they exist in the blood, their rapid mobilization at any point threatened with disease, injury, or invasion of germ cells offers no problem. They can squeeze between the cells of a capillary wall and so, escaping from the blood stream, can engage the enemy in the tissues of the body; they are often called **'wandering cells.'**

Five varieties of leucocytes exist (fig. 464, b, c, d, e, f) and many of them are phagocytic, i.e., capable of engulfing and digesting foreign material. Each variety seems to select its foe so that while the neutrophiles, for example, engage the germs of acute infections, and the monocytes clear up the battleground after the victory, the lymphocytes are the active ones in many chronic diseases.

Leucocytes are much less numerous than erythrocytes, there being between 5000 and 8000 in a cubic mm of blood. All are fully formed nucleated cells and none has a long life—probably not more than a few days. They are produced in red bone marrow and in the spleen and other lymphoid tissues. As has been suggested, their numbers are notably increased in many conditions and the determination of what particular variety is increased often reveals the nature of the disease.

2. The **fixed cells**—large and phagocytic —differ from the mobile ones chiefly in that they remain at their posts in the tissues of the body and let the blood bring them their victims. Strictly speaking, they are not blood cells but are found in connective tissue, spleen, liver, lung, lymphoid tissue, and bone marrow. They are like certain leucocytes and they have been given a variety of names by different scientists—*histiocytes* (Gr. = tissue cells), *macrophages* (Gr. = big eaters) and *reticulo-endothelial cells*. They are sometimes said to constitute the **reticulo-endothelial system** and they are the cells that, in the spleen and liver, destroy effete erythrocytes and, out of their hemoglobin, produce bile pigment.

Blood Platelets. These are less than half the size of a red cell and are more numerous than white cells. There are about 250,000 of them in a cubic mm of blood and they are important agencies in blood clotting. Their origin is still under investigation.

Plasma

Blood plasma is a solution in water of crystalloids (e.g., various salts, sugars) and of colloids. The latter are the **plasma proteins** (albumin, globulin, and fibrinogen). Remaining in the blood stream because they normally cannot pass through the capillary wall (a semipermeable membrane), they exert an osmotic pressure that helps to attract water from the tissue fluids back to the blood stream. In addition, fibrinogen is important in the mechanism of blood clotting.

If the plasma proteins leak away from the capillaries as the result of some serious injury or damage to the tissues (e.g., a burn) the water cannot be returned to or retained in the circulation and the heart receives less and less blood to pump. The same effect is achieved, of course, by an acute hemorrhage. A vicious spiral is set up which results in a serious collapse of the circulation known as **shock.** Only large transfusions of whole blood or plasma, in addition to measures preventing further loss of plasma proteins, can avert death.

BLOOD VESSELS

Blood reaches the capillaries because the circulatory system is provided with a pump; this is the heart. From the heart, an ever-branching system of tubes conducts the blood to the capillaries; these tubes, the arteries and arterioles, vary in diameter from 30 mm to 0.3 mm; arterioles are just visible to the unaided eye. Venules and veins corresponding to arterioles and arteries conduct the blood back to the heart (fig. 465). Not all tissue fluid, however, makes its way back into the blood capillaries. Some enters yet another system of conducting tubules and these are known as **lymph vessels** or **lymphatics** (page 292). A few general remarks about arteries, arterioles, capillaries, venules, and veins will be offered before the heart is described and the principal vessels are named and located.

Arteries (fig. 466) are expansile tubes possessing three coats: (1) an exceedingly smooth and thin inner endothelial lining known as the **intima;** (2) a relatively thick intermediate coat, consisting largely of elastic fibers but partly of smooth muscle fibers, known as the **media;** (3) an outer fibrous coat, the **externa** (or *adventitia*).

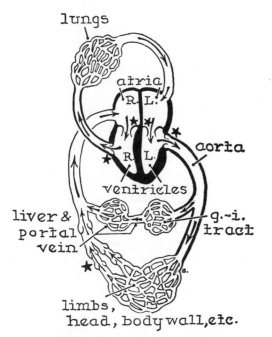

lungs

atria
R. L.

aorta

R. L.

ventricles

liver &
portal
vein

g.-i.
tract

limbs,
head, body wall, etc.

Fig. **465**. Scheme of systemic, pulmonary, and portal circulations. Arrows show direction of blood flow; stars indicate sites of one-way valves.

It may be observed that, because elastic tissue is such a notable feature of its wall, an artery largely responds passively to the pressure of the contained blood. Elastic tissue is lost in old age when arteries tend to become stretched, tortuous, and hardened.

Arterioles possess a structure similar to that of arteries but the emphasis is on muscle tissue in the media rather than on elastic tissue. On structural grounds alone, therefore, one expects to find active—not passive—changes in their caliber, one of the conspicuous features of arterioles. When it is further observed that it is to the arterioles especially that nerves are distributed, it is apparent that the phenomena of **vaso-constriction** and **vaso-dilation** are special functions of the arterioles. It is by this change in caliber of the arterioles that the quantity of blood delivered to a capillary field can be increased or diminished according to functional activity—and, sometimes, according to emotional activity, e.g., the blush of maidenly modesty, the blanching of fear, and the cold hands of apprehension.

An average **capillary** is not much more than a mm long and has a bore just large enough to permit the passage of red blood cells in single file. Because a capillary field is a network, many channels are available for the passage of blood through it, and not all channels need be open at a given time (fig. 468). As the blood is forced through the capillaries by the hydrostatic pressure in the arterioles, interchange between plasma and tissue field occurs, so that by the time the blood emerges from a capillary field it has delivered its oxygen and other products to the cells of the tissue and has received in exchange carbon dioxide and waste products. The collecting blood vessel is a venule.

Venules unite to form *veins*, and veins present certain features which distinguish them:

1. They possess coats similar to those possessed by arteries but the media is much thinner. Veins are thin-walled because the blood pressure in them is very low compared with that in the arteries (fig. 466).
2. Because the flow in veins is much more sluggish than it is in arteries, the veins are larger calibered so that they may return to the heart in a given interval of time a volume of blood equal to that leaving it. This increase in caliber alone is insufficient, so a single artery is often accompanied by a pair of veins known as its **venae comitantes** (see fig. 467). Further, deep to the skin there exist in several regions, particularly in the limbs,

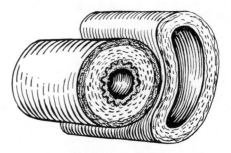

Fig. **466**. Schematic section of an artery (on the left) and its companion vein. Wall of artery is much thicker. Note three layers.

Fig. **467**. An artery with its venae comitantes.

large **cutaneous veins** unaccompanied by corresponding arteries.

3. Veins that have to return blood against the influence of gravity are equipped with **valves**. These, by breaking up the column of blood, relieve the more dependent parts of excessive pressure and encourage the flow of venous blood toward the heart (fig. 516 on page 284). The venous congestion that the hot and tired feet feel at the end of a busy day is relieved by reclining with the feet higher than the torso.

Lymph capillaries and **lymph vessels (lymphatics)** offer another special pathway for the return of tissue fluid to the blood stream. They are distinguished by certain special characters and will be considered after the heart and blood vessels have been described (see page 292).

<h3 style="text-align:center">HEART</h3>

General

The heart is a muscular pump with two duties to perform: (1) it must pump venous blood to the lungs, so that the red blood cells may exchange their cargoes of carbon dioxide for new cargoes of oxygen; (2) it must pump this oxygenated blood, received from the lungs, to all parts of the body, In consequence, the heart is a double pump whose two parts work in unison. The right side receives the venous blood and pumps it to the lungs; the left side received the oxygenated blood from the lungs and pumps it to the body at large. The circulation to and from the lungs is called the lesser or **pulmonary circulation** (pulmo, L. = lung); the circulation to and from the body at large is called the greater or **systemic circulation.**

Each side of the heart consists of: (1) a 'receiving' chamber, the **atrium** (L. =

ante-chamber), which pumps its blood through an orifice guarded by a valve into: (2) a 'discharging' chamber, the **ventricle** (L. = little belly), which pumps its blood through an orifice also guarded by a valve into an artery. The artery leading out of the right ventricle is called **pulmonary artery,** that out of the left ventricle, the **aorta** (see fig. 469).

The walls of the heart consist of three layers. The outermost layer is a thin, serous membrane known as **epicardium;** it is part of the serous pericardium to be described shortly. The middle layer, known as the **myocardium,** consists of special cardiac muscle and is responsible for the heart's ability to pump. The innermost layer, known as **endocardium,** lines the myocardium and is a very delicate endothelial membrane continuous with the intima (endothelium) of the blood vessels. Besides these three layers, there exists a fibrous framework for the heart (often referred to as its 'skeleton') in the form of tough rings which, in essence, surround the orifices of the heart; they give attachment to the heart muscle and to the delicate valves that guard the orifices.

The **cardiac muscle** is disposed in a complex manner which can be described only briefly here. Big bundles of muscle are attached to the fibrous framework at each end but run a spiral or a circular course in the walls of the chambers. When they contract they reduce the size of the cavities and 'squirt' the blood out through the orifices.

Control of contraction of cardiac muscle is exercised through the vagus nerve

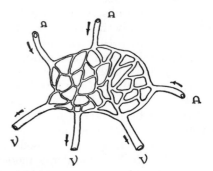

Fig. **468**. Arterioles (A) open into a capillary network (or 'bed') which is drained by venules (V).

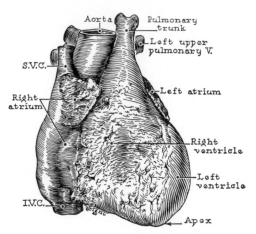

Fig. **469**. Normal heart and its great vessels directly from front.

ately contracts and expels this blood. Thus, at an average rate of 70 beats a minute, at least four liters of blood are pumped into the general circulation in a minute. Some athletes have highly efficient hearts which are said to expel more than twice the average amount at each beat. As a result, the pulse rate of such athletes under normal conditions may be markedly slower than average.

Pericardium

The heart is enclosed in a double-walled serous sac, known as the **pericardium** (Gr. = around the heart), which is similar to pleura or peritoneum (fig. 471). The two

(which retards) and the sympathetic nervous system (which accelerates). These act on a discrete lump of special tissue in the wall of the right atrium—*sinu-atrial or S-A node*—to regulate the strength and rate of the heart-beats (fig. 470). The S-A node is known as the 'pace maker' of the heart. From it, at the rate of about 70 per minute, an impulse (whose nature is quite complex) spreads over the heart, resulting in a contraction. The spreading impulse is picked up and quickly relayed by another node —*atrioventricular or A-V node*—through a prolongation (*A-V bundle*) down to the ventricles. At the upper edge of the muscular interventricular septum the bundle divides into a right and a left branch. These spread and rebranch to the right and left ventricular myocardium (fig. 470). The atria contract just before the ventricles and they both relax completely just before the next impulse arrives.

Because the atria have less work to do than the ventricles, their walls contain less muscle and are thinner. Because the right ventricle has to pump blood only to the lungs whereas the left ventricle has to pump blood to all parts of the body, the wall of the right ventricle contains less muscle and is thinner than that of the left ventricle; the wall of the left ventricle is more than 1 cm (½″) thick.

At each contraction of an atrium about 60 cc of blood are pumped into the ventricle which immedi-

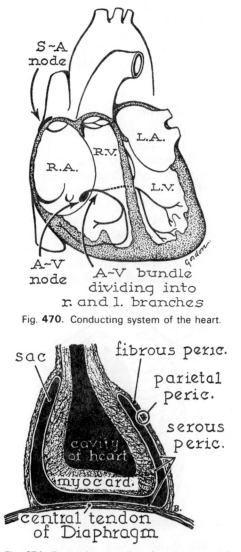

Fig. **470**. Conducting system of the heart.

Fig. **471**. Parts of pericardium (semi-schematic).

layers of pericardium, **parietal** and **visceral,** are separated from one another only by a minimal amount of serous fluid sufficient to keep their smooth contiguous surfaces moist and glistening.

The pericardial sac is literally a bag whose walls are a continuous sheet of fibrous tissue. This fibrous sac—the **fibrous pericardium**—adheres to the upper surface of the central tendon of the diaphragm and loosely encloses the heart. Its inner surface is lined by the parietal serous pericardium.

By the protection the pericardium affords it, the heart moves and beats in a frictionless environment. The visceral pericardium is closely applied to the myocardium where it is also known as epicardium; it resembles the serous coat of a digestive organ.

Position of Heart

The heart—about the size and shape of a clenched fist—occupies a central position in the thoracic cavity (fig. 472). It lies behind the body of the sternum and in front of the middle four thoracic vertebrae (T. 5, 6, 7, and 8). It lies, however, much nearer the front than the back of the thorax and much more to the left than to the right.

During development (fig. 473), the heart undergoes rotation so that its right side is carried forward and its left side backward

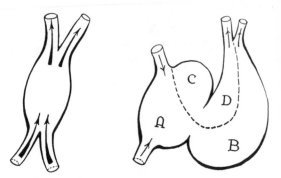

Fig. **473**. The heart develops from a dilated blood vessel. A and C are receiving chambers; B and D, discharging chambers.

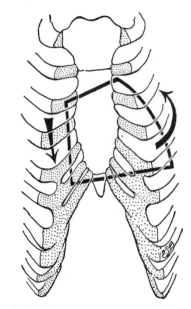

Fig. **474**. Outline of adult heart. Arrows indicate direction of rotation of heart to left during development.

(fig. 474). In consequence, the dispositions of its chambers are as shown in figs. 475–477. The following relations should be noted.

1. The right atrium alone forms the right border.
2. The right ventricle occupies most of the anterior surface and forms all but the extremities of the inferior border.
3. The left ventricle forms the left border except at the extreme upper end which is formed by a small part (the auricle) of the left atrium. Only a

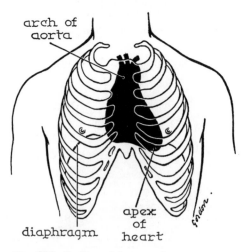

Fig. **472**. Outline of heart and great vessels in normal position.

narrow strip of the left ventricle can be seen in front. At the lower end of this strip lies the apex of the heart.

4. The left atrium lies almost entirely behind.
5. The upper border of the heart is ill defined but said to be formed by the two atria.

Surfaces of Heart

For purposes of description, the surfaces of the heart are referred to as the **sternocostal** (or *anterior*) surface, **diaphragmatic** (or *inferior*) surface, **base** (or *posterior surface*), and *pulmonary* or *left* surface. The anterior surface is overlapped by the lungs (particularly the left lung) as their thin anterior borders try unsuccessfully to meet one another in front of it. The name diaphragmatic surface is self-explanatory. The posterior surface is well-named as such, but its official name, *base*, is confusing unless one realizes that it is 'opposite' the *apex* of the cone-shaped heart; it consists almost entirely of the left atrium. The full and rounded left surface, consisting entirely of left ventricle, buries itself, as it were, in a marked concavity in the medial surface of the left lung.

The surface groove that marks the separation of atria from ventricles is the **coronary** (or **atrioventricular**) **sulcus;** it is more vertically than horizontally disposed. The surface grooves (one on the anterior surface, the other on the inferior surface)

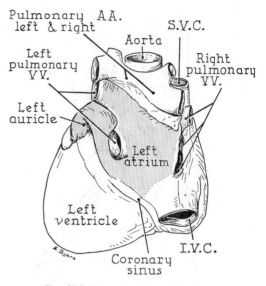

Fig. **476**. Posterior view of heart

that mark the position of the thick muscular **interventricular septum** are the **anterior** and the **posterior** (really **inferior**) **interventricular sulcus** (figs. 475, 476).

When the heart contracts it straightens, as it were, on the great vessels from which it hangs, until the apex of the heart taps the chest wall at a spot in the fifth left intercostal space some 9 to 10 cm (3½") from the midline. The effect of this tapping is, in the living, visible on the exposed chest wall, except when the apex strikes the fifth rib rather than the muscles of the space.

Right Atrium

The right atrium receives the venous blood returning from all parts of the body except the lungs. If the chamber is opened from in front (fig. 478), a large vein, the **superior vena cava,** will be seen entering it vertically from above; a still larger one, the **inferior vena cava,** will be seen entering it vertically from below. The superior vena cava returns blood from everything above the diaphragm (lungs excepted), the inferior vena cava from everything below the diaphragm. The two veins are in line with one another, and their right borders are in line and are continuous with the right border of the heart. The posterior wall of the chamber between the veins is smooth

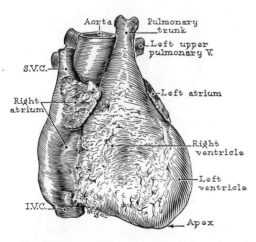

Fig. **475**. Normal heart, unopened, front view.

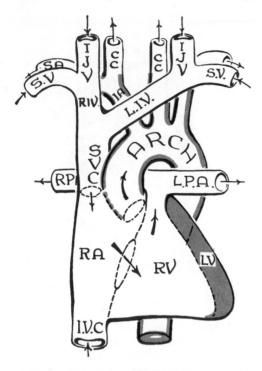

Fig. **477**. Course of blood indicated by arrows. I. V. C. = inferior vena cava; S. V. C. = superior vena cava; R. A. = right atrium; R. V. = right ventricle; R. P. A. and L. P. A. = right and left pulmonary arteries; L. V. = left ventricle, R. I. V. and L. I. V. = right and left innominate (brachiocephalic) veins; S. A. and S. V. = subclavian artery and vein; I. J. V. = internal jugular vein; C. C. = common carotid artery; I. A. = innominate artery (brachiocephalic trunk).

and gives the appearance of being a continuation of the veins, which indeed, developmentally, it is.

The superior vena cava has no valve, but the inferior vena cava has, on its left, a narrow crescentic fold of endothelium, inappropriately called its *valve*. This fold sweeps upwards to outline an oval depression in the posteromedial (interatrial) wall of the chamber known as the **fossa ovalis,** whose significance is appreciated only when the circulation before birth is understood.

In the fetus, the inferior vena cava returned blood not only from the lower half of the body but also from the placenta (page 10) and, since this placental blood contained oxygen and food products necessary to the life of the fetus, it had to be transferred at once to the left side of the heart for delivery to all parts of the body. It could not be pumped to the lungs, since before birth they were not functioning as respiratory organs. The fossa ovalis indicates the site of the fetal **foramen ovale,** and through this foramen the blood from the inferior vena cava passed directly to the left atrium (figs. 479, 480). It closes at birth and then the blood passes through the atrioventricular opening into the right ventricle and is pumped to the lungs.

If a foramen ovale remains open it allows unoxygenated blood to mingle with oxygenated blood and one of several types of 'blue baby' results. The severity of the condition depends on the size of the persisting foramen; when large it is incompatible with life.

To the left of the opening of the inferior vena cava, the **coronary sinus** (a 'vein') opens into the right atrium. It has the caliber of a pencil and it returns blood from the walls of the heart itself which, of course, needs a blood supply. The mouth of

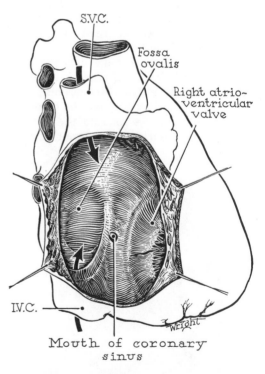

Fig. **478**. Interior of right atrium in right anterior view. Arrows in superior and inferior venae cavae.

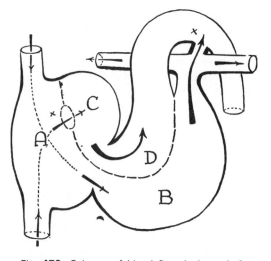

Fig. **479**. Scheme of blood flow in heart before birth. Non-functioning lungs are by-passed at two places marked x. The first is an oval foramen between right atrium (A) and left atrium; the second is a large arterial duct between pulmonary artery and aortic arch. Both close at birth or shortly after (cf., fig. 480).

orifice which is toward the back and right of the ventricle. The blood, pumped by the ventricle into the **pulmonary trunk** which lies above and in front, has to take a U-shaped course between entrance and exit. The chamber itself is crescentic in cross-section since, to the left, the muscular **interventricular septum** bulges into the right ventricle and reduces the size of the chamber (see fig. 481).

The walls of the chamber are rough and trabeculated because of the existence of many coarse bundles of muscle fibers which are in the form of ridges and bridges known as the **trabeculae carneae** (L. = little beams or rafters of flesh). There also project from the walls nipple-like **papillary muscles** to whose apices are attached delicate tendinous cords **(chordae tendineae);** these resemble the cords of a parachute and are attached at their other ends

the coronary sinus is more or less guarded by a flap-valve which is closed when the atrium contracts.

The **right atrioventricular** or **tricuspid orifice** lies anteriomedially and is large enough to admit the tips of three fingers. Through it the blood is pumped to the right ventricle. The tricuspid **valve** will be described with the right ventricle.

If the anterior wall of the right atrium is inspected, a vertical muscular ridge, the **crista terminalis,** will be seen to pass from the front of the superior vena cava to the front of the inferior. To the left of the ridge the wall is rough. The site of the ridge is indicated on the outer surface of the atrium by a groove, the **sulcus terminalis.** Parallel muscle ridges run from the crista transversely to the left, the whole having the appearance of a comb (hence the name **pectinate muscles**). The appendage at the extreme upper left corner of the appendage at the extreme upper left corner of the chamber is known as the **auricle** of the right atrium because of a fancied resemblance to a dog's ear; the pectinate muscles extend into it. The auricle is shown but not labeled in figures 475, 478, and 481 (in which it is retracted by a loop of string).

Right Ventricle

The right ventricle receives blood from the right atrium through the tricuspid

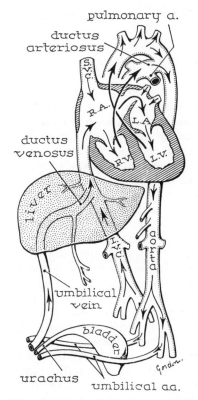

Fig. **480**. Circulation of blood in fetus. Arrows indicate direction of main blood flow. Note arrow passing from right to left atrium through (unlabeled) foramen ovale (cf., fig. 479).

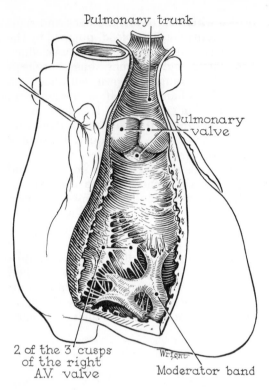

Pulmonary trunk

Pulmonary
valve

2 of the 3 cusps
of the right
A.V. valve

Moderator band

Fig. **481**. Interior of right ventricle and pulmonary trunk.

to the margins of delicate and transparent endothelial flaps known as **cusps** and comparable to the silk of a parachute. The cusps are attached by their bases to the fibrous ring that surrounds the tricuspid orifice. When the ventricle contracts, the cusps come together and occlude the opening; the contraction of the papillary muscles tenses the cords and prevents the cusps from suffering the same fate as that suffered by an umbrella blown inside out on a windy day. Such a catastrophe occurs only very rarely.

Below the opening of the pulmonary trunk, the ventricular wall is smooth. Because it is shaped like a funnel, it is known as the **infundibulum** or **conus.** The entrance to the pulmonary trunk is guarded by the **valve of the pulmonary trunk** (*pulmonary valve*) which consists of three semilunar valvules or cusps whose bases are fixed to the fibrous ring at the base of the artery. These valvules are the walls of little pockets between them and the vessel

wall. As the blood rushes out between the valvules, it compresses them against the wall of the vessel so that they offer no obstruction to the stream. When the ventricle relaxes, blood, attempting to re-enter the ventricle, fills the pockets behind the valvules and forces them together until they completely occlude the exit (figs. 481–483).

Left Atrium

The left atrium receives oxygenated blood from the lungs via the **pulmonary veins.** These veins, therefore, are unlike veins in general in that they contain oxygenated blood. As a rule, four pulmonary veins enter the left atrium, two from each lung. The left atrium, which presents few features of interest, is essentially a chamber resulting from the confluence of pulmonary veins. To the left, lies a little **auricle** which 'peeks' around the left border of the heart (fig. 475); it is the only part whose interior is not smooth. Since the left atrium (almost alone) forms the posterior surface (or base) of the heart, it opens forward and into the back of the left ventricle. The opening and its double-flapped valve are known as **left atrioventricular** or **mitral orifice** and **valve.** The name *mitral* was

Fig. **482**. Pulmonary valve with its three semilunar valvules closed, from above.

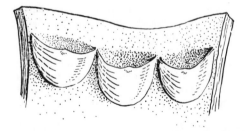

Fig. **483**. Pulmonary valve consists of three semilunar cusps or valvules. Here the artery is split open longitudinally and laid flat.

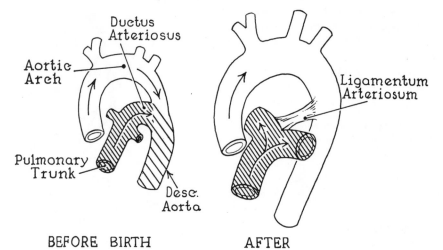

Fig. **484**. After birth the ductus arteriosus, through which blood from the right ventricle bipasses the lungs in the fetus, closes off and turns into a ligament.

given to this valve because the two cusps together suggested the shape of a bishop's miter.

Left Ventricle

The left ventricle is thick-walled and is circular in cross-section. As in the right ventricle, the blood must take a U-shaped course between the entrance behind and the exit above. In its other features too, the left ventricle resembles the right. It is similarly equipped with muscular ridges and bridges; it possesses papillary muscles, tendinous cords, and cusps. The cusps are two in number (hence the obsolete name 'bicuspid') and their bases are attached to the fibrous ring surrounding the mitral orifice—the entrance to the chamber.

The **aorta** is the great artery leaving the left ventricle above, and it lies in close apposition with the right side of the pulmonary trunk and somewhat behind it. It is guarded at its entrance by the **aortic valve,** exactly similar to the valve of the pulmonary trunk (fig. 483).

Pulmonary Trunk and Arteries

As the pulmonary trunk leaves the heart, it spirals up the left side of the aorta until it divides in a T-shaped manner in the concavity of the aortic arch (fig. 477). The limbs of the T are known as the **right** and

left pulmonary arteries and they run horizontally, at the summit of the heart and across the fronts of the bronchi, to their respective lungs. They lie above the level of the pulmonary veins. The right pulmonary artery passes below the aortic arch and behind the termination of the superior vena cava.

Pulmonary Circulation. Each pulmonary artery, upon entering the hilum of its lung, branches in company with the bronchial tree and provides companion arteries for each bronchus and bronchiole. Just as the bronchioles finally end as alveoli the arterioles end as capillary networks which enclose the air spaces. From these capillaries the return flow is by a venous 'tree'—the veins finally forming into two large pulmonary veins for each lung, and these carry oxygen-laden blood to the left atrium.

The left pulmonary artery is connected to the posterior end of the aortic arch by a short but stout fibrous cord known as the **ligamentum arteriosum** (arterial ligament). **Before birth,** the ligament was an arterial duct (fig. 484); its significance lay in the fact that any blood that failed to pass from the right atrium through the foramen ovale to the left atrium entered the right ventricle; but, instead of going via the pulmonary artery to the non-functioning lung, it short-circuited the lung by passing out of the pulmonary artery through the **ductus arteriosus (arterial duct)** and so into the aorta. It was the superior vena caval blood especially that took this course (fig. 480). At birth, the ductus arteriosus becomes occluded and, in the adult, is recognized as the ligament.

Persistence of the ductus arteriosus (producing another type of 'blue baby') is an abnormality which, fortunately, can be remedied surgically. Various groups of developmental abnormalities also occur *within* the heart (e.g., improper development of septa and of valves) which are now amenable to surgical repair.

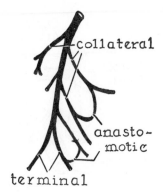

Fig. **485**. Types of branches of an artery

SYSTEMIC ARTERIES

General

Before the systemic arteries are described, it is well to observe that by far the most important duty falling to the lot of the arteries is to supply blood to the muscles. It is, therefore, perhaps a little unfortunate that these important vessels are so often dismissed in a rather cavalier fashion as **'muscular branches.'** It should be remembered that the chief duty of many *named* vessels is also to supply muscles, the name merely giving an indication—often in somewhat vague terms—of where the vessel is to be found. It is never the name that is important, it is the territory or field of supply. The intelligent student will work out these fields of supply for himself, by regarding each artery as he meets it as an irrigation system flooding with blood the surrounding territory. Since fields of supply overlap, one must expect any given artery to **anastomose** (join) with others by communicating branches (fig. 485). An anastomosis provides a secondary channel along which blood may pass to an area. The chief value of this is that temporary obstruction of an artery by 'kinking' or pressure does not necessarily deprive the area beyond of its blood supply. Therefore, the biggest and most important anastomotic arteries must be located where there is the greatest mobility and danger of obstruction, e.g., arteries to the intestines and arteries crossing major joints.

While, in general, it is the named branches that will be chiefly described, it will be taken for granted that the student is familiar with the fact that **the (often omitted) unnamed muscular branches are important;** sometimes they will be referred to in order that their existence may not be forgotten.

Branches arising from any artery before its termination are called **collateral branches;** when an artery ends by dividing into two it is said to **bifurcate** and the resulting two arteries are called the **terminal branches.**

AORTA

The aorta (fig. 486) is the great arterial trunk issuing from the upper end of the left ventricle; it has the diameter of a garden hose (about 3 cm) and the shape of a curved-handled cane, the long 'shaft' running down on the vertebral column to the level of the fourth lumbar vertebra where it bifurcates into its terminal branches. On its way it has many important collateral branches. Because it is very important, the aorta is divided along its length into four parts. These will be described now in general terms. Their branches will be discussed later, area by area.

Beyond the aortic valve the first 5 cm (2″) of the aorta run upward and to the right within the pericardial sac and are known as the **ascending aorta.** This ascending aorta has the pulmonary artery in front and to the left of it, and is itself in front and to the left of the superior vena cava; thus these three great vessels are 'in echelon.'

As the aorta leaves the pericardial sac it bends to the left and arches backward in contact with the left side of the end of the trachea and over the root of the left lung. This part is known as the **arch of the aorta** and it ends when the vessel reaches the left side of the fourth or fifth thoracic vertebral body; the arch, therefore, begins and ends at the same horizontal level (see fig. 487).

The third part of the aorta, known as the **descending aorta** (*descending thoracic aorta*), is about 20 cm (8″) long and lies in bare contact with the left sides of thoracic vertebral bodies five to 12; it may, by its contact, actually flatten the sides of these bodies.

Leaving the thorax between the two crura of the diaphragm, the aorta—now

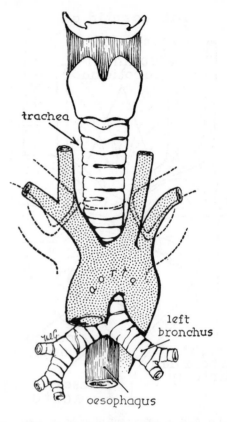

Fig. **487**. Aorta reaches vertebral column by arching backward over root of left bronchus and in contact with left side of trachea and esophagus. From summit three large branches arise in quick succession—brachiocephalic, left common carotid, left subclavian.

known as the **abdominal aorta**—lies almost in the midline and immediately in front of the first four lumbar vertebral bodies. Its terminal branches resulting from its Λ-shaped bifurcation are the **right** and the **left common iliac artery;** the division occurs on the fourth lumbar vertebra and at a level about 2 cm (¾″) below the umbilical plane. The aorta, as might be expected, gets progressively a little smaller in caliber from above downward.

Development. See figure 488.

BRANCHES OF AORTA

The branches of each of the four parts of the aorta—**ascending, arch, descending** and **abdominal**—will be described in turn.

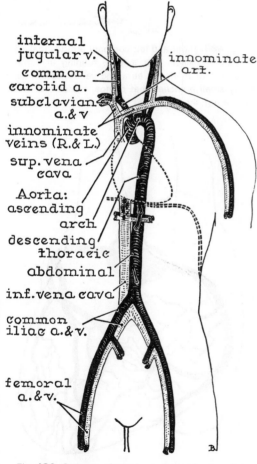

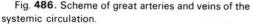

Fig. **486**. Scheme of great arteries and veins of the systemic circulation.

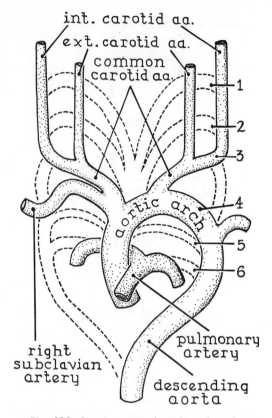

int. carotid aa.

ext. carotid aa.

common carotid aa.

1

2

3

aortic arch

4

5

6

right subclavian artery

pulmonary artery

descending aorta

Fig. **488**. Development of aortic arch and great vessels. Fleetingly in the embryo there are six bilateral arches. From the left fourth, the aortic arch develops; from the right fourth the brachiocephalic and right subclavian arteries; from the left sixth, the pulmonary trunk and ductus arteriosus (which later becomes the ligamentum arteriosum).

BRANCHES OF ASCENDING AORTA

Coronary Arteries

At the very beginning of the ascending aorta, where lie the three semilunar valvules of the aortic valve, the wall of the vessel is dilated in the form of three bulbous swellings known as the **aortic sinuses** (sinus, L. = a hollow). In the arterial wall of two of these sinuses (and therefore under shelter of a valvule) is the mouth of a **coronary artery**—a right and a left. Blood can enter these arteries only when the left ventricle is relaxed and the valvules do not shelter the mouths.

The **coronary arteries** are the only branches of the ascending aorta but they are of great importance since they supply blood to the muscle of the heart. Fortunately, they anastomose liberally with each other. The word coronary means a crown and an arterial crown encircles the heart in the atrioventricular sulcus. From this crown or circle, an arterial loop runs in the interventricular sulci. Reference to figure 489 will make obvious this arrangement and the distributions of the two vessels.

Scheme

Fig. **489**. Right and left coronary arteries make an anastomotic circle and a loop, running in surface grooves that mark out atria and ventricles of heart. Many small branches that plunge into cardiac muscle are not shown (cf., fig. 490).

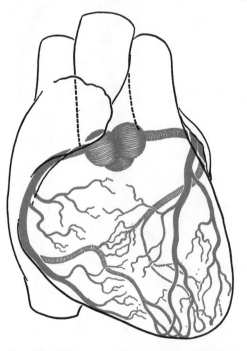

Fig. **490**. Right and left coronary arteries (semi-schematic).

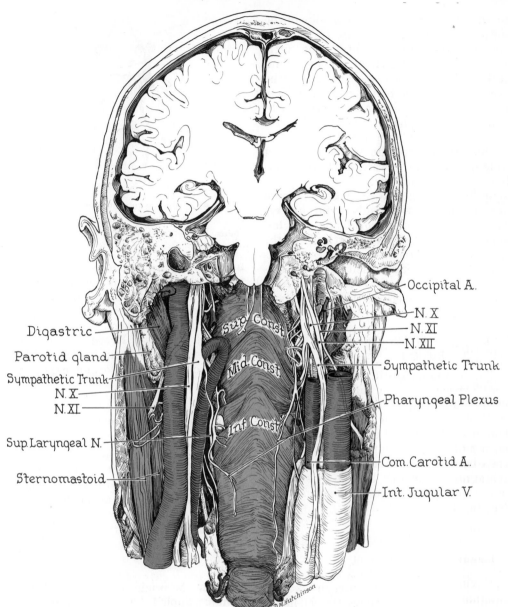

Fig. **491**. Great vessels and nerves in carotid sheath (from behind) on either side of pharynx.

From the arterial circle, branches run over all parts of the surface of the heart, both atria and ventricles (see fig. 490).

From the loop, branches, arranged like the steps of a ladder, plunge into the thick interventricular septum.

Thrombosis (or blood clot) in a coronary artery is a frequent cause of sudden heart failure and death.

BRANCHES OF AORTIC ARCH

From the summit of the arch of the aorta three large vessels arise in rapid succession. In order from front to back they are: (1) the **brachiocephalic trunk** (still widely called the *innominate artery*); (2) the **left common carotid artery;** (3) the **left sub-clavian artery.** The innominate (L. =

unnamed) artery is to the right side of the body what the other two are to the left. Together, the three branches of the aortic arch supply the region of the head and neck, and the upper limbs. That is why the new term *brachiocephalic trunk* is widely accepted for the innominate artery (brachium = arm; cephale = head). The student should know both names.

The **brachiocephalic trunk** is a short wide stem passing upward and to the right. It ends behind the right sternoclavicular joint by dividing into the **right common carotid** and the **right subclavian artery.** The **left common carotid** and the **left subclavian artery** ascend vertically on the left side of the trachea and the esophagus until they lie behind the left sternoclavicular joint. Thereafter, the courses and distributions of the vessels are the same on both sides.

CAROTID ARTERIES; THEIR BRANCHES AND DISTRIBUTION

The **common carotid artery** mounts the neck vertically on the lateral side of the trachea and esophagus; it has no collateral branches. At the level of the upper border of the thyroid cartilage, the artery ends by bifurcating into the **internal** and the **external carotid artery.** The two terminal arteries at first lie side by side and only gradually diverge from one another (see fig. 493).

Internal Carotid Artery

Having no branches in the neck, the internal carotid artery runs up the side of the pharynx (fig. 491) to enter the skull by running through a sinuous tunnel in the petrous part of the temporal bone. It enters the middle cranial fossa at the apex of the petrous bone at the side of the body of the sphenoid. Running forward to the anterior clinoid process, it makes a hair-pin bend upward and backward to pierce the membranes (or *meninges*) that surround the brain (fig. 492). Here the internal carotid artery gives off a tiny, but vitally important, branch before ending by dividing into two large branches to the brain.

The tiny branch is the **ophthalmic**

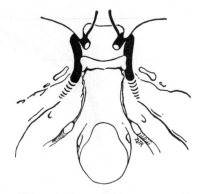

Fig. **492**. Hair-pin turn of internal carotid arteries in middle cranial fossa just before piercing dura mater. Note ophthalmic branches.

artery which accompanies the optic nerve to the orbital cavity and supplies the eye.

The larger branches, **anterior** and **middle cerebral arteries**, supply most of the cerebral hemisphere.

Blood Supply of Brain. An account of the arteries to the brain is given with the discussion of that important organ (page 309).

External Carotid Artery

The external carotid artery supplies almost everything in the neck and head regions except the contents of the cranial and orbital cavities. However, since it begins relatively high up in the neck, the external carotid artery calls, as it were, upon the subclavian artery to look after the lower part of the neck as that vessel crosses the root of the neck on its way to the upper limb.

From its origin (at the level of the 'Adam's apple') the external carotid artery passes up the neck to the region of the temporomandibular joint where, at the neck of the mandible, it ends by dividing into its two terminal branches, the superficial temporal artery and the maxillary artery. It will suffice to present a list of the branches of the external carotid artery and to note briefly the principal distribution of each (fig. 493).

1. **Superior thyroid artery.** Arising *just below* the greater horn of the hyoid bone, it descends medially to supply

the upper part of the thyroid gland and to give a branch to the larynx.

2. **Lingual artery.** Arising *at* the greater horn of the hyoid bone, it runs horizontally forward to enter the oral cavity for the supply of the tongue.

3. **Facial artery.** Arising *just above* the greater horn of the hyoid bone, it runs, somewhat tortuously, upward and medially to cross the lower border of the mandible two fingers'-breadths in front of the angle, where it can be felt pulsating. Entering the face, the artery is directed toward the medial angle of the eye. It supplies the facial region.

4. **Occipital artery.**

5. **Posterior auricular artery.** These two vessels arise from the back of the parent stem and supply the structures of the scalp behind the ear.

6. **Ascending pharyngeal.** This is a slender artery running up the wall of the pharynx to supply it.

7. **Superficial temporal artery.** This

enters the scalp in front of the ear and can be seen in elderly people as a tortuous vessel on the side of the head.

8. **Maxillary artery.** This is the larger of the two terminal branches and it passes deep to the neck of the mandible to enter the infratemporal (i.e., the masticatory) region.

The numerous branches of the maxillary artery supply the following structures.

a. The external acoustic meatus and the outer surface of the eardrum.

b. The teeth. An **inferior alveolar** (*dental*) **artery** enters a canal in the mandible; **superior alveolar** (*dental*) **arteries** enter minute canals in the maxilla.

c. The meninges or coverings of the brain. The important **middle meningeal artery** enters the cranial cavity by passing upward through the foramen spinosum situated just lateral to the foramen ovale. Once this artery has entered the skull it and its branches (not seen in fig. 493) run in grooves on the deep surface of the brain-case deep to the area of origin of Temporalis muscle. A severe blow in the temporal region may rupture the middle meningeal artery in its bony groove and the resulting hemorrhage or blood clot requires emergency surgical treatment.

d. The muscles of mastication by many **muscular branches.**

e. The mucous membrane of the nose, pharynx, and palate. Having passed into the pterygomaxillary fissure and into the deeply placed pterygopalatine fossa, these last branches accompany the several branches of the maxillary nerve.

SUBCLAVIAN ARTERY; BRANCHES AND DISTRIBUTION

The subclavian artery is the vessel of the upper limb or, rather, this is the name given to the first part of the continuous arterial channel which reaches the elbow before bifurcating into terminal branches. This main channel changes its name when it enters the axilla (axillary artery) and when it enters the arm (brachial artery) (fig. 494).

The first part of the channel, the **subclavian artery** (fig. 495) after passing (left side) or beginning (right side) behind the sternoclavicular joint, arches lateralward and upward at the root of the neck in front of the apex of the lung and behind the Scalenus Anterior. Here, it crosses the

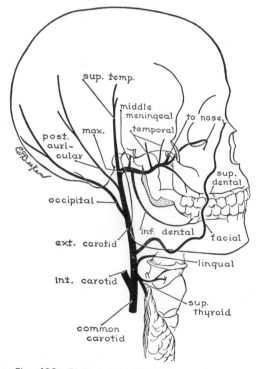

Fig. **493**. Right external carotid artery and its distribution. Middle meningeal artery enters cranium.

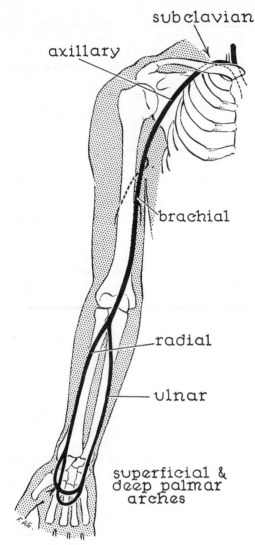

Fig. **494**. Main arterial channels of the upper limb (branches omitted for clarity).

costal cartilages and one finger's-breadth away from the lateral border of the sternum. As it crosses successively the anterior ends of the upper six intercostal spaces, it gives small branches to each space, the **anterior intercostal arteries.** The internal thoracic divides when it reaches the lower end of the sternum. One terminal branch, the **musculophrenic artery,** courses along the costal origin of the diaphragm, supplying that muscle and the lower intercostal spaces. The other terminal branch, the **superior epigastric artery,** passes near the xiphoid process into the abdominal wall behind the Rectus Abdominis.

2. **The vertebral artery** ascends the neck by entering the foramen in the transverse process of the sixth cervical vertebra; it threads each succeeding transverse process in turn and finally enters the foramen magnum on the side of the brain stem. It next moves to the front of the brain stem where it meets its fellow to form with it a single midline trunk, the **basilar artery.** The vertebral arteries supply branches to the upper part of the spinal cord (the aortic intercostals supply the lower part) and they (and the basilar) supply the brain stem and

upper surface of the first rib and then runs beneath the clavicle.

The distribution of the branches—with, perhaps, the partial exception of one—is *not* to the upper limb but to areas that, in development, were closely related to the upper limb bud.

Three of the four branches of the subclavian artery are clustered near the summit of its arch (fig. 495).

1. **The internal thoracic (int. mammary) artery** descends vertically into the thorax where it is in contact with

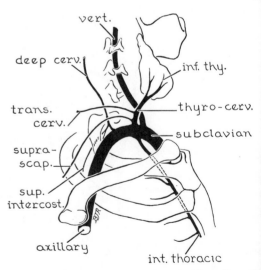

Fig. **495**. Right subclavian artery and its collateral branches.

the cerebellum, and the posterior pole of the cerebrum. The contribution of the vertebral arteries to the *arterial circle* is shown in figure 557 on page 309.

3. The thyrocervical trunk is a short stem dividing into three branches.

a. The **inferior thyroid artery,** running a curved path, upward and medialward to supply the lower half of the thyroid gland.

b. The **transverse artery of the neck,** running from front to back round the root of the neck to reach the vertebral border of the scapula at its end.

c. The **suprascapular artery,** running lateralward behind the clavicle to (and above) the upper border of the scapula near the coracoid process. It and (b) aid in the supply of the many muscles clothing the scapula. They are shown in figure 495.

4. The costocervical trunk provides the 'aortic' intercostal arteries for the first and second intercostal spaces, the aorta itself not reaching high enough to provide vessels for these spaces. It provides also a branch, the *deep cervical artery*, which passes up the neck behind the cervical transverse processes. This branch helps to supply muscles at the back of the neck.

ARTERIES OF UPPER LIMB

AXILLARY ARTERY; COURSE AND BRANCHES

The **axilla** (page 52 and fig. 241) is a pyramidal space filled with loose areolar tissue. Its apex or entrance from the neck lies behind the middle of the clavicle and its base is the skin of the armpit. It possesses three muscular walls: (1) an anterior, consisting of the pectoral muscles in the same coronal plane as the clavicle and coracoid process; (2) a medial, consisting of the Serratus Anterior clothing the upper ribs; (3) a posterior, consisting of muscles clothing the front of the scapula. At the bicipital sulcus on the humerus the anterior wall meets the posterior.

The axillary artery traverses the axilla surrounded by the bundle of nerves of the brachial plexus and two or three accompanying veins (venae comitantes). It and its companions are wrapped close together

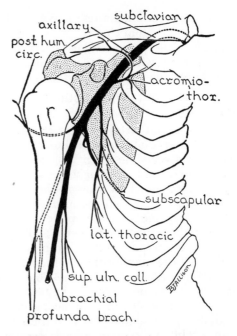

Fig. **496**. Axillary artery (right) becomes brachial artery when it reaches humerus. Profunda brachii is largest collateral branch of brachial artery.

in a sheath of fibrous tissue known as the *axillary sheath.* Because this sheath and a large quantity of fat in the axilla are both removed in a dissection, the student must beware of getting a false picture of the relationships.

The axillary artery provides a branch for each wall (fig. 496).

1. The thoraco-acromial (acromio-thoracic) artery arises just below the clavicle and supplies the anterior wall by radiating branches.

2. The lateral thoracic artery descends on the side of the thorax and supplies the medial wall. A large branch supplies the female breast.

3. The subscapular artery, the largest branch, descends along the axillary border of the scapula and supplies the posterior wall, sending a large branch (*circumflex scapular*) to the dorsal surface of the scapula.

4. The posterior humeral circumflex artery runs backward below the capsule of the shoulder joint and encircles the neck of the humerus with the axillary (or circumflex) nerve to supply the Deltoid chiefly.

The branches of the axillary artery anastomose freely with one another and with branches of the subclavian artery.

BRACHIAL ARTERY; COURSE AND BRANCHES

The brachial artery enters the upper arm on the medial side of the humerus and in close association with the principal nerves of the limb. Throughout its course, it is accompanied by the median nerve running in a groove on the medial side of the Biceps. It moves to the front of the arm at the midlength of the humerus and, from there to its termination, it occupies the midline of the limb, resting on the Brachialis behind. At the bend of the elbow, the brachial artery lies between the tendon of the Biceps laterally and the median nerve medially. Immediately distal to the elbow joint, it bifurcates into its two terminal branches, the **radial artery** and the **ulnar artery,** which will be described shortly (figs. 496, 497).

In the upper part of the arm the radial nerve lies behind the brachial artery and the ulnar nerve lies to its medial side. When these nerves diverge from the artery each is accompanied by a branch.

1. The **profunda brachii artery**—the largest collateral branch of the brachial—arises high up in the limb and runs with the radial nerve behind the humerus. It is the blood supply for the Triceps. Terminal branches run dowhward in front of and behind the lateral epicondyle and anastomose freely with arteries ascending from the forearm. This *collateral circulation* is discussed below.

2. The long and thread-like **superior ulnar collateral artery** runs with the ulnar nerve down the back of the medial side of the arm and behind the medial epicondyle; it contributes to the supply of the flexor muscles arising from that epicondyle.

3. A small **inferior ulnar collateral** (*supratrochlear*) **artery** arises from the brachial a little above the elbow and also runs to the flexor muscles.

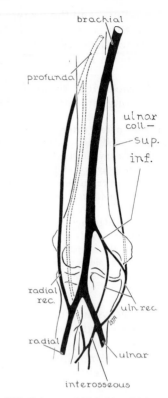

Fig. **497**. Scheme of (right) brachial artery and the anastomoses around elbow.

Collateral anastomoses around the elbow joint are relatively rich and important because they provide by-passes for the blood if the brachial artery is occluded at the elbow. They occur between the three branches of the brachial artery described above and recurrent branches (direct or indirect) of the radial and ulnar arteries (fig. 497).

ARTERIES OF THE FOREARM AND HAND

The radial artery runs down the lateral (radial) side of the front of the forearm, supplying the muscles along its course. It is sheltered by the border of the Brachioradialis but is otherwise close to the surface. It reaches the wrist at the base of the thumb where its pulsation is readily felt. Leaving the radius, it runs downward and backward through the 'snuff-box' to the proximal end of the space between metacarpals 1 and 2. Plunging through the first Dorsal Interosseus which fills this space, it enters the palm very deeply. It is principally concerned with supplying blood to the thumb and index finger, and in the formation of the *deep palmar arch* (fig. 498).

The ulnar artery makes its way deeply to the medial (ulnar) side of the front of the forearm. It gradually becomes more superficial until, in the lower half of the forearm, it is sheltered only by the tendon of the Flexor Carpi Ulnaris. It enters the hand superficially, just lateral to the pisiform bone, and protected only by skin and fascia. It is principally concerned in the formation of the *superficial palmar arch* (see below and fig. 498).

Near its origin, the ulnar artery gives off a short branch, the **common interosseous artery,** which splits at once into an **anterior** and a **posterior interosseous artery** and these descend, one on the front, the other on the back, of the interosseous membrane (fig. 497). The anterior one supplies the deep flexor muscles; the posterior one supplies the extensor muscles on the back of the forearm.

The superficial palmar arch is the larger of the two arches and it runs across

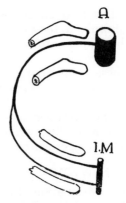

Fig. **499**. Scheme of arteries in an intercostal space. Those from aorta (A) anastomose with those from internal thoracic or mammary (I. M.) (see fig. 494).

the palm, at the level of the outstretched thumb, protected only by skin and palmar aponeurosis. It is derived from the ulnar artery and is completed laterally by a branch from the radial artery (fig. 498).

The deep palmar arch is a little proximal to the superficial one but lies on a much deeper plane; it runs across the palm just distal to the bases of the metacarpal bones. It is derived from the radial artery and is completed medially by a deep branch of the ulnar artery (fig. 498).

Both arches provide **palmar branches** that run forward and unite at the clefts between the fingers. Each common stem thus formed then divides to provide **digital arteries** to the contiguous sides of the fingers. Thus each finger has a (palmar) digital artery running down on each side of the fibrous flexor sheath that encloses the flexor tendons. These arteries end by forming a network of small vessels at the finger tips. The scheme of the blood supply of the hand is shown in figure 498.

BRANCHES OF DESCENDING AORTA

Eleven pairs of arteries arise from the back of the descending (thoracic) aorta. The first nine of these are **posterior intercostal arteries** for spaces three to 11; they anastomose with anterior intercostal arteries (fig. 499). Because the tenth pair lies below the 12th rib, each must be called a **subcostal artery** rather than an intercos-

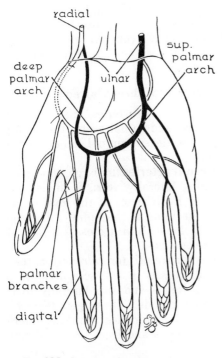

Fig. **498**. Scheme of arteries of hand

tal artery. The 11th pair, known as the **superior phrenic arteries,** supplies the back of the upper surface of the diaphragm.

Besides these, there arise a considerable number of small vessels for the adjacent viscera. These are a pair of **bronchial arteries** for the nourishment of the bronchial tree, and many little branches for the esophagus and the pericardium.

BRANCHES OF ABDOMINAL AORTA

The direct continuation of the descending thoracic aorta, the abdominal aorta, bifurcates at its lower end into its terminal branches, the **common iliac arteries** (which will be discussed later). During development, the aorta supplied: (1) paired branches to the walls of the abdomen; (2) paired branches to the three paired glands—suprarenal (adrenal) glands, kidneys, and testes (or ovaries); and (3) unpaired arteries which ran directly forward to the longitudinally dis-

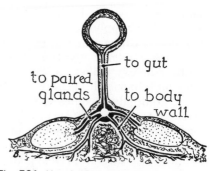

Fig. **501**. Unpaired branches of aorta to digestive tract come from front of aorta and run forward; paired branches to paired organs come off sides; paired arteries to wall come off farthest back and always lie behind others.

posed digestive tract. This basic fetal arrangement persists in the adult with minor modifications (see figs. 500, 501).

1. Branches to Walls of Abdomen

Five pairs of arteries arise from the abdominal aorta. The first pair, the **inferior phrenic arteries,** supply the inferior surface of the diaphragm. The remaining four pairs, the **lumbar arteries,** supply the posterior abdominal wall. All of the posterior intercostal arteries and several lumbar arteries send branches into the intervertebral foramina and contribute to the supply of the spinal cord and its nerve roots and coverings.

2. Branches to Three Paired Glands

Three pairs of arteries arise from the *sides* of the abdominal aorta; they supply the kidneys, the suprarenal glands, and the sex glands.

The **renal arteries** are large, short stems proceeding laterally at right angles from the aorta. Each arises at the level of, and enters, the hilus of the kidney. The right one is the longer and it crosses behind the inferior vena cava.

The **suprarenal** (or *adrenal*) **arteries** are small vessels that arise a little higher than the renals and run laterally to the suprarenal glands. These glands receive blood also by branches from the inferior phrenic arteries and from the renal arteries.

The **testicular arteries (ovarian** in fe-

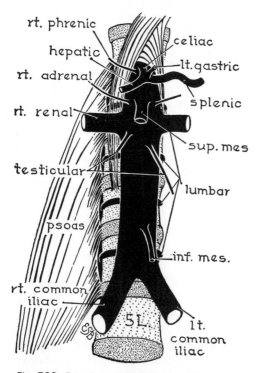

Fig. **500**. Branches of abdominal aorta: five pairs to wall; three pairs to paired glands; three unpaired to digestive tract; one pair—terminal (common iliac).

male) arise from the aorta a little below the renal arteries. Each is a long tenuous vessel that had to accompany its migrating organ while still arising near the site where that organ originally lay. Each descends obliquely downward and lateralward behind the peritoneum of the posterior abdominal wall.

The **testicular artery** enters the deep inguinal ring and, as a constituent of the spermatic cord, traverses the anterior abdominal wall. It enters the scrotum and supplies the testis.

The **ovarian artery** has not so far to go. It reaches the ovary situated on the side wall of the pelvic cavity and, after supplying the ovary, passes medialward beneath the uterine tube (which it supplies) and so reaches the side of the uterus. It anastomoses with the **uterine artery** to be described shortly.

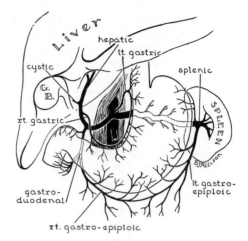

Fig. **503**. Scheme of celiac artery and its branches (see text).

3. Branches to Digestive Tract and Unpaired Organs

Three important and unpaired arteries arise from the *front* of the abdominal aorta and are devoted to the supply of the digestive tract, its accessory organs, and the spleen. They are: (1) the **celiac trunk** (celiac *artery*); (2) the **superior mesenteric artery**; (3) the **inferior mesenteric artery.**

Celiac Trunk and Branches

The celiac (coeliac) artery (figs. 502, 503) is a very short trunk with a large diameter, resembling the hub of a wheel from which three spokes radiate. It arises from the aorta as soon as that vessel has emerged from the thorax and it is straddled by the meeting of the two crura of the diaphragm. The three 'spokes' of this 'hub' at once diverge from one another. One runs upward and to the left; a second runs horizontally to the left; a third runs to the right.

a. The **left gastric artery** is the smallest; it runs upward to the esophageal end of the stomach, and descends along the lesser curvature. It supplies the lower end of the esophagus and part of the stomach (fig. 503).

b. The **splenic** (or **lienal**) **artery** is the largest; it runs a serpentine course to the left along the upper border of the pancreas to which it gives **pancreatic**

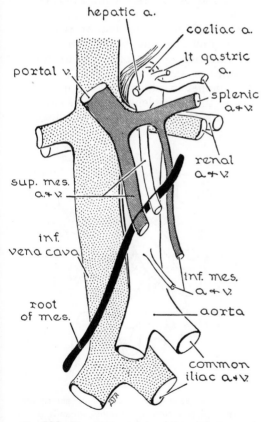

Fig. **502**. The great vessels of the abdomen and their relative positions.

branches. From it, several **short gastric arteries** are given off to the fundus of the stomach, and its several terminal branches enter the spleen at the hilus. A large branch, the **left gastrolienal ligament,** reaches the greater curvature of the stomach (via gastrosplenic omentum) in the vicinity of the spleen; it runs from left to right along the greater curvature of the stomach. Besides supplying the stomach, it supplied branches to the greater omentum.

c. The **common hepatic artery** is intermediate in size; it runs downward and to the right to reach the pyloric end of the stomach. Turning sharply upward at the first part of the duodenum, it runs in the free edge of the lesser omentum to the porta of the liver. It divides into **right** and **left hepatic arteries** which enter the liver for the nourishment of that organ; one of these branches, usually the right hepatic, gives a branch (the **cystic artery**) to the gall bladder.

Near the point where the common hepatic artery turns sharply up, it gives off a small branch, the **right gastric artery,** which supplies the pylorus and meets the left gastric along the lesser curvature of the stomach. In the same vicinity it gives off a large branch, the **gastroduodenal artery,** which, after giving off several direct branches to the duodenum, passes behind the first part of the duodenum. It gains the right extremity of the greater curvature of the stomach along which it courses (as the **right gastro-epiploic artery**) to its junction with the left gastro-epiploic artery. On its way, the gastroduodenal artery supplies a **superior pancreaticoduodenal branch** for the duodenum and the head of the pancreas.

In summary, the celiac artery through its three branches supplies the end of the esophagus, the stomach, duodenum, pancreas, spleen, gall bladder, and liver.

Superior Mesenteric Artery

This large artery (figs. 502, 504) is larger in caliber than a pencil. It arises immediately below the celiac artery where, between the origins of the two arteries, the aorta is crossed by the splenic vein. It then runs from the aorta acutely downward, and gradually curves to the right toward the cecum. From its origin, it runs behind the neck of the pancreas and across the front of the lower part of the head of the pancreas and the front of the third or horizontal part of the duodenum. The artery then enters the root of the mesentery near its upper end (fig. 502).

Branches. *The superior mesenteric artery supplies all of the small intestine (part of the duodenum excepted) and the right half of the large intestine or colon;* its many branches may be divided into **jejunal** and **ileal branches** arising from its left side, and **colic branches** arising from its right side.

a. The **jejunal** and **ileal arteries** are 10 to 15 in number and, between the layers of the mesentery, they form with one another a series of anastomotic loops. These, and the manner in which they succeed in supplying the entire length of the small intestine, are depicted in figure 504, where they

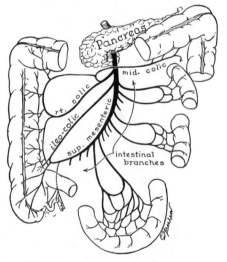

Fig. **504.** Scheme of superior mesenteric artery. and its branches. Note anastomotic arcades. (Intestinal branches = jejunal and ileal arteries.)

ileo-colic a.

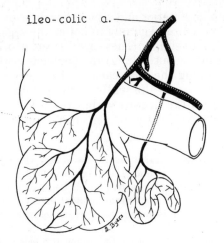

Fig. **505**. Ileocolic artery and its branches to cecum and appendix.

are labeled collectively as 'intestinal branches.'

b. Figure 505 depicts the **ileocolic artery** and its distribution to the termination of the ileum, to the cecum, to the appendix, and to the beginning of the ascending colon.

c. The **right** and **middle colic arteries** supply the ascending colon, hepatic flexure, and transverse colon. They must therefore, arise early from the superior mesenteric artery. They are shown in figure 504 as arising from a common stem—a very frequent occurrence. Just as frequent is a common origin for the ileocolic and right colic branches. At the left colic flexure the middle colic artery anastomoses with the highest branch of the inferior mesenteric artery (fig. 506).

Inferior Mesenteric Artery

The inferior mesenteric artery (fig. 506) is the smallest of the three vessels to the digestive tract. It arises from the front of the aorta about 4 cm (1½″) above the bifurcation; its origin is obscured by the third part of the duodenum which just succeeds in reaching low enough to cover it. The artery runs acutely downward and somewhat to the left; it crosses the left common iliac artery to enter the pelvis. The continuation of the vessel descends

behind the rectum which it supplies and is known as the **superior rectal artery.**

Within the abdomen, the inferior mesenteric artery gives off (*superior*) **left colic** and **sigmoid** (or *lower left colic*) **arteries** whose courses to the splenic flexure, descending colon, and sigmoid (pelvic) colon, are shown in figure 506.

COMMON ILIAC ARTERIES

From the bifurcation of the aorta, each **common iliac artery** takes a direction toward the midpoint of the inguinal ligament. One-third of the way along this path each common iliac artery bifurcates into **internal** and **external iliac arteries** (figs. 507, 508). The external iliac artery continues the line of the common iliac and skirts

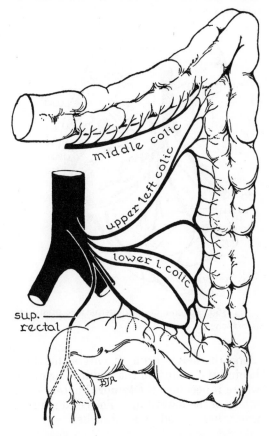

Fig. **506**. Scheme of inferior mesenteric artery. Superior rectal branch supplies most of rectum. Note anastomosis with middle colic branch of superior mesenteric artery at splenic flexure.

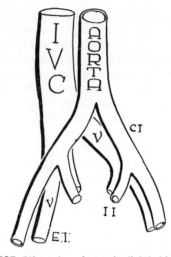

Fig. **507**. Bifurcation of aorta is slightly higher than beginning of inferior vena cava. Note relationships of veins to arteries.

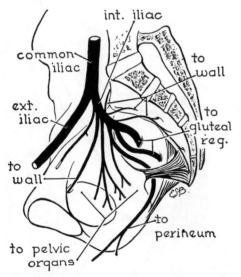

Fig. **508**. Scheme of right internal iliac artery and its branches.

along the pelvic brim. It enters the thigh behind the middle of the inguinal ligament where it becomes the **femoral artery**—the artery of the lower limb. The common iliac artery has no collateral branches.

INTERNAL ILIAC ARTERIES

Each internal iliac artery enters the pelvic cavity in front of the sacro-iliac joint (fig. 508). It is usually described as dividing

into an anterior and a posterior division. It is more useful simply to notice the regions it has to supply. These are: (1) the **gluteal region** or buttock; (2) the **perineal region** including the anal canal and the external genitalia; (3) the **walls of the pelvic cavity;** (4) the **organs** within the pelvic cavity: (a) urinary; (b) generative; (c) digestive.

1. The gluteal region receives two vessels which pass out of the greater sciatic foramen. They are the **superior gluteal artery** for the upper part of the region, and the **inferior gluteal artery** for the lower part, including the upper part of the back of the thigh.

2. The perineal region lies below the pelvic diaphragm, viz., the Levator Ani. The artery of the region is the **internal pudendal artery.** It escapes from the pelvis and enters the perineum by passing around the ischial spine. It is the main vessel for the region and, in consequence, devotes itself particularly to the external genitalia and to the anal canal.

3. The wall of the false pelvis is supplied by the **iliolumbar artery** in series with the lumbar arteries. The wall of the true pelvis is supplied by:
   a. The **lateral sacral artery** running vertically down the sacrum (it is often double).
   b. The **obturator artery** which, after supplying the inside of the lateral wall, passes through the obturator foramen to supply the adductor muscles of the thigh.

4. Small and variable vessels supply the organs within the pelvis. [These, with many companion veins and nerves, and enclosing fascia, form the posterior limit of a potential space which lies between the bladder and the side wall of the pelvis, and is below the peritoneum. This, the *retropubic space* (of Retzius), is continuous from one side to the other behind the symphasis pubis.]
   In the **male,** these arteries are:
   a. **Superior** and **inferior vesical,** for the bladder, prostate gland, seminal vesicle, and vas deferens.
   b. **Middle rectal,** variable vessels to the muscular coats of the rectum.
   In the **female,** in addition:
   c. **Uterine,** a large artery reaching the vagina at its upper part and then running tortuously up the side of the uterus to anastomose with the ovarian artery in the broad ligament.

EXTERNAL ILIAC ARTERIES

The continuation of the common iliac artery, the external iliac (figs. 508, 509) courses along the brim of the true pelvis.

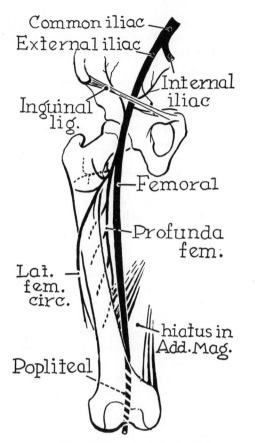

Fig. **509**. Scheme of arteries in thigh

Behind the middle of the inguinal ligament and separated from the head of the femur by the Psoas tendon, each external iliac artery enters the thigh, changing its name to the femoral artery.

Just before it escapes into the thigh it gives off two branches to the anterior abdominal wall.

1. **The deep circumflex iliac artery,** skirting along the inguinal ligament laterally, and supplying the lower and lateral half of the abdominal wall.
2. **The inferior epigastric artery,** running upward and medialward to enter the lower part of the sheath of the Rectus. Within the sheath it anastomoses with the superior epigastric artery.

ARTERIES OF LOWER LIMB

FEMORAL ARTERY

Course

The femoral artery supplies the lower limb. Entering the thigh in the midline in front, it descends vertically until 10 cm (4″) above the knee it meets the femur which descends the thigh obliquely. The artery then passes backward through an opening in the Adductor Magnus and on the medial side of the femur to become the **popliteal artery.** Throughout its femoral course it is entirely medial to the bone and rests behind on the adductor muscles (figs. 509–511).

Branches

Because the chief obligation of the femoral artery is to look after the back of the thigh as well as the front, most of its branches behave as does the main vessel itself, i.e., they perforate the adductor muscles at the insertions of those muscles into the shaft of the femur.

A little below the inguinal ligament the femoral artery provides two small vessels which turn upward (recur) to supply the skin of the lower half of the abdominal wall; it also supplies small vessels to the scrotum or labium majus.

In the middle of the femoral triangle (fig.

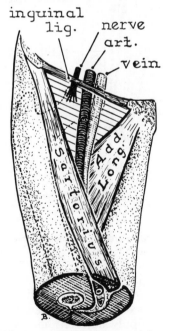

Fig. **510**. Boundaries and three chief contents of (right) femoral triangle. Also note position of femoral artery and vein deep to Sartorius in the cross-section of the midthigh.

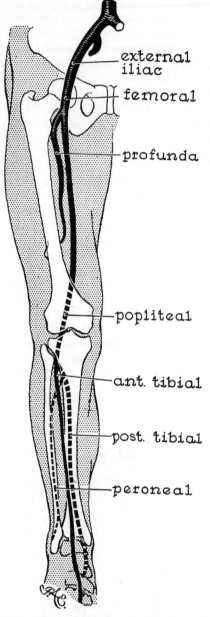

Fig. **511**. Main arterial channels of the lower limb (branches omitted for clarity).

inferior gluteal artery; it contributes to the supply of the hip joint and looks after the upper part of the back of the thigh.

Small but important branches from this and the next artery run along the neck of the femur to supply the head. If they are torn in a fracture of the femoral neck, the bone of the head may die.

2. **Lateral femoral circumflex artery.** This is the only important branch that runs laterally. It is the great muscular branch to the muscle clothing the femur—the Quadriceps. It passes laterally and provides branches which ascend or descend along the whole length of the lateral side of the thigh from hip to knee.

3. **The profunda femoris artery.** This is the largest branch of the femoral artery and it follows the course of the main vessel but on a deeper plane. It provides blood to the back of the thigh by means of a series of three or four **perforating branches** which pass close to the femoral shaft through the insertions of the adductors. It ends by itself acting as a final perforating branch.

POPLITEAL ARTERY

The popliteal artery, the continuation of the femoral, is the artery at the back of the knee; it is deeply placed against the back of the femur and the capsule of the knee joint. It traverses the popliteal space somewhat obliquely, since it begins on the medial side of the femur, but makes for the interval between the massive tibia and the slender fibula. It ends at this interval by dividing into posterior and anterior tibial arteries (figs. 512, 513).

The popliteal artery gives a series of branches, the genicular arteries, to the knee joint. These encircle the lower end of the femur and the upper end of the tibia and the fibular (fig. 512), providing branches to the adjacent muscles and joint.

ARTERIES OF LEG AND FOOT

The **posterior tibial artery,** the larger of the two terminal branches of the popliteal, descends the back of the leg deep

510), three important branches arise from the femoral; they may arise independently or by means of a common stem (fig. 509). All three supply adjacent muscles. They are:

1. **Medial femoral circumflex artery.** This vessel passes backward above the adductors and, gaining the back of the thigh high up, anastomoses with the

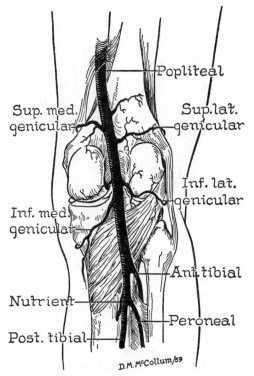

Fig. **512**. Popliteal artery at back of right knee (semischematic.)

The **plantar arteries** are the principal vessels of the foot; they lie in the sole after entering the sole deep to the belly of Abductor Hallucis. The **medial plantar artery** is mainly devoted to the muscles of the great toe, while the **lateral plantar artery** supplies everything else. They are shown schematically in figure 515.

SYSTEMIC VEINS

All the systemic veins, except the superior and inferior venae cavae, contain a series of bicuspid **valves** which allow the

to the Gastrocnemius and Soleus but superficial to the deep flexors. A large branch, the **peroneal (or fibular) artery,** descends on the back of the fibula and is devoted principally to the important Flexor Hallucis Longus. As the posterior tibial artery descends, it gives off many muscular branches. Having supplied the back of the ankle joint, the artery enters the foot behind the medial malleolus where it divides into **medial and lateral plantar arteries** for the structures in the sole of the foot.

The **anterior tibial artery** passes forward above the interosseous membrane and runs down the front of the leg on that membrane. It gives branches to the muscles on the front of the leg and crosses the front of the ankle joint in the midline. Entering the dorsum of the foot as the relatively small **dorsalis pedis artery,** it supplies that region and ends by piercing the posterior part of the space between the first and second metatarsal bones, where it anastomoses in the sole with the plantar arteries (fig. 514).

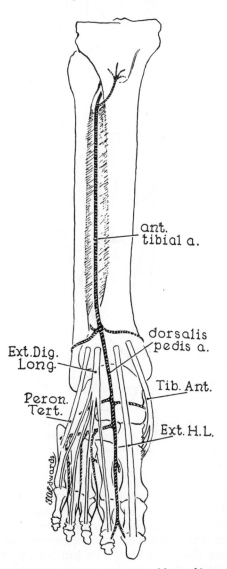

Fig. **513**. Course of main artery of front of leg and dorsum of foot.

Fig. **514**. Anterior tibial artery becomes the dorsalis pedis a. which supplies branches to dorsum of foot and toes.

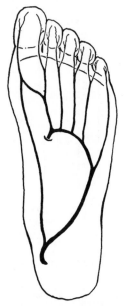

Fig. **515**. Posterior tibial artery enters sole of foot from behind medial malleolus and splits into two plantar arteries (medial and lateral). Note anastomoses.

blood to flow toward the heart but not in reverse. The two cusps are thin, semilunar, pocket-forming flaps of endothelium (intima) which can meet and so prevent backward flow (fig. 516). The pulmonary and portal systems of veins have no valves.

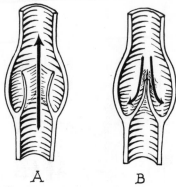

Fig. **516**. Blood flows in one direction only through a venous valve—toward the heart. A. Bicuspid valve open. B. Closed by back-pressure.

The systemic veins* drain blood from the tissues of the body as rivers drain water from the countryside. It is, therefore, proper to speak of **tributaries** rather than branches. The *tributaries of a vein correspond to the branches of its companion artery*, so that it would be redundant to discuss them in detail.

The **plan to be followed,** therefore, is to discuss briefly the veins returning blood from: (1) the head and the neck; (2) the upper limb; (3) the thorax; (4) the lower limb; (5) the pelvis and abdomen. Particular attention will be paid to those veins having no companion arteries or following independent courses.

The portal system of veins will be given separate consideration (page 290).

The popular concept that arteries contain 'pure' blood and veins 'impure' blood is an erroneous one, since it is the venous blood entering the heart that is rich in everything except oxygen necessary to the life of the cells of the body. It is better simply to speak of blood either as oxygenated or as unoxygenated.

VEINS OF THE HEAD AND NECK

Deep Veins

Cranial Venous Sinuses. Within the cranial cavity, the brain is covered by three fibrous membranes known as the meninges. These are described on page 310 where it is pointed out that in certain regions large and important veins known as

---

* Vena, L. = vein.

cranial venous sinuses course between two layers of the *dura mater*—the outer of the three *meninges* (Gr. = membranes). These sinuses drain the blood from the cranial cavity and the more important of them (shown in fig. 517 and 518) are:

1. **The superior sagittal sinus** lying in the upper edge of the *falx cerebri*—the sagittal dural partition intervening between the two cerebral hemispheres. It begins in the midline at the front. By receiving the many **superior cerebral veins** that drain blood from the surface of the cerebral hemispheres, it gains in size as it proceeds backward in the median plane immediately beneath the bones of the cranial vault. At the back of the cranial cavity, it ends on the occipital bone by turning (usually) to the right and running horizontally lateralward as:

2. **The right transverse sinus.** This runs horizontally in the attached margin of the *tentorium cerebelli*—the tented dural partition that roofs the posterior cranial fossa thereby separating the cerebrum from the cerebellum. When the sinus reaches the base of the petrous temporal bone it leaves the tentorium and, as the **sigmoid** (Gr. = S-shaped) **sinus**, descends to the jugular foramen.

    The left transverse sinus has a

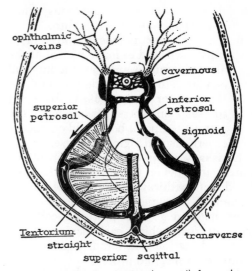

Fig. **518**. Venous sinuses (opened) from above (semischematic). Arrows show direction of flow of blood.

similar course but it usually begins by receiving:

3. **The straight sinus** (fig. 518) is a midline sinus running backward along the line of meeting of falx cerebri and tentorium cerebelli. At its anterior end it receives the **great cerebral vein** (of Galen) from the interior of the brain and—

4. **The inferior sagittal sinus,** which runs in the lower edge of the falx.

5. **The cavernous sinus** is an extensive sponge-like, blood-filled, venous space, lying on the side of the body of the sphenoid. Through it course the internal carotid artery, and the third, fourth, fifth, and sixth cranial nerves. These give a cavernous effect to the cross-sectional appearance of this pool of blood, hence its name. The sinus is a very important one because of its many relations and connections. Venous channels unite it, front and back, to its fellow of the opposite side; it receives in front **ophthalmic veins** from the orbital cavity, which usually communicate with facial veins. This is a potential channel for the spread of infection to the interior of the brain-case from unwisely treated infections

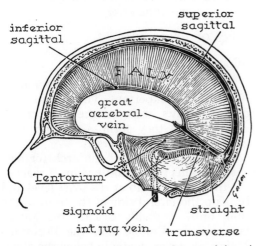

Fig. **517**. Venous sinuses in falx cerebri and tentorium cerebelli (semischematic).

of the face. The cavernous sinus drains backward (fig. 518) into:

6. **The inferior petrosal sinus.** This runs downward to make its exit at the jugular foramen and join the internal jugular vein just outside the skull.

7. **The superior petrosal sinus.** It courses along the superior edge of the petrous temporal bone, and unites the cavernous and sigmoid sinuses.

Emissary veins connect certain of the sinuses to veins on the outside of the skull by direct pathways through the bones. Certain of the larger ones leave (variable) large holes, e.g., the posterior condylar foramen, just behind each occipital condyle; and the mastoid foramen, above and behind each mastoid process.

Internal Jugular Vein. Within the cranial cavity, all veins lead to the **internal jugular vein.**†

Beginning at the jugular foramen, this large vein descends the neck in company first with the internal, then with the common, carotid artery. Vein and artery (with the vagus nerve lying behind and between them) are bound together in the fascial **carotid sheath;** they lie on the side of the pharynx and esophagus and in front of the transverse processes of the cervical vertebrae (fig. 491). In the neck, the vein receives tributaries which, in general, correspond to the branches of the external carotid artery. At the root of the neck behind each sternoclavicular joint, the internal jugular vein meets the subclavian vein to form the right or the left **brachiocephalic vein.**

Superficial (Cutaneous) Veins of Head and Neck

External Jugular Vein (fig. 519). Descending in the neck along a line which extends from a point a little behind the angle of the mandible to the middle of the clavicle, is a superficial vein known as the **external jugular vein.** It is formed by the union of a vein from the posterior part of the face (**retromandibular**) with one

---

† Jugular, not an ancient word, was coined from jugulum, L. = the hollow part of the neck above the clavicle.

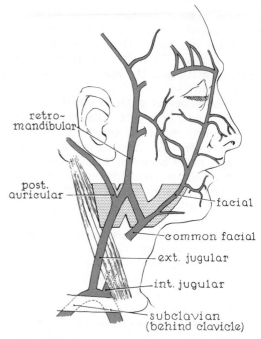

Fig. **519**. Main subcutaneous veins of the face and neck form a 'W.'

draining the scalp above and behind the ear (**posterior auricular**).

Having crossed the surface of the Sternomastoid, the external jugular vein receives large tributaries at the root of the neck, as well as the **anterior jugular vein** from the front of the neck. It then turns deeply to end in the **subclavian vein.** *Like other cutaneous veins, it is unaccompanied by a corresponding artery.*

VEINS OF UPPER LIMB

Superficial Veins

The cutaneous veins of the upper limb (fig. 520) are important. They begin, as a **dorsal venous arch,** on the back of the hand, where they form a pattern that varies from person to person.

At the lateral end of the arch (or network), begins the **cephalic vein.** It reaches the front of the limb at the wrist level and runs up the lateral side of the front of the limb until it finally reaches the groove between the Deltoid and the Pectoralis Major. Swinging medially below the clavicle, it plunges deep at the noticeable infraclavicular fossa near the lateral end of the clavicle and ends by emptying into the **axillary vein.** Most of the course of the cephalic vein often may be seen in the living.

At the medial side of the dorsal venous arch, begins the **basilic vein.** It runs up the medial side of the forearm and reaches the front of the limb a little below the elbow region, where it and its several tributaries are often quite conspicuous. At the bend of the elbow it receives a communication from the cephalic vein, and midway up the arm it turns deeply to run with the brachial artery. It finally becomes the main component of the axillary vein.

Deep Veins

The arteries of the limb are accompanied by veins having similar names and needing no independent discussion. However, it may be observed that: (1) the axillary vein lies on the medial side of its companion artery; (2) it is often double and with cross-connections; (3) the axillary vein becomes the **subclavian vein** which lies in

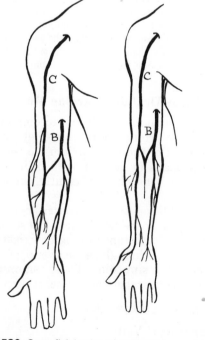

Fig. **520.** Superficial veins of upper limb—cephalic (C) and basilic (B)—form variable pattern at elbow.

the concavity of its companion artery; (4) where the subclavian vein crosses the first rib it is separated from its artery by the insertion of the Scalenus Anterior; (5) in many regions an artery may have two or three veins—called its *venae comitantes* (L. = accompanying veins)—clinging to its sides like ivy (fig. 467 on page 258).

Origin of names. It is interesting to note that the words cephalic, basilic, and saphenous (a vein in the lower limb to be discussed shortly) are three of the few legacies left to anatomy by the great Arab physicians who kept the flame of Greek medicine burning while Europe was passing through the Dark Ages. The European translators of the arabic manuscripts rendered *al-kifal* (whose exact meaning is not known) into cephalic (Gr. = pertaining to the head), and *al-basilic* (= the inner vein) into basilic. *Al-safin*—meaning 'secret'—became saphenous, which was derived from a Greek word that actually means the opposite of secret, i.e., visible or apparent.

VEINS IN THORAX

In the thorax are found the following:
1. Superior vena cava and its tributaries,
2. Inferior vena cava,
3. Azygos vein and its tributaries,
4. Veins draining the walls of the heart.

1. Superior Vena Cava and Tributaries

The right and left brachiocephalic (or innominate) veins join to form the superior vena cava (fig. 477, page 262) which also receives the azygos vein—the final pathway for the drainage of the chest wall (fig. 521). Since the brachiocephalic veins drain head, neck, and upper limbs, it is correct to say that the superior vena cava drains everything (except the lungs and heart) above the diaphragm.

Behind the sternoclavicular joint of each side, the **subclavian vein,** having already received the **external jugular vein,** meets the **internal jugular vein** to form with it the **brachiocephalic (innominate) vein.** Thus, there are two brachiocephalic veins (right and left) but only one brachiocephalic artery (lying on right).

The principal veins of the thorax lie on the right side since they all must ultimately return their blood to the right atrium. Therefore, the left brachiocephalic vein must cross to the right; this it does by running obliquely behind the upper half of the manubrium and across the roots of the three great arteries that arise from the aortic arch. It joins the right brachioce-phalic vein half way down the right border of the manubrium.

The **superior vena cava,** then, is formed halfway down the right border of the man-ubrium. It descends for about 6 cm (2½″) and enters the right atrium. Halfway along its course, it receives the **azygos vein** (fig. 521).

2. Inferior Vena Cava

This enormous swollen vein‡ barely ap-pears in the thorax because it drains into the right atrium almost immediately after piercing the central tendon of the dia-phragm. It will be described with the veins of the abdomen.

3. Azygos Vein and Tributaries

The **azygos vein** (Gr. = unpaired) is a longitudinal stem (fig. 521) ascending ver-tically on thoracic vertebral bodies; it lies to the right of the descending thoracic aorta. It begins in the abdomen often by the union of lumbar veins and it usually communicates with the inferior vena cava or the right renal vein. It becomes an important pathway for the return of blood from the lower half of the body when the inferior vena cava is, for any reason, ob-structed.

The azygos vein receives **intercostal veins** from both sides of the chest wall; these are the companion veins of the poste-rior (aortic) intercostal arteries. Gaining in size as it proceeds upward, the azygos vein,ultimately as thick as a pencil, finally comes forward by arching over the root of the right lung and so reaches the back of the superior vena cava. They azygos arch is comparable in position, though not in size, to the aortic arch over the root of the left lung.

Because the azygos vein ascends no higher than the root of the lung, the veins draining the upper intercos-

‡ *Cava* is probably an early Latinist's error of translation for *koile,* Gr. = swollen, for otherwise it is ridiculous, since it means 'hollow.'

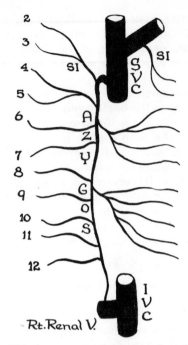

Fig. **521.** Azygos vein drains thoracic wall (inter-costal veins) and empties into S.V.C. (S.I. = superior intercostal vein.)

tal spaces on the right side unite to form the **right superior intercostal vein,** which descends to join the azygos just previous to its arch. The upper intercostal spaces on the left side are drained by the **left superior intercostal vein,** which, finding the azygos inaccess-ible to it, runs along the lateral side of the aortic arch and empties into the left brachiocephalic vein (fig. 521).

4. Veins Draining Cardiac Muscle

The veins of the heart accompany the coronary arteries but, since they drain to the back of the right atrium via the **coro-nary sinus** (a vein in spite of its name), they are somewhat differently named (see fig. 522).

In addition to the arteries, capillaries, and veins of the heart there is a secondary and less important system of irrigation of the walls through a network of small spaces, the **sinusoids.** These communicate with the chambers of the heart directly.

VEINS OF LOWER LIMB

Superficial Veins

There are two cutaneous veins in the lower limb which, like those in the upper limb, begin at the extremities of a **dorsal venous arch.** This arch lies on the dorsum of the foot a little proximal to the roots of the toes. From the lateral end of the arch runs the less important **small (or short) saphenous vein** which proceeds no higher than the knee level; from its medial end runs the conspicuous and important **great (or long) saphenous vein** which proceeds right up to the root of the limb (fig. 523).

The **small saphenous vein** courses proximally along the lateral side of the foot, behind the lateral malleolus, and up the posterolateral side of the calf; it turns deep in the lower part of the popliteal region to join the popliteal vein.

The **great saphenous vein** courses along the medial side of the foot, in front of the medial malleolus, and up to the posteromedial side of the calf to a point a hand's breadth behind the medial border of the patella. It then runs up the medial side of the thigh and, gradually coming to the front, turns deep 4 cm (1½″) inferolateral to the pubic tubercle; there it runs through an oval opening (the fossa ovalis) in the deep fascia; it empties into the front of the femoral vein. The great saphenous vein is important because of the frequency with which it becomes varicosed by a breaking down of its valves. On its long journey, it usually communicates with the small saphenous vein and it receives tributaries from the medial side of the thigh. The great saphenous vein is frequently visible in the living throughout most of its long course.

Fig. **523**. Great (or long) saphenous vein

Deep Veins

The deeper veins of the lower limb correspond to the arteries already described. It may be noted that: (1) the **popliteal vein** is intimately bound to its companion artery by a very dense fibrous sheath; (2) in the lower part of the thigh the **femoral vein** lies behind its artery; (3) at the root of the thigh the femoral vein lies medial to its artery; (4) on passing behind the inguinal ligament the femoral vein becomes the external iliac vein.

VEINS OF PELVIS AND ABDOMEN

Iliac Veins

The courses and tributaries of the **external, internal, and common iliac veins** correspond closely to the courses and branches of their companion arteries. They are shown in figure 507 in which it is to be observed that: (1) in general, the large veins lie within the angles of their arteries; (2) the end of the left common iliac vein passes behind the beginning of the right common iliac artery where it may be com-

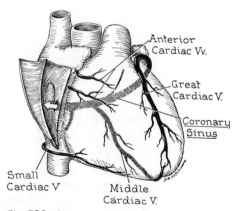

Fig. **522**. Coronary sinus on the back of the heart emptying into right atrium, and cardiac veins (semi-schematic).

pressed and where a clot may form; (3) the beginning of the inferior vena cava (formed by the two common iliac veins) lies at a lower level than the end of the abdominal aorta, and to its right side.

Inferior Vena Cava

This enormous vein (fig. 524) is the largest blood vessel in the body; it is about 3.5 cm (1¼″) in diameter. It runs up the right sides of the lumbar vertebral bodies and behind the peritoneum of the posterior abdominal wall. Its upper 5 cm (2″) lie buried in the back of the liver and, on leaving the liver, the vein at once pierces the diaphragm and the pericardium which is adherent to the diaphragm. It enters the right atrium at the lower right corner of the heart having a thoracic course of no more than about 1 cm; a bit of its right aspect may be seen below the lower right corner of the pericardial sac.

In the abdomen, the inferior vena cava receives tributaries that correspond to the

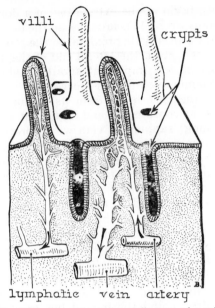

Fig. **525**. Three-dimensional diagram of a microscopic piece of small intestine. Blood is liberally supplied to all the blood capillaries of the villi. The villi are drained by both veins and lymphatics.

lumbar arteries, the renal arteries, the suprarenal arteries, and the testicular or ovarian arteries.

Because the inferior vena cava is on the right, the left renal vein is longer than the right renal vein and it reaches the inferior vena cava by passing across the front of the aorta in the acute angle between the aorta and its superior mesenteric branch. For the same reason, the left suprarenal and left testicular (ovarian) veins empty into the left renal vein.

A most significant fact about the inferior vena cava is that it does NOT receive veins from the digestive tract, the pancreas, or the spleen. The veins that drain these parts constitute the portal system of veins to be discussed below.

PORTAL SYSTEM

The mucous membrane of the small intestine, from duodenum to ileum, owes its velvety feeling to the myriads of villi that project from its surface into the lumen of the gut. These villi are microscopic, blind, finger-like processes resembling the pile on a carpet. Each villus (fig. 525) contains a minute arteriole, venule, and lymph

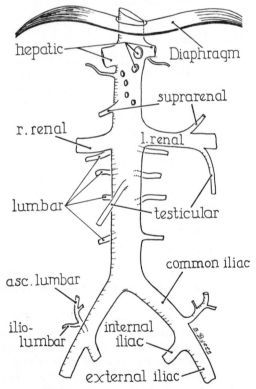

Fig. **524**. Scheme of inferior vena cava and its main tributaries.

vessel, with their associated capillaries. The arteriole nourishes the cells of the villus; the venule receives the products of digestion absorbed by the blood capillaries; the lymph vessels receive emulsified fat absorbed by the lymph capillaries.

The venules, thus laden with digestive products, unite with one another to form veins and these veins are the tributaries of the superior mesenteric vein.

Portal Vein and Its Tributaries

Venous blood from the spleen and pancreas via the **splenic (lienal) vein** joins the stream from the mesenteric veins. The **superior mesenteric vein** meets the splenic vein behind the neck of the pancreas; the resultant large vein is known as the **portal vein.** The **inferior mesenteric vein**—much longer than the corresponding artery and not closely associated with it—joins either the splenic or superior mesenteric veins, or the angle of union between the two. Veins corresponding to the gastric and gastro-epiploic arteries drain the stomach and empty, either directly into the portal vein or into the superior mesenteric vein. There is, therefore, no vein comparable to the celiac artery (fig. 526).

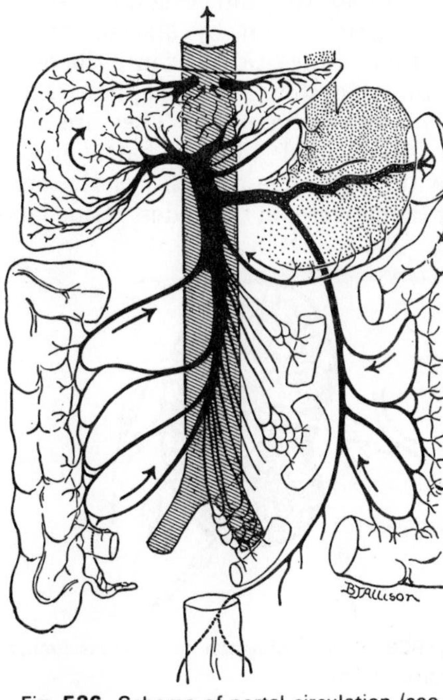

Fig. **526.** Scheme of portal circulation (see text).

The portal vein runs with hepatic arteries and bile ducts in the free edge of the lesser omentum; with them it enters the porta of the liver. It is by this means that the liver receives blood from the digestive tract (from stomach to rectum), from the pancreas, and from the spleen.

The portal vein is quite large, being 2 cm (¾″) in diameter and about 10 cm (4″) long. On reaching the liver, it divides into right and left branches and throughout its subsequent division is always accompanied by a branch of the hepatic artery and of the bile duct.

Circulation in Liver and Hepatic Veins

It has already been observed that the terminal branches of the portal vein are known as **interlobular veins.** From them, the blood runs in spaces between plates of liver cells to the center of the lobule. The liver cells absorb the glucose brought to it from the digestive tract, alter it to glycogen, and store it.

According to the traditional concept, the center of a lobule is occupied by a venule.§ This venule leaves the lobule and unites with other venules to form a **sublobular vein** (fig. 527). These veins unite to form a tributary of an **hepatic vein.** The hepatic veins finally become quite large and end by emptying into the inferior vena cava while that structure lies buried in the back of the liver; thus the hepatic veins—of which there are about two large and several small ones—lie entirely within the liver substance.

The last veins received, therefore, by the inferior vena cava are the hepatic veins. Thereafter, the blood has only to journey to the lung for oxygen when, on its return, it is rich in all the materials needed for the nourishment of the cells of the body.

It is this circulation of blood from the digestive tract, pancreas, and spleen, to

§ Recently, these have been shown to be plates or sheets of cells rather than cords. Further, the traditional concept of the lobule has been modified; in the newer view, the 'interlobular' branches of the portal vein lie in the middle of the unit called an *acinus.* Around the periphery of an acinus are the tributaries of the hepatic veins (the 'central' veins of the traditional lobule) which drain the liver.

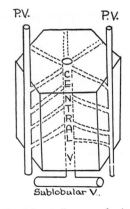

P.V.   P.V.

Sublobular V.

Fig. **527**. Traditional diagram of microscopic liver lobule. Blood passes from branches (P.V.) of portal vein to central vein that eventually drains into hepatic veins. Arteries and ducts omitted for clarity.

the liver that is known as the **portal circulation.**

Porto-Systemic Anastomoses

At the esophagus and at the rectum there are veins that drain into the portal system and others that drain into the systemic veins. The anastomoses between such veins, by becoming greatly enlarged, allow by-passing of the liver when this organ develops *cirrhosis*, a condition that obstructs the flow of blood through the liver.

LYMPHATICS

Lymph is the colorless fluid drained from the tissue fluid by lymph capillaries (lympha, L. = pure spring water).

Lymph vessels (lymphatics) offer an alternative pathway for the return of tissue fluid to the blood stream. They possess several distinguishing characters.

1. They are small, delicate, thin-walled vessels that tend to run side by side as *leashes of vessels* rather than to unite to form large trunks as do blood vessels. They are not seen normally in classroom dissections.
2. They are equipped with innumerable *valves.*
3. The flow of lymph is very sluggish but is increased in activities (such as muscular exercise) which 'massage' the vessel walls.

4. They are principally associated with skin, membranes (mucous, serous, and synovial), and glands.
5. They are absent where blood vessels are absent and are not found in the nervous system (including the eye).
6. Along their courses are found **lymph nodes** (fig. 528) through which the lymph must pass. Lymph nodes resemble the spleen in structure, and are filters (but, of course, they contain no blood).
7. After passing through at least one and often more than one node, the lymph is finally emptied into the venous blood stream at the root of the neck. The principal return is into the left brachiocephalic vein or into one of the two veins that unite to form it (left subclavian or left internal jugular). The vessel so ending is the **thoracic duct;** a smaller and less important **right lymphatic duct** enters the right brachiocephalic vein in similar fashion (fig. 529).

Lymph Nodes

Lymph nodes are extremely important, for in them many battles are fought and many victories won. Infecting organisms or foreign particles that by disease, injury, inspiration, or ingestion, get into the body, enter the ubiquitous lymph capillaries and are conveyed to the nodes. They are, as it were, shepherded to the battle-ground where the defensive agencies of the body await them. Here, in the nodes, reside large

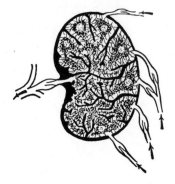

Fig. **528**. Section of lymph node (semischematic; magnified × 25).

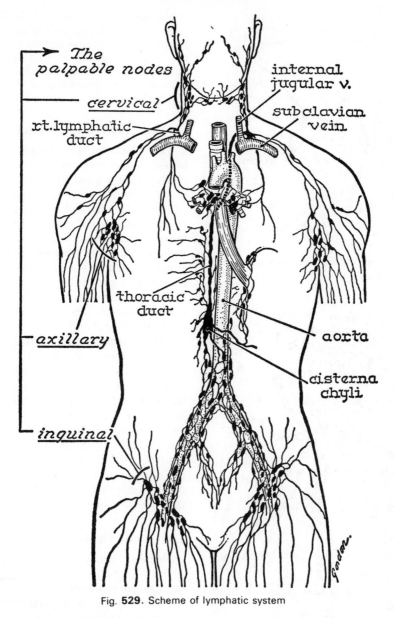

Fig. **529**. Scheme of lymphatic system

numbers of those phagocytic **reticulo-en-dothelial** or **scavenger cells** which have already been noted (page 256). These cells attack and digest bacteria or engulf and retain foreign particulate matter.

Lymph nodes vary in size from a pinhead to a large bean. They are found along the courses of the principal lymph vessels and are particularly numerous at the roots of the limbs—armpit and groin (axillary and inguinal nodes); at the side of the neck

(cervical nodes); and along the courses of the abdominal aorta and its branches.

Structure (fig. 528). Lymph vessels or lymphatics enter a node at several points on its periphery as afferent vessels. After percolating through sinusoids (spaces) in the node, the filtered lymph emerges via a (usually) single efferent vessel situated at the hilus of the node. In its passage, the lymph loses little, if any, of its volume but it carries off the **lymphocytes** (a type of

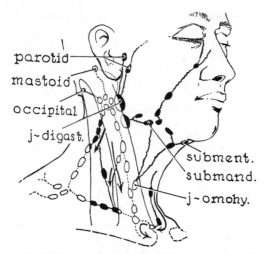

Fig. **530**. Lymph nodes and drainage of neck (modified after Rouvière).

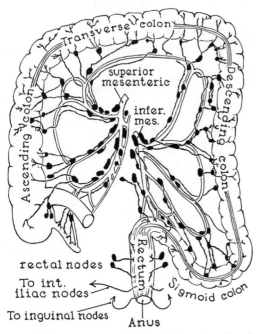

Fig. **531**. Lymphatic drainage and lymph nodes of large intestine.

whte blood cell) that are produced in the lymph nodes just as the blood carries off those that are produced in the spleen. Efferent lymph vessels unite to form lymph trunks and these return the lymph to the venous stream in the manner already indicated.

The swelling and tenderness exhibited by the axillary lymph nodes when the hand has an infection are evidences of increased activity of these nodes, as they strive to prevent bacteria and toxins from entering the blood stream. Similarly, the large lymph nodes in the root of each lung are often converted in city dwellers into black masses of gritty carbon.

Cisterna Chyli and Thoracic Duct

The lymph collected by the lymph capillaries of the intestinal villi is known as **chyle**. After a meal, the milky appearance of chyle is due to the emulsified fat with which it is charged. The lymph vessels from these villi empty by **intestinal trunks** into a reservoir known as the cisterna chyli. The **cisterna chyli** is about 5 cm (2″) long and about 1 cm (½″) broad; it lies to the right of the abdominal aorta (on lumbar vertebrae two and three) just below the diaphragm. From its upper end proceeds the thoracic duct.

The **thoracic duct**, pale, thin-walled, and about the size of a drinking straw, ascends on the right side of the lower thoracic vertebral bodies between the descending thoracic aorta and the azygos vein. At the level of the back of the aortic arch (T.5), it crosses obliquely to the left side behind the esophagus. It continues upward, lying to the left of the esophagus, and, at the root of the neck, it arches laterally behind the left carotid sheath to reach the left brachiocephalic vein (see also fig. 529).

Lymphatic Drainage

Because cancers and infections spread along lymphatics it is important in clinical studies to know the routes along which certain organs and parts are drained. Below is a brief summary of the more significant routes (see also figs. 529–531).

Pharynx. The lymphatics pass to a series of nodes in the neck related chiefly to the great vessels. Some are palpable when enlarged.

Upper Limb and Breast. There are many lymph nodes in the axilla which form the first 'stop' for secondary cancer from the breast and infections from the hand, forearm, and arm. Unfortunately, there are other, deeper routes for spread of cancer of the breast into the thorax.

Lungs. The lymphatics of the lungs drain to nodes in the hilus and thence to the thoracic (or to the right lymphatic) duct.

G-I Tract. The stomach, intestines, and related organs are drained along lymphatics and through nodes lying in the mesenteries and omenta with the vessels supplying these organs (fig. 531). These nodes are finally drained into the cisterna chyli.

Testis and Ovary. These organs, though far removed from their site of development (near the kidney), are drained by their lymphatics (which run with their veins) to nodes in the region of the renal vein.

External Genitalia and Lower Limb. The inguinal nodes, found just below the inguinal ligament (and often easily felt), drain the whole lower limb and the skin of the external genitalia (scrotum but not testis; labia majora et minora).

11

Nervous System

The duties of the nervous system are to receive information with regard to changes occurring in the environment of the body and to initiate and regulate appropriate responses. The nervous system is affected not only by the external environment but also by events occurring in the various parts of the body of which it is a part, i.e., by the internal environment. Changes in the external environment are consciously appreciated; changes in the internal environment quite often are not. When an event occurring in the environment affects the nervous system, that event is said to be a **stimulus.**

Neuron

The structural unit of all tissues is the cell. This is as true for the nervous system as it is for any other system, and the structural unit about to be described is simply a highly specialized cell that has concentrated on the properties of *irritability* and *conductivity*. Just as a muscle cell is designed to contract and do work, so a nerve cell is designed to receive stimuli and to conduct impulses.

In the nervous system, the cell or whole structural unit is called a **neuron.** A neuron is a nerve cell, including all its parts. The presiding executive of the neuron is the **cell body** but, unlike most other cell

bodies, that of the neuron has prolongations of its substance—sometimes single, sometimes very numerous—known as the nerve processes or fibers.

Nerve processes are of two kinds: (1) numerous, short, and branched processes, known as **dendrites** or **dendrons** (Gr. = tree), whose duties are always to conduct impulses toward the cell body; (2) a single, very long, and delicate process, known as an **axon** (Gr. = axis), whose duty is to conduct impulses away from the cell body.

All efferent ('motor') neurons possess the typical equipment of numerous dendrites and a single axon (fig. 532). The equipment of afferent ('sensory') neurons, however, requires further explanation.

The processes of most sensory neurons are reduced in number to one or two, such neurons being known as *unipolar* and *bipolar* (fig. 533). Of these the unipolar neurons are much the commoner.

The single process of a unipolar cell is properly called an axon. It bifurcates into two processes which are in line with each other, forming a T. Impulses are carried toward the cell body by the *peripheral process or fiber* (which therefore functions like a dendrite). *The central process or fiber* carries impulses away from the cell body and toward the central nervous system. The two processes of a bipolar cell,

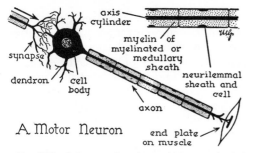

A Motor Neuron

Fig. **532**. Scheme of a (motor) neuron (greatly magnified). Inset: part of axon for detail.

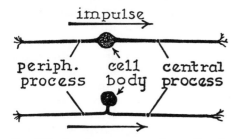

Fig. **533**. A unipolar neuron (lower figure) and a bipolar neuron.

although they spring direct from the cell body, are named in the same manner.

In all parts of the peripheral nervous system, bundles of fibers constitute the gross structure known as a **nerve** (fig. 534). Most nerves contain both afferent and efferent fibers and are called **mixed nerves;** some contain only afferent fibers and are called **sensory nerves;** nerves going to muscles are called **motor nerves** even though they contain both efferent fibers for the muscles and afferent fibers (proprioceptive) from muscles, tendons, and joints.

It is customary to restrict the use of the term *nerve* to the peripheral nervous system and to utilize other terms for bundles of fibers lying entirely within the central nervous system. The distinction is a useful one because a nerve, as has been seen, is made up of fibers of differing functions; some are efferent, some are afferent, and among both of these are fibers with differing duties. In the central nervous system a sorting out occurs, and fibers become bundled together according to the exact functions they serve; bundles of fibers serving the same function are called **tracts.**

In this connection, it should be noted here that tracts bear compound names. The first part of the name indicates where the impulse begins, the second where it ends. Thus the name 'corticospinal' reveals that the impulse begins in some part of the cortex (the outer gray matter of the brain) and ends somewhere in the spinal cord.

The **structure of fibers** varies as do their functions. Some are of large caliber, others of small. When fully equipped, an axon possesses two coats: (1) an inner thick one of a fatty substance known as **myelin**—the axon is therefore said to be **myelinated** or **medullated** (myelos, Gr. = medulla, L. = marrow); (2) an outer thin, nucleated, membranous one known as the **neurilemmal sheath** (lemma, Gr. = skin) (see fig. 532).

Nerve Impulse

When a nerve receives an adequate stimulus, a 'message' or nerve impulse passes along the nerve. A great deal is known about nerve impulses but their exact nature is still controversial. Although many fanciful comparisons have been made with electric currents and with the burning of a train of gunpowder, these are not especially informative, and the beginner is well advised to think of impulses simply as 'messages' passing along nerve fibers.

Types of Nerves

The nerves concerned with conveying impulses toward the central nervous sys-

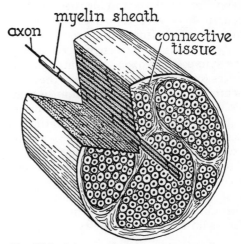

Fig. **534**. Scheme of a short segment of a peripheral nerve. Note that the delicate nerve fibers are protected by considerable connective tissue.

tem are called **sensory** or (better) **afferent nerves.**

An end-organ receiving a stimulus is also called a **receptor.** Any structure that makes a response to an impulse from the nervous system is called an **effector.** Effectors are stimulated by nerves which convey impulses to them from the central nervous system; these nerves are called **motor** or (better) **efferent** (L. = out-carrying) **nerves.** Two varieties exist according to whether they are destined for the voluntary muscles on the one hand, or for involuntary (smooth) muscle, cardiac muscle, or glands on the other.

Synapse. An impulse reaches its ultimate destination by traveling over a series or chain of neurons. It resembles a verbal command relayed by a series of soldiers who meet at a series of junction points. These junction points are where the terminal branches of an axon come in contact with the dendrites or cell bodies of the succeeding neuron. Each junction is known as a **synapse** (Gr. = a touching together). The accepted convention for showing a synapse in a diagram is to draw a bifurcation at the end of the axon and, lying within this bifurcation, a black dot representing the cell body of the 'receiving' neuron (fig. 535). This is only a diagram; in fact the synapse usually consists of a tiny end-bulb flattened against the cell body or dendrites of the receiving neuron. Each neuron may accommodate thousands of such synapses from many sources.

By producing electrochemical changes at the synapse, the axon stimulates the dendrites (or the cell body direct) of the second neuron; the resulting new impulse that is set up travels over this second neuron, and so the process may be repeated until a final destination is reached. The synapse is of great interest to the physiologist. Here, it is desirable merely to point out that there is no continuity of structure at a synapse and that synapses permit impulses to cross them in one direction only; they are, therefore, said to possess **polarity.**

Fig. **535.** Traditional diagram of synapse

For most afferent impulses, only the first neuron in a chain is found in the peripheral nervous system; for all efferent impulses to voluntary muscles, only the axon of the last neuron is found in the peripheral nervous system. For efferent impulses to cardiac muscle, smooth muscle, and glands, the greater part of the last two neurons are found in the peripheral nervous system; this is an important distinction and will be discussed later, but it is one reason why the autonomic nervous system, of which these two neurons are a part, usually receives independent consideration.

It is perhaps realized by now that the effects produced by an impulse depend, not on the nature of the impulse—which is obscure—but upon the nature of the receptor or effector. The exact nature of electricity is likewise, obscure, but the same current can be used to light a bulb, ring a bell, or boil a kettle; so it is with the nerve impulse.

Inhibition

The nervous system is capable of producing effects that are of either a 'plus' or a 'minus' nature. In other words it can either **stimulate** or **inhibit.** Perhaps the great function of the 'highest' part of our nervous system, 'the gray matter,' is its ability to inhibit activity. Everyone is familiar with the unfortunate effects produced on a person when the inhibiting power of the brain is temporarily overcome by alcohol.

Stimuli and Receptors

A stimulus received from outside the body is called an *exteroceptive* stimulus; one received from within the body is called an *interoceptive* stimulus. Interoceptive stimuli may originate in the muscles, tendons, and joints and give information as to their states of tension and position; these are called *proprioceptive*. Similar stimuli may originate in the organs (viscera) and give information as to their activities; these are *visceroceptive*.

Exteroceptive stimuli are of many varieties and some require special and elaborate anatomical structures to enable the nervous system to receive them. Any apparatus, simple or complex, situated at the end of a nerve and designed to receive a stimulus is called an **end-organ.**

The varieties of exteroceptive stimuli that the nervous system is capable of receiving, are interpreted as: *smell, sight, hearing, taste, touch, pain,* and *temperature.* For the stimuli interpreted as pain there are no special end-organs since the naked ends of nerve fibers themselves can receive directly the adequate stimulus. Almost all other stimuli seem to require special end-organs for their reception. For sight and hearing the end-organs are the very complex anatomical structures, the eye and the internal ear. Taste is received by the taste buds of the tongue; smell by special olfactory cells; and an assortment of microscopic end-organs deals with the varieties of touch which range from a light 'cotton wool' type to one in which pressure involves some deformity of the skin (see fig. 536).

Proprioceptive stimuli also are received by special end-organs. So-called **muscle-spindles** and **tendon-spindles** are capable of registering the state of contraction of muscle fibers and the state of tension of tendons. A highly complex structure within the internal ear is required to register posture and the proprioceptive stimuli resulting from disturbances of equilibrium. Naked nerve endings receive stimuli from the joints, and from lining membranes and secretory organs.

Classification of Sensation

A. **Exteroceptive**
   (1) Smell
   (2) Sight
   (3) Hearing
   (4) Taste (and chemoreception)
   (5) Touch (and pressure)
   (6) Pain and temperature
B. **Interoceptive**
   (1) Proprioceptive
     (a) Equilibrium
     (b) Tendons
     (c) Joints
     (d) Muscles
   (2) Visceroceptive—internal organs

Exteroceptive stimuli generally reach consciousness and, at the discretion of the individual, evoke, or do not evoke, re-

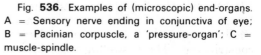

Fig. **536.** Examples of (microscopic) end-organs. A = Sensory nerve ending in conjunctiva of eye; B = Pacinian corpuscle, a 'pressure-organ'; C = muscle-spindle.

sponses, Proprioceptive stimuli, giving information as to the changing states of muscles, tendons, and joints, are ordinarily received at a subconscious level but are vital to the proper performance of co-ordinated and voluntary muscular activity. Visceroceptive stimuli from the organs, etc., are also ordinarily received at a subconscious level and they evoke responses which are generally involuntary and over which the individual has no control. We may become painfully aware of such stimuli when they become excessive.

The operations of the nervous system bear certain similarities to those of a modern army. Headquarters of an army (the brain), kept informed by telephone and telegraph, radio and radar (various types of afferent nerves) of what is happening throughout the whole organization (the body), analyzes and integrates its knowledge, and issues its orders. These orders are sent to divisional or brigade headquarters (lower centers in the brain or spinal cord) whence, by telephone, etc. (efferent nerves), appropriate orders are sent to units in the field whereby the engines of modern warfare (the muscles, glands) spring into action.

In the nervous system the headquarters organization is known as the **central nervous system** and it consists of the brain and spinal cord. The parts of the nervous system that carry information to, and orders from, the central nervous system are known as the **peripheral nervous system.** Because of the special characteristics that they possess, the parts of the nervous system that are concerned with the *activation of smooth muscle, cardiac muscle,* and *glands* are known as the **autonomic nervous system.** This last is a part of the general organization and giving it a special name expresses a functional rather than a structural entity.

Development of the Brain

The upper end of the neural tube (page 12) becomes the brain; very early, it enlarges to take the form of three rounded swellings marked off from one another by constrictions. These swellings are known as the three **primary brain vesicles** and, by following the fate of each, it is not difficult to make out the principal parts of the brain (figs. 537, 538).

Forebrain

The front half of the foremost or rostral vesicle swells out on each side and, as these evaginations grow in size, their walls thicken also. So large do these evaginations become in man, that they finally completely overshadow the parts developed from the rest of the vesicles. They are known as the **cerebral hemispheres** and in each a cavity persists which communicates with the midline cavity from which it grew out; each lateral cavity is known as a **lateral ventricle.** Even this is not enough,

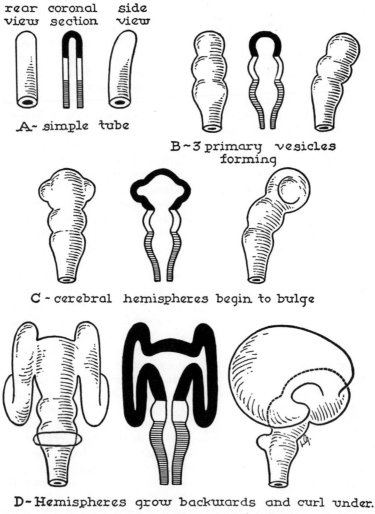

rear coronal side
view section view

A ~ simple tube

B ~ 3 primary vesicles
forming

C ~ cerebral hemispheres begin to bulge

D ~ Hemispheres grow backwards and curl under.
Cerebellum begins

Fig. **537**. Early development of brain. Series of three companion figures showing posterior view coronal section (in which black = forebrain, white = midbrain, and shaded = hindbrain) and view from right side.

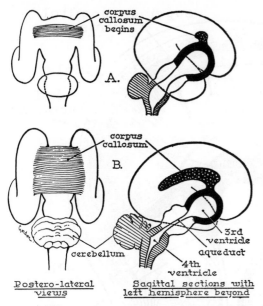

Postero-lateral views

Sagittal sections with left hemisphere beyond

Fig. **538**. Further development of brain (schematic). Hemispheres are pulled far apart to show corpus callosum developing. In sagittal section, cut surfaces are: black = forebrain, white = midbrain, shaded = hindbrain. A is at same stage as fig. 537-D. B is approaching adult condition.

for the hemisphere, continuing to grow, bends on itself in a U-shaped manner until its original hind end comes to be directed forward below its front end (see fig. 537).

The caudal half of the vesicle, the **diencephalon**, remains as a midline structure but with greatly thickened lateral walls, a relatively thin floor, and a roof so thin as to be almost deficient; its cavity is reduced to a slit-like space known as the **third ventricle.** The lateral walls consist of a pair of large and important nuclei* (cell masses), which ultimately are found buried deeply in the cerebral hemispheres; each is known as the **thalamus** and each is a great relaying station for incoming (afferent) impulses. Besides its other numerous connections each thalamus discharges to the cortex or gray matter of the cerebral hemisphere where afferent impulses are finally

* The word **nucleus** (L. = nut) is a little confusing. The student is probably accustomed to regarding it as the specialized central mass of a cell. In neurology, however, it is used to denote a collection of nerve cell bodies inside the central nervous system. A collection of nerve cells outside the CNS is called a **ganglion** (Gr. = swelling).

received. The floor of the region is known as the **hypothalamus** and is an important autonomic center for the regulation of the activities of smooth muscle, cardiac muscle, and glands.

Midbrain

The second vesicle is called the **midbrain.** It also remains in the midline and all its walls are thickened; through it run ascending (afferent) and descending (efferent) tracts, and it contains small but important nuclear masses. Its cavity is reduced to a narrow canal known as the **cerebral aqueduct.**

Hindbrain

The thickened floor of the lowest vesicle remains in the midline and, as the **pons,** is continuous rostrally (forward) with the midbrain; as the **medulla oblongata,** it continues below with the spinal cord. The side walls of the third vesicle are greatly thickened and meet one another across the roof to form the two **cerebellar hemispheres.** These are principally concerned with the reception of proprioceptive impulses and with the discharge of impulses designed for the co-ordination of muscular activities. They are solid structures since the cavity of the vesicle, known as the **fourth ventricle,** remains in the midline and intervenes between the cerebellum behind and the pons and medulla oblongata in front (figs. 538, 539).

The lower half of the medulla oblongata retains a tubular shape characteristic of the spinal cord below it.

The midbrain, pons, and medulla oblongata are collectively referred to as the **brain stem.**

TABLE 1

Vesicles	Parts of Brain	Cavity
First	Cerebral hemispheres	Lateral ventricles (r. & l.)
	Thalamus	Third ventricle
	Hypothalamus	
Second	Midbrain	Aqueduct
Third	Cerebellar hemispheres	
	Pons	Fourth ventricle
	Medulla Oblongata	

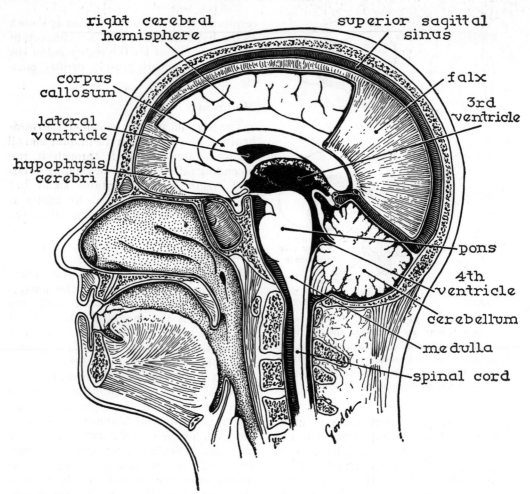

right cerebral hemisphere

superior sagittal sinus

corpus callosum

lateral ventricle

hypophysis cerebri

falx

3rd ventricle

pons

4th ventricle

cerebellum

medulla

spinal cord

Gordon

Fig. **539**. Brain *in situ* as seen in sagittal section of head. Window cut away in the midline falx cerebri reveals part of medial aspect of right cerebral hemisphere. The remaining parts of the brain are seen here in midline sagittal section.

CEREBRAL HEMISPHERES

Surfaces, Borders, Poles

Each cerebral hemisphere possesses three surfaces—a superolateral, a medial, and an inferior. The **superolateral surface** is large and convex and is protected by the overlying vault and sides of the cranial cavity. The **medial surface** is flat, lies close beside the median sagittal plane, and is separated from its fellow by the falx cerebri. It is separated from the superolateral surface by the **superior border** (see figs. 540–543). The **inferior surface** or **base** (figs. 544, 545) is on two levels, the anterior third being higher than the poste-

rior two-thirds. The anterior third rests on the floor of the anterior cranial fossa. The middle third rests on the floor of the middle cranial fossa. The posterior third is on the same level as the middle third since it rests—not on the floor—but on the roof of the posterior cranial fossa. This roof is the tentorium cerebelli (page 310); beneath it lies the cerebellum. The inferior surface is separated from the superolateral by the **inferior border;** it is separated from the medial surface by the **medial border** which in its middle third sweeps around the brain stem.

The full and rounded anterior extremity of the hemisphere is the **frontal pole.** The

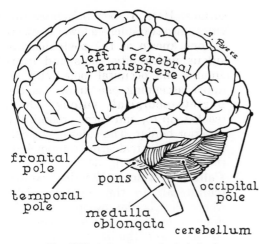

Fig. **540**. Left aspect of whole brain

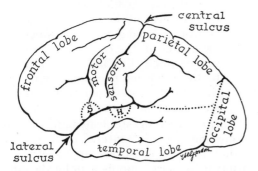

Fig. **541**. Lateral aspect of left cerebral hemisphere. Note centers for hearing (H) and speech (S) bordering lateral sulcus.

more pointed posterior extremity is the **occipital pole.** The rounded extremity on the inferior surface that occupies the front end of the middle cranial fossa is the **temporal pole;** it lies behind the orbital cavity.

Cortex of Cerebral Hemispheres

The outer surface of the cerebral hemisphere consists of a thin layer of gray matter known as the **cortex** (L. = bark). It is several millimeters thick and is made up principally of cell bodies whose axons run in the subjacent white matter. It must not be supposed that the cortex is of uniform texture and structure. Rather it varies from region to region in accordance with its varying functions. Indeed, the mapping of cortical areas according to function has, in recent years, been painstakingly under-

taken until today over 100 areas are recognized, each designated by a number, e.g., the precentral gyrus—to be presently noted —is known as 'area 4.'

The cortex is not smooth but is thrown into folds and furrows, a state of affairs that markedly increases the total surface area without demanding a greater brain volume. Each fold or rounded elevation is known as a **gyrus;** the intervening furrow—varying considerably in depth—is known as a **sulcus** or **fissure.** The gyri and sulci, while varying in minor detail from brain to brain, do not occur haphazardly but follow a recognizable pattern, so that it becomes possible (and desirable) to name some of them and particularly those occur-

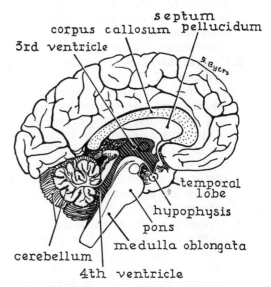

Fig. **542**. Medial view of left half of whole brain.

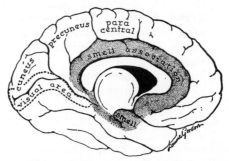

Fig. **543**. Medial aspect of left cerebral hemisphere. Dark cavity (lateral ventricle) is seen below the cut corpus callosum.

ring in, or separating, areas of known functional significance.

Sulci and Gyri of Hemispheres

Five sulci are particularly significant since they serve to locate gyri of known function:

1. **Lateral sulcus** (fig. 541). A deep sulcus that outlines and sweeps backward above the temporal pole, and then continues, on the superolateral surface (rather more than halfway down), almost horizontally backward. It marks the line along which the hemisphere became folded; if its lips be separated, a 'sunken' piece of cortex—the **insula**—will be seen in its depths. The sites of the **speech area** and of the **hearing area**—both closely associated with the lateral sulcus —are shown in figure 541.

2. **Central sulcus.** Starting very near the midpoint of the superior border it then runs obliquely downward and forward across the superolateral surface toward the lateral sulcus which it just fails to reach. Its importance lies in the fact that the gyrus in front of it—the **precentral gyrus**—is the **motor area,** while the gyrus behind it—the **postcentral gyrus**—is the **sensory area** (fig. 541).

3. **Sulcus cinguli.** A prominent sulcus on the medial surface that runs from front to back parallel to, and one gyrus above, the corpus callosum— the very thick plate of fibers that unites the two cerebral hemispheres (fig. 543). Below the sulcus is the **gyrus cinguli,** a **smell association area** and an **'emotional' area** that, today, is receiving increasing attention. [The cingulum (L. = girdle) is a bundle of fibers which runs in the gyrus and almost engirdles the corpus callosum.]

4. **Calcarine sulcus** (fig. 543). Starting as a short, deep sulcus on the inferior surface close behind the posterior extremity of the corpus callosum, the calcarine sulcus passes backward like a spur (L. = calcar) to the medial surface. Here it soon bifurcates (divides) into two sulci—a lower horizontal one known as the **postcalcarine sulcus** which often cuts and turns around the occipital pole, and an upper ascending one known as the **parieto-occipital sulcus** which reaches up to, and turns around, the superior border. The sides of the calcarine and postcalcarine sulci include the **visual area** (fig. 543).

5. **Collateral sulcus.** Running parallel (collateral) to the medial border where that border sweeps round the brain stem, this sulcus on the inferior surface marks the lateral limit of the **parahippocampal gyrus** that lies medial to it. At its front end a hook-shaped piece of the parahippocampal or hippocampal gyrus—the **uncus**—is the center for the conscious appreciation of **smell** (figs. 543, 544). [Medial to the (para-)hippocampal gyrus are the closely associated *hippocampal* and *choroidal sulci.*]

Cortical areas of established function are sometimes called **'centers'** and they are either for the conscious appreciation of afferent (sensory) impulses, or for the initi-

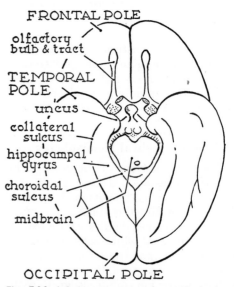

FRONTAL POLE

olfactory bulb & tract

TEMPORAL POLE

uncus

collateral sulcus

hippocampal gyrus

choroidal sulcus

midbrain

OCCIPITAL POLE

Fig. **544.** Inferior aspect of cerebral hemispheres. Optic chiasma, pituitary gland, and mammillary bodies which all lie below thalamus, and cut surface of midbrain are seen centrally.

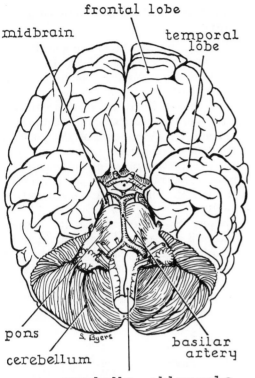

Fig. **545**. General inferior view of whole brain. Hypophysis has been removed and basilar artery and some nerves retained.

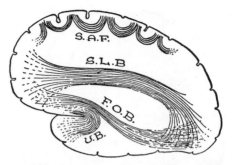

Fig. **546**. Association fiber bundles. S.A.F. = short association fibers; S.L.B. = superior longitudinal bundle; F.O.B. = fronto-occipital bundle; U.B. = uncinate bundle.

ation of efferent (motor) activities. Other areas of the cortex are sometimes called 'silent areas' to remind us—not of their lack of function—but of the inadequacy of our knowledge regarding them.

Lobes and Lobules of Hemispheres

Figures 542 to 545 illustrate the three surfaces of the hemisphere and show also how certain of the sulci mentioned are utilized to divide the hemisphere for descriptive purposes into **frontal, parietal, temporal,** and **occipital lobes.** On the medial surface three **lobules** are often recognized. These are the *cuneus, precuneus,* and *paracentral lobule;* their boundaries are sufficiently indicated in figure 543.

White Matter of Cerebral Hemispheres

The myriads of nerve fibers (axons) that go to make up the white matter of the cerebral hemisphere are classified as: (1) association fibers; (2) commissural fibers; (3) projection fibers.

Association fibers bring neurons of one part of the cortex of a hemisphere into communication with those of another part of the same hemisphere. Fibers nearest the cortex run from one gyrus to the next; they have merely to take a U-shaped course around the depths of the intervening sulcus. They are called **short association fibers.** On a deeper plane run fibers that pass from one region of the cortex of a hemisphere to a more distant region of the cortex of the same hemisphere. They are gathered into recognizable bundles and are known as **long association fibers.** The principal bundles are shown and labeled in figure 546.

Commissural fibers cross the midline from one hemisphere to the other. Reference to figure 538 reveals that at first the only pathway of communication between the two hemispheres lies at a very narrow region at the original front end of the brain. As the hemispheres grow back, this narrow region is drawn out and grows back also until, in the adult, a very conspicuous, compact, and thick plate of commissural fibers, known as the **corpus callosum,**† roofs over the lateral ventricles and the midline brain stem; it serves to keep all parts of the cortex of one hemisphere in communication with corresponding parts of the other. The long lines in figure 547

---

† *Callum*, L. = hard; also = a beam or rafter. Either meaning is appropriate to the corpus callosum.

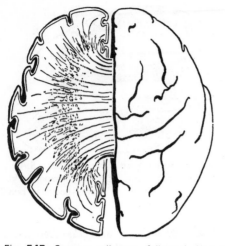

Fig. **547**. Corpus callosum followed into left hemisphere, viewed from above (semischematic).

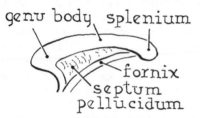

Fig. **548**. Parts of corpus callosum and its related structures in median sagittal section.

suggest the ultimate extent of the corpus callosum. Figures 539 and 548 show the appearance of the adult corpus callosum in sagittal section.

Projection fibers leave the cortex for lower centers or mount to the cortex from lower centers; they are projected out of, or into, the cortex. Just as the commissural fibers must run at right angles to the long association fibers, so the projection fibers must run at right angles to both.

Gathered from almost all parts of the cortex, the projection fibers must crowd together in order to enter the restricted territory of the brain stem. In the hemisphere they appear as a spray of fibers known as the **corona radiata** which, making its way through the fibers of the corpus callosum, reaches the base of the hemisphere. Here the fibers, narrowed to a compact mass, find a large nucleus of gray matter astride their path. This nucleus and how the projection fibers deal with it will be considered now.

Basal Nuclei, Thalamus, and Internal Capsule

The term **basal nuclei** (or **ganglia**) is used to designate masses of gray matter in the base of the cerebral hemisphere lateral to each thalamus. Together, two of these masses, the **lentiform** and **caudate nuclei**, are the **corpus striatum**. The **thalamus** (right and left) was noted as the greatest part of the *diencephalon* that took no part in the evagination resulting in the cerebral hemispheres, but remains as a midline structure. Its functions are discussed on page 319.

Corpus Striatum. Since the corpus striatum is that part of the basal nuclei in the base of the evaginated hemisphere, it necessarily accompanied the hemisphere in its subsequent folding. Figure 549 suggests how the original elongated form became a U-shaped one.

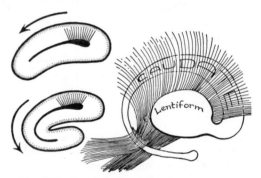

Fig. **549**. Development of internal capsule and corpus striatum which follows folding of hemisphere and gives final appearance as in right figure (see fig. 550).

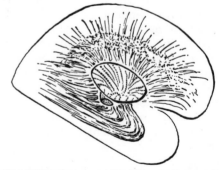

Fig. **550**. Diagram of right internal capsule (broken lines) deep to oval lentiform nucleus. Beyond edges of lentiform nucleus, corona radiata spreads to hemisphere.

In order to reach the restricted area of the brain stem, the projection fibers make their way through the fore part of the corpus striatum. They need not concern themselves with the hind part now lying in the temporal lobe. In their passage, the projection fibers split the corpus striatum almost completely into lateral and medial parts. Because the lateral part is shaped like a biconvex lens it is known as the **lentiform nucleus;** because the medial part is tail-like it is known as the **caudate nucleus.** A few fibers succeed in passing entirely lateral to the lentiform nucleus and join their fellows lower down, so that the nucleus is entirely surrounded by fibers that form a capsule for it. That is why the projection fibers in the region of the basal nuclei become known as the **internal capsule** (fig. 550) and the (much smaller) **external capsule.** Figure 551 indicates the appearance of the base of the hemisphere when cut through horizontally at the level of the internal capsule. It will be seen that the capsule, in such a section, consists of

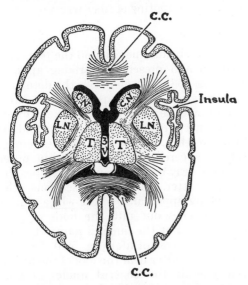

Fig. **551.** Horizontal section through forebrain (cerebral hemisphere and thalamus). Between right thalamus and left (T) lies the third ventricle which extends, as lateral ventricles, into hemispheres in front of caudate nucleus (C.N.). Internal capsule separates thalamus and caudate nucleus from lentiform nucleus (L.N.) which lies deep to insula in depth of lateral sulcus. Anterior and posterior fibers of corpus callosum (C.C.) also seen.

anterior limb, genu (knee), and posterior limb.

Because many important tracts are packed into the narrow confines of the **internal capsule,** the region is a very important one. For example, rupture of a small blood vessel in this region may cut off efferent tracts descending from the cortex and result in an apoplexy or stroke, a condition of *hemiplegia* or paralysis of one-half of the body.

BRAIN STEM AND CEREBELLUM

Midbrain

Having passed beyond the region of the basal nuclei, the descending projection fibers leave the cerebrum and enter the **midbrain.** Here, in the **cerebral peduncles** (L. = little feet), they appear on the anterior (ventral) surface of the midbrain as two broad bundles (the *bases pedunculi*) which approach one another and, distally, sink into the pons. The midbrain is a small but important region and is depicted in figures 538, 552, and 553. In it lie a pair of nuclei, which, because of their pinkish tinge, are known as the **red nuclei.**

The 'cavity' of the midbrain is the small canal known as the *cerebral aqueduct (of Sylvius)*. Surrounding it is a layer of gray matter containing important nuclei. On the dorsal aspect of the midbrain are two

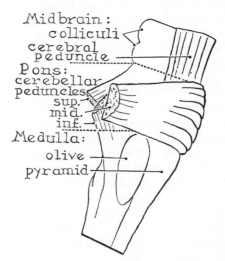

Fig. **552.** Brain stem from right side (semi-schematic).

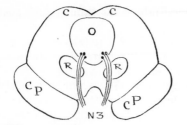

Fig. **553**. Cross-section of midbrain (schematic). C = colliculus; CP = basis pedunculi of cerebral peduncle; R = red nucleus; N3 = oculomotor nerves with cell bodies in gray matter around aqueduct.

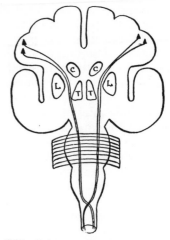

Fig. **554**. Scheme of corticospinal tracts (see text).

pairs of rounded eminences, the superior and inferior *colliculi* (L. = little hills) containing nuclei associated with sight and hearing.

Pons

The striking feature of the **pons,** and the one from which it gets its name, is the wide bridge of coarse bundles of fibers visible on its anterior aspect. These fibers cross at right angles to the general direction of the brain stem and narrow somewhat as they plunge laterally into the cerebellar hemispheres. As they enter the cerebellum they are known as the **middle cerebellar peduncles** (figs. 552, 555); deep to the transverse fibers of the pons lie the vertical projection fibers which, above, are the cerebral peduncles.

The *fourth ventricle* lies mostly in the pons (the remainder being in the medulla). The gray matter which forms a thick carpet for the floor of the fourth ventricle contains many important nuclei.

Medulla Oblongata

The medulla is the part of the brain stem immediately below the pons. It tapers somewhat below to become the spinal cord (fig. 552). On its ventral aspect, a pair of elongated rounded masses—the *pyramids*—lie separated by a midline groove. They carry the major efferent pathway of the central nervous system, the *corticospinal tract* (fig. 554). On each side of the medulla is a rounded mass appropriately named the *olive.*

The part of the fourth ventricle that occupies the medulla is roofed by a very thin membrane (known as the inferior velum). In the gray matter carpeting the floor of the cavity lie very important centers for the regulation of many vital activities such as respiration, body temperature, and heart rate. In the medulla, rearrangement and re-routing of fiber tracts occur.

Cerebellum

The cerebellum lies behind the pons and medulla and presents a striking appearance (figs. 538, 539, 555), but the details of nomenclature are confusing and of little importance because of the lack of exact knowledge in regard to function. Its folds or **folia** are thin, transverse, and numerous. Like the cerebrum, it has a thin outer cortex of gray matter deep to which lies the white matter. In sagittal sections, the branching appearance of the white matter as it reaches out into the folds of gray matter gives it the fanciful name of *arbor vitae* (L. = tree of life) (fig. 556). Deep in the white matter near the roof of the fourth ventricle lie the **central nuclei** of gray matter from which all the pathways exit from the cerebellum. The most conspicuous central nucleus is the **dentate nucleus,** so-called because of its shape and size (dens, L. = tooth).

Besides the **middle cerebellar peduncle,** the cerebellum possesses a **superior cerebellar peduncle** which runs upward to

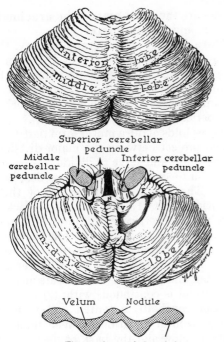

Fig. **555**. Cerebellum, from above and from below, with a scheme of flocculonodular lobe. Middle lobe (old terminology) is also known as the posterior lobe.

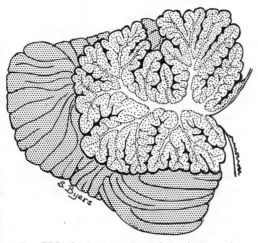

Fig. **556**. Sagittal section of cerebellum showing arbor vitae of white matter and cortical folia.

connect it to the midbrain, and an **inferior cerebellar peduncle** which runs upward into the cerebellum from the spinal cord (fig. 552).

The **functions** of the cerebellum are

discussed with Cranial Nerve VIII, on page 329.

BLOOD SUPPLY OF BRAIN

The two internal carotid arteries and the two vertebral arteries (which are branches of the subclavian arteries) contribute to a rich anastomosis at the base of the brain known as the **arterial circle** (of Willis). This is shown in figure 557, in which it is to be observed that the vertebral arteries devote themselves chiefly to the supply of the cerebellum and brain stem, and that from the arterial circle **three cerebral arteries** supply the cerebrum. In general, each of these three cerebral arteries supplies a surface and a pole. Thus: **anterior cerbral**—medial surface and frontal pole; **middle cerebral**—superolateral surface and temporal pole; **posterior cerebral** —inferior surface and occipital pole (figs. 558, 559).

From each of the three main cerebral arteries, several series of **central branches** are given off. These very slender but impor-

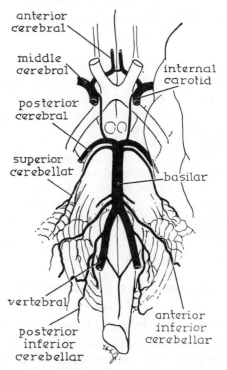

Fig. **557**. Arteries at base of brain. Note anastomoses forming circle of Willis.

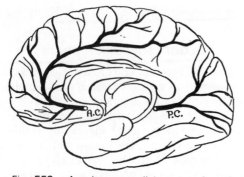

Fig. **558**. Arteries on medial aspect of cerebral hemisphere. A.C. = anterior cerebral; P.C. = posterior cerebral.

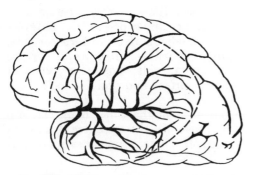

Fig. **559**. Middle cerebral artery supplies large area on lateral aspect of hemisphere enclosed by broken line.

tant branches are grouped in leashes; they plunge into the substance of the brain to supply deeper parts. It is one of these that may burst or become blocked in older people, a condition known as a 'stroke.'

The main branches at the base of the brain are intimately related to one or other of the 12 cranial nerves which also issue from the base of the brain. In consequence, a diseased artery may seriously interfere with the proper functioning of the nerve or nerves with which it is closely associated. Compare figures 557 and 587.

MENINGES AND CEREBROSPINAL FLUID

Meninges (Coverings of Brain and Cord)

The brain and cord possess for protection three complete membranous coverings known as meninges (L. = membranes).

These are the **dura mater,** the **arachnoid mater,** and the **pia mater.**‡

Dura Mater. In the cranial cavity, the dura mater acts not only as the strong, tough, fibrous, outermost covering for the brain, but also as the lining membrane on the inner surfaces of the bones (for which it is a periosteum).

Three folds of dura mater form partitions which project between the major divisions of the brain and serve to divide the cranial cavity into compartments (figs. 539, 560, 561). The two most important of these partitions are: (1) the **falx** (L. = sickle) **cerebri,** a vertical, midline (sagittal), sickle-shaped partition hanging between the two cerebral hemispheres. The anterior two-thirds of its lower border are free, but the posterior third meets at right angles and is continuous with: (2) the **tentorium cerebelli,** a tented and more or less horizontal partition between the cerebellum below and the back of the cerebral hemispheres above. The tentorium is attached round the upper margin of the posterior cranial fossa, but, of necessity, it has an aperture in the midline for the brain stem to descend to the foramen magnum. It, therefore, possesses an attached (peripheral) and a free (central) border. By dividing the cranial cavity into compartments, the dura mater gives valuable support to the brain, the partitions preventing 'shifting of the cargo' and so safeguarding the brain against injury.

Along certain lines—particularly in association with falx and tentorium—the dura contains large venous channels known as the cranial **venous sinuses.** These receive the venous drainage of the brain, and the more important ones are shown in figures 560 and 561.

For an account of the venous sinuses see page 284.

Arachnoid Mater. This is a very thin, delicate, transparent, and grayish mem-

---

‡ *Mater,* L. = mother (cf., the 'mother,' or membrane-like scum, of old vinegar or wine); *dura,* L. = tough; *arachnoid,* Gr. = 'spider-webby,' *pia,* L. = faithful.

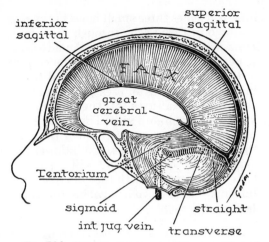
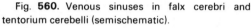

Fig. **560**. Venous sinuses in falx cerebri and tentorium cerebelli (semischematic).

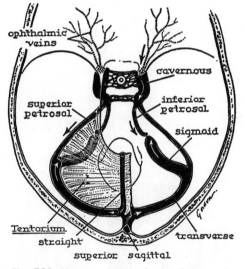

Fig. **561**. Venous sinuses (opened) from above (semischematic). Arrows show direction of flow of blood.

brane applied to the inner surface of the dura mater so that no true space exists between the two. Between the arachnoid mater and the pia mater, however, an irregular space exists—the fluid-filled **subarachnoid space.**

Pia Mater. Even more delicate than the arachnoid, the pia follows faithfully every twist and turn of the brain surface like a 'skin' while the arachnoid and dura do not.

Ventricles of Brain

The lumen or cavity in the developing brain persists as four, connecting, fluid-filled cavities called *ventricles* (figs. 562–565). The first two are deep in the cerebral hemispheres; they are the right and left **lateral ventricles** and they are filled with cerebrospinal fluid (CSF). Each opens by a small foramen into the anterior part of the midline **third ventricle,** which is little more than a vertical slit between right and left thalami. Joining the third ventricle to the fourth ventricle is a short tube in the midbrain called the **cerebral aqueduct** (of Sylvius). The **fourth ventricle,** in the back part of the pons and medulla, is mostly roofed by the cerebellum, and tapers below to the capillary-sized *central canal* of the spinal cord.

Cerebrospinal Fluid

In the subarachnoid space, and, therefore, bathed by cerebrospinal fluid, lie the principal arteries and veins of the brain. In order to supply the brain tissue, these vessels pierce the pia mater carrying tubular prolongations of it with them into the brain substance.

The subarachnoid space is most voluminous where the general contour of the brain is most irregular, i.e., at the base of the

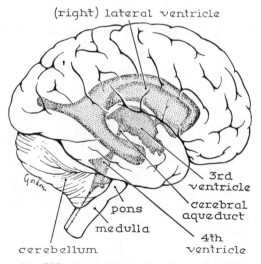

Fig. **562**. 'Transparency' to show outlines of cavities of brain.

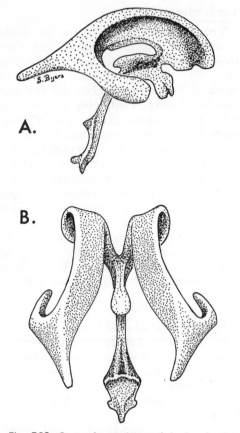

A.

B.

Fig. **563**. Cast of ventricles of brain. A, right view; B, superior view.

lum, it reaches three small holes in the roof of the fourth ventricle; through these foramina it escapes to the subarachnoid space. It next seeps upward, bathing the surface of the whole brain, and is returned to the blood in the venous sinuses by special little secretory structures of the arachnoid, known as **arachnoid villi.** These villi bulge into the venous sinuses through the dura that covers the sinuses. Cauliflower-like aggregations of arachnoid villi—particularly in association with the superior sagittal sinus—are known as **arachnoid granulations** (fig. 566).

Blockage of the escape of cerebrospinal fluid from the ventricles leads to the enlargement of the ventricular cavities at the expense of the brain tissue. Blockage of the circulation of the fluid in the subarachnoid space increases the pressure on the outer surface of the brain or spinal cord and may be just as damaging. Both conditions are known as hydrocephalus.

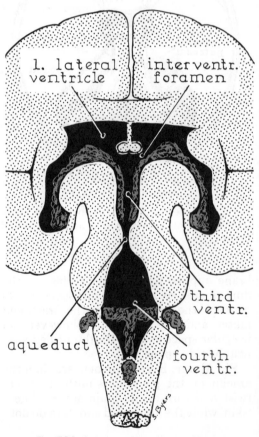

Fig. **564**. Scheme of ventricles of brain

brain where it rests on the floor of the cranium. In consequence, there are large spaces at the base of the brain, and, because they are filled with fluid, they are known as *cisterns*. These act as water-cushions or shock-absorbers for the brain.

The cerebrospinal fluid is derived from the blood, being secreted into all four ventricles in the interior of the brain. This is made possible by the fact that the walls or roofs of the ventricles are in certain places very thin (fig. 565); here, the lining membrane of the ventricles (called the *ependyma*) comes into contact with pia mater and blood capillaries. At these situations, tufts of modified capillaries hang into the ventricles. They are known as **choroidal plexuses,** and they secrete the cerebrospinal fluid into the ventricles.

The CSF flows slowly through the ventricular system until, below the cerebel-

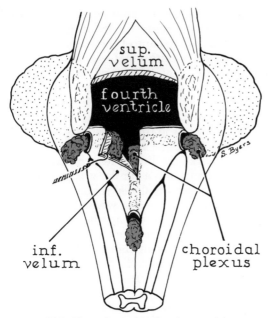

Fig. **565**. View into the fourth ventricle permitted by cutting away cerebellum. Choroidal plexus in this ventricle confined to inferior velum. Note three apertures out of which plexus protrudes and CSF escapes from ventricular system.

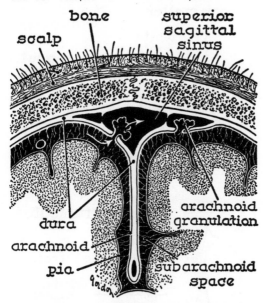

Fig. **566**. Semischematic coronal section of superior sagittal sinus and meninges. Note, in the fluid-filled subarachnoid space, the fine filaments joining pia to arachnoid mater. The arrows show direction of cerebrospinal fluid. It is returned to the blood in the sinus by the arachnoid granulations, which have ballooned through dura into the blood stream.

SPINAL CORD (SPINAL MEDULLA)

Before attempting to describe some of the more important afferent and efferent tracts or pathways, it is necessary to give a brief account of the spinal cord or medulla. Since a knowledge of the cord is essential to the understanding of the peripheral nerves, it will be considered somewhat more fully than may appear necessary. (Medulla, L. = marrow.)

Dura, arachnoid, cerebrospinal fluid in the subarachnoid space, and pia are all continued, at the foramen magnum, into the succession of vertebral foramina that goes to make up the vertebral canal. They therefore invest the spinal cord as they do the brain. At each intervertebral foramen the dura and arachnoid are prolonged out and fade away on the issuing spinal nerve.

An average vertebral foramen in a dried bone has the same diameter as that of an index finger, but the contained cord, which is considerably smaller, is not much thicker than a pencil. Surrounded by fluid, the cord is maintained in a central position by a series of tiny sawtooth-like 'ligaments' known as the **denticulate ligaments.** Alternating with the spinal nerves, the denticulate ligaments stretch on each side from the pia and are attached by their points to the dura. They, and the CSF, afford the cord protection comparable to that enjoyed by the brain (fig. 567).

In the fetus, the spinal cord was as long as the vertebral canal but, since it failed to keep pace with the growth of the canal, its lower end is found in the adult no lower than the level of the first or second lumbar vertebra. The cone-like tip is called the *conus medullaris.*

Cauda Equina. Originally, the roots of a spinal nerve had merely to pass directly lateralward in order to find their proper intervertebral foramen; in the adult, the lower roots have become increasingly lengthened and must descend considerable distances within the vertebral canal. Consequently, below the lowest limit of the spinal cord, the vertebral canal is occupied by a leash of long nerve roots; these are known collectively as the **cauda equina** (L. = horse's tail). The dura, arachnoid, subarachnoid space, cerebrospinal fluid, and pia reach as low as the second sacral vertebra. An unimportant terminal thread

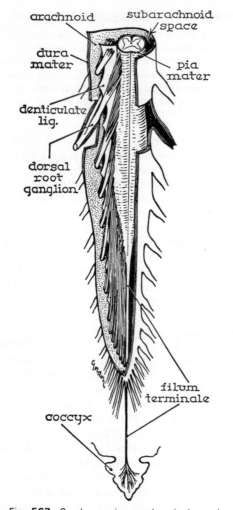

arachnoid — subarachnoid space

dura mater

pia mater

denticulate lig.

dorsal root ganglion

filum terminale

coccyx

Fig. **567**. Cauda equina and spinal cord exposed from behind. On right side, structures in subarachnoid space have been removed (semischematic).

(fig. 568) reveals an outer covering of white matter with a central mass of gray matter which is in the shape of the letter H. In the center is a minute canal—all that remains of the cavity of the neural tube. A deep groove runs down the anterior (ventral) aspect of the cord and always appears in a cross-section. A fine midline septum divides the white matter posteriorly into right and left halves. A lesser groove anterolaterally marks the line of attachment of the ventral roots, and a similar groove posterolaterally is for the dorsal roots.

The nerve cell bodies whose axons make up the anterior roots are in the anterior horn of the gray matter of the cord; they are called **anterior horn cells.** On the other hand, the bodies of the neurons whose axons make up the dorsal roots are outside the cord. These collections of cell bodies are known as the **spinal** (or *dorsal root*) **ganglia** and each spinal nerve possesses one; the ganglia are found, each in an intervertebral foramen, resting on a

of pia, known as the *filum* (L. = thread) *terminale*, extends from the conus medullaris to the back of the coccyx (fig. 567).

Structure of Cord and Spinal Nerves

Thirty-one pairs of **spinal nerves** are attached to the spinal cord. Each nerve is attached to its own half of the cord by two roots—a ventral (anterior) and a dorsal (posterior). The **ventral roots** are efferent (motor)—they 'issue from' the cord; the **dorsal roots** are afferent (sensory)—they 'enter' the cord.

A typical cross-section of the spinal cord

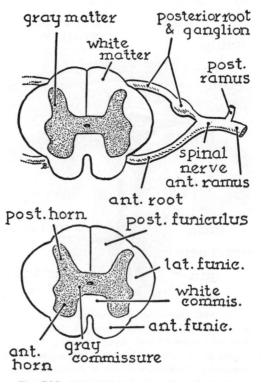

gray matter

posterior root & ganglion

white matter

post. ramus

spinal nerve

ant. ramus

ant. root

post. horn

post. funiculus

lat. funic.

white commis.

ant. funic.

ant. horn

gray commissure

Fig. **568**. Cross-sections of spinal cord and arrangement of a spinal nerve (semischematic).

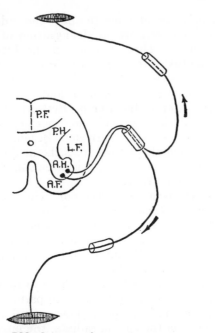

Fig. **569**. Scheme of two separate motor neurons shown with their bodies in anterior horn of gray matter.

The time spent in studying figures 568 to 572, learning the names, and understanding the connections indicated, will be repaid by the ease with which subsequent discussion can be followed. *Therefore from now on familiarity with these important diagrams will be assumed.*

EFFERENT OR MOTOR PATHWAYS

Just as most of us use a radio without knowing how it works, so we use our nervous systems. We turn a knob to select one of a dozen or more programs. Having made our choice, we turn other knobs in order to regulate tone and control volume.

In its balanced control of the muscles the nervous system also has two mechanisms. It selects its program, i.e., it decides what movements among a myriad possible ones it desires to perform. Having made its choice, it utilizes another mechanism to regulate the movement so that it becomes a smooth, coordinated, and purposeful one. Since both mechanisms impinge on the anterior horn cells whose axons provide the

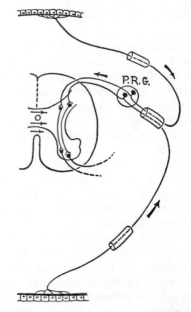

Fig. **570**. Scheme of two separate sensory neurons with cell bodies in spinal or posterior root ganglion (P.R.G.) and ending on internuncial (connector) neurons in post. horn. These, in turn, synapse with motor neurons, completing spinal reflex arc.

pedicle. Each cell body in the spinal ganglion is a part of a unipolar, afferent neuron having a T-shaped axon (fig. 570). The peripheral branch of the T is the afferent fiber of the spinal nerve; the central branch of the T enters the cord where its fate depends on its function. It will be followed in greater detail later.

Figures 568 to 575 are diagrammatic cross-sections of the cord with the names of the parts of the gray and of the white matter indicated. An afferent ('sensory') neuron and an efferent ('motor') neuron are shown forming a typical spinal nerve. The composite spinal nerve—consisting of many thousands of both afferent and efferent fibers—is formed just beyond the spinal ganglion, i.e., just outside the intervertebral foramen. It divides at once into a **ventral** (*anterior*) **ramus** (L. = branch) and a **dorsal** (*posterior*) **ramus,** each of which contains, of course, many afferent and many efferent fibers.

In addition, every spinal nerve contains fibers from neurons of the autonomic nervous system; these autonomic fibers will be discussed later.

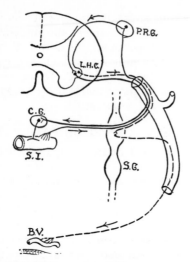

Fig. **571**. Scheme of a sympathetic reflex. Sensory neuron from an organ (S.I.) joins spinal nerve, has cell body in posterior root ganglion (P.R.G.), and synapses with neurons (L.H.C. or 'lateral horn cells') in spinal cord. These neurons, leave cord with spinal nerve, join a sympathetic chain ganglion (S.G.) where they either: (1) synapse with a fiber that travels with spinal nerve to an effector organ (B.V.); or (2) run through to celiac ganglion (C.G.) to synapse with a neuron that runs to effector organ, S.I.

only motor pathway to the muscles, the anterior horn cells are sometimes referred to as the **'final common pathway.'** Each mechanism will be briefly examined in turn.

Corticospinal Tract. This is the first mechanism and by it are activated the anterior horn cells concerned in the desired movement.

The *precentral gyrus* has already been identified as the motor area; in its cortex are large pyramidal cells (**Betz cells**) whose axons constitute the corticospinal tract (fig. 554); they run in the *corona radiata*, then in the *posterior limb of the internal capsule.* They pass out of the hemisphere in the *cerebral peduncle,* traverse the midbrain and pons, and are found on the anterior surface of the medulla oblongata in a vertical rounded column known as the pyramid. Each is separated from its companion of the opposite side by a midline groove. The lower part of this groove is obliterated by bundles of obliquely crossing fibers. Here most of the

axons cross at once to the other side in what is known as the **decussation of the pyramids.** They descend in the **lateral funiculus** of the cord (figs. 568, 575) and, at all levels of the cord, axons end by entering the gray matter and synapsing with **anterior horn cells.** Between cortex and anterior horn cell, no other neuron is involved in this pathway. Because the corticospinal tract occupies the medullary pyramid in its course, it is often called the **pyramidal tract.**

Extrapyramidal System. This is the second mechanism and involves many pathways each of which has something to contribute to the anterior horn cell whereby smooth and co-ordinated muscular action is possible.

Large cell bodies in the gyri in front of the motor area, and in certain parts of the parietal and temporal lobes, send their axons, via the corona radiata, the internal capsule, and the cerebral peduncle, to the pons in which are found scattered collections of nerve cells known as the **pontine nuclei.** Within these nuclei the axons end

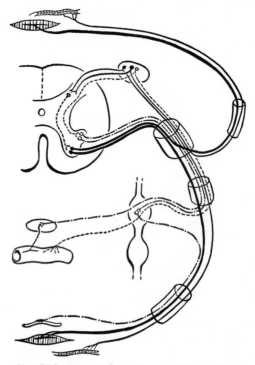

Fig. **572**. Composite of figures 568 to 571 indicates complexity of a spinal nerve.

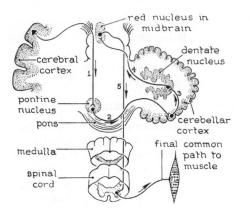

Fig. **573**. Scheme of extrapyramidal system

and from the pontine nuclei a second series of axons passes, via the **middle cerebellar peduncle,** to the opposite **cerebellar cortex.** These are the axons largely responsible for the coarse bundles of transverse fibers which are such a characteristic feature of the pons. The axons end around specialized cells (**Purkinje cells**) of the cerebellar cortex (see fig. 573).

The cerebellum, thus linked to the cerebrum and under its dominance, sends axons from the Purkinje cells to the **dentate nucleus**—a cellular mass in the depths of the cerebellar hemisphere. A fourth series of axons starts in the cells of the dentate nucleus and leaves the cerebellum, via the **superior cerebellar peduncle,** to pass upward (rostrally) to the midbrain where, after crossing to the opposite side, the axons end around cells in the **red nucleus.**

Axons from cells in the red nucleus quickly cross to the other side once more, and descending in the brain stem as the **rubrospinal tract,** enter the cord. The axons of this fifth series of neurons end around **anterior horn cells.** Thus pontine nuclei, cerebellum, dentate nucleus, and red nucleus are able to influence the anterior horn cell in ways that are not fully understood. The pathway is depicted schematically in figure 573 and, if one must give it a name, it will be necessary to call it the cerebro-ponto-cerebello-dentato-rubrospinal tract!

Besides this mechanism, which has other connections not overlooked by the neurolo-

gist, the cortex of the cerebrum can discharge directly to the red nucleus as the **corticorubal tract.** It can discharge also to the corpus striatum via the **corticostrial tract;** the corpus striatum can discharge to the red nucleus via the **striorubral** tract, and to a nucleus situated below the thalamus and known as the **subthalamic nucleus.** There are also important extrapyramidal fibers descending in the spinal cord from a network of cell bodies and fibers in the brain stem known as the **reticular formation** (*reticulospinal tract*).

It is scarcely necessary for the beginner to make himself familiar with all of these connections and they are given merely to illustrate the fact that even the simplest muscular activity requires a very complex nervous integration for its performance, so that the totality of impulses reaching the anterior horn cells from all these subsidiary centers ensures that the 'orders' for a given movement, issued to the anterior horn cells via the corticospinal tract, are carried out in a smooth co-ordinated and purposeful fashion. Disturbance by disease or injury of any part of the extrapyramidal mechanism results in loss, to varying degree, of the ability to carry out movements in a smooth and effortless fashion.

SPINAL REFLEX

We may now investigate the fate of the afferent neurons, the central processes of whose T-shaped axons enter the cord. It has been observed that all peripheral afferent neurons have their cell bodies in the **spinal** (*dorsal root*) **ganglia.** Figure 570 shows such a ganglion cell with its typical T-shaped axon forming with its fellows, the dorsal (post.) root of a spinal nerve. The peripheral process of the T-shaped axon contributes to the spinal nerve; the central process enters the cord and performs two duties: (1) it takes part in a spinal reflex; (2) it conveys information to the brain by pathways which vary according to the function served.

When the central process of a dorsal root ganglion cell enters the cord, it may divide into branches which turn at right angles and proceed upward or downward or which

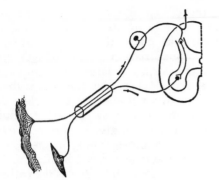

Fig. **574**. The basic functional unit of the nervous system—the simple reflex arc.

quickly enter the **posterior horn** of gray matter and end around a **posterior horn cell.** The posterior horn cell (a so-called **internuncial neuron**) sends a short axon through the gray matter to an anterior horn cell in the neighborhood; it thus completes an afferent-efferent circuit or arc; this is the simple spinal reflex arc (fig. 574).

The mechanism is an emergency one. For example, if a hot object is touched, there is an immediate withdrawal. The impulse travels into the cord which flashes out an impulse for the necessary response even before the information reaches the brain, which by then has little left to do but indulge in exclamatory responses.

It is upon the spinal reflex as a foundation that the most complex nervous system is built up. One may regard the anterior horn cell as being brought under the control in turn of sucessively 'higher' and 'higher' centers until at last a cortex comes into existence by means of which the power to exercise judgment (inhibition) over motor responses seems to be acquired.

AFFERENT OR SENSORY PATHWAYS

The spinal reflex having been discussed, it will be sufficient now to illustrate afferent pathways by describing briefly the classical concept of the paths taken by: (1) a **pain** axon; (2) a **touch** axon; (3) a **proprioceptive** axon; and (4) a **visceroceptive** axon. The pathway for **temperature** sense is the same as that for pain, and that for **pressure** is the same as that for touch. Although in recent years these ideas

have been subjected to well-deserved criticism by experimental scientists, nevertheless they serve the student well in understanding the mechanisms of most normal and abnormal functions.

1. Pain (and Temperature) Pathway

As with all sensory neurons, the unipolar cell bodies are in the spinal ganglia, and the central processes of the T-shaped axons enter the cord via the dorsal root (fig. 576).

The 'pain fibers' then turn upward. Each ascends a very short distance before turning into the gray matter and synapsing with a posterior horn cell. The axon of this second neuron arising from the posterior horn cell crosses in front of the central canal to the other side of the cord; it turns upward, forming with its fellows the **lateral spinothalamic tract** situated in the lateral funiculus of the cord (see fig. 575). This tract traverses the cord, medulla oblongata, pons, and midbrain to end in the thalamus.

2. Touch (and Pressure) Pathway

Each 'touch fiber' enters the cord via the dorsal root and runs into the posterior funiculus of the cord (fig. 576).

It now has alternative pathways: (1) the ascending branch may run up in the posterior funiculus for a variable distance and then turn into the gray matter to synapse with a posterior horn cell. The new axon crosses in front of the central canal to the

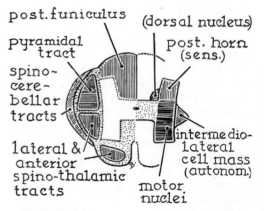

Fig. **575**. Schematic cross-section of spinal cord.

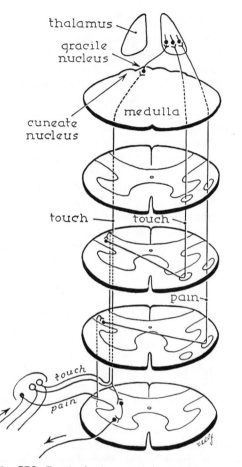

thalamus

gracile nucleus

medulla

cuneate nucleus

touch — touch

pain

touch

pain

Fig. **576**. Touch (and pressure) pathway and pain (and temperature) pathway. (Spinal reflex shown near bottom.)

other side and turns up in the **anterior (ventral) spinothalamic tract** (fig. 575); (2) the ascending branch may run up the whole length of its own side of the cord (in the posterior funiculus) and synapse with a cell in one of two centers in the medulla oblongata at the summit of the posterior funiculus and known as the **gracile** and **cuneate nuclei.** From this cell the new axon crosses in the medulla and, joining those that crossed lower down, proceeds to the thalamus. Thus for touch there is a pathway on the same side of the cord and one on the opposite side. That is why, in destruction of one-half of the cord, pain sense may be lost while touch sense is retained.

On reaching the medulla oblongata, the several exteroceptive pathways unite to form the **medial lemniscus.** The fibers that cross in the medulla are cal!ed the **(sensory) decussation of the medial lemnisci;** they then proceed upward to the thalamus with the other earlier-crossing fibers.

Thalamus

The thalamus is a great sensory center second in importance only to the sensory area of the cerebral cortex which is a 'newer' part of the brain. In man, the thalamus is under the dominance of the cortex to which it relays sensory impulses (fig. 551 on page 307).

Actually, the thalamus consists of a numerous collection of nuclei each with its own connections. Thus, its large **lateral nucleus** receives exteroceptive impulses from lower levels (via the medial lemniscus) and relays them to the postcentral gyrus, the great sensory (somaesthetic) center. Its **anterior nucleus** is a relay station on the smell pathway and discharges to the gyrus cinguli. The **lateral** and **medial geniculate bodies**—at the posterior end of the thalamus—are relay stations on the visual and auditory pathways, respectively; and a proprioceptive pathway exists whereby the thalamus can relay to consciousness sensations produced by movements of muscles, tendons, and joints.

3. Proprioceptive Pathway

a. To cerebral cortex. The 'proprioceptive fibers' enter the spinal cord with the 'touch fibers' and turn with some of them into the posterior funiculus where, mixed with the touch fibers, they ascend. Like the 'uncrossed touch fibers' they synapse with cell bodies in the **gracile** and **cuneate nuclei** of the medulla oblongata. Impulses are relayed to thalamic nuclei of the opposite side and then to the cerebral cortex (fig. 577).

b. To cerebellum. Collateral branches come off proprioceptive fibers as they enter the cord. These collaterals synapse with posterior horn cells which then send their

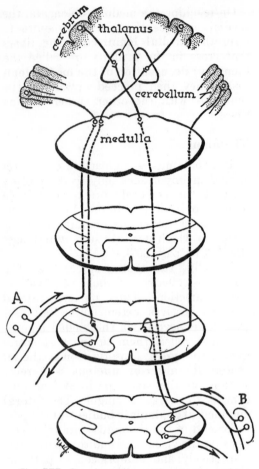

Fig. **577**. Proprioceptive pathway to cerebrum and to cerebellum; A. From upper limb. B. From lower limb of opposite side.

axons up the **spinocerebellar tracts** on the same side (fig. 577). Passing through the inferior cerebellar peduncle, these fibers end in the cerebellar cortex of the same side (fig. 577).

Because the posterior funiculi are so largely concerned in the conveyance of proprioceptive impulses, their destruction by a disease results in considerable disturbance of co-ordinated muscular activity (locomotor ataxia). Such a condition illustrates—not a motor disturbance—but a loss of vital information concerning posture, tone, tension, etc., in muscles, tendons, and joints; without this information intelligent motor responses cannot be performed smoothly and efficiently.

4. Visceroceptive Pathway

The visceroceptive axons can travel as afferent fibers with the autonomic nervous system or with the ordinary spinal nerves. Although they transmit impulses from the viscera, their cells are situated in the spinal ganglia (or their counterparts in the skull) and in no respect do they differ in anatomical structure or disposition from peripheral afferent neurons already discussed. They make up a considerable proportion of the myelinated fibers of the autonomic nervous system. The central processes of the visceroceptive cells in the spinal (post. root) ganglia enter the cord but the pathways up the cord for subsequent neurons are not established. However, they are known finally to reach the hypothalamus whence pathways exist to the (frontal) cortex.

AUTONOMIC NERVOUS SYSTEM

General

Smooth muscle, cardiac muscle, and glands have, in general, two sets of efferent neurons for the regulation of their activities. One is the **sympathetic** set; the other is the **parasympathetic** set. The two constitute the efferent or motor side of the autonomic nervous system and, together with the efferent visceroceptive neurons (described above), they provide for an autonomic reflex.

Since a good deal of confusion exists as a result of the lack of uniformity in the use of the word 'autonomic' table 2 shows how it is used in these pages.

TABLE 2

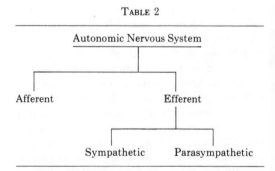

The decision for any given structure as to which of the two sets—sympathetic or parasympathetic—is stimulatory and which inhibitory depends on functional requirements and has been set forth realistically by W. B. Cannon in what he described as the 'fight or flight' mechanism. This is best explained by an illustration.

If one is awakened in the middle of the night by the sudden slamming of a door, a great many autonomic efferent impulses are set up by the stimulus. The pupils dilate, the hair stands up or at least a 'goose-flesh' condition of the skin occurs, the heart beats fast, blood is directed to the muscles which are tensed, respiration is increased. In short, the whole body is put on the alert and is prepared to do battle or to flee. At the same time those bodily functions that, for the moment, can be of no assistance, are suppressed. Digestion is slowed, the bladder wall musculature remains relatively relaxed, and the activities of the sex organs are inhibited. It is the sympathetic division of the autonomic nervous system that takes charge in an emergency and the parasympathetic that is in abeyance.

If, then, for any structure under the control of the autonomic nervous system the question is asked—"Does its activity contribute in an emergency to the defensive and offensive powers?" the answer informs us which of the two divisions is stimulatory and which inhibitory. The parasympathetic division is primarily concerned with those functional activities that operate when all is peaceful and quiet.

Both divisions are characterized by requiring two peripheral neurons (instead of one) for the conduction of the efferent impulse. In each case, the axon of the first neuron comes from a cell body in the central nervous system; it is known as the **preganglionic fiber** because it runs to a ganglion outside the CNS to synapse with the second neuron. The axon of this second cell is known as the **postganglionic fiber** and it is the 'final pathway' of the autonomic nervous system since it activates smooth muscle, cardiac muscle, or gland.

Sympathetic Division

Cell bodies of the first neurons of the sympathetic division are found only in the spinal cord, and only in a region that is restricted to between the first thoracic and the third lumbar segments. They are small cells lying lateral to the anterior horn cells in a column of cells, **lateral (intermediolateral) cell mass** or **column,** figs. 575, 578. These cells also may be called *lateral horn cells*. The preganglionic fibers issuing from these cells, because of the restricted area of the cord from which they come, are sometimes referred to as the **thoracolumbar outflow.**

Lying a little lateral to the vertebral bodies is a vertical series of ganglia connected to form a trunk or chain. It extends from the level of the firxt cervical vertebra down to the coccyx. The (paired) chain is known as the **sympathetic trunk** and its ganglia are called **paravertebral ganglia.** In figure 578 the right sympathetic trunk is represented from the lower cervical to the lower lumbar region.

The myelinated preganglionic fibers of the thoracolumbar outflow pass out of the cord with the ventral (or motor) nerve roots of the spinal nerves. They leave all the thoracic and the upper three lumbar spinal nerves to join the sympathetic trunk, running in little branches known as the **white rami communicantes**—white because myelinated fibers have a whitish color.

Some *preganglionic* fibers that reach the trunk synapse there with cells of the ganglion at the same level, as shown at level A in figure 578. Some travel up the chain to synapse with cell bodies in higher ganglia (as shown at levels B to B') while others travel down the chain (as shown at level C) to synapse.

Those *postganglionic* fibers destined for smooth muscle and glands *in regions that are supplied by spinal nerves*, i.e., those fibers that go to blood vessels, hair follicles, sweat and sebaceous glands in the body wall and in the limbs, run in the spinal nerves to those regions (e.g., levels A, B', and C in fig. 578). Thus there are short communicating nerves—**gray rami communicantes**—connecting the sympathetic chain at each and every level with the spinal nerves. These gray rami carry only postganglionic fibers and are gray because postganglionic fibers are relatively unmyelinated.

The *postganglionic fibers* to smooth

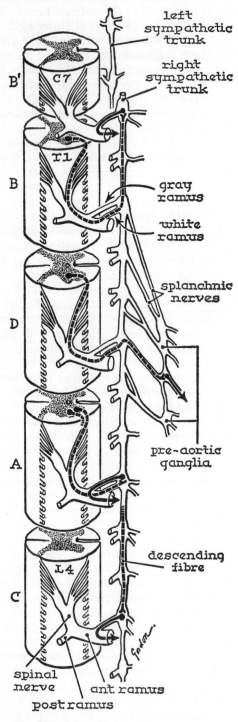

left
sympathetic
trunk

right
sympathetic
trunk

B'   C7

B   T1

gray
ramus

white
ramus

splanchnic
nerves

D

pre-aortic
ganglia

A

descending
fibre

L4

C

spinal
nerve   ant.ramus
post ramus

Fig. **578**. Scheme of sympathetic nervous system from the right side (see text).

muscle and glands *in the head region* arise from cells in the highest (and largest) ganglion—the *superior cervical ganglion* (not shown in fig. 578). These fibers climb up the arteries (internal and external carotid) and are distributed with them and their branches to their final destinations.

The *preganglionic* fibers that transmit impulses destined for the abdominal organs run the same course as any others as far as the sympathetic trunk, but instead of synapsing there, they sweep through and form nerves called *thoracic* (greater, lesser, and lowest or least) *splanchnic nerves* and *lumbar splanchnic nerves* (as shown at level D in fig. 578). These paired nerves run medially to an irregular series of ganglia around the roots of the main branches of the aorta—hence called **pre-aortic ganglia** (fig. 579). The largest and most important are the paired **celiac ganglia** at the root of the celiac artery. [The celiac ganglia and the many nerves that radiate from them like the rays of the sun are referred to as the 'solar plexus.'] The preganglionic fibers that form the splanchnic nerves synapse on the cell bodies that form the pre-aortic ganglia.

The *postganglionic* fibers going to *organs in the body cavities* run, in the main, with the blood vessels to those organs, and they cling to the blood vessel walls like ivy climbing up a tree. Although this is essentially true of abdominal organs, there are no pre-aortic ganglia in the thorax. The synapse between preganglionic fibers and the cell bodies of the postganglionic fibers takes place in the sympathetic trunk. The postganglionic fibers run as discrete nerves from the upper part of the trunk to the roots of the great vessels where they form *cardiac* and *pulmonary plexuses*. From these plexuses branches run to the heart and lungs.

Parasympathetic Division

The cells of the first neurons of the parasympathetic division are found in two widely separated regions of the central nervous system, viz., in the brain stem and in the sacral portion of the spinal cord. For this reason the parasympathetic nervous

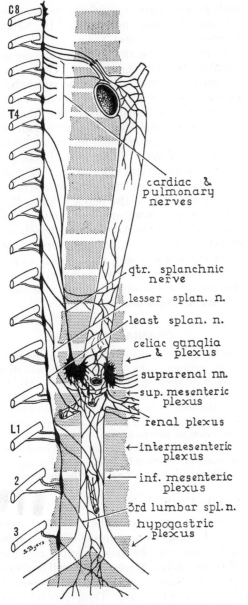

Fig. **579**. Sympathetic nerve supply to the viscera (semischematic).

C8
T4

cardiac &
pulmonary
nerves

qtr. splanchnic
nerve

lesser splan. n.

least splan. n.

celiac ganglia
& plexus

suprarenal nn.

←sup. mesenteric
plexus

renal plexus

L1

←intermesenteric
plexus

← inf. mesenteric
plexus

3rd lumbar spl. n.

2

hypogastric
plexus

3

along their courses—usually not far away from the structure to be supplied—the preganglionic fibers come to a **parasympathetic ganglion** and they end by synapsing with its cells. Postganglionic fibers arising from cells in such a ganglion carry the impulses to their terminations.

In the head region the structures receiving parasympathetic innervation are as follows:

1. Involuntary muscles of the eye concerned with the size of the pupil and the shape of the lens.
2. Glands in the mucous membrane of the mouth, palate, nose, and nasopharynx.
3. Lacrimal gland.
4. Salivary glands—submandibular, sublingual, parotid.

The great parasympathetic nerve for the organs in the thorax and in the abdomen is the **vagus nerve** (tenth cranial). The word vagus means 'wanderer' (cf., vagabond)

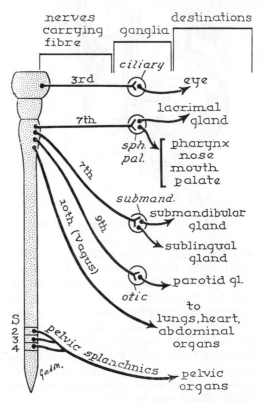

Fig. **580**. Scheme of parasympathetic nervous system (see text).

system in general is sometimes referred to as the **craniosacral outflow** (fig. 580).

The preganglionic fibers issuing from the brain stem in the head travel via certain of the cranial nerves. Fibers may travel for part of their course with one nerve and for part with another seeking only a convenient route to their destination. Somewhere

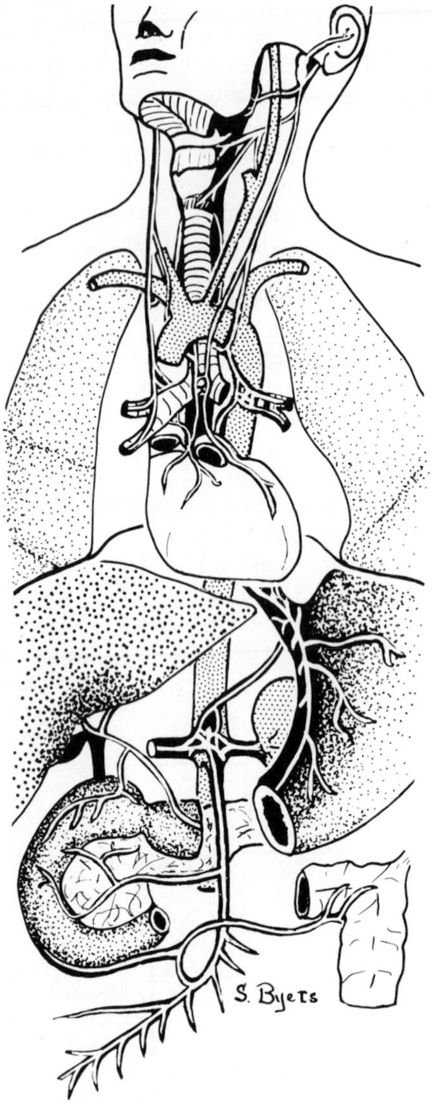

Fig. **581**. Scheme of vagus nerve (see text)

and the vagus in its wanderings supplies heart, lungs, digestive tract (and its associated organs), kidneys and, indeed, all except the organs in the pelvic cavity. The innumerable preganglionic fibers of the vagus end in the walls of the structures to be supplied by synapsing with nerve cell bodies in those walls. From these cells the impulse is conducted to adjoining smooth muscle and glands by microscopic postganglionic fibers (see fig. 581).

The parasympathetic nerves will be discussed when the cranial nerve with which

each is associated is under consideration. See under cranial nerves III, VII (Nervus Intermedius), IX, and X.

As noted above, the pelvic organs are not supplied by the vagus nerve. Their parasympathetic nerves, the **pelvic splanchnic nerves,** are branches of sacral spinal nerves two, three, and four. The fibers in the pelvic splanchnic nerves have their cell bodies in the (*intermedio*)*lateral cell mass* of the sacral part of the cord. These preganglionic fibers run in nerves S. 2, 3, and 4 and then, forming the pelvic splanchnics, run to the walls of the pelvic organs. Here they synapse with neurons having very short axons that end on smooth muscle and glands. The pelvic splanchnic nerves stimulate the bladder and rectal walls to contract, and because they cause erection of the penis (or clitoris), they are also called the **nervi erigentes.**

CRANIAL NERVES

Before discussing the 12 cranial nerves it is profitable to construct a table showing the groups of structures in the head region requiring innervation. The composition of any one of the cranial nerves can then be determined by enquiring how many and which of these groups it supplied. Table 3, shown on page 325, is intended for reference.

It will be convenient to discuss, first the nerves and pathways in the brain for smell, sight, hearing, and equilibrium, then the remainder of the cranial nerves.

Smell—Cranial Nerve I

Over an area about the size of a fingernail, the mucous membrane of the extreme upper part of the nasal cavity contains **olfactory cells.** These are the cell bodies of the first neurons and they act almost as their own end-organs. Their central processes (axons) pierce the narrow, thin, bony roof of the nose (the cribriform plate of the ethmoid) and, piercing the meninges, enter the **olfactory bulb** which rests on the nasal roof. The olfactory bulb appears merely as the slightly swollen anterior extremity of the **olfactory tract** (see fig. 544) but within it are the cell bodies (mitral cells) of the second neurons in the chain. From these cell bodies axons run backward in the olfactory tract to end in the cortex of the **uncus.** This is the forepart of the hip-

TABLE 3

No.	Names	Efferent		Afferent		
		Striated muscles	Smooth muscles, cardiac muscle, glands	Skin	Mucous membranes, organs	Special senses
I	Olfactory					Smell
II	Optic					Sight
III	Oculomotor	Extrinsic muscles of eye	Muscles of lens and iris of eye (see V)			
IV	Trochlear	An extrinsic muscle of eye				
V	Trigeminal	Muscles of mastication	Carries the parasympathetic ganglia for III, VII and IX, which are rerouted via V	Skin of face Front of scalp	Teeth and mucous membrane of tongue ($\frac{2}{3}$), mouth and nose Also taste fibers from VII	Eye
VI	Abducent	An extrinsic muscle of eye				
VII	Facial	Muscles of expression	N. Intermedius Glands of mouth, nose, palate Lacrimal gland Submandibular and sublingual (see V)			N. Intermedius Taste—anterior $\frac{2}{3}$ of tongue
VIII	Vestibulocochlear (Auditory)					Hearing, equilibrium
IX	Glossopharyngeal	Stylopharyngeus muscle	Parotid gland (see V)		Mucous membrane of eardrum, middle ear, pharynx, tongue ($\frac{1}{3}$)	Taste—posterior $\frac{1}{3}$ of tongue
X	Vagus	Muscles of pharynx	Organs in neck, thorax, and abdomen	Skin of external acoustic meatus and eardrum	Organs in neck, thorax, and abdomen	Taste (?) (epiglottis)
XI	Accessory	Muscles of larynx (via X)				
XII	Hypoglossal	Extrinsic and intrinsic muscles of tongue				

pocampal gyrus and is the cortical center where smell is received and appreciated (fig. 543).

The cortical center is associated with the cortex of the remainder of the hippocampal gyrus and, by the long association bundle known as the **cingulum,** with the cortex of the **gyrus cinguli.**

An efferent pathway known as the **fornix** (fig. 548) leads from the cortex of the uncus. It finally reaches the **mammillary bodies** in the hypothalamic region whence it is relayed to the **thalamus.** The thalamus sends axons to the gyrus cinguli.

The importance of motor responses to the sense of smell is subsidiary in man but is transcendent in lower forms—witness the motor responses smell calls forth in a dog.

Sight—Cranial Nerve II

The lining of the interior of the eye is known as the **retina.** It is, in fact, a projection of the brain looking out upon the world. The remainder of the eye protects it, nourishes it, and brings objects to a focus upon its surface.

The retina contains many chains of three microscopic neurons. Each chain includes: (1) the neuron carrying the end-organ (**a rod or a cone**); (2) the **bipolar-cell** neuron; (3) the **ganglion-cell** neuron. Rods and cones are end-organs especially designed to be sensitive to light. Rods are more sensitive than cones and are about three times as numerous; they are used in lights of low intensity (twilight vision). Cones are used in full light and alone are sensitive to color. The three neurons are shown schematically in figure 582. When one looks directly at an object, a small specialized area of the retina, the **macula lutea** (yellow spot) in a *fovea* (pit) is used; there cones alone are present.

The central processes of the ganglion cells are the fibers of the **optic nerve** (II). The fibers of the optic nerve leave the eyeball at the 'blind spot'; this is situated a little medial to the macula which is at the posterior pole of the eyeball.

An object in the visual field, unless one looks directly at it, is seen by the retina of the temporal (lateral) half of one eye and by that of the nasal (medial) half of the other eye. It is necessary that the fibers conveying the impulse from a visual field

accompany one another to the same cortical center. In order to achieve this, the 'nasal' fibers from each retina cross, behind the eyeball, in what is known as the **optic chiasma** (see fig. 583). The fibers passing back from the chiasma are known as the **optic tract** and, whereas an optic nerve represents an eye, an optic tract represents a visual field.

The optic tract of each side encircles a cerebral peduncle (fig. 583) and its fibers end according to their function. A light stimulus can be effective in three ways.

1. It can **reveal the attributes** of the object seen.
2. It can **initiate reflex adjustments** of the eye mechanism according to the intensity of the light and the distance of the object seen. This is known as the **light reflex.**
3. It can **initiate reflex** (protective) **movements** of a widespread nature whereby the head and eyes, and indeed the whole body, can be turned away from the source of 'inimical' light. This is known as the **somatic reflex.**

1. **Visual path** (fig. 583). The fibers

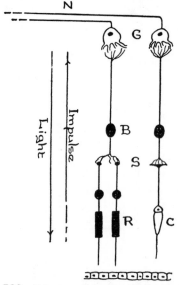

Fig. **582.** Scheme of composition of retina—chains of three neurons. Rods (R) and cones (C) are receptors. They synapse (S) with bipolar cells (B) which synapse in turn with ganglion cells (G) whose axons form fibers of optic nerve (N).

VISUAL PATH

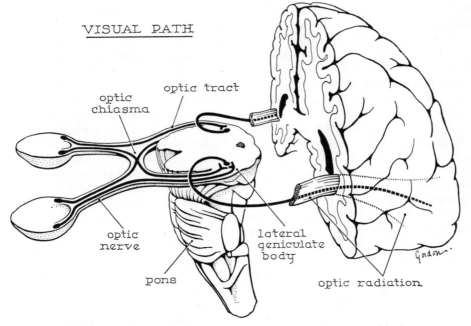

Fig. **583**. Scheme of visual pathway from retina to visual area of brain

concerned with the recognition of the object seen end in the **lateral geniculate body,** a small, ill-defined nucleus situated on the back of the brain stem under shelter of the overhanging posterior end of the thalamus; it is part of the thalamic complex. From the cell bodies in the lateral geniculate body, the axons sweep backward as the **optic radiation** (*geniculocalcarine tract*) ending in the cortex of the **calcarine sulcus** which reaches as far back as the occipital pole (see fig. 543). This cortical center is known as the **visual area.**

Destruction of the optic tract or the visual cortex results in blindness in the nasal half of one eye and the temporal half of the other.

2. Light reflex path (fig. 584). The optic nerves carry also fibers concerned with the quantitative measure of light entering the eye. These fibers do not end in the lateral geniculate body but enter the brain stem and pass to cells situated in the transition zone between thalamus and midbrain.

From cell bodies in this so-called *pretectal region*, axons pass to a small collection of cells situated farther forward at the front end of the motor cells of the third (oculo-

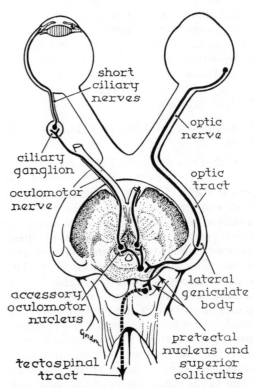

Fig. **584**. Scheme of light reflex pathway (from retina, through midbrain, to sphincteric muscle of iris) and somatic reflex pathway (see text).

motor) nerve (see page 330). This collection is called the **accessory oculomotor nucleus** (of Edinger-Westphal) and from its cells, axons pass out of the brain with the third nerve and by it are conducted to the little **ciliary ganglion** at the back of the eyeball. These axons are parasympathetic preganglionic fibers, and the ciliary ganglion is the site of origin of the postganglionic fibers which enter the eyeball and cause constriction of the pupil and an increased convexity of the lens for focussing on a near object.

3. **Somatic reflex path** (fig. 584). The fibers of the optic nerve concerned with the somatic reflex end in cells of the **superior colliculus.** The colliculi, two superior and two inferior, are four rounded swellings that go to make up the roof or dorsal aspect of the midbrain; they are sometimes referred to as the *corpora quadrigemina* (figs. 552, 553, 584). They are somatic cell stations and from the cells of the superior colliculi axons pass: (1) upward, as the **tectobulbar tract** to the cells of the motor cranial nerves of the eye (NN. III, IV, and VI); and (2) downward (in the cord), as the **tectospinal tract** to anterior horn cells. By these paths, widespread motor responses can occur reflexly in response, say, to a flash of lightning.

Hearing—Cranial Nerve VIII, Cochlear Part

Within the forepart of the petrous temporal bone lies the highly specialized **spiral organ** of Corti (fig. 633, page 357). It is an end-organ on the first neurons of the auditory pathway and is stimulated by sound waves. The cell bodies of these neurons constitute the *spiral ganglion* and it, too, lies within the petrous bone. The central processes make their exit by the internal acoustic meatus situated on the posterior surface of the petrous bone; the resulting nerve is known as the **cochlear nerve**—one division of the **vestibulocochlear (eighth cranial) nerve.** Almost at once, the cochlear nerve enters the brain stem at the lower border of the pons (figs. 585, 587).

Just deep to the site of entrance of the cochlear nerve, lie the **cochlear nuclei** around whose cells the axons of the first

neuron end. The axons of the second neuron cross deeply in the pons to the other side and turn upward through the midbrain as the **lateral lemniscus.** The lateral lemniscus can be seen as a plate of fibers on the lateral surface of the midbrain. The fibers end by synapsing with cell bodies in the **medial geniculate body** which also lies under shelter of the posterior end of the thalamus medial to the lateral geniculate body; it is a small but well defined collection of cells. From these cells, axons pass directly lateralward, beneath the lentiform nucleus, to end in the **hearing center** in the cortex of the temporal lobe (see fig. 541–H).

Just as the superior colliculus is a somatic reflex station on the visual path, so the **inferior colliculus** is a **somatic reflex station** on the auditory path. It possesses similar connections and, by its agency, reflex and widespread motor responses are made as a result of a sudden and unusual sound.

Equilibrium—Cranial Nerve VIII, Vestibular Part

Within the petrous temporal bone lie also the three semicircular canals (fig. 631, page 356). Postural changes bring about disturbances in the fluid and sense-organs and these disturbances set up impulses transmitted by the **vestibular nerve**— the other division of the **eighth cranial nerve.** The cells of the vestibular nerve constitute the **vestibular ganglion** which also lies within the petrous bone, and the central processes of these cells leave the petrous

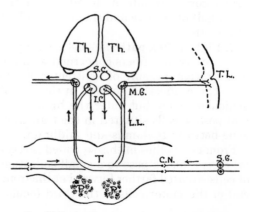

Fig. **585**. Auditory path in brain (see text)

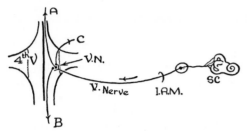

Fig. **586**. Vestibular path (see text)

bone at the internal auditory meatus and in company with the cochlear division (fig. 586).

The axons reach a series of well-recognized nuclei, known as the **vestibular nuclei,** lying in the medulla oblongata deep to the floor of the fourth ventricle; several paths are open to them.

1. They may proceed without interruption right to the cerebellar cortex, or they may be relayed to the cerebellar cortex by cells in the vestibular nuclei.
2. They may be relayed by cells in the vestibular nuclei: (a) upward to the motor cranial nerves especially those of the eye; (b) downward as the **vestibulospinal tract** to the anterior horn cells.

The vestibulocerebellar fibers enter the cerebellum via the inferior cerebellar peduncle and they inform that organ of postural changes. The pathways up the brain stem and down the spinal cord permit reflex motor adjustments of the head, eyes, and trunk in response to postural changes.

Functions of Cerebellum

These may now be briefly summarized.

1. It receives **proprioceptive impulses via the spinal cord** (spinocerebellar tracts) from tendons, joints, and muscles. These enter by the inferior cerebellar peduncles and end largely in the anterior lobe of the cerebellum (see figs. 552, 555).
2. It **receives proprioceptive impulses from the semicircular canals.** These enter by the inferior cerebellar peduncles and end largely in the flocculonodular lobe.
3. It is **under the dominance of the cerebral cortex** via the middle cerebellar peduncles which end in the very large posterior lobe.
4. Its efferent pathway, whereby it produced its motor effects, leaves by the superior cerebellar peduncles and is part of the **extrapyramidal mechanism** (see page 316).

Ganglia in Region of Head

It has been observed that the cell bodies of all peripheral sensory (afferent) neurons lie outside the central nervous system. This is as true for the cranial nerves as it is for the spinal, so that, associated with those cranial nerves that carry sensory fibers, are ganglia *comparable in many respects to spinal ganglia on dorsal roots.* Reference to table 4 will make clear that the nerves that posses such ganglia are V, VII (N. Intermedius), VIII, IX, and X. Moreover, just as the central processes of many spinal ganglia cells enter the cord and end around posterior horn cells, so the central processes of cells of these cranial ganglia enter the brain stem and end around what are called sensory nuclei or **nuclei of reception.** In general, impulses pass from these nuclei of reception to the thalamus and by it are relayed to the cerebral cortex.

It was earlier observed, too, that the preganglionic fibers carried by cranial nerves III, VII (N. Intermedius), IX, and X end by synapsing with neurons in parasympathetic ganglia (table 3, p. 325). From the ganglia there arise postganglionic fibers for the supply of smooth muscle and glands in the head region. These parasympathetic ganglia are, of course, 'motor' ganglia in the head region and are not to be confused with the sensory ganglia just noted. The names of the ganglia in the head region are given in table 4.

TABLE 4

Sensory Ganglia (Afferent)	Parasympathetic Ganglia (Efferent)
Trigeminal (V)	Ciliary
Geniculate (taste)	Sphenopalatine
(N. Intermedius)	Submandibular
Spiral } (VIII) Vestibular	Otic
Petrosal (IX)	All attached to nerve
Jugular (Superior) } (X) Nodose (Inferior)	V for convenience of distribution

Just as the nuclei of origin of the motor spinal nerves (the anterior horn cells) lie within the spinal cord, so the nuclei of origin of the motor cranial nerves lie within the brain stem. They extend from midbrain to medulla oblongata and exist for all the cranial nerves with the exceptions of I, II, and VIII, which have already been seen to be nerves of special senses. These nuclei of origin are under the dominance of the motor cortex and for them the corticospinal tract finds its counterpart in the so-called **corticobulbar tract.**

Origins of Cranial Nerves

The site of exit from the brain of each cranial nerve is known as its **superficial origin.** The site within the brain of its nucleus of reception or (and) nucleus of origin is known as its **deep origin.** Figure 587 shows the superficial origins of the cranial nerves.

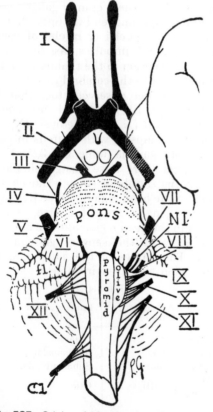

Fig. **587**. Origins of 12 pairs of cranial nerves from surface of brain. (NI = nervus intermedius, part of cranial VII.)

Cranial Nerves III, IV, and VI

These are three, highly specialized, cranial nerves entirely devoted to the extrinsic muscles of the eye. They are, therefore, motor, and are known as the **oculomotor nerve (III),** the **trochlear nerve (IV),** and the **abducent nerve (VI).** From their superficial origins these nerves cross the apex of the petrous temporal bone, lie in the cavernous sinus on the side of the body of the sphenoid, and enter the orbit through the superior orbital fissure. The trochlear and abducent supply one muscle each; the oculomotor supplies the remainder. As has been seen, the oculomotor also carries preganglionic fibers to the **ciliary ganglion** for the supply of the smooth muscles of the lens and iris.

Trigeminal Nerve—V

The large fifth cranial nerve is called the **trigeminal nerve** because it branches into three major divisions: the **ophthalmic,** the **maxillary,** and the **mandibular.** The ophthalmic and maxillary are entirely sensory, the mandibular is mixed (see figs. 588, 589).

The trigeminal is the sensory nerve of the face, the anterior half of the scalp, the mouth, the teeth, the anterior two-thirds of the tongue, and the nose; it is the motor nerve to the four muscles of mastication. Many of its terminal branches carry parasympathetic efferents.

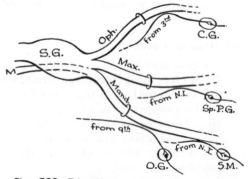

Fig. **588**. Trigeminal nerve (cranial V) has a sensory root with a ganglion (S.G.) and a motor root (M.) which joins mandibular division. Each division also carries parasympathetic fibers to synapse in related ganglia—ciliary (C.G.), sphenopalatine (Sp. P.G.), submandibular (S.M.), and otic (O.G.). (N.I. = nervus intermedius.)

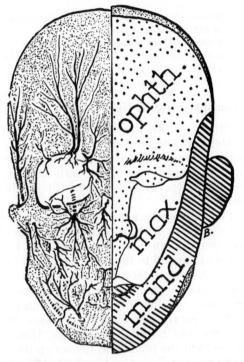

Fig. **589**. Scheme of distribution of three divisions of trigeminal nerve (Cranial V).

After leaving the ventral surface of the pons well laterally, the coarse-bundled trigeminal nerve passes forward to the large **trigeminal ganglion** (sensory) situated in a shallow bony depression at the apex of the petrous bone. The ganglion serves as a 'spinal (dorsal root) ganglion' for all sensory fibers of the nerve and from it the three divisions of the nerve radiate forward; they all lie in the lateral wall of the cavernous sinus.

The **ophthalmic division** enters the orbit through the superior orbital fissure. It divides into many branches.

1. Some travel along the orbital roof, turn up at the orbital margin, and supply the skin of the forehead and front half of the scalp. The largest of these is known as the **supraorbital nerve.**
2. One branch runs a circuitous course and supplies the ethmoidal air cells, the mucous membrane of the front half of the nasal cavity, and the skin of the nose.
3. A leash of branches (**ciliary nerves**) constitutes the great sensory supply of the whole eyeball. They pierce the eyeball around the entrance of the optic nerve.

The **maxillary division** passes through the foramen rotundum and then lies in the pterygopalatine fossa at the back of the face and in front of the pterygoid plates (see page 42). From this division branches pass:

1. Forward, along the floor of the orbit to appear on the face as the **infraorbital nerve;** it is a sensory nerve to the face.
2. Medialward, into the nose to supply the back part of the mucous membrane.
3. Downward, into the mouth to supply the mucous membrane of the palate and of the adjacent part of the pharynx (*palatine nerves*).
4. Into minute canals in the maxilla to supply the upper teeth and gums as well as the mucous membrane of the maxillary air sinus.

The **mandibular division** passes through the foramen ovale. It alone of the three divisions carries motor fibers and is the motor nerve of the masticatory muscles. Its sensory branches supply:

1. The skin of the face over the mandible and side of the head.
2. The mucous membrane of the floor of the mouth.
3. The lower teeth.
4. The mucous membrane of the anterior two-thirds of the tongue (*lingual nerve*).

Facial Nerve—VII; and Nervus Intermedius

The seventh or facial nerve (figs. 590, 591) is the motor nerve of the muscles of expression. Its superficial origin from the brain is so closely associated with that of

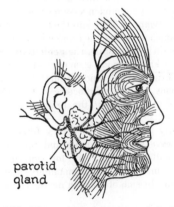

parotid gland

Fig. **590**. Five or six branches of facial nerve (cranial VII) radiate from parotid gland to supply facial muscles.

the eighth nerve that the two appear almost as one nerve. Furthermore, the facial nerve accompanies the eighth nerve into the interior of the petrous bone by entering the internal acoustic meatus. It tunnels the bone between the organs of hearing and equilibrium and, on reaching the lateral wall of the internal ear, turns downward at right angles near the posterior wall of the middle ear cavity to escape from its bony canal at the stylomastoid foramen. It next swings into the substance of the parotid gland and divides into numerous and radiating branches which reach from scalp to neck. It is entirely devoted to the facial and scalp muscles.

Nervus Intermedius. Table 3 (p. 325) implies that the facial nerve is by no means purely motor, as described above. It is there indicated as possessing afferent fibers from the tongue and efferent fibers for many glands. This apparent contradiction is explained by the fact that accompanying the facial nerve is a small mixed cranial nerve that has not yet been given numericial status; it has, unfortunately, been bundled descriptively, as it is bundled structurally with the facial nerve. It is known as the **nervus intermedius** and it carries taste fibers from the tongue (anterior two-thirds) and secretory fibers to many glands (see fig. 591).

The afferent **taste fibers** find it convenient to leave the tongue with the ordinary sensory nerve of the tongue, viz., the lingual nerve, a branch of the fifth nerve. They leave the fifth and join the seventh (facial) nerve by running in a connecting nerve called the *chorda tympani.* They join the facial nerve in the petrous bone just behind the tympanic cavity (middle ear) since their cell station, the **geniculate ganglion** (genu, L. = knee or bend), is on the facial nerve where it makes a right-angled bend within the interior of the petrous bone. The central processes of the cells in the geniculate ganglion enter the brain stem with the facial nerve.

Similarly the efferent **secretory fibers** (preganglionic) leave the facial nerve and run to ganglia (parasympathetic) which are for convenience attached to branches of the maxillary and of the mandibular division of the fifth nerve. These ganglia are:

1. **Pterygopalatine (or sphenopalatine) ganglion,** lying in the pterygopalatine fossa (page 42), whence postganglionic fibers run to the lacrimal gland and to the glands of the nose, mouth, palate, and pharynx.

2. **Submandibular ganglion,** lying beneath the mucous membrane of the floor of the mouth, whence postganglionic fibers run to the submandibular and to the sublingual gland.

Glossopharyngeal Nerve—IX

From its superficial origin by several fine rootlets on the side of the medulla, the ninth or glossopharyngeal nerve leaves the skull in company with the tenth and 11th nerves via the medial part of the jugular foramen.

The glossopharyngeal nerve is the great sensory nerve of the mucous membrane of the pharynx and all its extensions. Descending medial to the styloid process (fig. 592), the nerve accompanies the Stylopharyngeus (which arises from that process) and, having supplied that muscle, with it enters the pharynx between the superior and middle constrictors. Here it lies behind the tonsil and breaks up into a leash of small twigs which supply the mucous membrane of the pharynx, the tonsil, and the posterior third of the tongue.

Just after leaving the skull the nerve gives a branch for the supply of the mucous membrane of the middle ear cavity, the mastoid air cells, and the auditory tube.

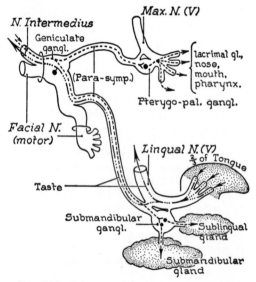

Fig. **591**. Scheme of facial nerve and nervus intermedius.

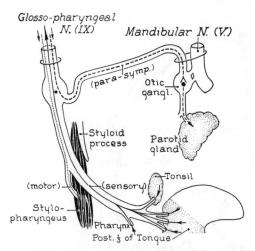

Fig. **592**. Scheme of glossopharyngeal (cranial IX) nerve; a mixed nerve with sensory, motor, and parasympathetic fibers. Twigs to middle ear and the carotid nerve are not shown. (Parasympathetic branch is the lesser superficial petrosal nerve.)

The glossopharyngeal nerve carries **taste fibers** centrally from the posterior third of the tongue. It also carries parasympathetic preganglionic fibers peripherally and these are relayed by the **otic ganglion** (situated just below the foramen ovale) to the **parotid gland.**

Two ganglia (superior and inferior) form small swellings on the nerve just after it leaves the skull. They are the cell station for all sensory branches of the glossopharyngeal, including taste.

A small but important little branch of the ninth nerve, known as the **carotid nerve,** ends on the wall of the internal carotid artery where the common carotid bifurcates. In the medulla is a center for the regulation of blood pressure and it is from the carotid nerve that this center receives its impulses.

Vagus Nerve—X

The superficial origin of the tenth cranial or vagus nerve is by rootlets on the side of the medulla in series with those of the ninth. Passing out of the skull through the jugular foramen with the ninth and 11th nerves, the vagus descends vertically on the side of the pharyngeal wall. Here it is wrapped up in the carotid sheath with the internal and common carotid arteries and the internal jugular vein (fig. 593).

At the root of the neck the *right vagus* crosses in front of the subclavian artery and enters the thorax behind the great veins; it is found on the side of the trachea

which conducts it to the back of the root of the right lung where it breaks up into the **right pulmonary plexus** (plexus, L. = network; plaiting) (fig. 594).

The *left vagus nerve* descends in the interval between the left common carotid and left subclavian arteries. It crosses the lateral side of the aortic arch and then passes behind the root of the left lung to lose its identity in the **left pulmonary plexus.** The vagi—the parasympathetic components of these plexuses—are responsible for constricting the bronchi.

From each pulmonary plexus several nerves pass to the esophagus to form and **esophageal plexus**—a network on that tube.

Shortly before piercing the diaphragm, the esophageal plexus gathers together as the **anterior (left)** and the **posterior (right) vagal trunks** and these—the only branches of the two vagi in the abdomen —pass to the corresponding surfaces of the stomach. The gastric branches supply the smooth muscles of the stomach wall and of its blood vessels; they also supply the gastric glands. Branches from the vagal trunks may be traced to the liver, duodenum, pancreas, spleen, and kidneys.

From this point on, the identities of the vagi are lost since they become inextricably intermingled with sympathetic fibers in the **celiac** (and subsidiary) **plexuses** that lie along the abdominal aorta and its branches.

Just below the skull, two swellings appear on the vagus. The smaller upper one is the **jugular (superior) ganglion** and the larger lower one is the **nodose (inferior) ganglion.** Both act as a 'spinal ganglion' for sensory fibers traveling with the vagus (page 329).

High up in the neck the vagus gives off: (1) a **pharyngeal branch** which, forming a plexus on the wall of the pharynx with fibers from the glossopharyngeal nerve, supplies most of the muscles of the pharynx and soft palate; (2) a **superior laryngeal branch** which has a sensory branch to the mucous membrane of the larynx and epiglottis, and a motor one to the Cricothyroid.

A **recurrent** (or **'inferior'**) **laryngeal nerve** comes off the vagus. On the *left side,* it arises in the thorax and recurs by hooking around the aortic arch; on the *right*

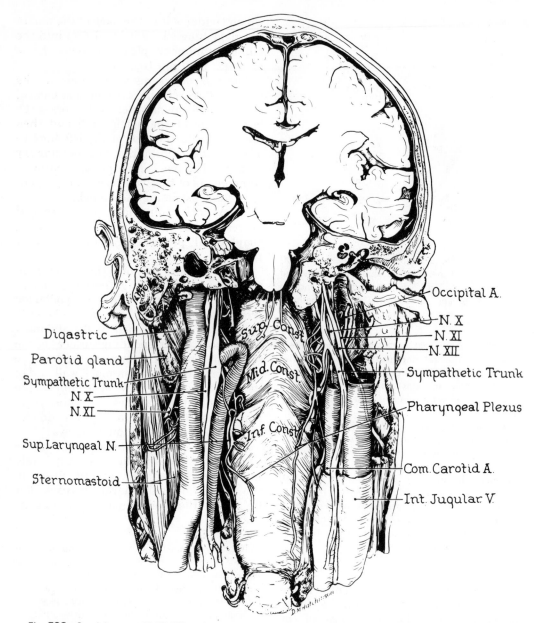

Fig. **593**. Cranial nerves X, XI, XII and sympathetic trunk running with great vessels on side of pharynx (seen from behind).

side, it arises in the root of the neck and recurs by hooking around the right subclavian artery. Both ascend in the angle between the trachea and the esophagus to reach the larynx where they are the motor nerves for the laryngeal muscles. They are in danger in an operation for goiter; should they be cut, the muscles of the larynx are paralyzed and the voice is lost (fig. 594).

Cardiac branches arise from the vagus in both the neck and the thorax. They are responsible for slowing the heart beat.

In summary, the vagus is both a sensory and a parasympathetic nerve for the heart,

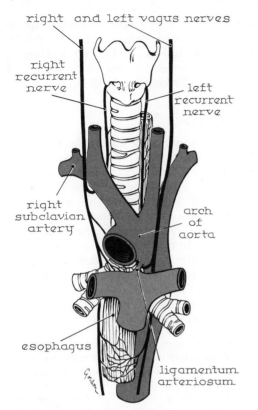

Fig. **594**. Course and relationships of vagus and recurrent laryngeal nerves.

('spinal accessory') that descends in the neck to supply the Sternomastoid and the Trapezius. Figure 595 depicts the nerve schematically (see also fig. 596).

Hypoglossal Nerve—XII

The 12th cranial or hypoglossal nerve arises by rootlets from the front of the medulla and leaves the skull via the hypoglossal canal. It joins the ninth, tenth, and

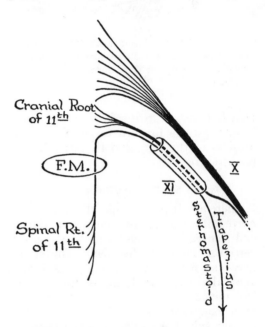

Fig. **595**. Scheme of accessory (XI) nerve. F.M. = foramen magnum.

respiratory tract, digestive tract as far as the transverse colon, the liver, pancreas, spleen, and kidneys. It is also the motor nerve to the pharyngeal and laryngeal muscles.

Accessory Nerve—XI

The 11th cranial or accessory nerve is a peculiar one. Part of it, the so-called **cranial root,** belongs to the vagus which it soon joins to be distributed chiefly to the voluntary muscles of pharynx, soft palate, and larynx.

Part of the nerve, the so-called **spinal root,** consists simply of nerve fibers from the upper five segments of the spinal cord which have entered the foramen magnum and have left the skull in company with the cranial portion as the definitive accessory nerve. Those fibers that arise from the cord gather here to form a separate motor nerve

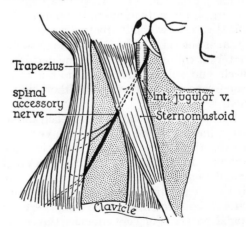

Fig. **596**. Right accessory nerve crossing posterior triangle.

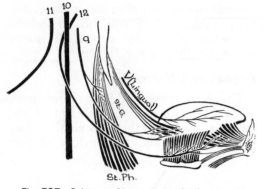

Fig. **597**. Scheme of hypoglossal (XII) nerve. Also seen: nerves IX, X, XI, and a sensory branch of V, the lingual n., to tongue. St. G. = Styloglossus; St.Ph. = Stylopharyngeus.

11th but at the greater horn of the hyoid bone it swings forward to supply all the muscles of the tongue both extrinsic and intrinsic (fig. 597).

SPINAL NERVES

There are 31 pairs of spinal nerves. Each nerve is formed by the union of its **dorsal** and **ventral roots** just outside the intervertebral foramen (fig. 598). The spinal nerve so formed at once divides into a **ventral** and a **dorsal ramus.** Each ramus is a mixed nerve consisting of fibers from both roots (see figs. 567–572).

Each spinal nerve gets its name from the part of the vertebral column where it emerges from the vertebral (neural) canal. In turn, it gives its name to the part of the spinal cord to which its roots are attached. Because the spinal cord is not as long as the vertebral column (page 313), the vertebral levels and spinal cord levels do not coincide—especially below the midthoracic region.

Except in the cervical region, each spinal nerve has the same name as the vertebra above, thus there are 12 thoracic, five lumbar, five sacral, and one coccygeal spinal nerves. In the cervical region, the first nerve emerges *above* the first vertebra (atlas) and so there are eight cervical spinal nerves. For convenience, the terms cervical, thoracic, lumbar, and sacral are often abbreviated to **C., T., L.,** and **S.,** respectively, and 'spinal nerve' is under-

stood (e.g., C. 8 = eighth cervical spinal nerve). The student using these abbreviations must make clear that he is referring to spinal nerves, in view of the fact that these designations are used also for vertebrae.

Dorsal Rami

A typical dorsal ramus passes backward between the transverse processes and, on reaching the territory of the intrinsic or native muscles of the back, divides into a medial and a lateral branch; these supply the muscles of the back and the skin over them, piercing and ignoring any muscles of the upper limb girdle that may have migrated to the region. The dorsal rami are segmentally disposed and are found from the back of the head to the coccyx. Generally, they are much smaller than the corresponding ventral rami.

The following additional facts are of interest.

1. The territory of the posterior rami lies, in general, between the posterior angles of the ribs of the two sides.
2. The posterior ramus of C. 1, unlike most posterior rami, is much larger than the anterior ramus. It supplies muscles in the suboccipital

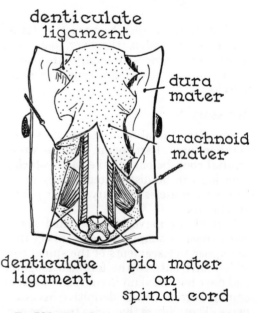

Fig. **598**. Dorsal view of a segment of spinal cord in its meningeal coverings showing rootlets forming spinal nerves. The row of dorsal rootlets are cut short.

region and is known as the **suboccipital nerve.**

3. The posterior ramus of C. 2, also larger than the anterior ramus, is known as the **great occipital nerve.** After contributing to the supply of the muscles of the suboccipital region, it supplies the skin of the back of the head as high as the vertex.

Ventral Rami

The ventral rami of the spinal nerves are, as a rule, larger and more important than the dorsal ones. They supply not only the skin and muscles of the trunk (viz., those of the front of the vertebral column, the oblique, and the rectus musculature) but also the whole of the limbs. They tend to form plexuses (nerve networks) whereby more than one spinal nerve can contribute to the formation of a definitive nerve, e.g., the radial nerve is composed of fibers from spinal nerves C. 5, 6, 7, 8, and T. 1.

CERVICAL PLEXUS

The anterior rami of C. 1, 2, 3, and 4, after issuing in the neck, unite by loops to form the **cervical plexus.** At its origin, the plexus is hidden by the Sternomastoid.

Cutaneous branches radiate from the middle of the posterior border of Sternomastoid. Thus, branches pass: (1) upward, to the skin over the angle of the jaw and the back of the ear; (2) medialward, to the skin of the front of the neck; (3) downward, to the skin over the side of the neck and the front of the thorax as far as the level of the second rib and reaching from sternum to acromion. These nerves are depicted and named in figure 599.

Motor branches supply the 'strap' muscles of the neck, the muscles in front of the cervical vertebrae, the Levator Scapulae, and the scalene muscles, as well as contributing branches to the Trapezius and Sternomastoid.

Phrenic Nerve. An important branch of C. 4, with contributions from C. 3 and 5, is known as the phrenic nerve and, descending in front of the Scalenus Anterior, it enters and traverses the thorax to reach the diaphragm; it is the great motor nerve for that important muscle and, in addition, it is sensory to the full thickness of the diaphragm.

BRACHIAL PLEXUS

The ventral rami of C. 5, 6, 7, 8, and T. 1 issue, in the lower part of the neck, between the Scalenus Anterior and the Scalenus Medius. They at once form the **brachial plexus.** This runs obliquely downward and lateralward and passes over the first rib and behind the middle third of the clavicle to enter the axilla: it is the whole nerve supply for the upper limb. The manner of formation of the plexus, its parts, and its collateral and terminal nerves are shown schematically in figures 600 and 601.

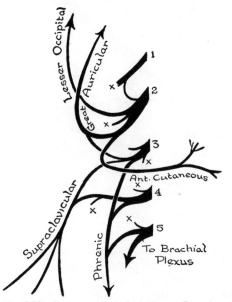

Fig. **599**. Scheme of cervical plexus. Branches to muscles of neck marked X.

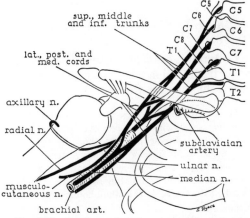

Fig. **600**. Course and main relationships of right brachial plexus (semischematic).

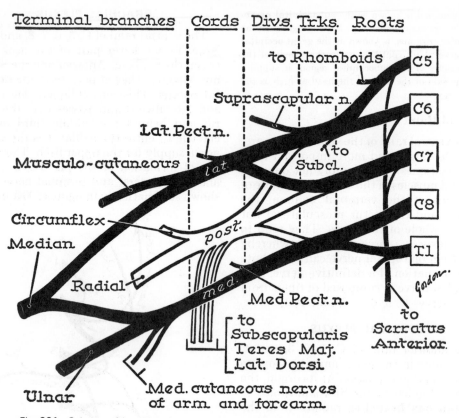

Fig. **601**. Scheme of brachial plexus, its parts, and its branches (right side, from in front)

It will suffice to give a brief account of each collateral and terminal nerve in turn.

For descriptive purposes it is usual and useful to designate the parts of the plexus as the **roots** (the ventral rami themselves), the **trunks,** the **divisions,** and the **cords.**

Collateral Branches of Roots and Trunks

Collateral branches usually appear at sites from which they can most readily reach the muscles they supply.

Thus, from the roots of the plexus twigs are given to the Longus Cervicis on the cervical vertebral bodies, and to the two muscles between which the roots issue, viz., Scalenus Medius and Scalenus Anterior.

Similarly, a branch from the back of C.5 can at once pass down the vertebral border of the scapula; it is called the **nerve to the Rhomboids** (and Levator Scapulae).

The **nerve to Serratus Anterior (long**

thoracic nerve) descends from the backs of the upper three roots (C. 5, 6, 7) and runs vertically down the chest wall on the lateral surface of the muscle, which is applied to the upper eight ribs; by so doing it can supply each digitation of the Serratus Anterior as it meets it.

The **suprascapular nerve** (C. 5, 6) springs from the upper trunk, crosses the root of the neck behind the clavicle, and reaches the notch on the upper border of the scapula near the caracoid process. It first supplies the Supraspinatus; then, at the neck of the scapula, it supplies the shoulder joint; and, passing into the infraspinous fossa, it supplies the Infraspinatus.

A small but important twig from the same trunk supplies the Subclavius (C. 5, 6).

No collateral branches arise from the divisions of the plexus, and the remaining branches are all from the cords; they are found in the axilla.

Collateral Branches of Cords

From the **lateral cord,** the **lateral pectoral nerve** supplies the Pectoralis Major.

From the **medial cord,** the **medial pectoral nerve** supplies the Pectoralis Minor and contributes to the supply of Pectoralis Major. Just where it divides into its terminal branches (see below), the medial cord gives off two (long) cutaneous branches whose names sufficiently proclaim their area of supply. They are the *medial cutaneous nerve of the arm* and the (larger) **medial cutaneous nerve of the forearm.**

The **posterior cord** gives off collateral branches for the three muscles of the posterior wall of the axilla—Subscapularis, Teres Major, and Latissimus Dorsi.

The Five Terminal Branches

Figure 601 illustrates the composition of the terminal branches. Each cord bifurcates, but, since a branch of the lateral cord joins a branch of the medial cord to form the median nerve, there are only five terminal branches of the plexus. The median nerve is said to have a *lateral* and a *medial root.* The musculocutaneous nerve is the other terminal branch of the lateral cord, while the ulnar nerve is from the medial cord. The radial and circumflex nerves are the terminal branches of the posterior cord. All of these nerves are extremely important and persons who deal with patients with nerve injuries must be able to visualize figure 601 clearly.

1. The **musculocutaneous nerve** (C. 5, 6, 7) supplies the muscles of the front of the arm (fig. 602). It pierces the Coracobrachialis, runs between the Biceps and the Brachialis, supplies all three, and becomes cutaneous 2 to 3 cm (1″) above the elbow, on the lateral side of the Biceps. It is now known as the **lateral cutaneous nerve of the forearm** and it supplies the skin of the lateral half of the forearm, both front and back, as far down as the root of the thumb.

2. The **median nerve** (C. 5, 6, 7, 8; T. 1), formed from the lateral and medial cords of the brachial plexus, is the great 'flexor' nerve of the upper limb. It accompanies the axillary and brachial arteries but has no branch until it reaches the elbow region;

there it supplies the following flexor muscles arising from the medial epicondyle: Pronator Teres, Flexor Carpi Radialis, Palmaris Longus, and Flexor Digitorum Superficialis (fig. 602).

As it passes through the cubital fossa—the region of the bend of the elbow—the median nerve gives off a deep branch, the **anterior interosseous nerve,** which supplies the Flexor Pollicis Longus,

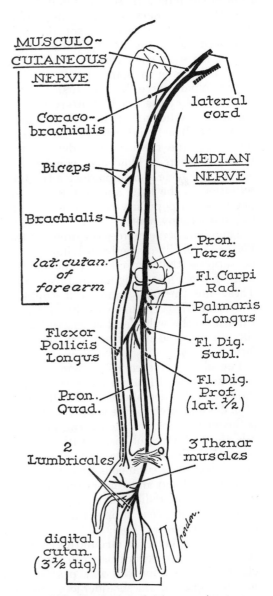

Fig. **602.** Distribution of right musculocutaneous and median nerves (semischematic).

the lateral half of the Flexor Digitorum Profundus, and the Pronator Quadratus.

The median nerve passes deep to the Pronator Teres and traverses the forearm by clinging to the deep surface of the Flexor Digitorum Superficialis.

It enters the palm (having supplied the skin of the palm) with the long tendons and supplies the three thenar muscles, the two lateral Lumbricals, and the skin on the front of the lateral three and a half digits.

3. The **ulnar nerve** (fig. 603) might well be entitled the 'accessory flexor' nerve. It has been called the nerve of fine movements, since its chief concern is with the small muscles of the hand.

The ulnar nerve lies at first on the medial side of the brachial artery but, at about the midlength of the arm, it passes backward. It is found behind, and in contact with, the medial epicondyle of the humerus where a blow may compress it against the bone and so produce a tingling sensation in the little and ring fingers—a phenomenon popularly referred to as 'hitting one's funny bone.'

The ulnar nerve pierces the origin of the Flexor Carpi Ulnaris and now lying deep to it, supplies that muscle and the medial half of the Flexor Digitorum Profundus. These are the only muscles it supplies above the wrist. It runs with the Fl. Carpi Ulnaris to the wrist, and enters the palm superficially and in contact with the pisiform bone.

The ulnar nerve ends by dividing into: (1) a **superficial branch** for the supply of the skin of the medial one and a half digits; (2) a **deep branch** for the supply of the hypothenar muscles, the medial two Lumbricals, the Adductor Pollicis, and all the Interossei. A little above the wrist the ulnar nerve gives off a cutaneous branch for the back of the hand and fingers sharing that supply with the radial nerve as, in front, it shares the cutaneous supply with the median nerve (Figs. 603, 604).

4. The **axillary** (or **circumflex**) **nerve (C. 5, 6)** runs backward between the muscles of the posterior wall and, passing below the capsule of the shoulder joint (which it supplies), gives a branch to the Teres Minor; it ends by supplying the Deltoid and the skin over it (figs. 605, 606).

By thus encircling the neck of the humerus it justifies its older name. However, it is now known as the **axillary nerve** because it is confined to the general region of the axilla.

Quadrangular Space. The axillary nerve, in leaving the axilla, passes backward through a 'space' between the capsule of the shoulder joint above, surgical neck of humerus laterally, tendon of Teres Major below, and long head of Triceps medially. The 'space' is actually a potential one since all the boundaries are in contact with each other.

5. The **radial nerve** (C. 5, 6, 7, 8; T. 1) is

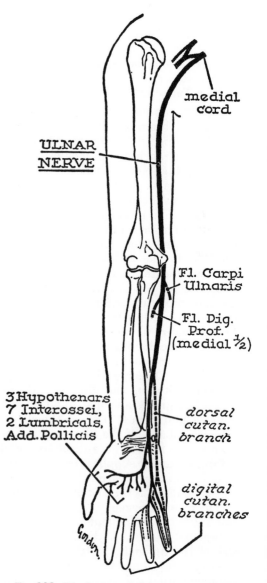

medial cord

ULNAR NERVE

Fl. Carpi Ulnaris

Fl. Dig. Prof. (medial ½)

3 Hypothenars
7 Interossei,
2 Lumbricals,
Add. Pollicis

dorsal cutan. branch

digital cutan. branches

Fig. **603**. Distribution of right ulnar nerve (semi-schematic).

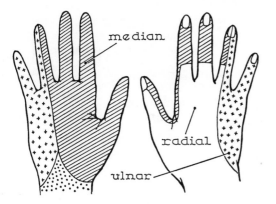

Fig. **604**. Usual cutaneous nerve supply of hand

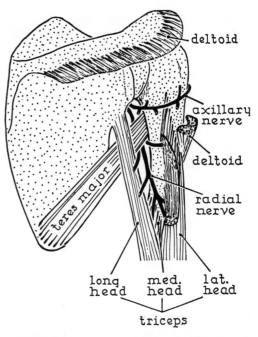

Fig. **606**. Course and relations of axillary and radial nerves in the shoulder region.

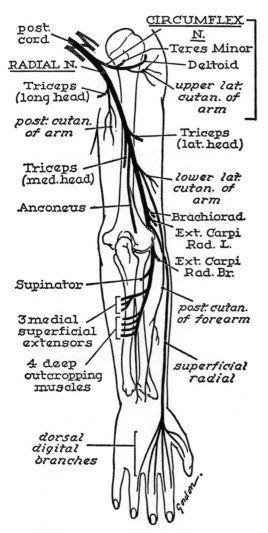

Fig. **605**. Distribution of right radial and circumflex (axillary) nerves (semischematic, from behind).

the largest nerve of the upper limb and it is the great 'extensor' nerve (figs. 605, 606).

As the continuation of the posterior cord, the radial nerve leaves the axilla behind the axillary artery and lies in front of the long head of the Triceps. A little farther down the arm, the radial nerve passes obliquely lateralward to run behind the humerus in its spiral groove, deep to the Triceps. During this course, the radial nerve gives one or two branches to each of three heads of the Triceps; also branches to the Anconeus, the skin of the back of the arm, and the skin of the back of the forearm (*posterior cutaneous nerve of forearm*).

On the lateral side, above the elbow, the radial nerve comes forward to the front of the limb and lies between the Brachioradialis and the Brachialis.

As it crosses the front of the elbow joint on its lateral side, the radial nerve supplies the muscles arising from the lateral supracondylar ridge. These are Brachioradialis, and Extensor Carpi Radialis Longus. The nerve then divides into: (1) a long **superficial branch** for the skin of the back of the hand (fig. 604); (2) the **deep branch** or **posterior interosseous nerve.** This latter first supplies Extensor Carpi Radialis Brevis; then it supplies and tunnels through Supinator on the lateral side of the radius and so regains the back of the limb, where it supplies the medial superficial extensors (Extensor Digitorum (Communis), Extensor Digiti Minimi, and Extensor Carpi Ulnaris) and the deep out-cropping muscles (Abductor Pollicis Longus, Extensor Pollicis Longus, Extensor Pollicis Brevis) and Extensor Indicis; it ends by supplying the wrist joint.

INTERCOSTAL NERVES

On issuing from their intervertebral foramina, the **ventral rami of the thoracic spinal nerves** enter the thorax as the **intercostal nerves.** Each runs under the shelter of the subcostal groove situated on the lower border of the rib with which it numerically corresponds. The last of the series (12th), being below the 12th rib, is not intercostal but is the **subcostal nerve.**

Each nerve courses between and supplies the intercostal muscles of its space; at the side of the sternum each of the upper six nerves ends by becoming cutaneous near the midline. Halfway around the chest each gives off a large **lateral cutaneous branch** for the supply of the skin of the thoracic and of the abdominal wall (fig. 607).

The following features are to be noted:
1. Most of T. 1 joins the brachial plexus and its intercostal part is small.
2. The lateral cutaneous branch of T. 2 (and sometimes of T. 3) enters the upper limb as the **intercostobrachial nerve** and supplies the skin of the axilla.
3. The lateral cutaneous branches of T. 7–12 inclusive, descend to supply the skin of the anterior abdominal wall and that of the upper front part of the gluteal region.
4. The ventral rami of T. 7–10, on reaching the front ends of their spaces, pass behind the upturned costal cartilages and enter the abdominal wall to supply the muscles of the anterior abdominal wall and the skin near the midline of the abdomen.

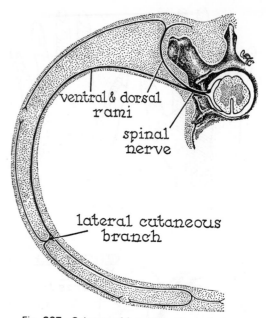

Fig. 607. Scheme of branches of a typical spinal nerve.

5. The ventral rami of T. 11 and T. 12 escape into the abdominal wall at the open ends of their spaces; they supply the lower part of the abdominal wall.

Dermatomes and Myotomes

The human torso shows evidence of its segmental development in that its nerve supply is by an orderly series of spinal nerves. The strip-like area of skin supplied by one pair of spinal nerves is known as a dermatome (Gr. = skin-slice) and these dermatomes have been mapped by different investigators by varying techniques and with somewhat varying results. In figure 608, it will be noted that there is a regular sequence of dermatomes on the limbs, too, because the limbs are outgrowths from the torso in the cervical and lumbosacral regions. Physicians and other persons dealing with patients having nerve injuries must be familiar with the general pattern of the dermatomes.

Although the segmental origin of skeletal muscles and their nerves is less apparent, the *myotomes* (Gr. = muscle-slices), which parallel the dermatomes in general distribution, can be traced and are important, too. In reviewing the section on muscles, it

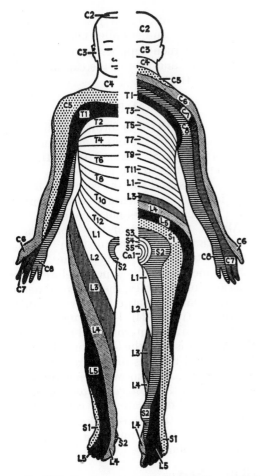

The six named branches of the lumbar plexus are: the ilio-hypogastric, ilio-inguinal, lateral femoral cutaneous, genitofemoral, femoral, and obturator nerves. The last two are much larger and more important than the others.

The **first lumbar nerve** divides into two branches, both in series with the intercostal nerves. They are the **ilio-hypogastric** and the **ilio-inguinal nerves;** having crossed the muscles of the posterior abdominal wall they course in the lowest part of the anterior abdominal wall supplying skin and muscles immediately above the inguinal ligament. The lateral cutaneous branches of the subcostal nerve and iliohypogastric nerve reach the gluteal region; the ilio-inguinal nerve escapes through the inguinal canal to supply the skin on the upper, medial side of the thigh.

Three branches of the lumbar plexus pass behind the inguinal ligament. Two of them can be noted briefly. (1) The **lateral cutaneous nerve of the thigh** (lateral femoral cutaneous) declares by its name its area of distribution; (2) the **genitofemoral nerve** supplies the skin on the medial side of the upper part of the thigh and the adjacent scrotal wall or labium majus, and the Cremaster muscle. (3) The **femoral nerve** is large and has an extensive distri-

Fig. **608**. Dermatomes: the strips of skin supplied by the various levels or segments of the spinal cord. (After Keegan, modified.)

will be noted that we have named the segments of the spinal cord supplying the more significant muscles (e.g., C. 5 and 6 for Deltoid).

LUMBAR PLEXUS

The ventral rami of L. 1, 2, 3, 4 nerves (fig. 609) form the **lumbar plexus.** As they issue from lumbar intervertebral foramina, they encounter the origin of the Psoas which, it will be recalled, arises from the sides of the lumbar vertebrae and their discs. The lumbar plexus is formed in the substance of the Psoas; its branches issue from that muscle. Where these branches lie on the posterior abdominal wall they supply its muscles.

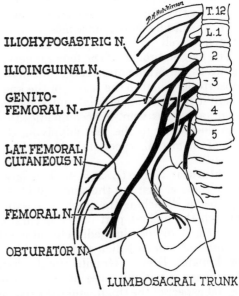

Fig. **609**. Lumbar plexus and lumbosacral trunk

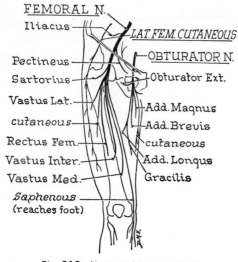

Fig. **610**. Nerves of front of thigh

bution (fig. 610). As soon as it enters the thigh behind the midpoint of the inguinal ligament, the femoral nerve (L. 2, 3, 4) breaks up into a leash of nerves for the supply of: (a) the muscles of the front of the thigh; (b) the skin of the front of the thigh (lateral, intermediate, and medial cutaneous branches) and of the medial side of the leg as far as the base of the great toe (saphenous nerve); (c) the hip and knee joints.

The muscles it supplies are: Pectineus, Sartorius, Rectus Femoris, and the Vasti. Each Vastus receives a separate branch near its middle and sometimes a lower branch as well.

The longest branch of the femoral nerve, known as the **saphenous nerve,** runs in the subsartorial canal to the region of the knee where it becomes superficial and joins the great saphenous vein. It accompanies this vein downward to the foot supplying the skin of the medial aspect of the leg and foot (fig. 610).

Obturator Nerve (L. 2, 3, 4). This large nerve runs, at first, on the side wall of the true pelvis. Reaching the thigh by passing through the highest part of the obturator foramen, it divides immediately into: (1) an *anterior branch* which supplies Adductor Longus, Brevis, Gracilis (and often Pectineus), a variable area of skin of the medial aspect of the thigh, and the capsule

of the hip joint; (2) a *posterior branch* which pierces and supplies Obturator Externus, and then supplies part of Adductor Magnus, and part of the capsule of the knee joint (fig. 610).

SACRAL PLEXUS

The sacral plexus is formed on the front of the sacrum by the union of the anterior rami of L. 4, 5; S. 1, 2, 3, 4 (fig. 611); thus, L. 4 contributes to two plexuses.

The *fifth sacral* and the *coccygeal nerves* are, in man, rudimentary and unimportant, because he does not have a tail.

Almost the whole of the broad sacral plexus narrows down to form a huge branch—the sciatic nerve. Other important branches are the superior gluteal and inferior gluteal nerves, the pudendal nerve, and the (parasympathetic) pelvic splanchnic nerves (page 324). Lesser branches run to the Obturator Internus and its Gemelli, the Piriformis, and the Quadratus Femoris. Levator Ani receives direct branches from S. 3, 4, and 5. A cutaneous branch from S. 2 and 3, called the *posterior cutaneous nerve of the thigh*, duplicates the course of the sciatic nerve (described below) down the back of the thigh, but at a much more superficial plane.

Gluteal Nerves

The *superior* gluteal nerve (L. 4, 5; S. 1) leaves the pelvis through the greater sciatic

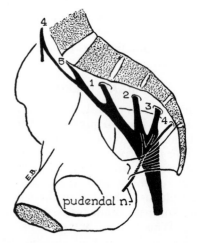

Fig. **611**. Scheme of right sacral plexus

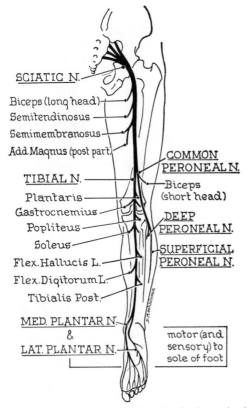

SCIATIC N.

Biceps (long head)
Semitendinosus
Semimembranosus
Add.Magnus (post part)

COMMON
PERONEAL N.

TIBIAL N.

Biceps
(short head)

Plantaris
Gastrocnemius
Popliteus
Soleus

DEEP
PERONEAL N.

SUPERFICIAL
PERONEAL N.

Flex.Hallucis L.
Flex.Digitorum L.
Tibialis Post.

MED. PLANTAR N.
&
LAT. PLANTAR N.

motor (and
sensory) to
sole of foot

Fig. **612**. Sciatic nerve and its distribution to back of lower limb.

nerve in the body (fig. 612). It gets most of its fibers from L. 4, 5; S. 1, 2, and 3. Escaping through the greater sciatic foramen below Piriformis, it lies on the back of the ischium deep to the lower part of Gluteus Maximus (fig. 613). The sciatic nerve then runs vertically downward in the midline of the back of the thigh deep to the long head of Biceps Femoris, and it gives branches to the hamstrings. Reaching the upper limit of the popliteal space still in the midline, it divides into two terminal branches.

Although the sciatic nerve is said to divide in the popliteal fossa, one can—by pulling the popliteal nerves apart—split the sciatic nerve easily into a *tibial division* (larger and medial) and a *peroneal (fibular) division*. The motor nerves to Semitendinosus, Semimembranosus, and the long head of Biceps Femoris come from the tibial division; the nerve to the short head of Biceps is from the peroneal division.

Common Peroneal Nerve

Large and important, the *common peroneal nerve* (L. 4, 5; S. 1, 2) runs obliquely laterally as it descends; it first takes the

foramen above Piriformis and turns forward to supply Gluteus Medius, Gluteus Minimus, and Tensor Fasciae Latae. The *inferior* gluteal nerve (L. 5; S. 1, 2) leaves below Piriformis, plunges into the middle of the deep surface of Gluteus Maximus, and supplies this huge muscle.

Pudendal Nerve

Branches from the anterior rami of S. 2, 3, and 4 form the pudendal nerve which leaves the pelvis and enters the perineum by turning around the ischial spine; it is destined for the muscles of the external genitalia, for the Sphincter Ani and Sphincter Urethrae, and for the skin of the perineum and back of the scrotum in the male (labia in female).

Sciatic Nerve and Its Branches

Students are usually amazed at the great diameter of the sciatic nerve, the largest

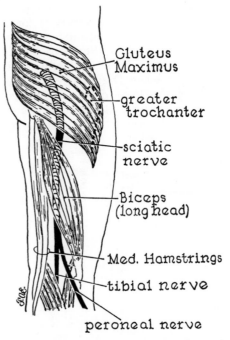

Gluteus
Maximus

greater
trochanter

sciatic
nerve

Biceps
(long head)

Med. Hamstrings

tibial nerve

peroneal nerve

Fig. **613**. The course and chief superficial relations of the (right) sciatic nerve, from behind.

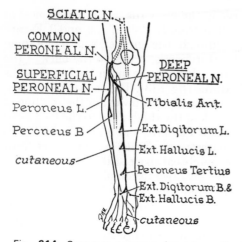

Fig. **614**. Common peroneal (lateral popliteal) nerve distribution.

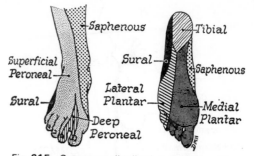

Fig. **615**. Cutaneous distribution of nerves of foot

tendon of the Biceps for its guide and then spirals around the neck of the fibula, where it can be felt and rolled by the palpating finger (fig. 614). It divides into two branches: one, known as the **superficial peroneal** (or *superficial fibular*) **nerve,** is for Peroneus Longus and Peroneus Brevis and, farther down, it supplies the skin of the front of the leg and of the dorsum of the foot; the other, known as the **deep peroneal** (or *deep fibular*) **nerve,** is for the anterior muscles of the leg and foot (Extensor Digitorum Longus, Peroneus Tertius, Tibialis Anterior, Extensor Hallucis Longus, and Extensor Digitorum Brevis). It runs down the front of the interosseous membrane and ends as a cutaneous nerve for the skin 'between' the first and second toes (figs. 614, 615).

Tibial Nerve

The *tibial nerve* (L. 4, 5; S. 1, 2, 3), which is very large, continues the line of the sciatic nerve and bisects the popliteal space vertically. It passes into the back of the leg (with the posterior tibial artery) deep to Soleus. It gives branches to the muscles of the back of the knee and leg (Gastrocnemius, Soleus, Plantaris, Popliteus, Flexor Digitorum Longus, Flexor Hallucis Longus, and Tibialis Posterior).

Plantar Nerves (fig. 612). Finally, the tibial nerve divides behind the medial malleolus into two terminal branches which enter the sole of the foot with the plantar vessels and are known as the medial and lateral plantar nerves; between them they look after the intrinsic muscles as well as the skin of the sole. The *medial plantar nerve* resembles the median nerve of the hand in that it supplies the muscles of the great toe and is sensory to the medial three and one-half digits. The *lateral plantar nerve* resembles the ulnar nerve in that it supplies the remaining muscles of the sole and the lateral one and one-half digits.

The **sural nerve** and the *communicating sural nerve* are cutaneous branches of the tibial and common peroneal nerves, respectively, and they supply the back of the leg and the lateral margin of the foot (see also fig. 615).

12

The Eye

The eyeball is a highly specialized end-organ on the nerve of vision. It is a spherical body about 2½ cm (1″) in diameter situated in the forepart of the orbital cavity. It lies just within the orbital margin where the cavity is most capacious; it is maintained in position by semi-fluid fat around and behind it, and by the tension of the ocular muscles.

ORBIT

The **medial (nasal) walls** of the **orbit (orbital cavities)** are parallel to one another and are exceedingly thin; they help to enclose ethmoidal air cells. The **lateral walls** are set at 90° to one another and are quite thick at the orbital margin. The margin, easily felt through the skin, is rather oblong in shape. The lateral walls do not reach as far forward as the medial and, in consequence, the eye is most vulnerable from the lateral side.

The thin **roof** of the orbit separates the eye from the frontal lobe of the brain. The **frontal air sinus** often extends well laterally above the eye in the orbital margin. It may be so extensive as to split the whole roof of the orbit into two lamellae of bone.

The **floor,** also thin, separates the eye from the large **maxillary air sinus.** Behind the **apex** of the orbital cavity lies the **sphenoidal air sinus** which, therefore, is an important relation to structures entering the cavity.

Along the roof course nerves and vessels destined for the forehead and scalp. Along the floor (or sunk into it) course nerves and vessels destined for the face.

The orbital cavity has variously been described as shaped like a pear, pyramid, or cone. At the apex lie the two entrances—the **optic canal** and the **superior orbital fissure.**

The axes of the two cavities (**orbital axes**) are at an angle of 45° to one another; the axes of the two eyes (**visual axes**) which—when one looks straight forward—pass horizontally from front to back through the middle of the eyes, are parallel to one another and at an angle of 22½° to the orbital axes. The ends of each visual axis are at the anterior and posterior **poles** of the eye; the **equator** bisects the eye into anterior and posterior halves. These relations are shown in figure 616; their significance will be appreciated when the muscles are discussed.

EYELIDS AND LACRIMAL APPARATUS

Before birth, the eye is completely separated from the exterior by a fibrous sheet or curtain attached at its periphery to the orbital margin. This fibrous sheet is called the **orbital septum** or the **palpebral** (eyelid) **fascia.** Shortly before birth its central part dissolves so that the fascia separates into upper and lower parts which become the fascial frameworks of the lids; the slit between them is the **palpebral fissure.** In the kitten this separation is not completed until after birth so the kitten is born blind.

Thickenings of the palpebral fascia form the **tarsi** (*tarsal plates*) which give rigidity to the lids (fig. 617). The outline of the *upper tarsus* is shaped like a D laid on its side— ◠ ; the *lower tarsus* is a bar or strip.

The tarsi contain within them **tarsal (Meibomian) glands** whose mouths open on the lid margin; by their oily secretion they 'waterproof' the margins. If the tears are excessive they break across this barrier and overflow onto the face.

The outer surfaces of the tarsi are cov-

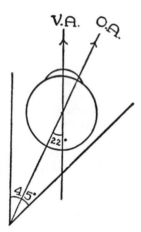

Fig. **616.** Visual axis and orbital axis compared

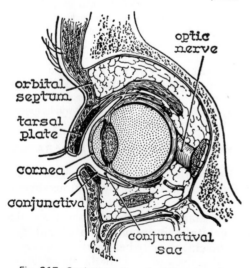

Fig. **617.** Sagittal section of orbit and eyelids

ered with muscle and skin of the lid; the inner surfaces are lined with a membrane known as **conjunctiva** which is intimately adherent to the tarsi. The conjunctiva is modified skin and it not only lines the inner surfaces of the lids but also is reflected on to the eyeball. The resulting conjunctiva-lined 'space' that lies between the lids and the eyeball is known as the **conjunctival sac** (fig. 617). At the margin of the cornea—the foremost, transparent segment of the globe—the conjunctiva is continuous with surface cells of the cornea.

The **lacrimal gland** is situated just inside the upper lateral margin of the orbit (fig. 618). Its several ducts pierce the conjunctiva, and their secretions, the **tears,** wash across the front of the eye. The tears finally find their way into the **naso-lacrimal duct** (to be presently described) which drains into the inferior meatus of the nose. In order to cope with excessive lacrimal secretion, the child first blinks rapidly to wash the tears across the eye; he then sniffs vigorously to encourage the flow down the duct; he 'bursts into tears' when he gives up the struggle.

The **lid margins** are thick and flat in their lateral four-fifths, and in front of the openings of the tarsal glands are the roots of several rows of **cilia** or eyelashes equipped with sebaceous glands; an infection of one of these sebaceous glands is the common stye. The medial one-fifth of the lid margin is rounded and possesses neither tarsal glands nor cilia. It is rounded because within it lies a little canal or **canalic-**

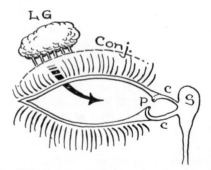

Fig. **618.** Scheme of lacrimal apparatus. L.G. = lacrimal gland; arrow = direction of tears to punctum (P), canaliculi (C), and lacrimal sac (S) which leads down to nasal cavity via duct.

ulus, whose tiny mouth or **punctum** is on the summit of a little elevation or **papilla,** visible where the two parts of the margin meet (fig. 618).

The canaliculi drain the tears and lead medially to the **lacrimal sac** which occupies a fossa at the front end of the medial wall of the orbit. The sac drains via the **nasolacrimal duct** lodged in a bony canal whose orbital opening lies just within the orbital margin where floor meets medial wall. Lacrimal gland, canaliculi, sac, and duct are known as the lacrimal apparatus and are shown schematically in figure 618.

The **function** of the lids and of the tears is to protect the delicate eye from minor —but serious—mechanical and chemical irritants, and from excessive light.

OCULAR MUSCLES

At the apex of the orbital cavity lies the **optic canal** through which the **optic nerve**—the nerve of vision—and the **ophthalmic artery**—the artery for the orbital contents—enter the orbit. Just lateral to the optic canal lies the medial end of the slit-like **superior orbital fissure,** through which pass all other vessels and nerves for the eye and for parts beyond the orbit. (Also see fig. 619.)

Attached to the bones, and so surrounding the optic canal and the medial end of the superior orbital fissure, is a fibrous **anulus,** ring or cuff. From this cuff and the bone nearby arise five ocular muscles and one muscle which raises the upper lid. They are flat little ribbons of muscles (rather pointed at their origins and wider as they pass forward) that form a cone as they diverge to their insertions; their actions are influenced by the fact that they do not pull from a point quite in line with the visual axis, as figure 620 shows.
The six muscles arising from the cuff are:
1. **Levator Palpebrae Superioris** or the muscle that raises the upper lid. It is the highest muscle in the cavity and is triangular. As it passes forward it spreads out until, at the orbital margin, it occupies the whole width of the cavity; it pierces the palpebral fascia and, splitting into lamellae like the

pages of a book, is inserted into the skin, the upper tarsus, and the conjunctiva of the upper lid, all of which it therefore raises. So as to safeguard against the muscle pulling the upper lid into the orbital cavity, the margins

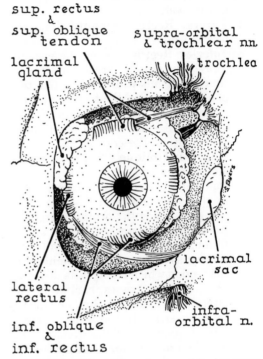

Fig. **619**. Right orbit and some of its important contents.

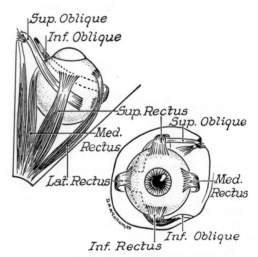

Fig. **620**. Muscles of (right) eye from above and from in front (semischematic).

of the muscle in front are fixed to the medial and lateral sides of the orbital cavity; these attachments act as *check ligaments*.

2. **Superior Rectus**—above the eyeball and below the Levator.
3. **Medial Rectus**—running along the medial wall.
4. **Lateral Rectus**—running along the lateral wall.
5. **Inferior Rectus**—below the eyeball. These four muscles are all inserted by thin, short but wide, ribbon-like tendons; these insertions are into the **sclera**—the tough, outer, fibrous coat of the eye—just behind its forward continuation, the *cornea*.
6. **Superior Oblique.** This muscle runs along the angle between the medial wall and the roof of the orbit; it lies immediately above the Medial Rectus. Just inside the orbital margin its rounded tendon passes through a little fibrous ring attached to the wall. This ring acts as a pulley or **trochlea** for the tendon which changes its direction abruptly and passes backward and lateralward below the Superior Rectus to its insertion in the upper postero-lateral quadrant of the eyeball. This muscle and **Inferior Oblique** are the only two muscles inserted into the back part of the sclera. The trochlea is readily palpable by the exploring thumb.

One muscle only arises from the front part of the orbital cavity; it is the **Inferior Oblique.** It arises from the orbital floor at a spot vertically below the trochlea or pulley of the Superior Oblique; this site is just lateral to the nasolacrimal duct. The muscle runs backward and lateralward below the Inferior Rectus and reaches an insertion near that of the Superior Oblique.

Motor Nerves. The Superior Oblique is supplied by the fourth or trochlear nerve; the Lateral Rectus by the sixth or abducent nerve; all the others (including the Levator) are supplied by the third or oculomotor nerve.

The **sensory nerve supply** of the eyeball is the ophthalmic division of the fifth (trigeminal) cranial nerve.

The muscles are shown in figure 620 and their actions in figure 621.

Actions of Eye Muscles. Some authors consider each muscle independently. It is perhaps simpler and certainly more useful to observe the following: (1) the Medial

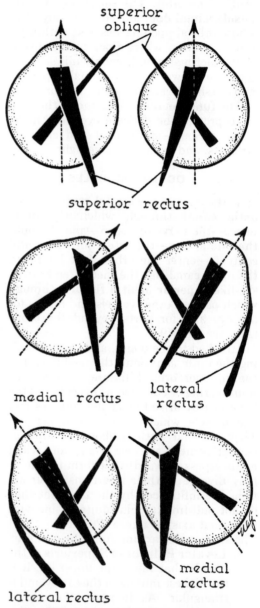

Fig. **621.** When eyes are turned to one side (Lateral Rectus of one eye, Medial Rectus of other), muscles that are brought in line with visual axis (S. and I. Recti of one eye, Obliques of other) become elevators and depressors.

Recti turn the eyes 'in'; the Lateral Recti turn the eyes 'out'; (2) when the eye is turned 'in' the Superior Oblique and Inferior Oblique lie nearly in the visual axis; hence S.O. depresses, I.O. elevates. When the eye is turned 'out' the Superior Rectus and the Inferior Rectus lie nearly in the visual axis; hence S.R. elevates, I.R. depresses; (3) with eyes right, the left S.O. co-operates with the right I.R. and the left I.O. with the right S.R. With eyes left, the left S.R. co-operates with the right I.O. and the left I.R. with the right S.O. Notice that the muscles that lie at right angles to the visual axis can produce rotation.

Imbalance of the power or length of these muscles results in various types of squint.

Fascia. Each ocular muscle, like muscles everywhere else, is contained in a fascial envelope. These are continuous from muscle to muscle, forming a hollow cone of fascia. This *muscular fascia* is continuous with a loose sheath of fascia on the eyeball, the *vagina bulbi*, which is attached to the sheath of the optic nerve and, in front, reaches the margin of the cornea.

Therefore, the eyeball rests in a fascial socket on which it can slightly glide. In all major excursions, however, eyeball and socket roll on the retrobulbar fat. When an eye has to be removed, the fascial socket is carefully preserved in order that it may serve as a socket for the artificial eye.

THE EYEBALL

The eyeball (*bulbus oculi*) is almost ideally spherical with a diameter of about $2\frac{1}{2}$ cm (1″). Owing to a forward bulge of its cornea, it is slightly longer in the antero-posterior diameter than in either the transverse or the vertical (fig. 622).

The eyeball possesses three coats, an outer protective, an intermediate nutritive, and an inner visual. They are known respectively as: the **fibrous coat (sclera and cornea)**; the **vascular coat (choroid, ciliary body, and iris)**; and the **internal (*nervous*) coat (retina)**.

The interior of the eye is filled in order from front to back with: (1) a weak solution of salt known as the **aqueous humor**; (2) the **lens;** (3) a jelly-like substance, the **vitreous** (L. = glassy) **body** (or *vitreous humor*).

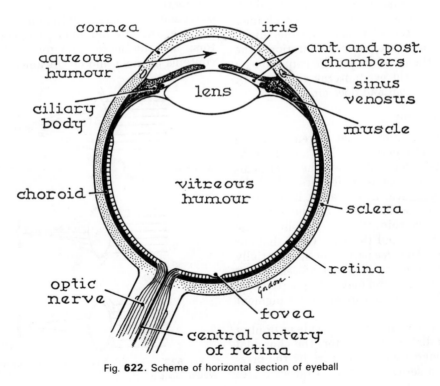

Fig. **622**. Scheme of horizontal section of eyeball

Light rays, before they reach the retina, must traverse cornea, aqueous, lens, and vitreous; these are known as the **refracting media** of the eye and of them the cornea is the most important.

Fibrous Coat

The protective fibrous coat of the eyeball is thick and unyielding. Its opaque posterior five-sixths is called the **sclera** (Gr. = hard); the transparent anterior one-sixth is called the **cornea.** The two are continuous with one another—the region of transition being called the *corneosclerotic (sclerocorneal) junction.*

Just medial to the posterior pole, the sclera is pierced by the fibers of the optic nerve at the *optic disc* or *blind spot* of the eye.

The surface of the cornea is part of a sphere of smaller radius than that of the sclera; in other words, the cornea possesses a more pronounced curvature than does the sclera.

The region of the *corneosclerotic junction* is important, for within it lies an encircling canal or sinus, *the sinus venosus sclerae* (or canal of Schlemm) (fig. 622). Between the canal and the aqueous humor (see below), the sclera is permeable, allowing aqueous to seep from the interior of the eye into the canal. Since the canal joins a vein outside the eye (anterior ciliary vein), the aqueous can and does circulate.

Vascular Coat

When the fibrous coat is removed, the eye and optic nerve look like a dark grape on its stalk; for that reason the intermediate vascular coat is sometimes called the *uveal tract* (uva, L. = grape). It possesses three parts which, from behind forward, are: the **choroid** (Gr. = skin-like), the **ciliary body,** and the **iris** (Gr. = halo or rainbow). All three parts consist essentially of great numbers of blood vessels (veins predominating) and nerves, supported by a loose fibrous stroma in which lie pigment cells.

Over a zone about 2 mm wide in front of and parallel to the equator, the choroidal coat is increased in thickness by the presence of a series of about 70 raised ridges

known as **ciliary processes;** they are vascular processes and constitute a major part of the **ciliary body.** The ciliary processes secrete (and possibly help to absorb) the aqueous. To the ciliary body are attached delicate fibrous strands which anchor the lens and maintain its position (fig. 622).

Ciliary Muscle. In the depth of the ciliary body adjacent to the sclera, lies the involuntary **ciliary muscle** which regulates the shape of the lens. When the muscle contracts, the fibrous strands that anchor the lens become looser and so allow the lens to bulge or become fatter. The resulting increase in curvature of the lens surface brings the images of near objects to a focus on the retina. The mechanism is known as **accommodation.**

The most anterior portion of the vascular coat is the iris, which is the diaphragm of the camera (fig. 623).

The **iris** springs from the front of the ciliary body and is a delicate, permeable, vascular, and disc-like structure with a central aperture, the **pupil.** It floats on the front of the lens. Within the iris are circular and radial muscle fibers which, by their contractions, alter the size of the pupil and so regulate the amount of light admitted to the eye; the circular fibers, the constrictors, are the more numerous. The iris contains pigment and on the amount pres-

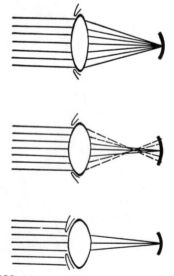

Fig. **623**. Iris cuts off light rays that edges of lens would focus poorly.

ent depends the color of the eye. The eye of the newborn is blue because it contains little pigment. It may remain blue or, by the acquisition of pigment, may gradually become brown.

Retina

The retina or visual coat (along with the optic nerve) is an outgrowth of a part of the brain which, in structure, it closely resembles. The three neurons composing it have been discussed (page 326). Here it should be pointed out that the rod and cone layer of the retina lies adjacent to the choroid, pigment alone intervening. In order to reach the rods and cones light must first traverse the thickness of the retina. Light traverses the retina in one direction; the nerve impulse set up by the entering light traverses it in the reverse direction (fig. 582).

Contents of Eyeball

The **aqueous humor** secreted by the ciliary body occupies the space behind the cornea and in front of the lens. The part of the space in front of the iris is called the **anterior chamber;** the part behind the iris and in front of the lens is called the **posterior chamber.** *A common mistake is to regard the posterior chamber as the region occupied by the vitreous body.* Aqueous can pass from the posterior chamber to the anterior (or the reverse) through the pupil. Its drainage via the sinus venosus (or canal of Schlemm) has been noted.

The **vitreous body** is a gel. It resembles a rather poorly set jelly and it occupies the large space behind the lens, its surface being applied to the nerve fiber layer of the retina.

The crystalline **lens** is a bi-convex transparent body about 1 cm in diameter and ½ cm thick. Its situation is by now apparent. To its equator are attached delicate fibrils from the ciliary body. The effect of their normal tension is to maintain the lens somewhat flat. The effect of their relaxation has been described.

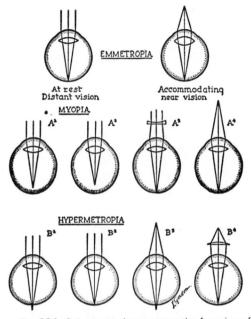

Fig. **624**. Scheme to demonstrate the focusing of parallel rays of light (distant vision) and diverging rays (near vision) by cornea and lens in normal (emmetropia), myopia, and hypermetropia. Correcting lenses added at A^3 and B^4.

The lens may become opaque in old age, a condition known as **cataract;** the sight is restored by surgical removal of the lens.

Visual Defects

Eyeballs vary somewhat in length. If the anteroposterior diameter is either unduly long or unduly short, muscular effort is required to focus objects on the retina. Such muscular effort causes eye strain. A long eye is short-sighted or **myopic;** it requires a concave lens for its correction. A short eye is long-sighted or **hypermetropic;** it requires a convex lens for its correction. With increasing age, one's eyes tend to become long-sighted, so that after middle life 'reading glasses' are usually a necessity.

Figure 624 shows the visual defects resulting from long and short eyes and how these defects are corrected. Sometimes the surface of the cornea is not the segment of a perfect sphere—it may have a greater curvature in the vertical than in the transverse plane, or the reverse. This condition is known as *astigmatism* and it can be corrected with a compensatory lens.

13

The Ear

The organ of hearing is composed of three parts, the **external ear,** the **middle ear or tympanic cavity,** and the **internal ear.** With the last is closely associated the organ of equilibrium which will therefore be considered at the same time. The close association of hearing and equilibrium gives us the new official name for the whole complex: the **vestibulocochlear organ,** reflecting the new name of the nerve (the eighth cranial) which supplies it.

External Ear

The **auricle** of the external ear is an appendage attached to the side of the head and primarily designed to catch sound waves. Its use in this respect is apparent in lower animals capable of cocking their ears but, in man, the appendage is more ornamental than useful.

The auricle is made up of a single piece of **elastic cartilage** with a dependent fatty mass, the **lobule;** the whole, of course, is covered with thin skin. The shape and parts of the auricle are shown in figure 625. A tubular extension reaches inward as the cartilaginous portion of the **external acoustic meatus** (ear passage), and is fixed to the outer rim of the bony meatus. The muscles of the human auricle, being practically functionless, are unimportant.

External Acoustic Meatus (figs. 626, 627). This canal is 2 to 3 cm (1″) long and leads medially toward the middle ear cavity deep in the petrous temporal bone. Its lateral third is cartilaginous and its medial two-thirds are bony.

In life the passage is barred by a (fibrous) **tympanic membrane** or **'eardrum'** set obliquely across it; the drum separates the external acoustic meatus from the tympanic cavity and is set so that its outer surface faces forward, downward, and lateralward. [Perhaps this direction is the most advantageous one to catch sounds from the ground in front of an animal advancing cautiously on 'all fours.']

Modified skin lines the meatus and the outside of the membrane; it possesses hairs, sebaceous glands, and sweat glands. The secretions of the sweat glands are known as **cerumen** or ear wax.

Middle Ear (Tympanic Cavity)

The tympanic cavity is shaped like a red blood cell with a 1 cm (½″) diameter. It is filled with air and the tympanic membrane sepafates this air from the air in the external auditory meatus (figs. 626, 627, 629).

The back part of the petrous bone and the mastoid process are honeycombed by **mastoid air cells** which communicate with the middle ear cavity via a restricted opening called the **aditus** to the **antrum.**

The function of the organ of hearing is to

convert sound vibrating the tympanic membrane into a stimulus for nerve impulses conducted by the cochlear nerve.

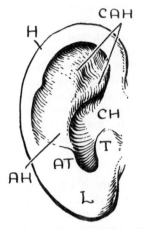

Fig. **625.** The auricle. Helix (H), lobe (L), and tragus (T) are most obvious. (CH = crus of helix; CAH = crura of antihelix (AH); AT = antitragus).

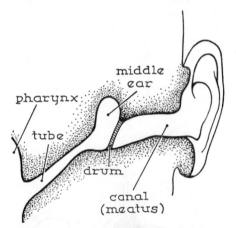

Fig. **626.** Scheme of air passages of the ear

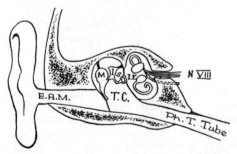

Fig. **627.** Scheme of ear passages. E.A.M. = external auditory meatus; T.C. = tympanic cavity (or middle ear) containing ossicles (M, I, and S); I.E. = internal ear. (Ph.T. = pharyngotympanic.)

But between membrane and nerve lie the width of the tympanic cavity and part of the petrous bone, for the tympanic membrane (because of its oblique set) forms most of the lateral wall of the tympanic cavity, while the nerve issues (into the posterior cranial fossa) from the short internal acoustic meatus whose mouth is situated on the posterior surface of the petrous bone. It is necessary, therefore, to enquire: (1) how the stimulus crosses the tympanic cavity; (2) how, in the interior of the petrous bone, it reaches the nerve.

Ossicles of Ear

A chain of three minute bones known as the **ossicles** of the ear stretches across the tympanic cavity from lateral to medial wall (fig. 627). The outermost of the three is called the **malleus** because it looks like a mallet or hammer. Its handle (long process) is attached along its length to the inner surface of the tympanic membrane, the lower end reaching the center of the membrane. Its head lies above the membrane in the epitympanic recess (figs. 628, 629) where it is suspended by a short ligament; this head articulates by a little synovial joint with the body of the intermediate ossicle, the incus (fig. 630).

The word **incus** means anvil but the ossicle looks more like a miniature bicuspid tooth with its two roots widely divergent. Its *short crus* (L. = leg) is attached by a ligament to the back wall. Its *long crus* hangs free in the tympanic cavity and at its

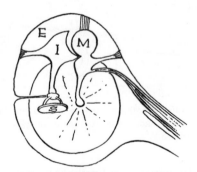

Fig. **628.** Ossicles and membrane viewed from within. Note Tensor Tympani muscle coming from auditory tube and attached to malleus (M). I = incus, S = 'foot-piece' or stapes. (E = epitympanic recess.)

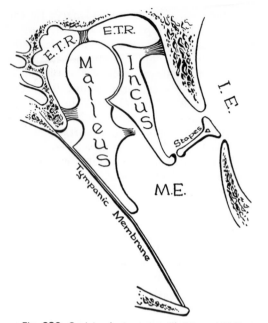

Fig. **629**. Ossicles (enlarged × 7). I.E. and M.E. = internal and middle ears. Stapes fits into oval window.

Fig. **630**. The three ossicles of the middle ear viewed from medial side. (Corner of postage stamp indicates size.)

bent tip carries the head of the innermost ossicle, the stapes.

The word **stapes** means stirrup and that is exactly what this ossicle resembles. Its

foot-piece or *base* fits snugly in a little **oval window** in the medial wall of the middle ear; the window is called the **fenestra vestibuli** and the base of the stapes is held in place by a circumferential ligament made of elastic fibers. Vibrations of the tympanic membrane are carried by the ossicles to this base which is alternately thrust farther into and partially withdrawn from the fenestra vestibuli. The movements of the ossicles can be controlled or 'damped' by two little muscles, one attached to the malleus (*Tensor Tympani*) the other to the stapes (*Stapedius*).

Internal Ear

Imagine yourself small enough to stand in the tympanic cavity. Then pull the stapes out of its window in order to look inside the petrous bone. At once there would be a gush of fluid called **perilymph** which, until released, filled a system of caverns and tunnels hewn out of the solid 'rock' of the petrous bone and known as the **bony labyrinth** (see fig. 631).

Looking through the fenestra vestibuli after the flood has subsided, one would see a little room or chamber, the **vestibule.** Leading forward from it, immediately beside the window through which one is

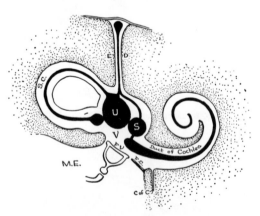

Fig. **631**. Internal ear (semischematic). Membranous labyrinth (black) lies in bony labyrinth. U. = utricle; S. = saccule; S.C. = a semicircular canal; V. = vestibule; M.E. = middle ear. Stapes in fenestra vestibuli (F.V.) or oval window. F.C. = fenestra cochleae or round window. C. of C. = canal of cochlea. (E.D. = endolymphatic duct whose sac lies under dura mater.)

gazing, is the opening of a tunnel, which is part of the cochlea (described below). After running forward for a short distance, the tunnel spirals around a central core of bone called the **modiolus** of the cochlea.

The **cochlea** (L. = snail) forms a unit that is both structurally and functionally different from the vestibule with which it is connected. The tunnel which takes two and one-half turns, is only the upper compartment of a bony canal that is divided into upper and lower compartments, partly by an *osseous* (bony) *spiral lamina* or ledge projecting from the modiolus as a thread projects from the shaft of a screw, and partly by a fibrous membrane (known as the *basilar lamina*) which stretches from the edge of the bony lamina to the outer wall of the canal (fig. 632). The mouth of the upper compartment or **scala** (L. = stair-case) **vestibuli** is the one we looked into from the vestibule.

At the tip of the osseous spiral lamina the two compartments are continuous with one another so that the ascending scala vestibuli becomes the descending **scala tympani.** How this happens can be understood if one remembers that he can run his finger along the upper surface of the thread of a screw to its tip whence the finger will be guided by the thread to return along the lower surface. The scala tympani is so named because it leads directly to the tympanic cavity at a **round window,** the **fenestra cochleae** which lies a little below the fenestra vestibuli. It is closed by a membrane which, if ruptured, would doubtless allow the remainder of the perilymph to escape.

Floating within the bony labyrinth is the **membranous labyrinth** whose existence until now has been ignored. A little membranous sac of fluid known as the **saccule** lies in, but by no means nearly fills, the front part of the vestibule; from the saccule a fine membranous tube, the **cochlear duct,** runs into the scala vestibuli, spiraling with it until it reaches the tip of the modiolus where it ends blindly. The duct of the cochlea rests partly on the osseous but mainly on the fibrous (basilar) spiral lamina; it is much smaller than the scala vestibuli whose lateral wall it hugs and is in all 2 to 3 cm (about 1″) long. Saccule and duct are filled with fluid known as **endolymph** (figs. 631, 633).

Running from end to end inside the duct

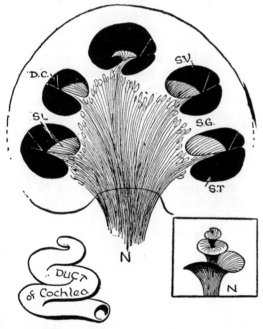

Fig. **632.** Section through modiolus. N. = nerve (see also inset); S.G. = spiral ganglion; S.V. and S. T. = scala vestibuli and scala tympani; D.C. = duct of cochlea; S.L. = spiral lamina.

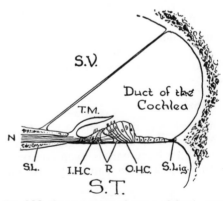

Fig. **633.** Cross-section of organ of Corti—greatly enlarged—resting on floor of duct of cochlea. S.V. and S.T. = scala vestibuli and scala tympani partly separated by spiral lamina of bone (S.L.) with contained nerves (N) that run to organ. Parts of organ: inner and outer hair cells (I.H.C. and O.H.C.), supporting rods (R), and tectorial membrane (T.M.). (S.Lig. = spiral ligament.)

and along its floor is the **spiral organ** (of Corti). This is the highly specialized end-organ receiving the stimuli of sound waves and setting up impulses transmitted by the fibers of the cochlear nerve. Figure 633 is a highly magnified cross-section of the duct and the organ.

The **modiolus,** the bony core around which the two scalae and the duct and the cochlea are wound, contains minute bony canals (see fig. 632). In these canals lie bipolar nerve cells whose peripheral axons run in still smaller branchings of the canals until they emerge through many openings at the very edge of the osseous spiral lamina. The nerve fibers thus can reach the organ of Corti via the basilar lamina (fig. 633). The central processes of the bipolar cells run to the bottom of the modiolus which itself forms the floor (fundus) of the internal acoustic meatus. These central processes constitute the **cochlear nerve.**

The **perilymph** (which fills the bony labyrinth and surrounds the membranous labyrinth) is agitated by movements of the base of the stapes and, since fluid is incompressible, the inward thrust at the fenestra vestibuli results in an outward bulging of the membrane closing the fenestra cochleae. The agitation of the perilymph is, of course, transmitted to the **endolymph** inside the membranous labyrinth and the spiral organ is stimulated. The cells in the modiolus receiving the stimulus constitute the **spiral ganglion;** the central processes of these cells, the cochlear nerve, convey the impulse to the brain.

Equilibrium

The back part of the bony labyrinth, now to be explored, houses the organ of equilibrium. Three bony canals each comprising two-thirds of a circle lead backward out of

the vestibule and finally return to it. One ought, therefore, to see the six openings of these three canals but, because two of them share at one end a common opening, only five openings appear. The three canals lie at right angles to one another as do the floor and two adjacent walls of a room. They are named the **lateral** (horizontal), the **anterior,** and the **posterior semicircular canals** (fig. 631).

The vestibule also houses a membranous sac, the **utricle,** which, like the saccule, is much smaller than the bony space in which it lies. The two sacs are joined together by a fine membranous duct. Into the utricle open the mouths of three **membranous semicircular ducts** which float in the bony canals; they, too, are much smaller in caliber than their bony counterparts.

Perilymph surrounds the utricle and the ducts as it does the saccule and the cochlear duct; **endolymph,** in the same way, occupies their interior.

Movements of the head disturb the endolymph in the utricle and saccule, and the semicircular ducts. This stimulus sets up an impulse in the **vestibular nerve** whose endings reach the utricle and dilated ends of the ducts through numerous foramina in the bone. The cell bodies of the vestibular nerve form a ganglion which lies in the internal acoustic meatus where the central processes of the neurons are bundled with the cochlear nerve as the **vestibulocochlear or eighth cranial nerve.**

Note: Two little 'safety valves' exist, one for the perilymph, the other for the endolymph. That for the perilymph is the *perilymphatic duct* (or *aqueduct of the cochlea*); that for the endolymph is the *endolymphatic duct* running in the bony aqueduct of the vestibule. They are indicated in figure 631.

14

Endocrine (Ductless) Glands

An endocrine gland is one that produces an internal secretion, i.e., a secretion carried off in the venous blood stream; it, therefore, does not need a duct. If it produces also a secretion carried off by a duct it is both an exocrine and an endocrine gland. This is the state of affairs in such organs as the testis and the ovary whose exocrine secretions have been described; their endocrine secretions are essential to the maintenance of many secondary sex characters. The pancreas also has an exocrine and an endocrine secretion (see page 218). The internal secretion of the pancreas is insulin.

A hormone is a chemical substance produced in one part of the body and carried by the blood stream to influence the activities of another part. The secretions produced by endocrine glands are hormones.

Hypophysis Cerebri (Pituitary Gland)

The hypophysis (fig. 634) is a small gland about the size of a pea hanging by a stalk, the infundibulum, from the floor of the third ventricle of the brain. Just behind the optic chiasma, it rests, like an egg in its nest, on the deeply concave body of the sphenoid where, except for its stalk, it is completely enclosed by dura mater.

It was popularly, but quite erroneously, supposed to produce the glairy mucous secretion that escaped from the nose, and was, in consequence, called the **pituitary gland,** a name which is now recognized as an official alternative.

The gland is made up of two lobes, an anterior and a posterior, entirely different from one another and, functionally, little if at all related; they are developed from different sources.

The anterior lobe (or **adenohypophysis**) forms most of the gland; it developed from the pharyngeal roof. It produces several hormones which, because of their influencing other endocrine glands, have caused the hypophysis cerebri to be referred to as the **'master gland.'** The anterior lobe is the source of the growth hormone and of one that, by influencing the sex glands, is responsible for male or female characteristics. Disturbances of the lobe result in dwarfism, giantism, infantilism, excessive obesity, and other manifestations associated with growth and sex.

The posterior lobe (or **neurohypophysis**) is a downgrowth from the brain and produces **'pituitrin,'** a secretion which stimulates the uterus to contract. The same or related hormones regulate the production of urine by the kidneys and in-

359

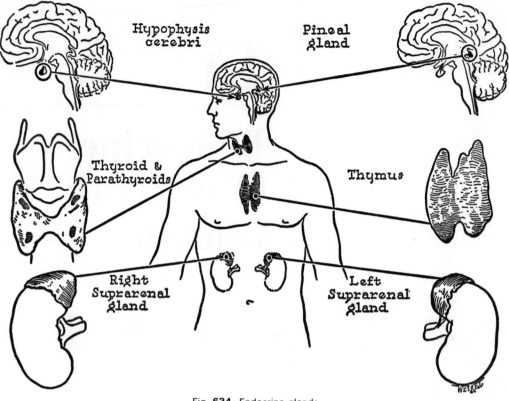

Fig. **634**. Endocrine glands

fluence the production of insulin by the pancreas. This lobe also affects blood pressure. It is possibly associated with sympathetic and (or) parasympathetic functions; its hollow stalk leads to the (third) ventricular cavity whose floor, the hypothalamus, has been noted as a sympathetic and a parasympathetic center.

Because the situation of the gland brings it into close association with the optic and other nerves of the eye, tumors of the gland may, by pressure, cause serious visual and other ocular disturbances.

The blood supply of the hypophysis is from small branches of the right and left internal carotid arteries which are close neighbors. The supply, like that of all endocrine glands, is rich. Of particular interest in recent years is its 'portal system' of veins, which are believed to carry the secretions first to the hypothalamus before the blood is returned to the systemic veins that drain that part of the brain.

Pineal Gland or Body

The pineal body, so called because in shape and appearance it resembles a miniature pine cone (fig. 634), is deeply placed within the brain. It projects backward from the back of the roof of the third ventricle. Here it lies between the posterior edge of the corpus callosum and the superior colliculi of the midbrain. Little is known about it and its status as an endocrine gland is open to challenge. It has been supposed to represent the vestige of a third or parietal reptilian eye, which is fairly well developed in the Sphenodon. In fishes it is truly glandular, emptying its secretions into the third ventricle, but in man its function remains obscure.

To the radiologist the pineal body is a landmark for deeply placed structures in the cranium because, in x-ray photographs, it often shows up as an opacity caused by calcium deposits that occur in it with advancing years.

Thyroid Gland

The thyroid lies in the neck (fig. 634). Its greatest length is about 5 cm (2'') and its breadth slightly more. It is made up of two **lobes** joined across the second and third tracheal rings by an **isthmus.** Each lobe partly hugs its side of the trachea and esophagus (fig. 635) and has the carotid sheath lateral to it. Its pointed upper pole reaches and partly covers the lower part of the thyroid cartilage; its more rounded lower pole partially embraces the third and fourth tracheal rings. The infra-hyoid, or strap, muscles almost completely hide it from view except in the midline (fig. 212 on p. 128).

The thyroid gland is reddish-brown in color because of its profuse blood supply. When it is cut it is found to consist of irregular masses of tissue. On closer examination these are found to be collections of microscopic sacs (follicles) filled with a colloid containing the thyroid hormone. From this store, the hormone is released steadily into the blood stream, and under special conditions, a more rapid release can occur without complete depletion of the hormone.

Blood Supply. It is profusely supplied via four arteries: the two **superior thyroid arteries** from the external carotids descend to the upper poles; the two **inferior**

thyroid **arteries** from the subclavians ascend to the lower poles. Corresponding veins drain into the internal jugular vein but a large **inferior thyroid vein** (veins) descends on the front of the trachea to enter the left brachiocephalic vein.

These **pretracheal veins** are surgically important in approaching the trachea from in front when the operation of tracheotomy has to be performed. The operation is an emergency opening of the trachea in order to save a victim threatened with choking from an air passage being blocked above.

Function. It produces an extremely important iodine compound known as **thyroxin.** This is a hormone concerned with a great many bodily functions and, when produced in excessive quantities, disturbs the normal rate at which the cells of the body function. This disturbance of the **basal metabolic rate** (BMR, as it is called) is a notable feature of the several varieties of goiter. Congenital deficiency or absence of the thyroid gland results in a mentally defective dwarf known as a **cretin.** The condition is alleviated by the administration of thyroxin.

Four Parathyroid Glands

The parathyroid glands lie inside the back of the thyroid capsule (fig. 635). There are two on each side and they are small ovoid bodies; yet, they are extremely important since they regulate the relative amounts of calcium in the blood and in the bones. Certain distressing diseases in which the bones become soft or extremely brittle are the result of malfunctioning of the parathyroids.

Thymus Gland

The thymus gland lies behind the manubrium sterni, and in front of the great vessels above the heart. It is essentially an elongated bi-lobed gland whose period of greatest functional activity is in fetal life. After birth it gradually retrogresses until, at puberty, it usually consists merely of two elongated fatty masses reaching down to the pericardium and with little thymic tissue remaining. It resembles a lymphoid organ and no hormone has yet been iso-

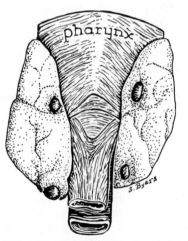

Fig. **635.** The thyroid gland—here seen from behind—wraps partway around trachea and esophagus. On the posterior surface of each lateral lobe are two parathyroid glands.

lated from it so, like the pineal gland, its status is questionable (fig. 636).

The thymus is an important production center for lymphocytes before puberty. Its life-long functions in controlling *immune responses* and *general* lymphocyte production are vital, but they lie beyond the scope of this book. It is, in turn, controlled by the sex hormones and adrenal cortical hormones (see below). In babies, it may be large enough to press upon the trachea and cause respiratory difficulties.

Paired Suprarenal (Adrenal) Glands

The suprarenal or adrenal glands, as their name implies, are closely associated in position with the kidneys; however they have no particular functional relationship to them. The right suprarenal gland surmounts the upper pole of its kidney, wedged in the triangular space between the

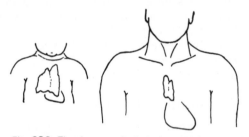

Fig. **636**. The thymus of a baby is much larger and more active than an adult's.

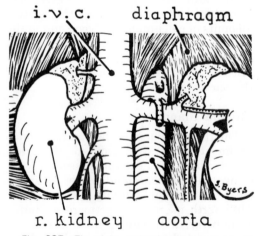

Fig. **637**. The right and left suprarenal glands capping the kidneys on the posterior wall of the abdomen.

kidney and the inferior vena cava. The left gland is crescentric in outline and fits on the medial border of its kidney between upper pole and hilus (figs. 635, 637). Both organs are fragile and are embedded in fat.

The suprarenal gland has two distinct parts, the outer cortex, and the inner medulla.

The **cortex** (L. = bark) forms a thick outer layer of the gland. It produces a group of hormones—among them cortisone—which are undergoing intensive investigation. Some are essential to life because they regulate various metabolic processes, e.g., salt metabolism, production of sex hormones, and production of collagen fibers of fibrous tissue throughout the body.

The **medulla** (L. = marrow) of the adrenal is closely associated in developmental origin with the sympathetic ganglia. It is in the medullary part of the gland that adrenaline is made. **Adrenaline** (*epinephrine*) produces the same general effects as are produced by stimulation of the sympathetic division of the autonomic nervous system (page 321). It floods into the circulation in time of emergency ('fight or flight').

Chromaffin Bodies

Scattered along the line of the sympathetic chain and the abdominal aorta are many almost microscopic bodies identical in structure with the suprarenal medulla. They are known as **para-aortic bodies** and **paraganglia or chromaffin bodies.** Like the suprarenal medulla they are associated with the activity of the sympathetic system.

The two *para-aortic bodies* are largest at about the age of three when they measure only about 1 cm in length. They atrophy and disappear by the mid-teens. The *paraganglia*, which are close associates of the sympathetic trunk and are quite tiny, atrophy to microscopic size in the adult. The chromaffin cells (so named because they are easily stained by yellow-colored chromic acid) secrete adrenaline just as the suprarenal medulla does.

Carotid Bodies

In the cleft between the internal and external carotid arteries there lies a small irregular brownish mass, the carotid body. To it run branches of the sympathetic trunk and of the glossopharyngeal and vagus nerves which transmit impulses from it to the brain stem (page 333).

The paired carotid bodies are chemoreceptors which respond to changes in the oxygen and carbon dioxide levels of the blood. They are therefore *not* endocrine glands. In very real sense, they are the opposite type of organ because they respond to the circulation of substances in the blood rather than release hormones.

Pancreas

As noted before, the pancreas has both an exocrine and an endocrine function. Scattered among the larger clumps of glandular tissue that pour their exocrine secretions through the pancreatic duct into the duodenum are small *islets* of cells that release the hormone *insulin* into the blood stream. This hormone was the first to be isolated. Insulin is vital to the proper metabolism of carbohydrates and its production is partly regulated by the hypophysis. The injection of the hormone into patients suffering from *diabetes* is a life-saving treatment. The discovery of insulin by Banting and Best in 1921 makes a stirring story that should be known by every student of biology and medical sciences.

Testes and Ovaries

The gonads not only produce the spermatozoa and ova, which may be considered as exocrine, but they have an important endocrine function as well. Specialized cells release the male and female sex hormones. These regulate the sexual function of the adult person and determine the secondary sexual characteristics including all aspects of maleness and femaleness.

In women, the monthly cycle of ovulation followed by menstrual bleeding is regulated by a balance of hormones produced by the ovaries and other endocrine glands, especially the hypophysis. These various hormones interact in such a way as to cause: (1) a sloughing off of the uterine mucosa if the ovum is not fertilized by a sperm within a few days of its release; or (2) an embedding and protection of a fertilized ovum in the mucosa.

The placenta of a developing embryo itself produces hormones which interact with hormones of the ovaries and hypophysis to prevent premature expulsion of the child. In recent years, much has been learned of these various influences. Some of the hormones have been isolated and purified and are now used for therapy.

15

Post-Natal Growth and Development

Weight and Height

At birth, the child weighs about 3½ kg (or 7½ lbs) and measures about 50 cm (or 20″) from the top of the head to the heels.

In the next 20 years, the weight increases about 20 times but this increase occurs irregularly. Birth weight is usually tripled in the first year, quadrupled in the second year, and averages about 15 kg (or 33 lbs) at three years of age. At puberty, girls tend to weigh somewhat more than boys, but during the late 'teens, boys rapidly surpass girls as they show a late spurt of growth (fig. 638).

The height also increases irregularly. The baby increases in length by 50 per cent the first year, but it takes four years in all to double the crown-heel stature. Although children seem to continue shooting up, their relative growth slows down after the first year until just before and during puberty when there is a pronounced spurt. This spurt begins and ends earlier in girls than in boys. Growth in height then tapers off to end finally at 16 to 18 years in girls and 19 to 21 in boys (see fig. 639).

Body Surface

Although the surface area of the body increases from birth (2500 sq. cm or 400 sq.

inches) to maturity by seven times, greater increase of body weight results in an actual proportional fall in surface area compared with bulk. The relative decrease in body surface is significant because the extent of the surface area influences metabolic processes including heat production and loss, sweating, and so on. The newborn has a relatively large area of skin from which to lose heat, making a baby very vulnerable to temperature changes in the surroundings, especially since the heat-regulating centers in the brain are not yet functioning adequately.

Parts of Body

Head. At birth the head is conspicuously large in comparison with the rest of the body (fig. 639). Its circumference is about 35 cm (14″) which is at least equal to the circumference of the trunk. Most of the volume of the head is occupied by the brain, the face being quite tiny. The jaws are small and the paranasal air sinuses have not yet invaded the bones of the face. With the growth and eruption of the deciduous teeth in the first two years of life, the face rapidly increases in size and the jaw becomes more prominent.

The cranial circumference increases by about 50 per cent over the first five years but very slowly thereafter. The circumfer-

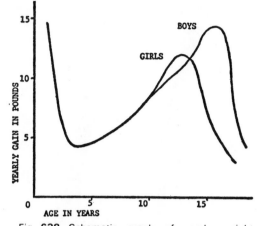

Fig. **638**. Schematic graph of yearly weight gain for boys and girls (based on Holt).

ence of the adult head is not much greater than it was in childhood, suggesting that growth in wisdom is not accompanied by volumetric growth in the brain.

Trunk. The torso or trunk always accounts for about one-half the stature. Its pelvic part contributes a much greater proportion after its relatively more rapid growth with sexual maturity. In a little baby, the relative importance and size of the thorax is obvious (fig. 639).

Vertebral Column. At birth, the baby has only one curvature for cervical, thoracic, and lumbar areas, with its concavity forward. In the first year, the cervical curvature develops in the opposite direction. After the child learns to stand the

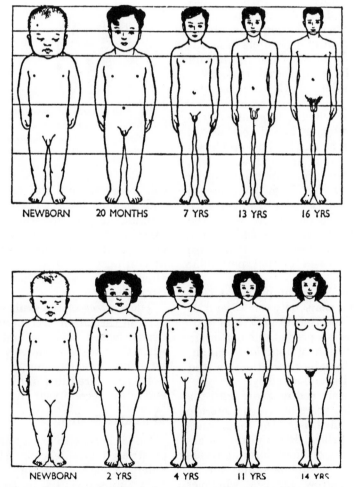

Fig. **639**. Proportions of various parts of body from birth to adolescence. (From R. W. B. Ellis, Child Health and Development, J. & A. Churchill Ltd.; by permission.)

lumbar curve appears and increases over several years (fig. 55 on page 33).

Limbs. Until the second year, the lower limbs are about the length of the upper ones, but thereafter they increase in length and weight more rapidly. In the adult the lower limbs are substantially longer and heavier than the upper limbs and account for about one-half the total stature and almost one-third of the body weight. The upper limbs always account for less than a tenth of the total body weight and, of course, they do not contribute to stature.

SYSTEMS

Skeletal System

The major changes after birth have been alluded to in Chapter 2. Growth of bones from primary and secondary centers (or epiphyses) has been described. At birth, all of the important primary centers of ossification, except for most of the carpal and tarsal bones, have appeared (fig. 640). These last develop slowly over several years.

The first epiphysis, the large one at the lower end of the femur, has almost always begun to ossify at birth. The epiphyses begin to ossify in a predictable pattern beginning with the large important ones in the first year of life. They continue to appear in descending order of importance and size throughout childhood. When the cartilaginous epiphyseal plate which separates an epiphysis from the shaft also ossifies, growth ceases in that particular bone.

The orderly fusion of epiphyses is concentrated in the years between 15 and 20 and spells the cessation in growth of stature. Again, there is a predictable pattern in the times for fusion of individual epiphyses, modified by minor variations related to sex, race, health, and heredity.

All bones contain **red marrow** until the age of about five years at which time the marrow in the cavities of the long bones begins to turn yellow. Red marrow is actively hemopoietic or blood-forming tissue while the yellow marrow is simply fat. In general, red marrow persists wherever there is cancellous bone, i.e., in the cubi-

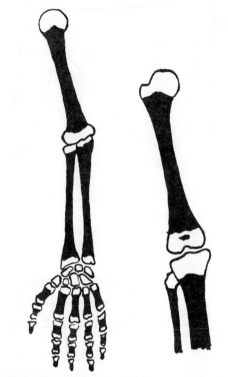

Fig. **640**. Extent of ossification at birth of arm and hand and of femur and knee region. Black represents bone; clear represents cartilage. Note secondary ossification (epiphysis) in lower end of femur.

cal, irregular, and flat bones and at the ends of long bones. These are the blood-forming organs. In adult life more red marrow is converted to fat. In an emergency, if the red marrow is challenged by hemorrhage, it can quickly become more abundant to replace blood cells in the circulation.

The *sutures* of the skull which are wide open in the infant become more and more firmly locked together until middle age when they are progressively fused together by bone. The fontanelles or *fonticuli* or 'soft spots,' are filled in by bone before the end of the second year (fig. 641).

Muscular System

One is born with one's total complement of striated muscle cells. However, these enlarge greatly in response to use. In the baby, muscular tissue contributes only one-quarter of the body weight, but its

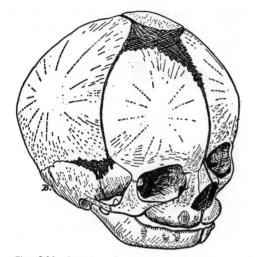

Fig. **641**. Anterior fontanelle and sutures of cranium at birth.

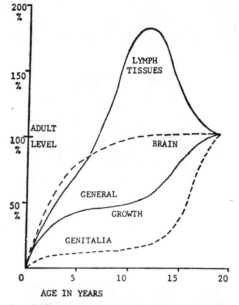

Fig. **642**. Schematic graph showing how different tissues develop at different rates at different ages (based on Scammon *et al.*).

progressive increase in volume and weight brings it up to two-fifths in the average adult. This means that a man weighing 70 kg (155 lbs) has about 28 kg (62 lbs) of muscle which began as a total of only 1 kg (2 lbs) in babyhood. With no sign of cellular increase, there is a 30-fold increase in bulk and an even more dramatic increase in strength.

Nervous System

Like muscular tissue, nervous tissue probably shows no cellular increase of neurons after birth. Nonetheless, the brain

doubles its weight in the first year. At six months it reaches half its adult weight. It continues to enlarge rapidly until six years when its growth is almost complete (see fig. 642). Final adult weight is reached by about ten years of age. Much of this growth is due to myelination of nerve fibers with resultant increase in bulk and in functional capacity. Although the spinal cord about

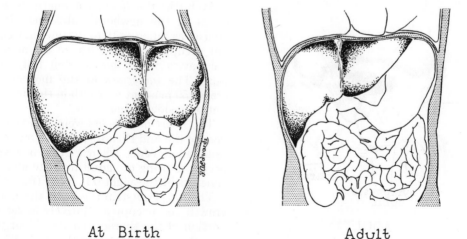

At Birth   Adult

Fig. **643**. At birth the size of the liver compared to other abdominal contents is much greater than it is later.

doubles in length to 45 cm (18″) in the adult, it fails to keep up with the somewhat faster growth of the vertebral column.

Therefore its tip comes to lie relatively higher and higher and finally ends at L.2 vertebra.

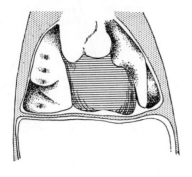 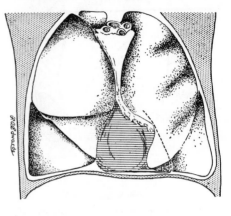

AT BIRTH   ADULT

Fig. **644**. Size relationships of lungs and heart compared in infant and adult. In infant note also the relatively large thymus gland just above heart.

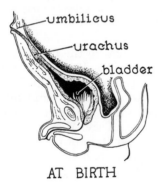

AT BIRTH

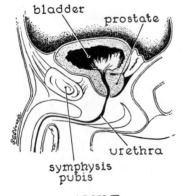

ADULT

Fig. **645**. The changing position of the urinary bladder.

Organs

The digestive, respiratory, and urinary organs, and the heart increase rapidly in size and weight in the infant and the young child. In middle childhood, their rate of growth slows up, spurts ahead again just before puberty, and slows once more during adolescence. In contrast to these structures, the sex organs remain dormant until puberty; then their period of greatest growth to adult proportions occurs over a brief year or two (fig. 642).

At birth, the *liver* is a relatively huge organ in the newborn's abdomen. It increases as much as 10 to 12 times in weight, but occupies less and less of the abdominal cavity which grows even faster (fig. 643). The *intestines* of the newborn increase 50 per cent in length in the first year but only slowly thereafter.

The *heart* doubles its weight in the first year of life and then its growth slows down until adolescence when a great new spurt occurs to bring the adult heart to about 12 times its birth weight (cf., liver) (fig. 644).

The *lungs* outstrip the heart throughout growth and occupy a relatively larger proportion of the thoracic cavity. Their total growth is about 20 times (fig. 644).

Of special interest are the slow changes

in the *larynx* up to puberty. The size doubles in boys but increases much less in girls. Puberty induces further changes in the larynx associated with the changing of the voice.

The most significant change in the *urinary system* is the gradual sinking of the bladder to a retropubic position from a high-riding position in the lower abdomen for the first several years. Increase in size of the pelvic cavity is the underlying cause (fig. 645).

Skin and Hair. The changes associated with puberty have been referred to in Chapter 5. Throughout life, there is a fluctuation in the amount of the subcutaneous fatty layer which varies with the individual. It is sparse in the several months before full term so that premature babies are wizened in appearance and light in weight. At birth, as much as one-quarter of a full-term baby's weight is due to subcutaneous fat.

Lymphoid Tissue

All lymphoid tissues, e.g., tonsils, thymus, and the lymph nodules in the gut, are prominent in childhood. They keep up with the general growth until puberty and then show a pronounced regression. There is actually less total lymphoid tissue in an adult than in a child of 10 to 12 years of age

(fig. 642). The thymus and tonsils atrophy and may almost disappear.

FACTORS INFLUENCING GROWTH

Many variable factors, normal and abnormal, play upon the growing child so that there is a substantial range of differences even in what is normal among individuals at any age. The most important underlying factors are heredity and race. They operate chiefly through the agency of the endocrine glands. These glands may, in turn, by reason of normal or abnormal functioning, produce variations in any population in spite of similar heredity.

Environmental influences play an important role in development. Malnutrition or privation, periods of severe or chronic diseases, and even the seasons of the year, make their mark on the growing child. Certain specific diseases have very specific influences, especially on skeletal growth, e.g., tuberculosis, rickets, and scurvy. Finally, everyone is familiar with the influence of lack of exercise on muscular and bone development and the reverse. It must be remembered that exercise cannot increase the number of muscle cells, which remain constant. As noted before, the increase in muscular bulk results from the growth in size of the muscle cells in response to use.

16

Regional Anatomy

The bulk of this book is devoted to individual chapters describing the systems of the body. This systematic approach is of fundamental usefulness to most students. However, in some regions of the body, parts of the various systems are so intermingled that a regional approach provides considerable clarity to the understanding of the structural organization. This chapter is devoted to selected regions that have been found to have the greatest practical significance to most students. Repetition is avoided by reference to other sections of the book where major structures have been described in detail.

HEAD AND NECK

Face

The bones of the face and jaw are described on pages 35 to 45. Clothing these bones are the facial muscles of expression (page 131) supplied by the facial nerve (cranial VII) (page 331) and the muscles of mastication (page 130) supplied by the motor branches of the mandibular nerve (cranial nerve V³) (page 331). The skin of the face is supplied by the three divisions of the trigeminal nerve—ophthalmic (V¹), maxillary (V²), and mandibular (V³). These are described on page 331.

The many branches of the facial nerve radiate from the margins of the **parotid gland** (fig. 646) deep to the muscles they supply. The **parotid duct** runs forward horizontally across the masseter and pierces the buccinator muscles that line the cheek (fig. 646).

The main vessels of the face are the **facial artery and vein.** On yourself palpate the pulsation of the artery where it enters the face by crossing the body of the mandible at the anterior border of the masseter. Thence the artery weaves its way among the muscles, giving them branches, until it reaches the medial angle of the eye. The vein follows a straighter course to the same point.

The **external nose** and **external ear** are described on pages 223 and 354), respectively, the **eyelids** on page 347.

The **orbit** and main **contents** are described on pages 347 to 353. The **ophthalmic artery** enters the orbit with the **optic nerve** (cranial nerve II) from behind and supplies both the eyeball and the other contents. The **ophthalmic veins** drain backward into venous sinuses within the cranial cavity through the superior orbital fissure. They anastomose also with the facial veins. In addition to the motor nerves (cranial nerves III, IV, and VI) the ophthalmic division (V³) of the trigeminal

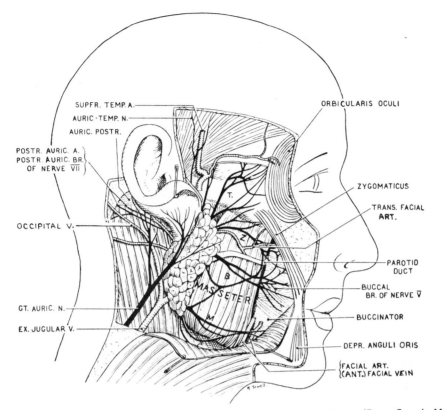

Fig. **646**. Dissection of side of face showing parotid gland and its neighbors. (From *Grant's Method of Anatomy*.)

nerve (cranial nerve V) enters the orbit through the superior orbital fissure. The motor nerves (fig. 647) enter the extraocular muscles but the ophthalmic nerve is sensory to the cornea; its other terminal branches supply the forehead and upper eyelid (fig. 648).

Interior of Cranium

The **bony interior** and **foramina** of the base of the skull are described on pages 38 to 43. The main content is the **brain** which is described, along with the **cranial nerves, meningeal coverings,** and **blood supply** in detail on pages 301 to 313.

The **intracranial course of the arteries and nerves** of the brain are indicated in figures 649 and 650. The clinically important **middle meningeal artery and vein** run in dura mater on the inner surface of the temporal region of the skull where a blow might rupture them (fig. 651). The

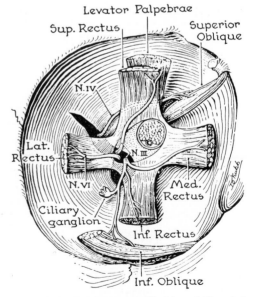

Fig. **647**. Cranial nerves III, IV, and VI entering orbit via superior orbital fissure (right orbit from in front). (From *Grant's Method of Anatomy*.)

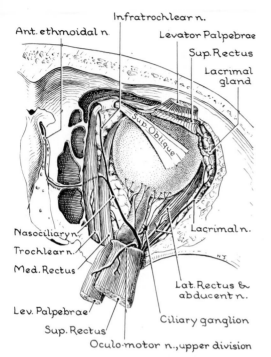

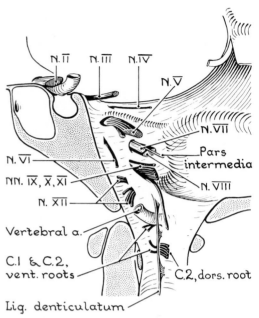

Fig. **648**. Dissection of right orbital cavity from above. (Bony roof, the floor of the anterior cranial fossa, removed.) (From *Grant's Method of Anatomy*.)

Fig. **650**. Cranial nerves piercing the dura mater to exit from the cranium. (From *Grant's Method of Anatomy*.)

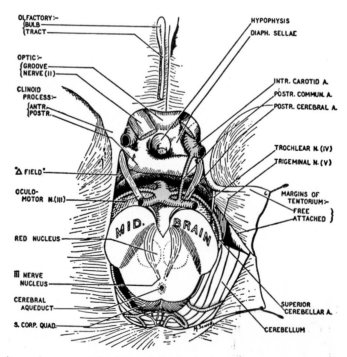

Fig. **649**. Arteries and nerves at the base of the brain. (Cerebral hemispheres removed.) (From *Grant's Method of Anatomy*.)

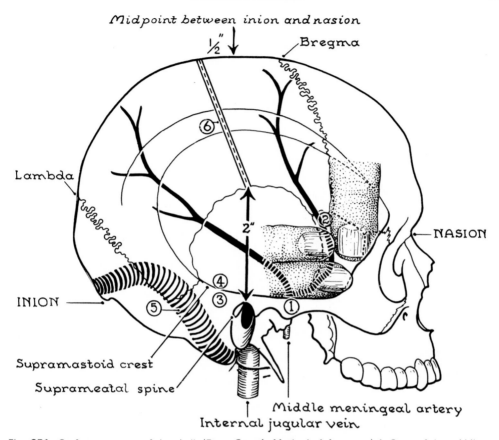

Fig. **651**. Surface anatomy of the skull. (From *Grant's Method of Anatomy*.) 1. Stem of the *middle menin-geal artery*, passing through the foramen spinosum, deep to the head of the mandible, which is readily located by palpation on opening and closing the mouth. 2. *Anterior* branch of the middle meningeal artery crossing the pterion. The *pterion* is the point where four bones meet (parietal, frontal, greater wing of sphenoid, and temporal squama). To locate it, place the thumb behind the frontal process of the zygomatic bone and two fingers above the zygomatic arch, and mark the angle so formed. This great landmark overlies anterior branch of the middle meningeal art., and stem of the lateral cerebral sulcus and, therefore, the insula, lentiform nu-cleus, internal capsule, etc. of the brain. 3. *Suprameatal triangle* lies below the supramastoid crest and be-hind the suprameatal spine; a hole drilled here enters the *mastoid antrum*. 4. A hole drilled above the supra-mastoid crest enters the *middle cranial fossa*. 5. *Lateral sinus*, i.e., transverse and sigmoid sinuses, passing from the inion to a point ¾'' (2 cm) behind the external acoustic meatus to become the internal jugular vein deep to the anterior border of the mastoid process. 6. *Central sulcus* of the cerebrum running from a point ½'' (about 1 cm) behind the mid inion-nasion point to a point 2'' (5 cm) above the external acoustic meatus.

complex course and relationships of the internal carotid artery as it enters the interior are shown in figure 652.

Neck

Anterior and Posterior Triangles

The soft structures of the neck are chiefly of or in its anterior or posterior triangles (figs. 653, 654). Partly deep to the Sternomastoid muscle and partly in a sub-

division of the anterior triangle—the ca-rotid triangle—are the great vessels and nerves in the carotid sheath (fig. 655). The carotid arteries are described on page 270; the internal jugular vein on page 286, and cranial nerves X, XI, and XII on pages 333 to 335.

In the midline are the trachea and esoph-agus leading down from the pharynx with the thyroid gland closely applied to them; the isthmus of the gland crosses the second

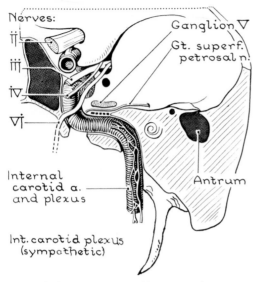

Fig. **652**. Internal carotid artery has many curves in its intra-osseous and intracranial course. Note relationships of cranial nerves. (From *Grant's Method of Anatomy*.)

and third tracheal rings (fig. 656). Most of the gland is covered by the infra-hyoid or 'strap' muscles (page 129).

Above the hyoid bone, the anterior triangle is subdivided into several smaller triangles or areas of interest to surgeons—(a) **diagastric (or submandibular) triangle** bounded by the mandible and two bellies of Digastricus; (b) floor or **diaphragm of the mouth,** paved by the Mylohyoid muscles. The tongue lies above these and its vessels and nerves are found in these regions.

The **posterior triangle** is important because it contains the brachial plexus and subclavian vessels *en route* to the axilla (fig. 657) as well as the spinal accessory nerve and beginning of the phrenic nerve. Except for the accessory nerve and many cutaneous nerves from the cervical plexus (page 337), all important structures are deep to the 'fascial carpet' that covers the muscular 'floor' of the posterior triangle.

Root of Neck

The root of the neck is a deep extension of the two major triangles (fig. 658). Lying above the inlet of the thorax, it is crowded by vitally important structures in transit.

Base of Skull and Pharyngeal Region

The bilateral area partly surrounding the upper end of the pharynx where it lies on the cervical vertebral bodies is packed with important structures running to and from foramina in the base of the skull (fig. 659). **Cranial nerves IX, X, and XI** emerge from the jugular foramen with the **internal jugular vein** to be joined by **nerve XII** and the **internal carotid artery** emerging through their own canals nearby. The air-carrying **auditory tube** runs through a bony canal to the middle ear just medial to the jaw joint; and deep to the mastoid process, the facial nerve (nerve VII) emerges from its bony canal to be immediately enveloped by the soft **parotid gland.**

The **infratemporal region** lies lateral to the pharynx deep to the ramus of the mandible; it contains the **muscles of mastication** (page 130) but also contains a large artery—the **maxillary artery** (page 271), the **pterygoid venous plexus,** and the branches of the mandibular nerve which emerges from the **foramen ovale** in the roof of the region (fig. 660). Its largest sensory branches are exposed here *en route* to the tongue (**lingual nerve**) and to the lower jaw and teeth (**inferior alveolar nerve**).

THORAX

The bony thoracic cage and ribs are described on pages 45 to 47, the diaphragm on

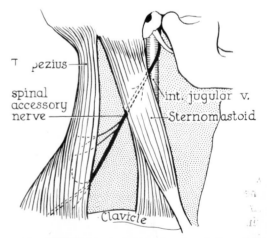

Fig. **653**. Right posterior triangle of the neck bounded by Sternomastoid and Trapezius muscles and clavicle.

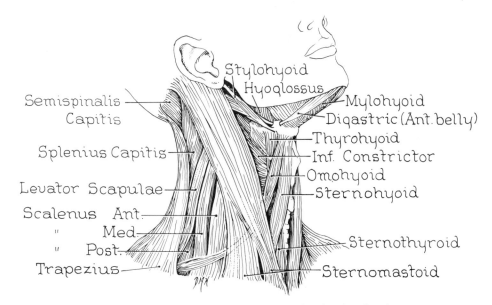

Fig. **654**. Muscles of posterior and anterior triangles of neck

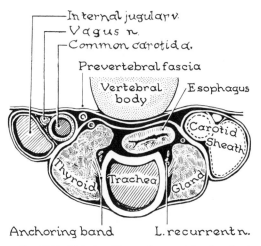

Fig. **655**. Relationships of carotid sheath in horizontal section. (From *Grant's Method of Anatomy*.)

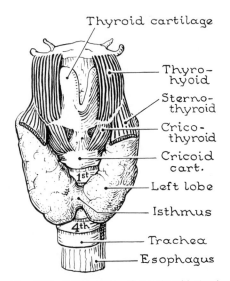

Fig. **656**. Relationship of the thyroid gland seen from the front with its coverings removed. (From *Grant's Method of Anatomy*.)

page 141 and intercostal muscles on page 136. Here we shall consider the general topographic anatomy of the thoracic content.

The thoracic cavity contains two pleural **sacs** and the intervening partition, the **mediastinum** (fig. 661, 662). Most of the important structures lie in the mediastinum or sacs but some lie behind the pleural sac just lateral to the partition—the right and left **sympathetic trunks** and the unpaired **thoracic duct** (lymph).

Mediastinum

The mediastinum extends from the root of the neck above to the central tendon of the diaphragm below. It is enclosed bilaterally by the mediastinal pleura, a portion of the parietal pleura. It is divided into four parts—the superior, anterior, middle, and posterior mediastina (fig. 663).

The **middle mediastinum** is synony-

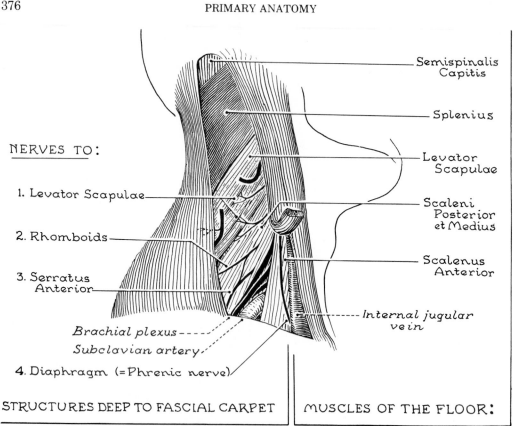

NERVES TO:

1. Levator Scapulae
2. Rhomboids
3. Serratus Anterior

Brachial plexus
Subclavian artery

4. Diaphragm (=Phrenic nerve)

STRUCTURES DEEP TO FASCIAL CARPET

Semispinalis Capitis

Splenius

Levator Scapulae

Scaleni Posterior et Medius

Scalenus Anterior

Internal jugular vein

MUSCLES OF THE FLOOR:

Fig. **657**. Floor of the right posterior triangle, great artery, and various nerves. (From *Grant's Method of Anatomy*.)

Phrenic n.

Scalenus { Anterior, Medius, Post.

Costo-cervical trunk
Trans. colli a.
Suprascapular a.
Thyro-cervical trunk
Int. thoracic a.

Subclavian v.
Int. jugular v.
(Innominate) Brachio-ceph. v.

Vagus n.
Recurrent laryngeal n.

C.6

Phrenic n.
Inf. thyroid a.
Thoracic duct

Contents of carotid sheath

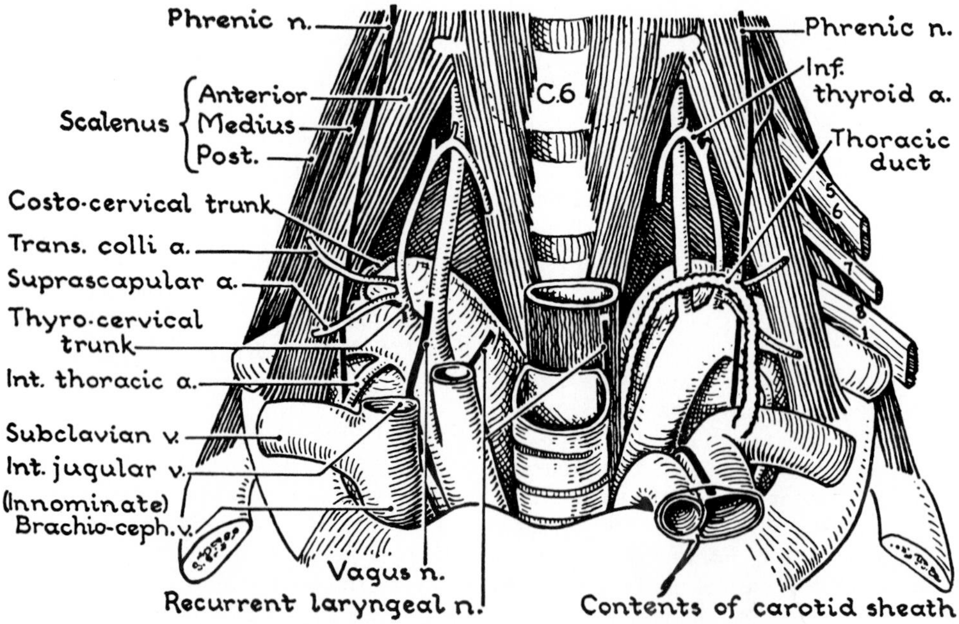

Fig. **658**. The root of the neck, from in front. (From *Grant's Method of Anatomy*.)

The following 26 illustrations are reproduced by permission from *Stedman's Medical Dictionary*, 23rd edition, 1976, published by The Williams & Wilkins Company.

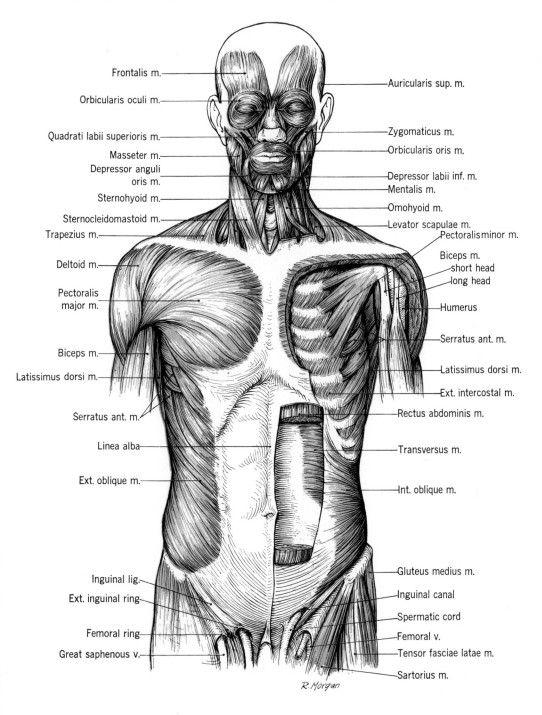

Frontalis m.
Orbicularis oculi m.
Quadrati labii superioris m.
Masseter m.
Depressor anguli oris m.
Sternohyoid m.
Sternocleidomastoid m.
Trapezius m.
Deltoid m.
Pectoralis major m.
Biceps m.
Latissimus dorsi m.
Serratus ant. m.
Linea alba
Ext. oblique m.
Inguinal lig.
Ext. inguinal ring
Femoral ring
Great saphenous v.

Auricularis sup. m.
Zygomaticus m.
Orbicularis oris m.
Depressor labii inf. m.
Mentalis m.
Omohyoid m.
Levator scapulae m.
Pectoralis minor m.
Biceps m.
short head
long head
Humerus
Serratus ant. m.
Latissimus dorsi m.
Ext. intercostal m.
Rectus abdominis m.
Transversus m.
Int. oblique m.
Gluteus medius m.
Inguinal canal
Spermatic cord
Femoral v.
Tensor fasciae latae m.
Sartorius m.

R. Morgan

PLATE 5
Muscles of Head, Neck, and Torso, Anterior View

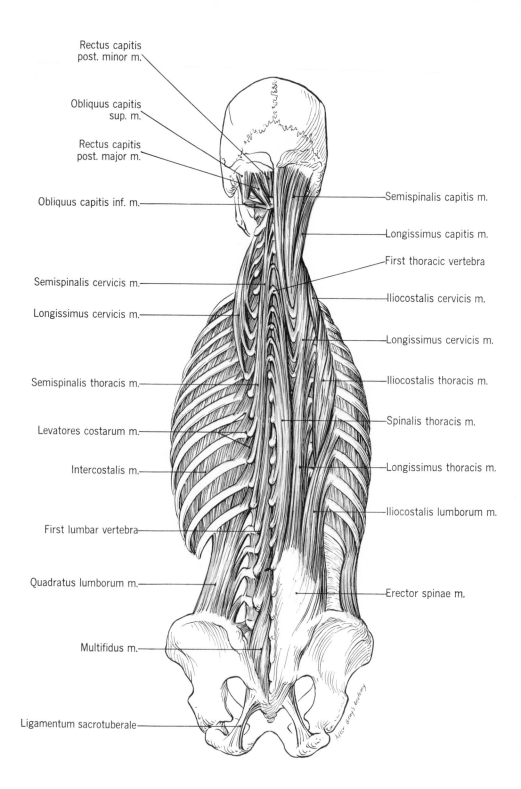

Rectus capitis
post. minor m.

Obliquus capitis
sup. m.

Rectus capitis
post. major m.

Obliquus capitis inf. m.

Semispinalis cervicis m.

Longissimus cervicis m.

Semispinalis thoracis m.

Levatores costarum m.

Intercostalis m.

First lumbar vertebra

Quadratus lumborum m.

Multifidus m.

Ligamentum sacrotuberale

Semispinalis capitis m.

Longissimus capitis m.

First thoracic vertebra

Iliocostalis cervicis m.

Longissimus cervicis m.

Iliocostalis thoracis m.

Spinalis thoracis m.

Longissimus thoracis m.

Iliocostalis lumborum m.

Erector spinae m.

PLATE 6
Muscles of Back Deep Dissection

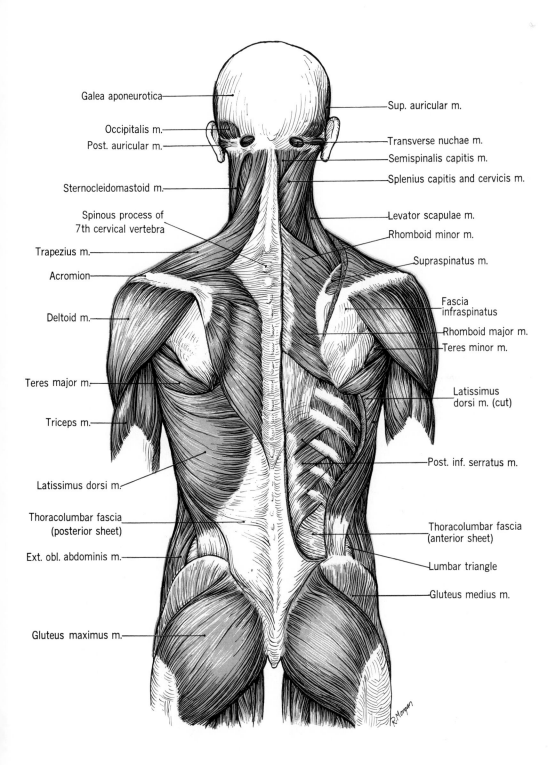

Galea aponeurotica

Occipitalis m.

Post. auricular m.

Sternocleidomastoid m.

Spinous process of
7th cervical vertebra

Trapezius m.

Acromion

Deltoid m.

Teres major m.

Triceps m.

Latissimus dorsi m.

Thoracolumbar fascia
(posterior sheet)

Ext. obl. abdominis m.

Gluteus maximus m.

Sup. auricular m.

Transverse nuchae m.

Semispinalis capitis m.

Splenius capitis and cervicis m.

Levator scapulae m.

Rhomboid minor m.

Supraspinatus m.

Fascia
infraspinatus

Rhomboid major m.

Teres minor m.

Latissimus
dorsi m. (cut)

Post. inf. serratus m.

Thoracolumbar fascia
(anterior sheet)

Lumbar triangle

Gluteus medius m.

PLATE 7
Muscles of Trunk, Posterior View

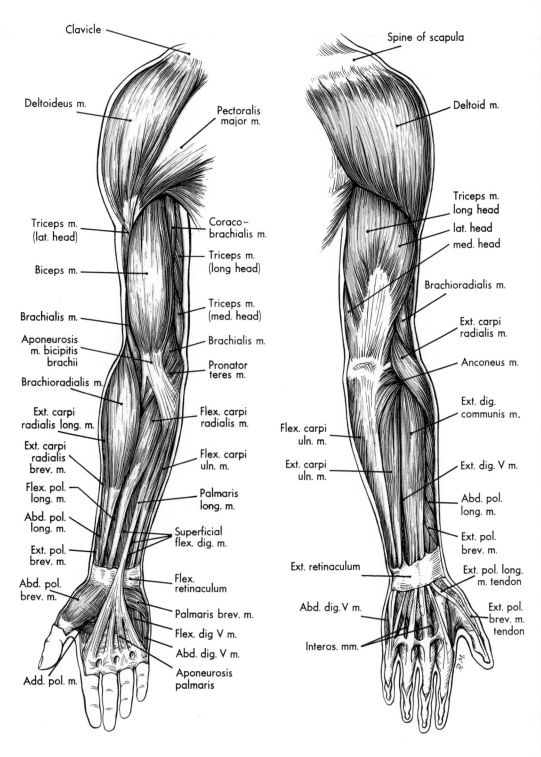

Clavicle

Deltoideus m.

Triceps m. (lat. head)

Biceps m.

Brachialis m.

Aponeurosis m. bicipitis brachii

Brachioradialis m.

Ext. carpi radialis long. m.

Ext. carpi radialis brev. m.

Flex. pol. long. m.

Abd. pol. long. m.

Ext. pol. brev. m.

Abd. pol. brev. m.

Add. pol. m.

Pectoralis major m.

Coraco-brachialis m.

Triceps m. (long head)

Triceps m. (med. head)

Brachialis m.

Pronator teres m.

Flex. carpi radialis m.

Flex. carpi uln. m.

Palmaris long. m.

Superficial flex. dig. m.

Flex. retinaculum

Palmaris brev. m.

Flex. dig V m.

Abd. dig. V m.

Aponeurosis palmaris

Spine of scapula

Deltoid m.

Triceps m. long head
lat. head
med. head

Brachioradialis m.

Ext. carpi radialis m.

Anconeus m.

Ext. dig. communis m.

Ext. dig. V m.

Abd. pol. long. m.

Ext. pol. brev. m.

Ext. pol. long. m. tendon

Ext. pol. brev. m. tendon

Flex. carpi uln. m.

Ext. carpi uln. m.

Ext. retinaculum

Abd. dig. V m.

Interos. mm.

PLATE 8
Superficial Muscles of Right Upper Limb

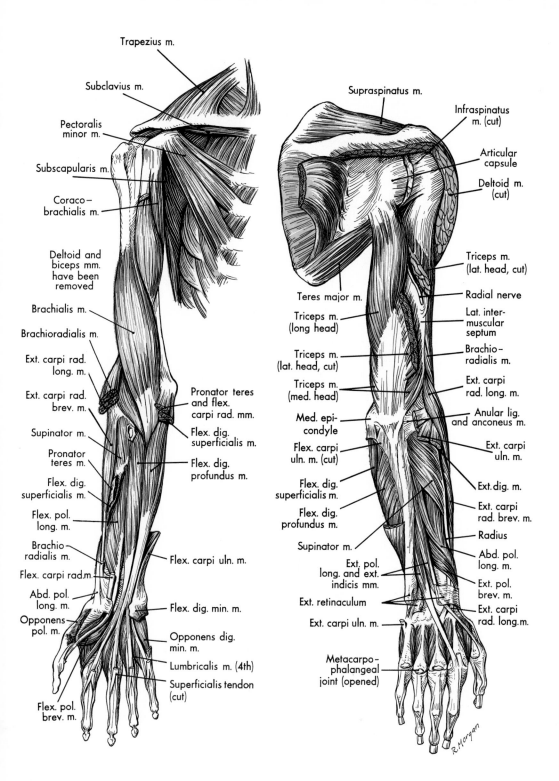

Trapezius m.

Subclavius m.

Pectoralis
minor m.

Subscapularis m.

Coraco-
brachialis m.

Deltoid and
biceps mm.
have been
removed

Brachialis m.

Brachioradialis m.

Ext. carpi rad.
long. m.

Ext. carpi rad.
brev. m.

Supinator m.

Pronator
teres m.

Flex. dig.
superficialis m.

Flex. pol.
long. m.

Brachio-
radialis m.

Flex. carpi rad.m.

Abd. pol.
long. m.

Opponens
pol. m.

Flex. pol.
brev. m.

Pronator teres
and flex.
carpi rad. mm.

Flex. dig.
superficialis m.

Flex. dig.
profundus m.

Flex. carpi uln. m.

Flex. dig. min. m.

Opponens dig.
min. m.

Lumbricalis m. (4th)

Superficialis tendon
(cut)

Supraspinatus m.

Infraspinatus
m. (cut)

Articular
capsule

Deltoid m.
(cut)

Triceps m.
(lat. head, cut)

Radial nerve

Lat. inter-
muscular
septum

Brachio-
radialis m.

Ext. carpi
rad. long. m.

Anular lig.
and anconeus m.

Ext. carpi
uln. m.

Ext. dig. m.

Ext. carpi
rad. brev. m.

Radius

Abd. pol.
long. m.

Ext. pol.
brev. m.

Ext. carpi
rad. long. m.

Teres major m.

Triceps m.
(long head)

Triceps m.
(lat. head, cut)

Triceps m.
(med. head)

Med. epi-
condyle

Flex. carpi
uln. m. (cut)

Flex. dig.
superficialis m.

Flex. dig.
profundus m.

Supinator m.

Ext. pol.
long. and ext.
indicis mm.

Ext. retinaculum

Ext. carpi uln. m.

Metacarpo-
phalangeal
joint (opened)

R. Morgan

PLATE 9

Muscles of Right Upper Limb, Deep Dissection

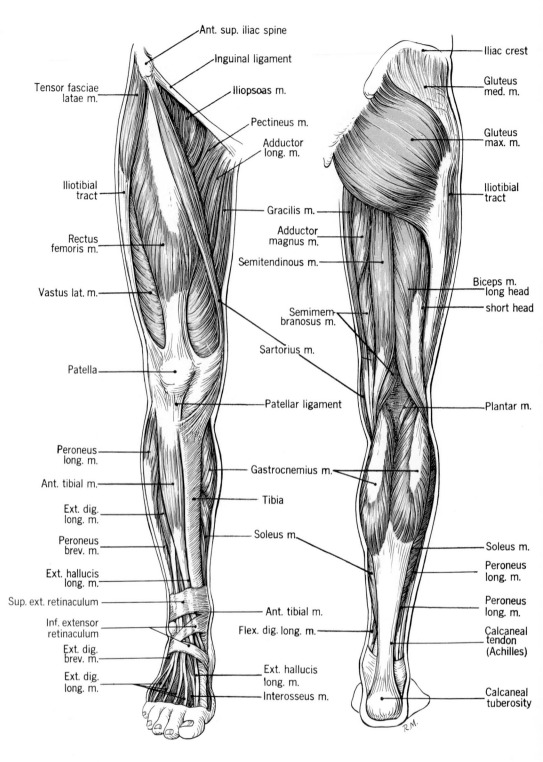

Ant. sup. iliac spine

Inguinal ligament

Iliopsoas m.

Pectineus m.

Adductor long. m.

Tensor fasciae latae m.

Iliotibial tract

Rectus femoris m.

Vastus lat. m.

Patella

Peroneus long. m.

Ant. tibial m.

Ext. dig. long. m.

Peroneus brev. m.

Ext. hallucis long. m.

Sup. ext. retinaculum

Inf. extensor retinaculum

Ext. dig. brev. m.

Ext. dig. long. m.

Gracilis m.

Adductor magnus m.

Semitendinous m.

Semimembranosus m.

Sartorius m.

Patellar ligament

Gastrocnemius m.

Tibia

Soleus m.

Ant. tibial m.

Flex. dig. long. m.

Ext. hallucis long. m.

Interosseus m.

Iliac crest

Gluteus med. m.

Gluteus max. m.

Iliotibial tract

Biceps m. long head

short head

Plantar m.

Soleus m.

Peroneus long. m.

Peroneus long. m.

Calcaneal tendon (Achilles)

Calcaneal tuberosity

R.M.

PLATE 10

Superficial Muscles of Right Lower Limb

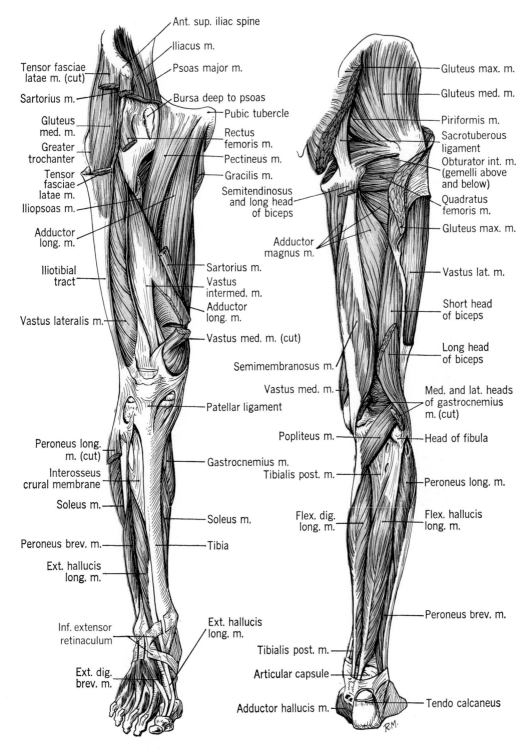

Ant. sup. iliac spine

Iliacus m.

Psoas major m.

Tensor fasciae latae m. (cut)

Sartorius m.

Gluteus med. m.

Greater trochanter

Tensor fasciae latae m.

Iliopsoas m.

Adductor long. m.

Iliotibial tract

Vastus lateralis m.

Peroneus long. m. (cut)

Interosseus crural membrane

Soleus m.

Peroneus brev. m.

Ext. hallucis long. m.

Inf. extensor retinaculum

Ext. dig. brev. m.

Bursa deep to psoas

Pubic tubercle

Rectus femoris m.

Pectineus m.

Gracilis m.

Semitendinosus and long head of biceps

Sartorius m.

Vastus intermed. m.

Adductor long. m.

Vastus med. m. (cut)

Patellar ligament

Gastrocnemius m.

Soleus m.

Tibia

Ext. hallucis long. m.

Gluteus max. m.

Gluteus med. m.

Piriformis m.

Sacrotuberous ligament

Obturator int. m. (gemelli above and below)

Quadratus femoris m.

Gluteus max. m.

Adductor magnus m.

Vastus lat. m.

Short head of biceps

Long head of biceps

Semimembranosus m.

Vastus med. m.

Med. and lat. heads of gastrocnemius m. (cut)

Popliteus m.

Head of fibula

Tibialis post. m.

Peroneus long. m.

Flex. dig. long. m.

Flex. hallucis long. m.

Peroneus brev. m.

Tibialis post. m.

Articular capsule

Adductor hallucis m.

Tendo calcaneus

PLATE 11

Muscles of Right Lower Limb, Deep Dissection

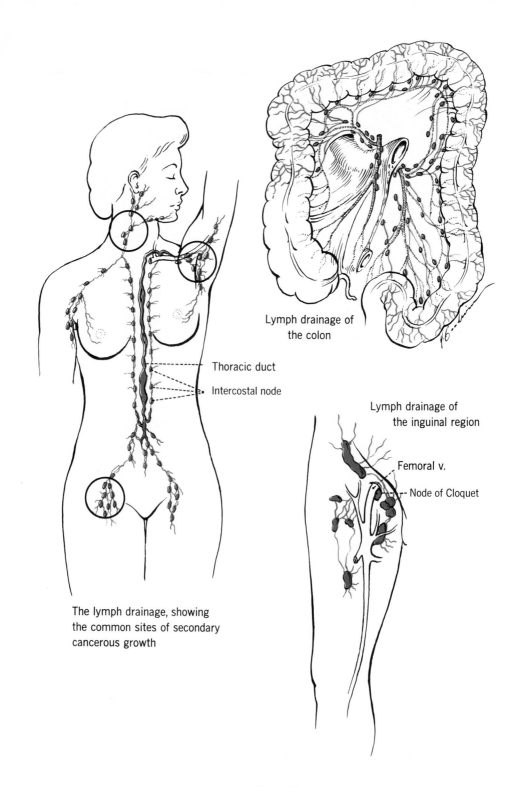

Lymph drainage of
the colon

Thoracic duct

Intercostal node

Lymph drainage of
the inguinal region

Femoral v.

Node of Cloquet

The lymph drainage, showing
the common sites of secondary
cancerous growth

PLATE 12
Lymphatic System

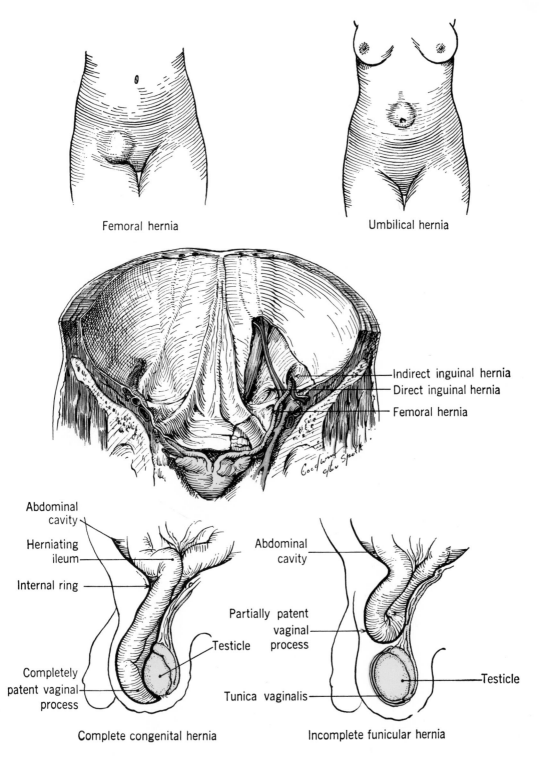

Femoral hernia

Umbilical hernia

Indirect inguinal hernia
Direct inguinal hernia
Femoral hernia

Abdominal cavity

Herniating ileum

Internal ring

Completely patent vaginal process

Testicle

Complete congenital hernia

Abdominal cavity

Partially patent vaginal process

Tunica vaginalis

Testicle

Incomplete funicular hernia

PLATE 13

Hernia of Anterior Abdominal Wall

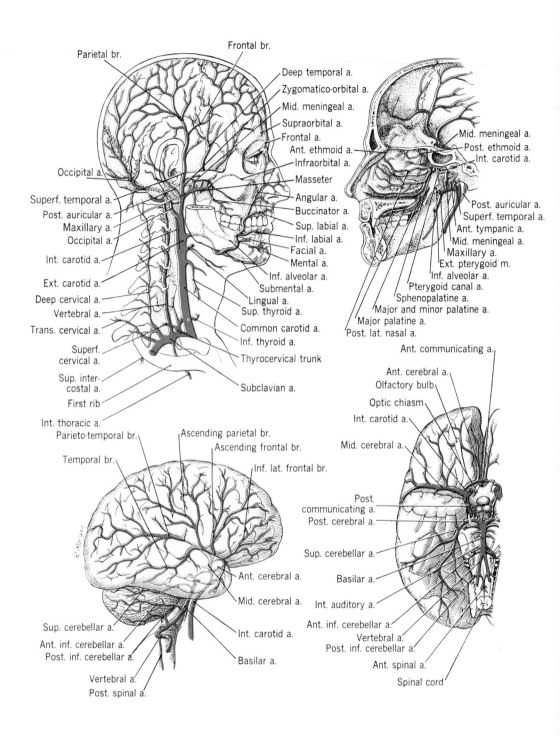

Parietal br.
Frontal br.
Deep temporal a.
Zygomatico-orbital a.
Mid. meningeal a.
Supraorbital a.
Frontal a.
Ant. ethmoid a.
Infraorbital a.
Masseter
Angular a.
Buccinator a.
Sup. labial a.
Inf. labial a.
Facial a.
Mental a.
Inf. alveolar a.
Submental a.
Lingual a.
Sup. thyroid a.
Common carotid a.
Inf. thyroid a.
Thyrocervical trunk
Subclavian a.

Occipital a.
Superf. temporal a.
Post. auricular a.
Maxillary a.
Occipital a.
Int. carotid a.
Ext. carotid a.
Deep cervical a.
Vertebral a.
Trans. cervical a.
Superf. cervical a.
Sup. inter-costal a.
First rib
Int. thoracic a.

Mid. meningeal a.
Post. ethmoid a.
Int. carotid a.
Post. auricular a.
Superf. temporal a.
Ant. tympanic a.
Mid. meningeal a.
Maxillary a.
Ext. pterygoid m.
Inf. alveolar a.
Pterygoid canal a.
Sphenopalatine a.
Major and minor palatine a.
Major palatine a.
Post. lat. nasal a.

Parieto-temporal br.
Ascending parietal br.
Ascending frontal br.
Temporal br.
Inf. lat. frontal br.

Ant. communicating a.
Ant. cerebral a.
Olfactory bulb
Optic chiasm
Int. carotid a.
Mid. cerebral a.

Post. communicating a.
Post. cerebral a.
Sup. cerebellar a.
Basilar a.
Int. auditory a.
Ant. inf. cerebellar a.
Vertebral a.
Post. inf. cerebellar a.
Ant. spinal a.
Spinal cord

Sup. cerebellar a.
Ant. inf. cerebellar a.
Post. inf. cerebellar a.
Vertebral a.
Post. spinal a.
Ant. cerebral a.
Mid. cerebral a.
Int. carotid a.
Basilar a.

PLATE 14
Arteries of Head and Brain

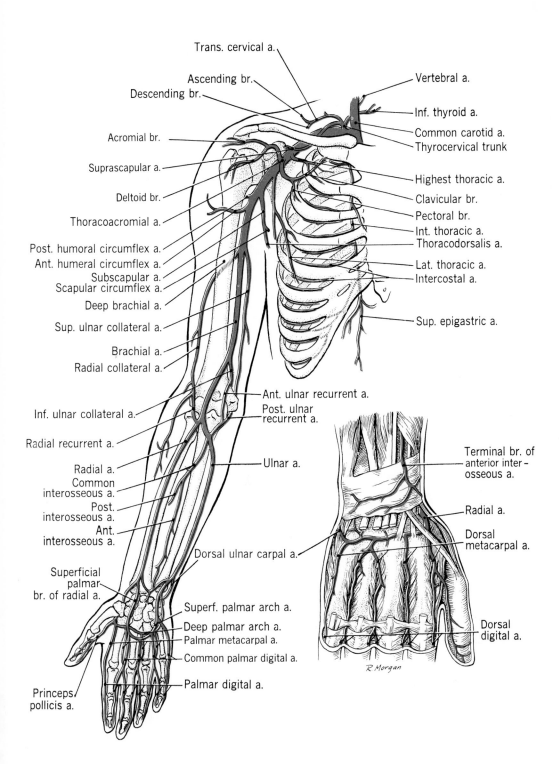

Trans. cervical a.

Ascending br.

Descending br.

Acromial br.

Suprascapular a.

Deltoid br.

Thoracoacromial a.

Post. humoral circumflex a.
Ant. humeral circumflex a.
Subscapular a.
Scapular circumflex a.
Deep brachial a.

Sup. ulnar collateral a.

Brachial a.
Radial collateral a.

Inf. ulnar collateral a.

Radial recurrent a.

Radial a.
Common
interosseous a.
Post.
interosseous a.
Ant.
interosseous a.

Superficial
palmar
br. of radial a.

Princeps
pollicis a.

Vertebral a.

Inf. thyroid a.

Common carotid a.
Thyrocervical trunk

Highest thoracic a.

Clavicular br.

Pectoral br.

Int. thoracic a.
Thoracodorsalis a.

Lat. thoracic a.
Intercostal a.

Sup. epigastric a.

Ant. ulnar recurrent a.
Post. ulnar
recurrent a.

Ulnar a.

Dorsal ulnar carpal a.

Superf. palmar arch a.
Deep palmar arch a.
Palmar metacarpal a.
Common palmar digital a.

Palmar digital a.

Terminal br. of
anterior inter-
osseous a.

Radial a.

Dorsal
metacarpal a.

Dorsal
digital a.

R. Morgan

PLATE 15
Arteries of Upper Limb and Chest

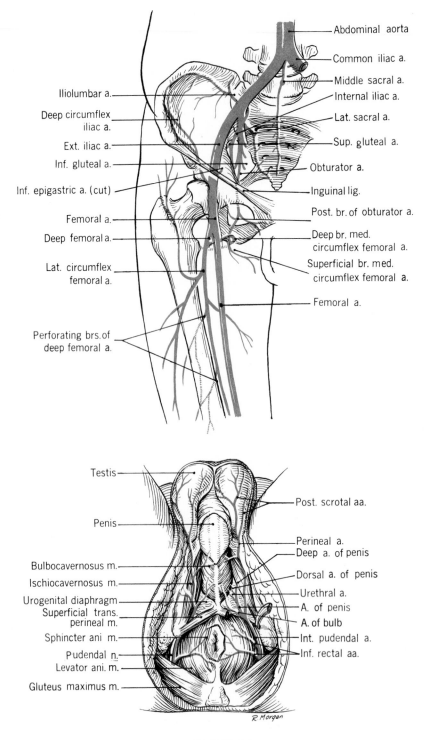

Abdominal aorta

Common iliac a.

Middle sacral a.

Internal iliac a.

Lat. sacral a.

Sup. gluteal a.

Obturator a.

Inguinal lig.

Post. br. of obturator a.

Deep br. med. circumflex femoral a.

Superficial br. med. circumflex femoral a.

Femoral a.

Iliolumbar a.

Deep circumflex iliac a.

Ext. iliac a.

Inf. gluteal a.

Inf. epigastric a. (cut)

Femoral a.

Deep femoral a.

Lat. circumflex femoral a.

Perforating brs. of deep femoral a.

Testis

Penis

Bulbocavernosus m.

Ischiocavernosus m.

Urogenital diaphragm

Superficial trans. perineal m.

Sphincter ani m.

Pudendal n.

Levator ani. m.

Gluteus maximus m.

Post. scrotal aa.

Perineal a.

Deep a. of penis

Dorsal a. of penis

Urethral a.

A. of penis

A. of bulb

Int. pudendal a.

Inf. rectal aa.

R. Morgan

PLATE 16

Arteries of Thigh and Pereneum

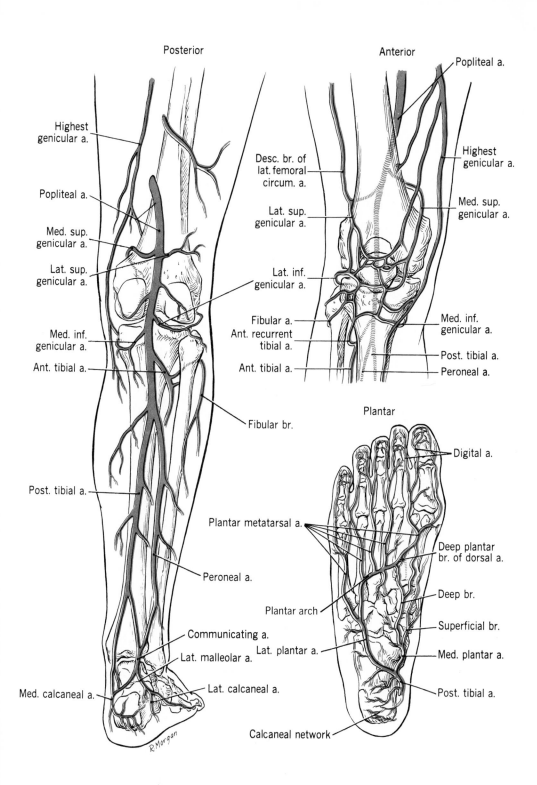

Posterior

Anterior

Popliteal a.

Highest genicular a.

Desc. br. of lat. femoral circum. a.

Highest genicular a.

Popliteal a.

Lat. sup. genicular a.

Med. sup. genicular a.

Med. sup. genicular a.

Lat. sup. genicular a.

Lat. inf. genicular a.

Med. inf. genicular a.

Fibular a.
Ant. recurrent tibial a.

Med. inf. genicular a.

Post. tibial a.

Ant. tibial a.

Ant. tibial a.

Peroneal a.

Fibular br.

Plantar

Digital a.

Plantar metatarsal a.

Deep plantar br. of dorsal a.

Post. tibial a.

Deep br.

Peroneal a.

Superficial br.

Plantar arch

Communicating a.

Lat. plantar a.

Med. plantar a.

Lat. malleolar a.

Post. tibial a.

Med. calcaneal a.

Lat. calcaneal a.

R.Morgan

Calcaneal network

PLATE 17
Arteries of Right Lower Limb in Relation to Bones

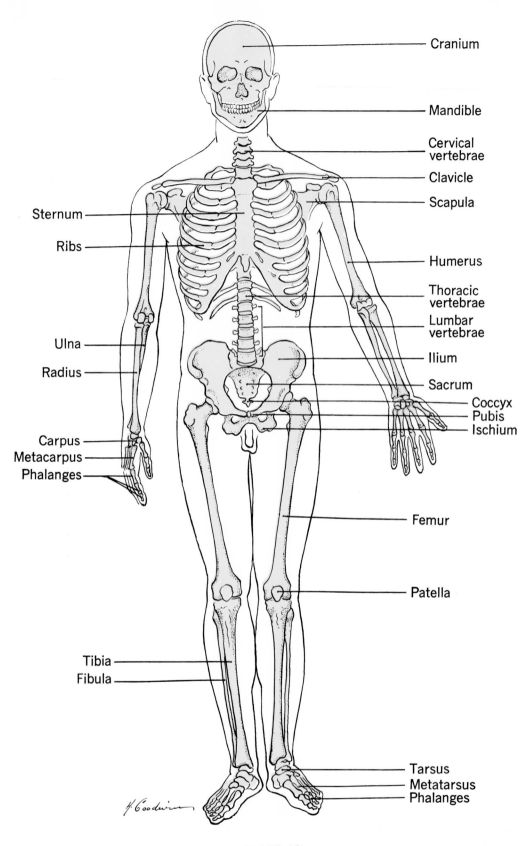

Cranium

Mandible

Cervical vertebrae

Clavicle

Scapula

Humerus

Thoracic vertebrae

Lumbar vertebrae

Ilium

Sacrum

Coccyx

Pubis

Ischium

Sternum

Ribs

Ulna

Radius

Carpus

Metacarpus

Phalanges

Femur

Patella

Tibia

Fibula

Tarsus

Metatarsus

Phalanges

PLATE 18

Human Skeleton, Anterior View

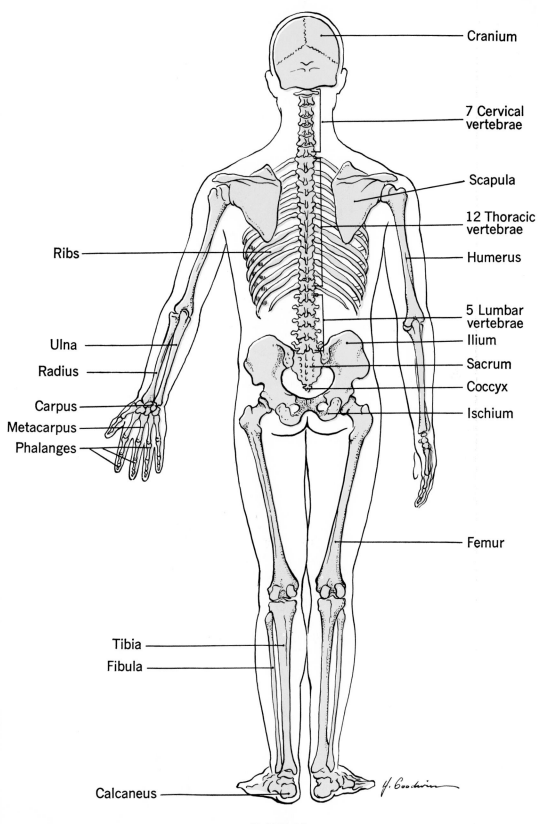

Cranium

7 Cervical
vertebrae

Scapula

12 Thoracic
vertebrae

Humerus

Ribs

5 Lumbar
vertebrae

Ilium

Ulna

Sacrum

Radius

Coccyx

Carpus

Ischium

Metacarpus

Phalanges

Femur

Tibia

Fibula

Calcaneus

PLATE 19
Human Skeleton, Posterior View

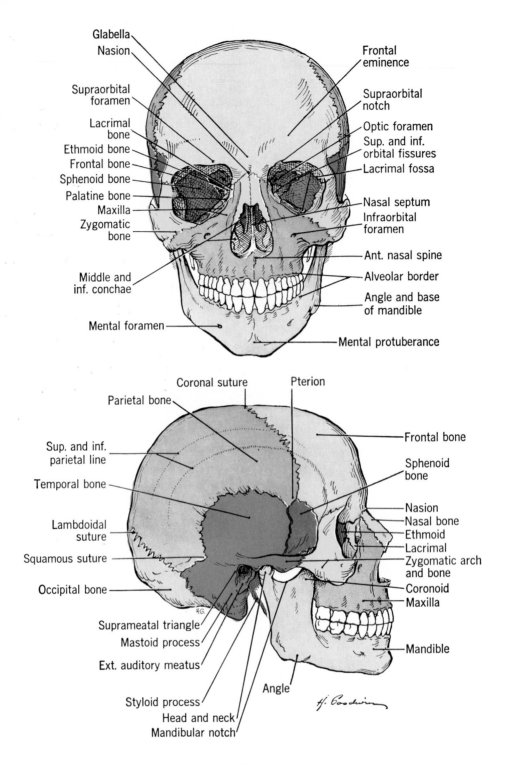

Glabella
Nasion
Frontal eminence
Supraorbital foramen
Supraorbital notch
Lacrimal bone
Optic foramen
Ethmoid bone
Sup. and inf. orbital fissures
Frontal bone
Lacrimal fossa
Sphenoid bone
Palatine bone
Nasal septum
Maxilla
Infraorbital foramen
Zygomatic bone
Ant. nasal spine
Middle and inf. conchae
Alveolar border
Angle and base of mandible
Mental foramen
Mental protuberance

Coronal suture
Pterion
Parietal bone
Frontal bone
Sup. and inf. parietal line
Sphenoid bone
Temporal bone
Nasion
Nasal bone
Lambdoidal suture
Ethmoid
Lacrimal
Squamous suture
Zygomatic arch and bone
Occipital bone
Coronoid
Maxilla
Suprameatal triangle
Mastoid process
Ext. auditory meatus
Mandible
Styloid process
Angle
Head and neck
Mandibular notch

PLATE 20
Norma of Skull, Anterior and Lateral Views

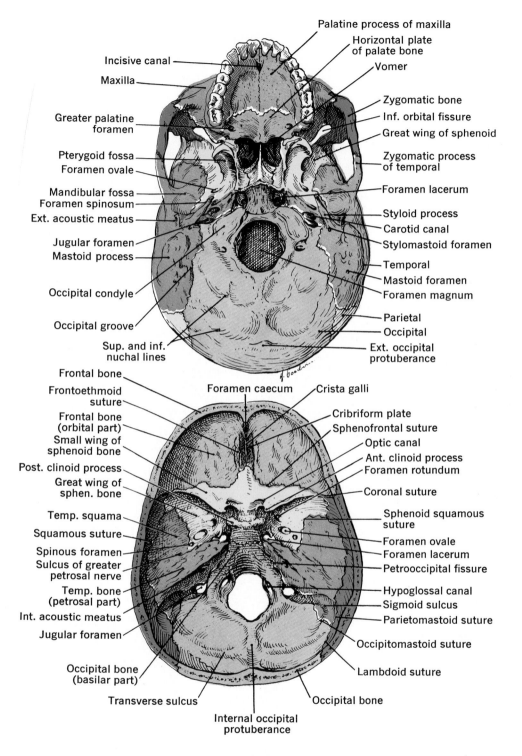

Incisive canal

Maxilla

Greater palatine foramen

Pterygoid fossa

Foramen ovale

Mandibular fossa
Foramen spinosum

Ext. acoustic meatus

Jugular foramen
Mastoid process

Occipital condyle

Occipital groove

Sup. and inf. nuchal lines

Palatine process of maxilla

Horizontal plate of palate bone

Vomer

Zygomatic bone

Inf. orbital fissure

Great wing of sphenoid

Zygomatic process of temporal

Foramen lacerum

Styloid process

Carotid canal

Stylomastoid foramen

Temporal

Mastoid foramen

Foramen magnum

Parietal

Occipital

Ext. occipital protuberance

Frontal bone

Frontoethmoid suture

Frontal bone (orbital part)

Small wing of sphenoid bone

Post. clinoid process

Great wing of sphen. bone

Temp. squama

Squamous suture

Spinous foramen

Sulcus of greater petrosal nerve

Temp. bone (petrosal part)

Int. acoustic meatus

Jugular foramen

Occipital bone (basilar part)

Transverse sulcus

Foramen caecum

Crista galli

Cribriform plate

Sphenofrontal suture

Optic canal

Ant. clinoid process
Foramen rotundum

Coronal suture

Sphenoid squamous suture

Foramen ovale

Foramen lacerum

Petrooccipital fissure

Hypoglossal canal

Sigmoid sulcus

Parietomastoid suture

Occipitomastoid suture

Lambdoid suture

Occipital bone

Internal occipital protuberance

PLATE 21

Base of Skull, External and Internal Views

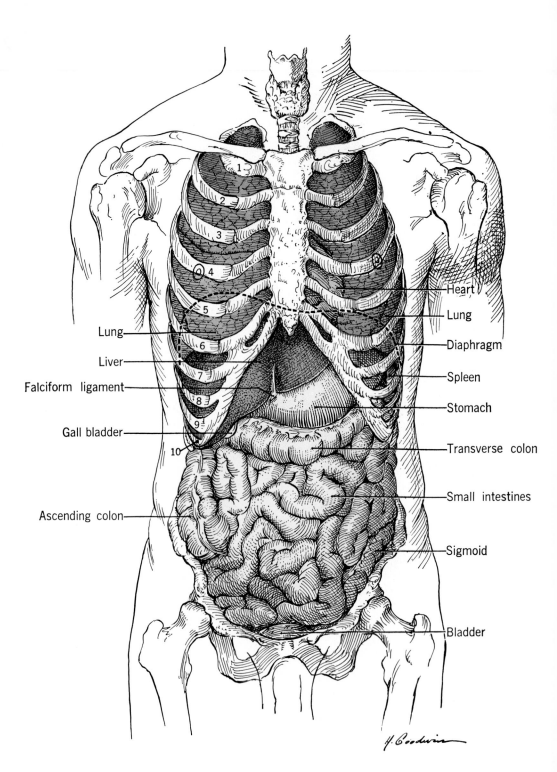

Lung

Liver

Falciform ligament

Gall bladder

Ascending colon

Heart

Lung

Diaphragm

Spleen

Stomach

Transverse colon

Small intestines

Sigmoid

Bladder

PLATE 22
Abdominal and Thoracic Viscera, Anterior View (After Pernkopf)

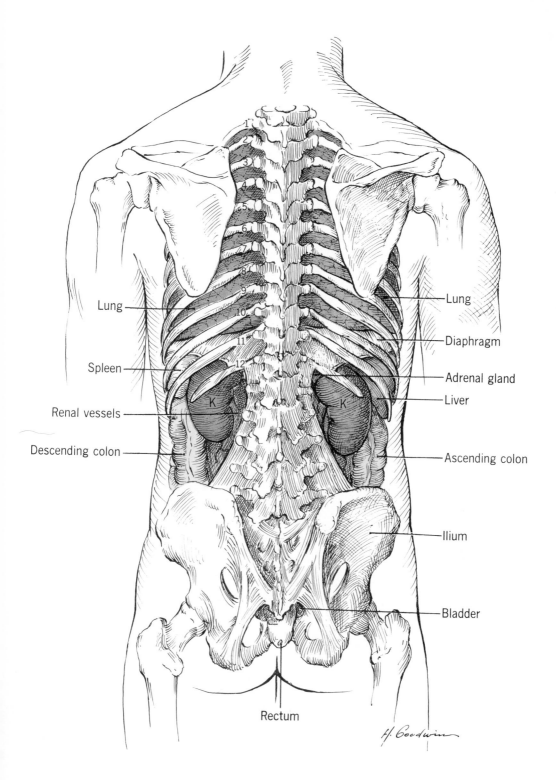

Lung

Lung

Diaphragm

Spleen

Adrenal gland

Renal vessels

Liver

Descending colon

Ascending colon

Ilium

Bladder

Rectum

H. Goodwin

PLATE 23
Abdominal and Thoracic Viscera, Posterior View (After Pernkopf)

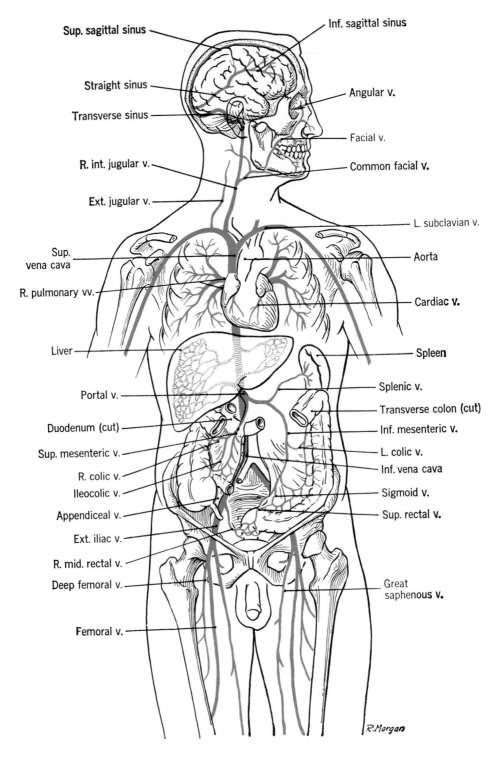

Sup. sagittal sinus

Inf. sagittal sinus

Straight sinus

Angular v.

Transverse sinus

Facial v.

R. int. jugular v.

Common facial **v.**

Ext. jugular v.

L. subclavian v.

Sup.
vena cava

Aorta

R. pulmonary vv.

Cardiac **v.**

Liver

Spleen

Portal v.

Splenic v.

Transverse colon (cut)

Duodenum (cut)

Inf. mesenteric **v.**

Sup. mesenteric v.

L. colic v.

R. colic v.

Inf. vena cava

Ileocolic v.

Sigmoid v.

Appendiceal v.

Sup. rectal **v.**

Ext. iliac v.

R. mid. rectal v.

Deep femoral v.

Great
saphenous **v.**

Femoral v.

R. Morgan

PLATE 24

Veins, Anterior View, Viscera Exposed

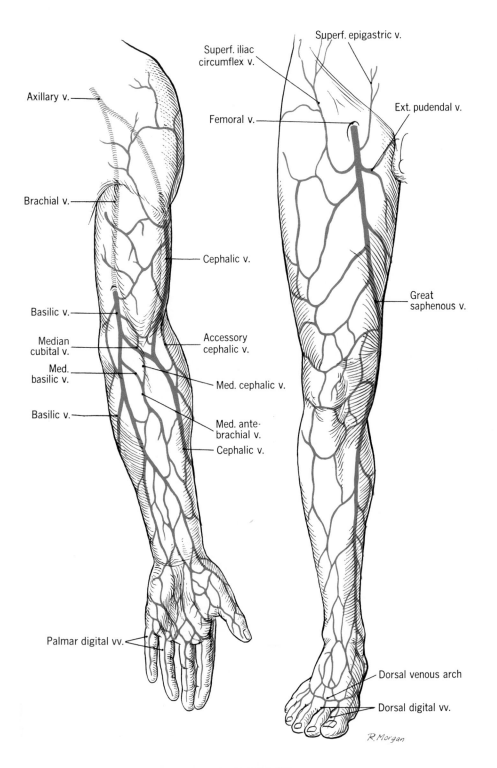

Axillary v.

Brachial v.

Basilic v.

Median cubital v.

Med. basilic v.

Basilic v.

Palmar digital vv.

Superf. iliac circumflex v.

Superf. epigastric v.

Femoral v.

Ext. pudendal v.

Cephalic v.

Accessory cephalic v.

Med. cephalic v.

Med. ante-brachial v.

Cephalic v.

Great saphenous v.

Dorsal venous arch

Dorsal digital vv.

R. Morgan

PLATE 25
Veins of Limbs

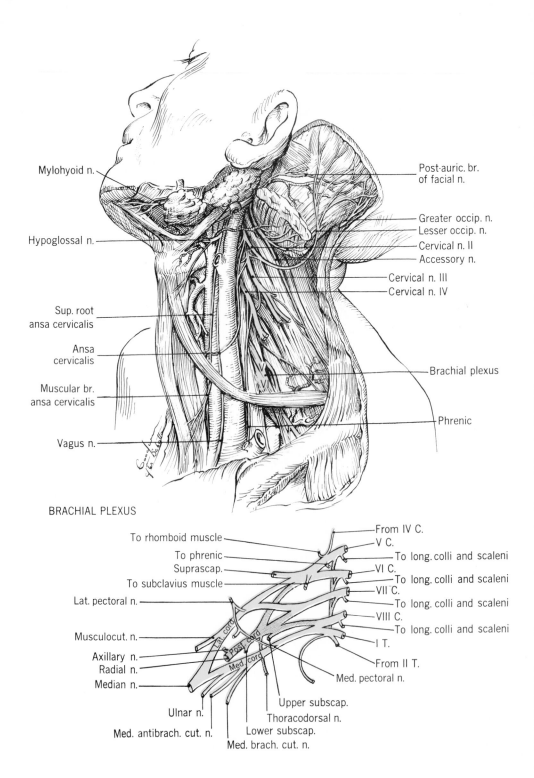

Mylohyoid n.

Hypoglossal n.

Sup. root
ansa cervicalis

Ansa
cervicalis

Muscular br.
ansa cervicalis

Vagus n.

Post-auric. br.
of facial n.

Greater occip. n.
Lesser occip. n.
Cervical n. II
Accessory n.
Cervical n. III
Cervical n. IV

Brachial plexus

Phrenic

BRACHIAL PLEXUS

To rhomboid muscle
To phrenic
Suprascap.
To subclavius muscle
Lat. pectoral n.

Musculocut. n.
Axillary n.
Radial n.
Median n.

Ulnar n.
Med. antibrach. cut. n.

From IV C.
V C.
To long. colli and scaleni
VI C.
To long. colli and scaleni
VII C.
To long. colli and scaleni
VIII C.
To long. colli and scaleni
I T.
From II T.
Med. pectoral n.

Lat. cord
Post. cord
Med. cord

Upper subscap.
Thoracodorsal n.
Lower subscap.
Med. brach. cut. n.

PLATE 26
Nerves of Neck and Axilla

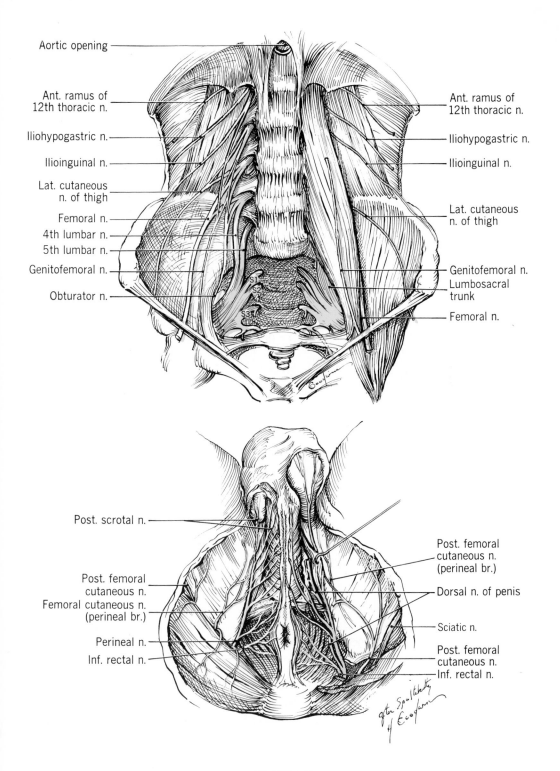

Aortic opening

Ant. ramus of
12th thoracic n.

Iliohypogastric n.

Ilioinguinal n.

Lat. cutaneous
n. of thigh

Femoral n.

4th lumbar n.

5th lumbar n.

Genitofemoral n.

Obturator n.

Ant. ramus of
12th thoracic n.

Iliohypogastric n.

Ilioinguinal n.

Lat. cutaneous
n. of thigh

Genitofemoral n.
Lumbosacral
trunk

Femoral n.

Post. scrotal n.

Post. femoral
cutaneous n.
(perineal br.)

Dorsal n. of penis

Post. femoral
cutaneous n.
Femoral cutaneous n.
(perineal br.)

Sciatic n.

Perineal n.

Inf. rectal n.

Post. femoral
cutaneous n.

Inf. rectal n.

PLATE 27
Nerves of Lumbar and Sacral Plexuses

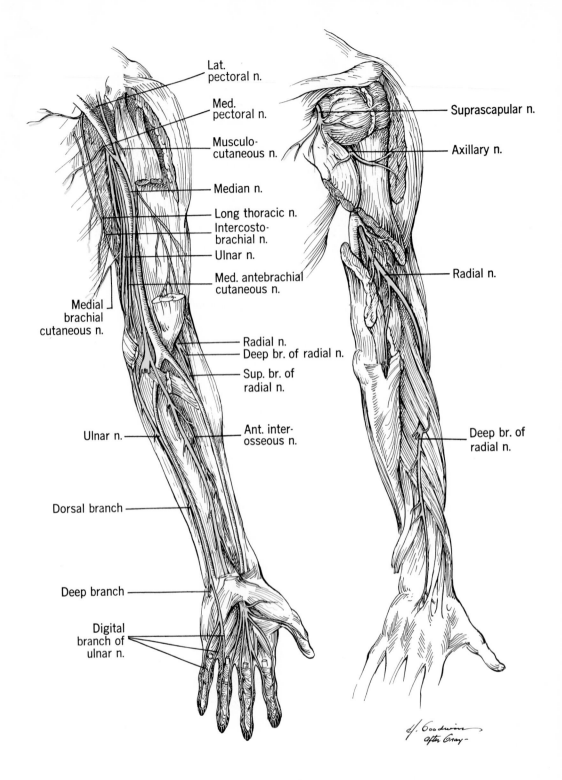

Lat. pectoral n.

Med. pectoral n.

Musculo-cutaneous n.

Median n.

Long thoracic n.

Intercosto-brachial n.

Ulnar n.

Med. antebrachial cutaneous n.

Medial brachial cutaneous n.

Radial n.
Deep br. of radial n.

Sup. br. of radial n.

Ulnar n.

Ant. inter-osseous n.

Dorsal branch

Deep branch

Digital branch of ulnar n.

Suprascapular n.

Axillary n.

Radial n.

Deep br. of radial n.

PLATE 28

Nerves of Superior Limb

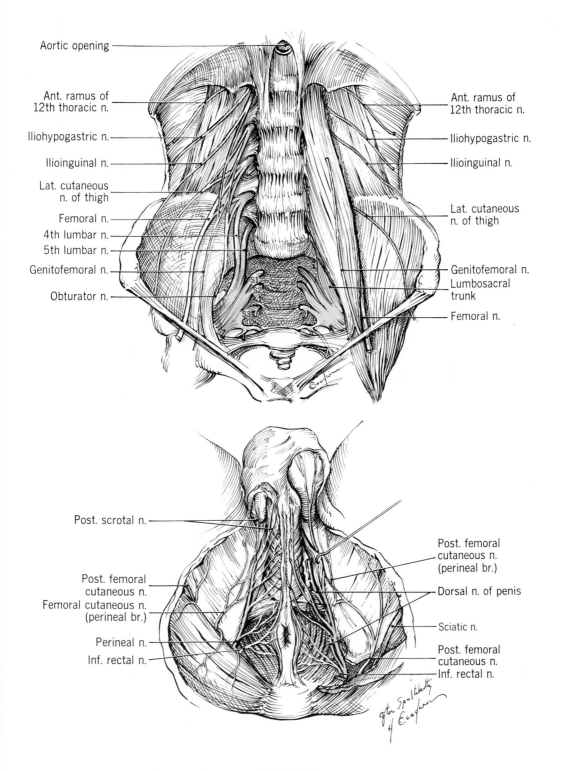

Aortic opening

Ant. ramus of
12th thoracic n.

Iliohypogastric n.

Ilioinguinal n.

Lat. cutaneous
n. of thigh

Femoral n.

4th lumbar n.

5th lumbar n.

Genitofemoral n.

Obturator n.

Ant. ramus of
12th thoracic n.

Iliohypogastric n.

Ilioinguinal n.

Lat. cutaneous
n. of thigh

Genitofemoral n.
Lumbosacral
trunk

Femoral n.

Post. scrotal n.

Post. femoral
cutaneous n.
(perineal br.)

Post. femoral
cutaneous n.
Femoral cutaneous n.
(perineal br.)

Dorsal n. of penis

Perineal n.

Sciatic n.

Inf. rectal n.

Post. femoral
cutaneous n.

Inf. rectal n.

PLATE 27
Nerves of Lumbar and Sacral Plexuses

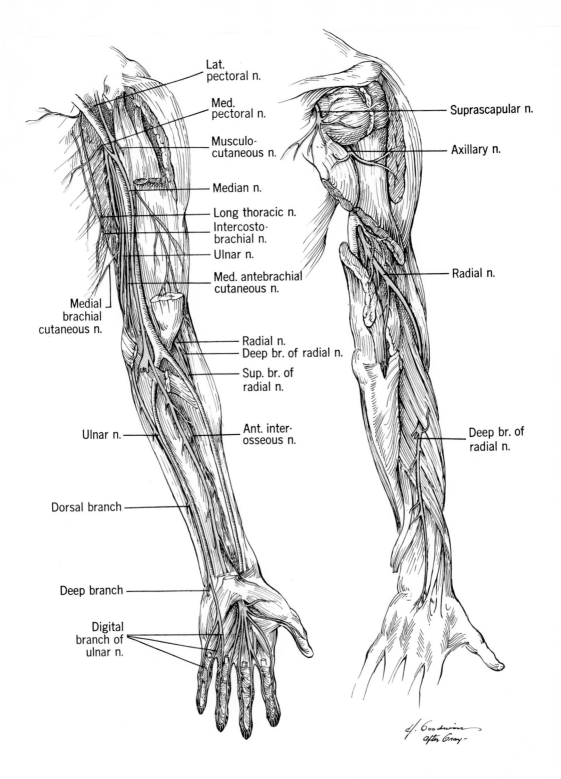

Lat. pectoral n.

Med. pectoral n.

Musculo-cutaneous n.

Median n.

Long thoracic n.

Intercosto-brachial n.

Ulnar n.

Med. antebrachial cutaneous n.

Medial brachial cutaneous n.

Radial n.

Deep br. of radial n.

Sup. br. of radial n.

Ulnar n.

Ant. inter-osseous n.

Dorsal branch

Deep branch

Digital branch of ulnar n.

Suprascapular n.

Axillary n.

Radial n.

Deep br. of radial n.

PLATE 28

Nerves of Superior Limb

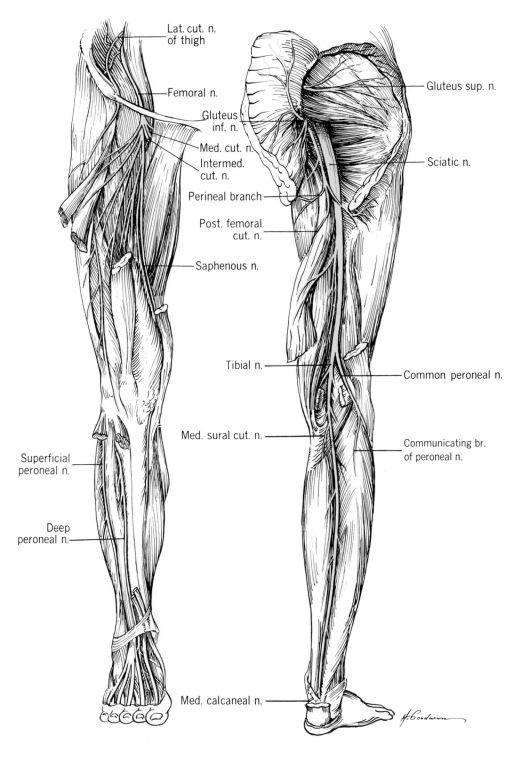

Lat. cut. n.
of thigh

Femoral n.

Gluteus
inf. n.

Med. cut. n.

Intermed.
cut. n.

Perineal branch

Post. femoral
cut. n.

Saphenous n.

Gluteus sup. n.

Sciatic n.

Tibial n.

Med. sural cut. n.

Superficial
peroneal n.

Deep
peroneal n.

Common peroneal n.

Communicating br.
of peroneal n.

Med. calcaneal n.

H. Goodwin

PLATE 29
Nerves of Inferior Limb

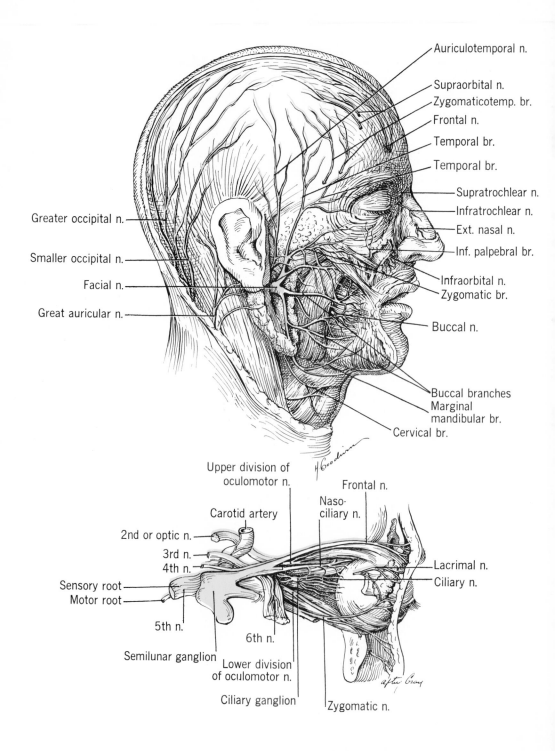

Auriculotemporal n.

Supraorbital n.
Zygomaticotemp. br.
Frontal n.
Temporal br.
Temporal br.
Supratrochlear n.
Infratrochlear n.
Ext. nasal n.
Inf. palpebral br.
Infraorbital n.
Zygomatic br.
Buccal n.

Greater occipital n.

Smaller occipital n.
Facial n.
Great auricular n.

Buccal branches
Marginal
mandibular br.
Cervical br.

Upper division of
oculomotor n.
Frontal n.
Naso-
ciliary n.
Carotid artery
2nd or optic n.
3rd n.
4th n.
Lacrimal n.
Ciliary n.
Sensory root
Motor root
5th n.
6th n.
Semilunar ganglion
Lower division
of oculomotor n.
Ciliary ganglion
Zygomatic n.

PLATE 30
Nerves of Head and Orbit

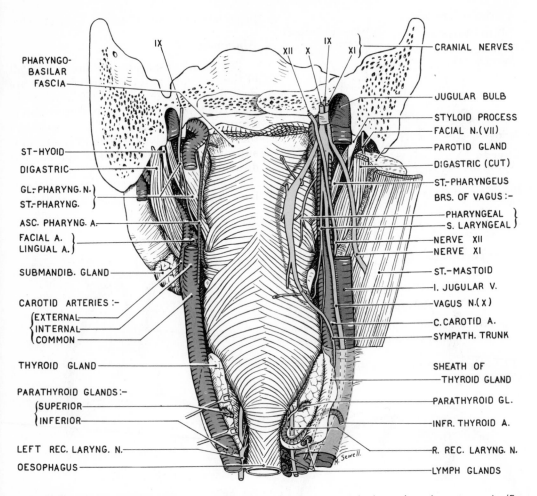

PHARYNGO-
BASILAR
FASCIA

ST-HYOID

DIGASTRIC

GL.-PHARYNG. N.
ST.-PHARYNG.

ASC. PHARYNG. A.

FACIAL A.
LINGUAL A.

SUBMANDIB. GLAND

CAROTID ARTERIES :-
  EXTERNAL
  INTERNAL
  COMMON

THYROID GLAND

PARATHYROID GLANDS :-
  SUPERIOR
  INFERIOR

LEFT REC. LARYNG. N.

OESOPHAGUS

IX   XII X   IX   XI

CRANIAL NERVES

JUGULAR BULB

STYLOID PROCESS
FACIAL N.(VII)
PAROTID GLAND

DIGASTRIC (CUT)

ST.-PHARYNGEUS

BRS. OF VAGUS :-

PHARYNGEAL
S. LARYNGEAL

NERVE XII
NERVE XI

ST.-MASTOID
I. JUGULAR V.
VAGUS N.(X)

C. CAROTID A.
SYMPATH. TRUNK

SHEATH OF
THYROID GLAND

PARATHYROID GL.

INFR. THYROID A.

R. REC. LARYNG. N.

LYMPH GLANDS

Fig. **659**. Posterior view of pharynx, last four cranial nerves, sympathetic trunk, and great vessels. (From *Grant's Method of Anatomy*.)

mous with the pericardial sac (page 259) and its contents are the heart, its vessels and the roots of the great vessels proceeding to and from the heart. Running vertically down between the fibrous pericardium and mediastinal pleurae are the right and left phrenic nerves (fig. 664).

The **anterior mediastinum** is an unimportant small cleft filled with areolar tissue and existing only because the anterior surface of the fibrous pericardial sac contacts the back surface of the body of the sternum.

The **posterior mediastinum** lies behind the pericardium and, below that level, behind the diaphragm down to T.12 vertebra. It contains the esophagus and de-

scending aorta continuing down from the superior mediastinum (fig. 665).

The **superior mediastinum** (fig. 665, 666) contains the great **arteries—aortic arch, brachiocephalic, left common carotid,** and **left subclavian;** the great veins draining the head and neck and limbs—**right and left brachiocephalic veins** and the beginning of the **superior vena cava** (fig. 667); the right and left **phrenic** and **vagus nerves;** the left **recurrent laryngeal nerve** running upward in the groove between trachea and esophagus; and, of course, the **trachea** and **esophagus,** with the **thoracic duct** on the left surface of the esophagus. Many **cardiac sympathetic** and **parasympathetic** nerves and **lymph**

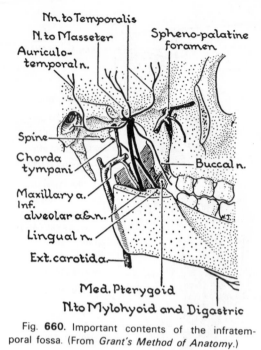

Fig. **660**. Important contents of the infratemporal fossa. (From *Grant's Method of Anatomy*.)

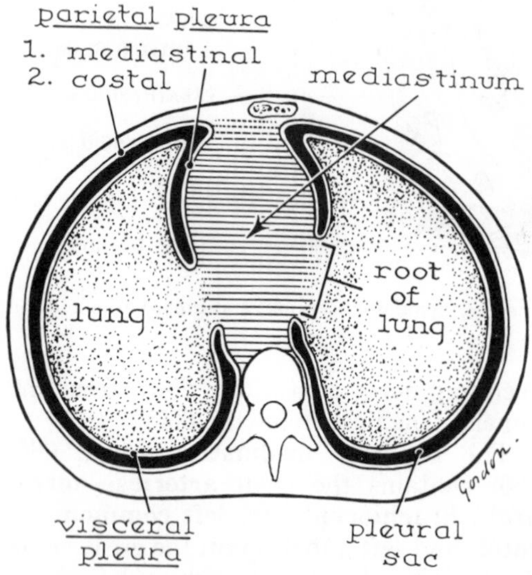

parietal pleura
1. mediastinal
2. costal

mediastinum

root of lung

lung

visceral pleura

pleural sac

Fig. **661**. Horizontal section of thorax

ABDOMEN AND PELVIS

The topographical anatomy of the abdomen and pelvis have been described in various sections of the book. The skeletal and muscular boundaries are described on

pages 60 to 64, and 137 to 143, respectively.

Abdomen

The **great vessels** on the posterior wall of the abdomen and their distributions are described on pages 276 to 279 and 287 to 288. The topography of the posterior wall is summarized in figure 668, which also shows the relationships of the **kidneys, ureters, suprarenal (adrenal) glands, spleen,** and **descending colon.**

The parts of the gastro-intestinal tract have been described on pages 207 to 214 and

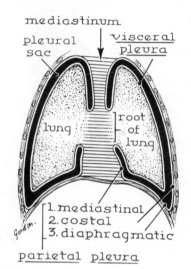

mediastinum

pleural sac

visceral pleura

lung

root of lung

1. mediastinal
2. costal
3. diaphragmatic

parietal pleura

Fig. **662**. Schematic coronal section of thorax

nodes and lymphatics are found around and between these structures.

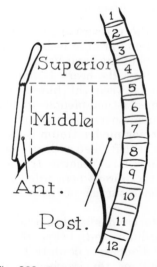

Fig. **663**. Divisions of mediastinum

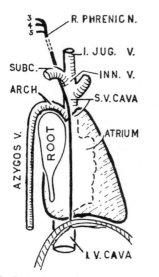

Fig. **664**. The right phrenic nerve runs sub-pleurally to pierce the diaphragm lateral to the I.V.C. (From *Grant's Method of Anatomy*.)

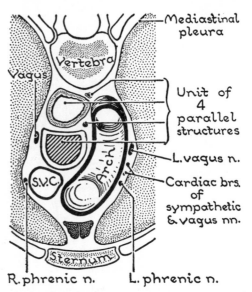

Fig. **666**. Horizontal section of superior mediastinum showing the relationships of aortic arch. (From *Grant's Method of Anatomy*.)

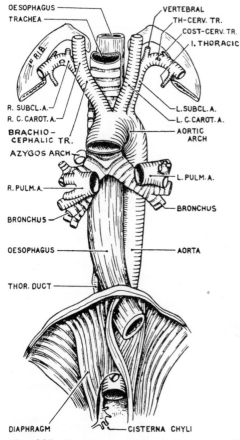

Fig. **665**. Esophagus, aorta, and great arteries in the thorax. (From *Grant's Method of Anatomy*.)

the liver on page 214. The stomach is moored to the abdominal walls at its two ends, i.e., at the point where the esophagus pierces the diaphragm and where the duodenum becomes retroperitoneal, a short distance beyond the pylorus; between these moorings it moves on its "bed" (fig. 669).

Pelvis

The pelvic bony topography is described on pages 62 to 64 and the muscular floor (**Levator Ani**) on page 142. The pelvis contains components of the urinary tract, generative (reproductive) tract in both sexes, and loops of small and large gut as well as the rectum. In its walls it has important vessels and nerves, some of which are transients—the most important being the **lumbosacral plexus** which forms the sciatic and other large nerves as it leaves the pelvis (fig. 670). The blood supply to the bladder and uterus as well as transient arteries form a 'sea' of blood around the pelvic organs (fig. 671).

The bladder and internal genitalia have been described on pages 241 to 242, 246 to 248, and 249 to 251. Figure 672 illustrates the important relationship of the uterus and vagina.

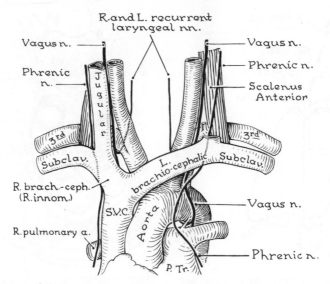

Fig. **667**. Courses of both phrenic and vagus nerves. (From *Grant's Method of Anatomy*.)

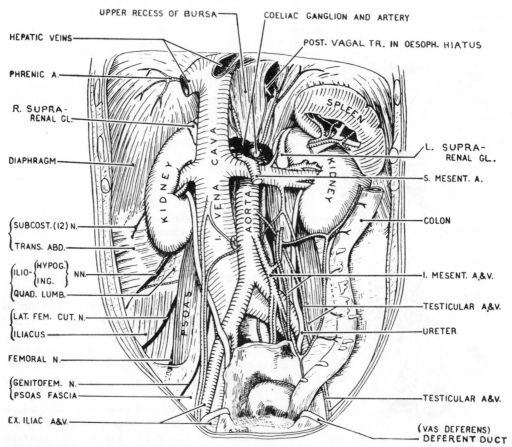

Fig. **668**. Dissection of posterior abdominal wall and its main retroperitoneal structures. (From *Grant's Method of Anatomy*.)

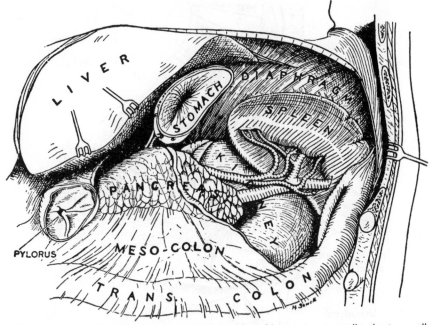

Fig. **669**. Stomach bed. (The tail of the pancreas in this subject was unusually short; usually it abuts against the hylus of the spleen.) (From *Grant's Method of Anatomy*.)

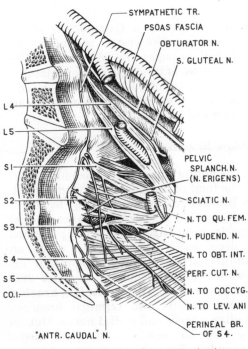

Fig. **670**. Left sacral and coccygeal plexuses. (From *Grant's Method of Anatomy*.)

Perineum. This diamond-shaped area, between the thighs and covered with skin, surrounds and includes the external genitalia and anus (figs. 673, 674). Details of the muscles are given on pages 214, 247 and 252, of the genitalia on pages 243 and 249, and of the anus on page 213. The **ischiorectal fossa** is the fat-filled space above the skin; it is roofed by the sloping **Levator Ani** muscle and bounded laterally by the bony pelvis covered with **Obturator Internus** (fig. 675).

LIMBS

The bones, joints, muscles, vessels, and nerves of the limbs have been thoroughly described in appropriate sections. Here we should consider only their composite arrangement in certain important areas.

Upper Limb

Axilla

The axilla is the pyramidal 'space' above the armpit (fig. 676). Its apex or entrance

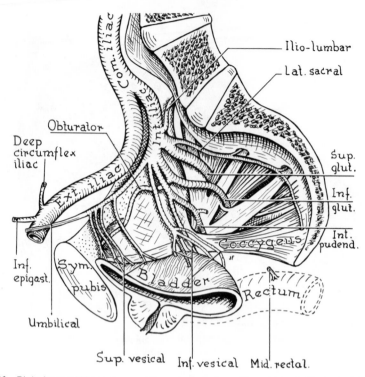

Fig. **671**. Right internal iliac artery and its main branches. (From *Grant's Method of Anatomy*.)

from the neck lies behind the middle of the clavicle and its base is the skin of the armpit. It possesses three muscular walls: (1) an anterior, consisting of the pectoral muscles in the same coronal plane as the clavicle and coracoid process; (2) a medial, consisting of the Serratus Anterior clothing the upper ribs; (3) a posterior, consisting of muscles clothing the front of the scapula. At the intertubercular (bicipital) sulcus on the humerus the anterior wall meets the posterior. The bones and muscles of the region are described on pages 49 to 53 and 143 to 152.

The axillary artery traverses the axilla surrounded by the bundle of nerves of the brachial plexus and two or three accompanying veins (venae comitantes). It and its companions are wrapped close together in a sheath of fibrous tissue known as the **axillary sheath.** Because this sheath and a large quantity of fat in the axilla are both removed in a dissection, the student must

beware of getting a false picture of the relationships. The vessels and nerves are described on pages 273, 287, and 337. The lymph nodes of this region, being related to lymph drainage of the breast, are especially important (page 294). The **female breast** is described on page 253.

Elbow Region

The **cubital fossa** is the triangular space at the front of the elbow, bounded laterally by Brachioradialis and medially by Pronator Teres. The apex is below, where the two muscles meet (fig. 677). Covering the fossa is deep fascia, superficial to which are **cutaneous nerves** and the clinically important **subcutaneous veins** (page 286). The contents are: **Biceps tendon; brachial artery** and its two terminal branches; **median nerve;** and **radial nerve.** Outside the fossa but nearby, behind the medial epicondyle, lies the **ulnar nerve.**

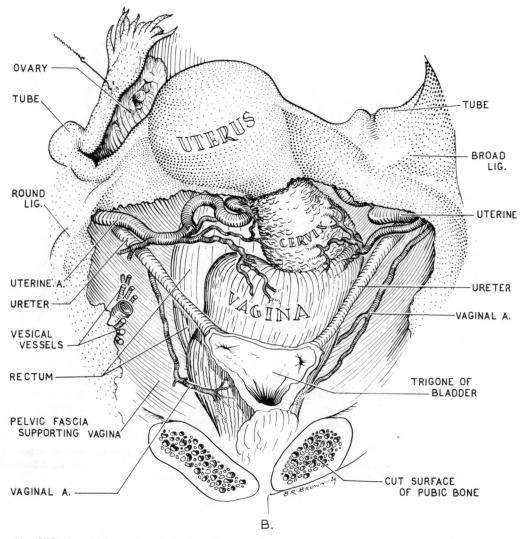

OVARY

TUBE

ROUND
LIG.

UTERINE A.

URETER

VESICAL
VESSELS

RECTUM

PELVIC FASCIA
SUPPORTING VAGINA

VAGINAL A.

UTERUS

CERVIX

VAGINA

TUBE

BROAD
LIG.

UTERINE

URETER

VAGINAL A.

TRIGONE OF
BLADDER

CUT SURFACE
OF PUBIC BONE

B.

Fig. **672**. Female internal genitalia seen from the front and with bladder removed. (From *Grant's Method of Anatomy*.)

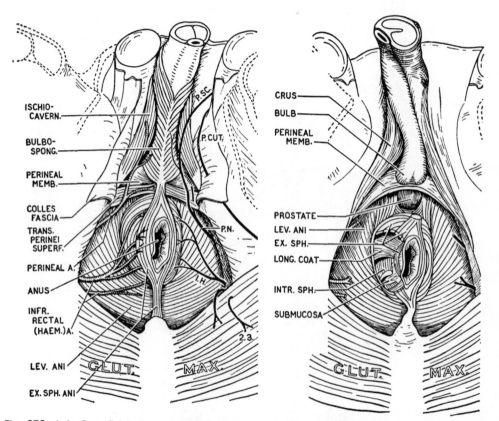

Fig. **673**. *Left*. Superficial dissection of male perineum. *Right*. Exposure of prostate and dissection of walls of the anal canal. I.H. = inferior rectal n.; C.N. = perineal n.; P.SC. = posterior scrotal n.; P.CUT. = perineal branch of posterior cutaneous n. of thigh. (From *Grant's Method of Anatomy*.)

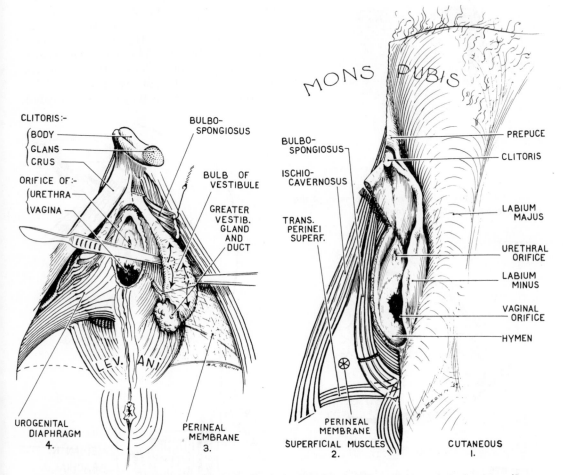

Fig. **674**. Successive dissections (1, 2, 3, 4) of the female genitalia and urogenital triangle.   (From *Grant's Method of Anatomy*.)

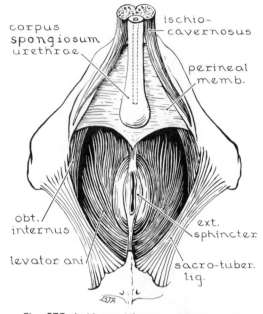

Fig. **675**. Ischiorectal fossae on the sides of the rectum with all their contents removed.

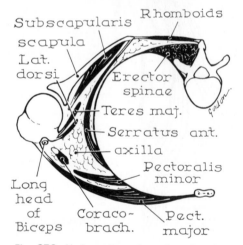

Fig. **676**. Horizontal section of walls of axilla (with great vessels and nerves omitted).

Front of Wrist and Hand

The structures you should be able to locate at the wrist are shown in figure 678. A superficial dissection is seen in figure 679. All of these structures have been individually described elsewhere. The **carpal tunnel** is the passageway behind the **flexor retinaculum** through which

crowd the **flexor tendons** to the fingers and thumb and the **median nerve.** The **ulnar nerve** does not traverse the tunnel although it is covered with a thick layer of fascia where it enters the palm, lateral to the **pisiform** bone.

Palmar Spaces. There are in the palm four closed fascial spaces. The thenar muscles occupy one, the **thenar space;** the hypothenar muscles occupy another, the **hypothenar space.** Between these two there is a large triangular **central space** that contains the tendons of the fingers. Its anterior wall is the palmar aponeurosis. Its posterior wall is formed by the three medial

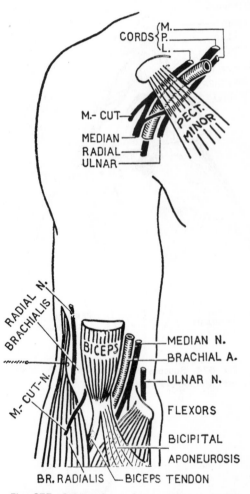

Fig. **677**. Cubital fossa (and posterior relationships of pectoralis minor tendon). (From *Grant's Method of Anatomy*.)

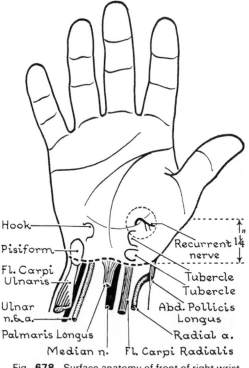

Hook

Pisiform

Fl. Carpi
Ulnaris

Ulnar
n.&a.

Palmaris Longus

Median n.

Recurrent 1¼
nerve

Tubercle
Tubercle

Abd. Pollicis
Longus

Radial a.

Fl. Carpi Radialis

Fig. **678**. Surface anatomy of front of right wrist.
(From *Grant's Method of Anatomy*.)

metacarpals, the palmar and deep transverse ligaments (fig. 680), and the fascia
covering the medial Interossei and the
Adductor Pollicis. Its side walls are the
backwardly turned edges of the palmar
aponeurosis which fuse with the thenar and
hypothenar fasciae. The fourth space is
placed between the Adductor Pollicis in
front and the two lateral intermetacarpal
spaces behind.

In the distal half of the palm, the central
space has eight subdivisions or tunnels, one
for each of the four pairs of long flexor
tendons and one for each of the four
Lumbricals and the companion digital
vessels and nerves. The septa separating
the tunnels are derived from the palmar
aponeurosis. The tendons of the Lumbricals prolong the spaces downward on to the
dorsum of the digits. It is by this 'lumbrical
route' that infection in the central palmar
space may spread to the dorsum of the
hand.

Lower Limb

Front of Thigh

The bony landmarks should be palpated
(fig. 681). The great saphenous vein is an
important structure to trace in a superficial dissection (fig. 682). It runs in the
subcutaneous fat up to where it pierces a
weak spot—the saphenous opening—in the
tough deep fascia, the **fascia lata.** Many
large lymph nodes surround it here; they
drain the limb and external genitalia (fig.
683).

Femoral Triangle. Bounded by the
Sartorius and the **Adductor Longus** this
triangle is important because it contains
the **femoral vessels** and **nerves** in front of
the hip joint—from which they are separated by the **Iliopsoas.** The vessels are
enclosed in the femoral sheath; the latter
has a medial compartment—the **femoral
canal**—loosely filled with fat, or fascia,
and a lymph node or two (fig. 684). A
femoral hernia is an (uncommon) evagination or bulge of the abdominal cavity
pushing down into this **canal.** It can be
confused with an inguinal hernia that occurs in the nearby inguinal canal (page
140).

Gluteal and Hamstring Regions

The gluteal (buttock) region has the
following important landmarks which are
described in appropriate sections of this
book.

1. Greater and lesser sciatic notches
   (page 61)
2. Ischial spine and tuberosity (page
   61)
3. Sacrotuberous and sacrospinous ligaments (page 64).

The 'lid' of the region is the **Gluteus
Maximus,** anterior to which lie **Gluteus
Medius and Minimus**—a good site for
intramuscular injections. The small muscles deep to the 'lid' are described on page
176 and shown in figure 685. The most
important structure here is the **sciatic
nerve** emerging below the midpoint of the
lower edge of the **Piriformis.** It is accessi-

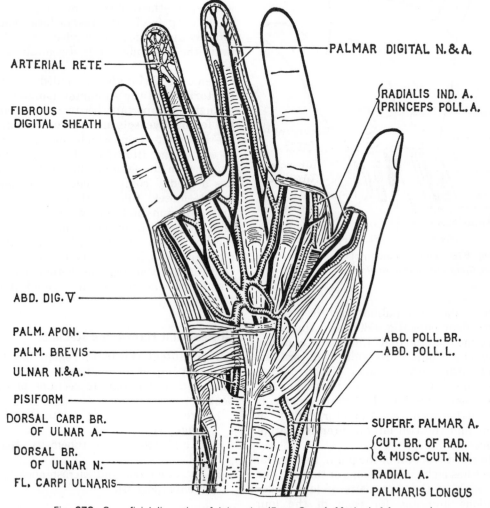

ARTERIAL RETE

FIBROUS
DIGITAL SHEATH

PALMAR DIGITAL N. & A.

RADIALIS IND. A.
PRINCEPS POLL. A.

ABD. DIG. Ⅴ

PALM. APON.

PALM. BREVIS

ULNAR N.&A.

PISIFORM

DORSAL CARP. BR.
OF ULNAR A.

DORSAL BR.
OF ULNAR N.

FL. CARPI ULNARIS

ABD. POLL. BR.
ABD. POLL. L.

SUPERF. PALMAR A.

CUT. BR. OF RAD.
& MUSC-CUT. NN.

RADIAL A.

PALMARIS LONGUS

Fig. **679**. Superficial dissection of right palm. (From *Grant's Method of Anatomy.*)

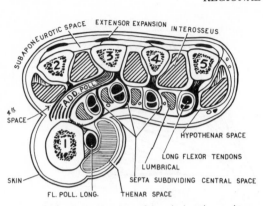

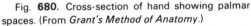

Fig. **680**. Cross-section of hand showing palmar spaces. (From *Grant's Method of Anatomy*.)

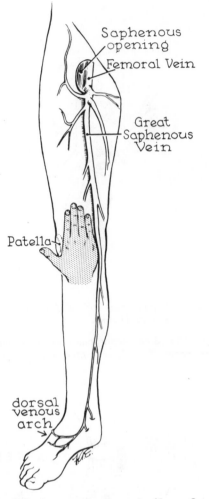

Fig. **682**. Great saphenous vein (From *Grant's Method of Anatomy*.)

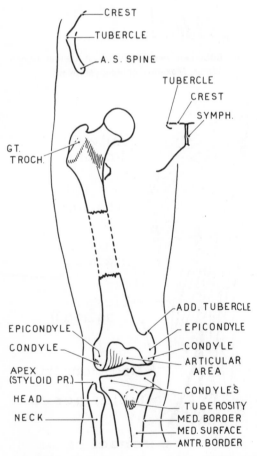

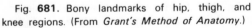

Fig. **681**. Bony landmarks of hip, thigh, and knee regions. (From *Grant's Method of Anatomy*.)

ble from the surface along the lower edge of **Gluteus Maximus** but immediately after it passes deep to the **Biceps Femoris** (fig. 686).

Also deep to Gluteus Maximus are the important structures shown in figure 687 and the back of the hip joint. Indeed, surgeons often use this route to operate on the joint.

The **hamstring region** is most important because it transmits the sciatic nerve down the middle of the back of the thigh. It leads to the **popliteal fossa.**

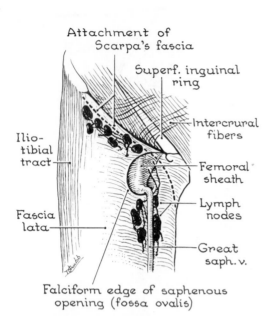

Fig. **683**. Great saphenous vein and inguinal lymph nodes. (From *Grant's Method of Anatomy*.)

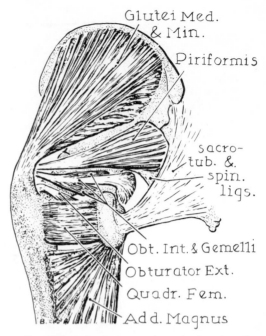

Fig. **685**. Left deep muscles of gluteal region (with Gluteus Maximus entirely removed).

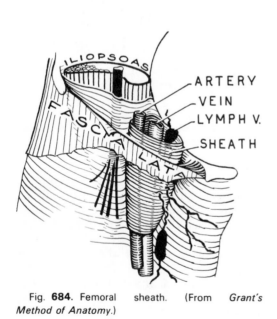

Fig. **684**. Femoral sheath. (From *Grant's Method of Anatomy*.)

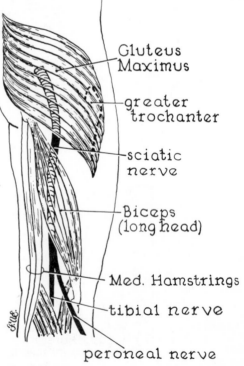

Fig. **686**. Course and chief superficial relationships of right sciatic nerve from behind.

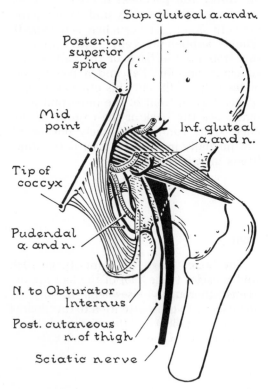

Fig. **687**. Important structures deep to the Gluteus Maximus. (From *Grant's Method of Anatomy*.)

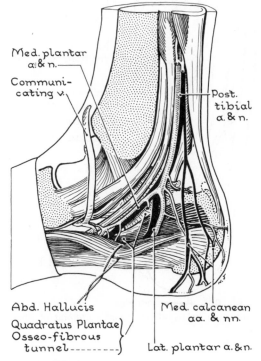

Fig. **689**. Door of right foot—structures passing deep to Abductor Hallucis. (From *Grant's Method of Anatomy*.)

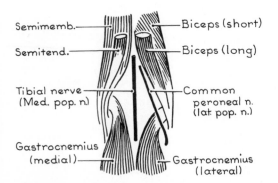

Fig. **688**. Boundaries of the popliteal fossa. (From *Grant's Method of Anatomy*.)

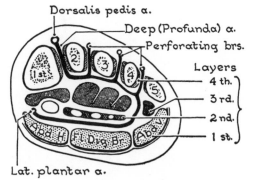

Fig. **690**. Schematic cross-section of foot, showing arteries, first between layers (1) and (2) and then between layers (3) and (4) of sole. (From *Grant's Method of Anatomy*.)

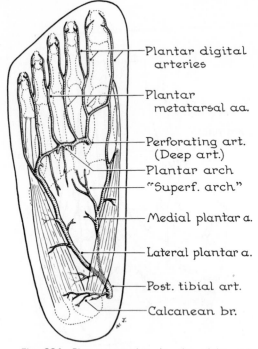

Fig. **691**. Plantar arteries in the right sole seen from below. (From *Grant's Method of Anatomy*.)

— Plantar digital arteries

— Plantar metatarsal aa.

— Perforating art. (Deep art.)

— Plantar arch

— "Superf. arch"

— Medial plantar a.

— Lateral plantar a.

— Post. tibial art.

— Calcanean br.

Popliteal Fossa

This diamond-shaped fossa transmits the **tibial and peroneal nerves** (fig. 688) and the **popliteal artery and its companion vein.** Joining the vein here is the **small (short) saphenous vein** which, prior to this, ran up the back of the calf in the subcutaneous fat. Fat and a few lymph nodes help to fill out this otherwise tight space. Just in front of the important contents, lies the back of the capsule of the knee joint; running obliquely downward and medially in its lower part is the **Popliteus** muscle (page 184).

Foot

The bones (page 70), joints (page 108), and muscles (page 195) of the foot are decribed thoroughly elsewhere. It only remains here to draw attention to the 'door of the foot' (fig. 689) and the scheme of the arterial supply of the sole (figs. 690, 691).

Index

393